Lecture Notes in Computer

T0230301

Commenced Publication in 1973
Founding and Former Series Editors:
Gerhard Goos, Juris Hartmanis, and Jan van Leeuwen

Editorial Board

David Hutchison
  Lancaster University, UK
Takeo Kanade
  Carnegie Mellon University, Pittsburgh, PA, USA
Josef Kittler
  University of Surrey, Guildford, UK
Jon M. Kleinberg
  Cornell University, Ithaca, NY, USA
Friedemann Mattern
  ETH Zurich, Switzerland
John C. Mitchell
  Stanford University, CA, USA
Moni Naor
  Weizmann Institute of Science, Rehovot, Israel
Oscar Nierstrasz
  University of Bern, Switzerland
C. Pandu Rangan
  Indian Institute of Technology, Madras, India
Bernhard Steffen
  University of Dortmund, Germany
Madhu Sudan
  Massachusetts Institute of Technology, MA, USA
Demetri Terzopoulos
  New York University, NY, USA
Doug Tygar
  University of California, Berkeley, CA, USA
Moshe Y. Vardi
  Rice University, Houston, TX, USA
Gerhard Weikum
  Max-Planck Institute of Computer Science, Saarbruecken, Germany

Anthony G. Cohn   David M. Mark (Eds.)

Spatial
Information Theory

International Conference, COSIT 2005
Elliottville, NY, USA, September 14-18, 2005
Proceedings

 Springer

Volume Editors

Anthony G. Cohn
University of Leeds, School of Computing
Leeds, LS2 9JT, UK
E-mail: a.g.cohn@leeds.ac.uk

David M. Mark
University at Buffalo, Department of Geography
105 Wilkeson, North Campus, Buffalo, NY 14261-0023, USA,
E-mail: dmark@geog.buffalo.edu

Library of Congress Control Number: 2005932208

CR Subject Classification (1998): E.1, I.2, F.1, H.2.8, H.1, J.2

ISSN   0302-9743
ISBN-10   3-540-28964-X Springer Berlin Heidelberg New York
ISBN-13   978-3-540-28964-7 Springer Berlin Heidelberg New York

This work is subject to copyright. All rights are reserved, whether the whole or part of the material is
concerned, specifically the rights of translation, reprinting, re-use of illustrations, recitation, broadcasting,
reproduction on microfilms or in any other way, and storage in data banks. Duplication of this publication
or parts thereof is permitted only under the provisions of the German Copyright Law of September 9, 1965,
in its current version, and permission for use must always be obtained from Springer. Violations are liable
to prosecution under the German Copyright Law.

Springer is a part of Springer Science+Business Media

springeronline.com

© Springer-Verlag Berlin Heidelberg 2005
Printed in Germany

Typesetting: Camera-ready by author, data conversion by Scientific Publishing Services, Chennai, India
Printed on acid-free paper   SPIN: 11556114   06/3142   5 4 3 2 1 0

Preface

This volume contains the papers presented at the "Conference on Spatial Information Theory", held in Ellicottville, New York in September 2005. COSIT 2005 was the 7th International Conference held under the COSIT name. When Andrew Frank and his colleagues organized the first COSIT conference on the island of Elba, Italy, in 1993, it represented the maturing of an international research community that had already met four or five times in the United States, Spain, and Italy. Of course, cognitive and computational approaches to space and spatial phenomena were not themselves new topics, but a context of providing theoretical underpinning for geographic information systems refocused some of these researchers and brought them up against practical and conceptual challenges. A second international symposium under the COSIT name, held in Semmering, Austria in 1995, established COSIT as a biennial conference series that continued at Laurel Highlands, Pennsylvania, USA (1997), Stade, Germany (1999), Morro Bay, California, USA (2001) and Ittingen, Switzerland (2003). A productive partnership with Springer's Lecture Notes in Computer Science has ensured that the papers from every COSIT meeting have been widely disseminated, and the COSIT community has contributed significantly to the development of Geographic Information Science, Geoinformatics and Spatial Information Theory in general.

This volume contains 30 papers, carefully selected from 82 manuscripts submitted for consideration. We believe that this is the largest number of papers ever submitted to a COSIT conference. Each submission was rigorously reviewed by three members of the program committee, and in cases where the reviewers disagreed on the quality of the paper or its appropriateness for COSIT, we initiated debates among the reviewers to resolve the disagreements. An acceptance rate of 37 percent meant that some solid papers were rejected, as well as some exciting new developments that the reviewers felt were not ready for publication in a fully refereed volume.

We believe that the authorship of the papers in this COSIT 2005 volume reflects both the continuity and strength of the series, and its openness to new blood and new ideas. The 30 accepted papers have a total of 58 authors. Of those, 26 have written or co-authored a paper in a previous COSIT volume, but 32 are appearing in their first COSIT volume (55 percent new). Of the 30 first authors, 18 are entirely new to the COSIT volumes (60 percent). Eleven of the papers have single authors, six of whom are appearing in a COSIT volume for the first time. Among the 19 co-authored papers, 11 had a mix of new and experienced COSIT authors, expanding the COSIT community through co-authorship; the remaining eight papers were equally split between four with all new authors and four with all authors having previously published in COSIT. The author list is also highly international, and the authors list current residences in 12 countries on four continents. Countries of residence of authors include USA (9 papers), Germany (7), UK (4), Canada (3), France (3), Australia (2), Austria (1), China (1), Israel (1), Italy (1), Japan (1), and The Netherlands (1). (The total adds up to more than 30 because 3 papers had multinational co-author lists.) By country of origin, the list would be even longer.

In addition to the refereed contributions, COSIT 2005 also featured two keynote presentations by Wolfram Burgard and Barbara Landau; abstracts of their talks can be found in this volume, and we thank them for their contributions to the conference. The conference also included a poster session one evening, and a Doctoral Consortium on the Sunday.

Like any other scientific conference, COSIT 2005 would not have been possible without the intellectual contributions and hard work of many people. We can thank only a few here. Most important are the authors and the Program Committee. The authors decided to share a significant piece of their academic lives with the COSIT community, and we thank them. Each PC member reviewed four or five papers, almost all within the short time frame available. Of the 246 reviews requested, only one failed to appear! We wish especially to thank the COSIT Steering Committee for providing advice on policy issues and some special cases in the review process. We also wish to thank Rich Gerber and his excellent START conference management software, which facilitated a very smooth review and manuscript handling process; we can certainly recommend it to other conference organizers! Finally, a long and incomplete list of others who contributed in many ways: Springer Lecture Notes in Computer Science and their staff for publishing the volume; Environmental Systems Research Institute (ESRI) for supporting the Doctoral Consortium, and Femke Reitsma for organizing and chairing it; Diane Holfelner, Linda Doerfler, LaDona Knigge, Pat Shyhalla, and others at NCGIA-Buffalo; Chip Day, Patti Perks, and other staff at the Holiday Valley resort and conference center; and Bruce Kolesnick and Mable Tartt Sumpter at the University at Buffalo Office of Conferences and Special Events for handling the conference registration. We thank Matt Duckham for continuing to maintain the www.cosit.info domain.

July 2005   Tony Cohn and David Mark

Organization

Program Chairs

Anthony G. Cohn University of Leeds, UK
David M. Mark　University at Buffalo, USA

Steering Committee

Anthony G. Cohn University of Leeds, UK   (Co-chair)
David M. Mark   University at Buffalo, USA   (Co-chair)
Michel Denis   Université de Paris-Sud, France
Max Egenhofer   University of Maine, USA
Andrew Frank   Technical University of Vienna, Austria
Christian Freksa University of Bremen, Germany
Stephen Hirtle   University of Pittsburgh, USA
Werner Kuhn   University of Münster, Germany
Benjamin Kuipers University of Texas, USA
Daniel Montello   University of California, Santa Barbara, USA
Barry Smith   University at Buffalo, USA
Sabine Timpf   University of Zurich, Switzerland
Barbara Tversky   Stanford University, USA
Michael Worboys University of Maine, USA

Program Committee

Gary Allen, USA
Thomas Barkowsky, Germany
John Bateman, UK
Brandon Bennett, UK
Michela Bertolotto, Ireland
Thomas Bittner, Germany
Mark Blades, UK
Gilberto Camara, Brazil
Roberto Casati, France
Eliseo Clementini, Italy
Helen Couclelis, USA
Matteo Cristani, Italy
Leila de Floriani, Italy
Matt Duckham, Australia
Geoffrey Edwards, Canada
Max Egenhofer, USA
Carola Eschenbach, Germany
Sara Fabrikant, USA
Andrew Frank, Austria
Christian Freksa, Germany
Mark Gahegan, USA
Anthony Galton, UK
Chris Gold, UK
Reg Golledge, USA
Christopher Habel, Germany
Kathleen Hornsby, USA
Chris Jones, UK
Marinos Kavouras, Greece

Markus Knauff, Germany
Werner Kuhn, Germany
Lars Kulik, Australia
Gerard Ligozat, France
Reinhardt Moratz, Germany
Bernhard Nebel, Germany
Dimitri Papadias, Hong Kong, China
Juval Portugali, Israel
Jonathan Raper, UK
Martin Raubal, Germany
Jochen Renz, Australia
Andrea Rodriguez, Chile
Christoph Schlieder, Germany
Michel Scholl, France
Barry Smith, USA
David Stea, USA
John Stell, UK
Holly Taylor, USA
Andrew Turk, Australia
Barbara Tversky, USA
David Uttal, USA
Laure Vieu, France
Stephan Winter, Australia
Michael Worboys, USA
Wai-Kiang Yeap, New Zealand
May Yuan, USA

Additional Referees

Maureen Donnelly
Stephen Hirtle
Dan Montello

Table of Contents

IV   Ontology and Spatial Relations

V   Spatial Reasoning

VI   Cognitive Maps and Spatial Reasoning

VII   Time, Change, and Dynamics

VIII   Landmarks and Navigation

IX   Geographic Information

X   Spatial Behavior

XI   Abstracts of Keynote Talks

Anchoring: A New Approach to Handling Indeterminate Location in GIS

Antony Galton and James Hood

University of Exeter, UK
{A.P.Galton, J.M.Hood}@exeter.ac.uk

Abstract. We describe a new approach to representing vague or uncertain information concerning spatial location. Locational information about objects in information space is expressed through various *anchoring relations* which enable us to state exactly what is known regarding the spatial location of an object without forcing us to identify that location with either a precise region in the embedding space or any precise mathematical construct from such regions, such as rough sets or fuzzy sets. We describe the motivation for introducing Anchoring, propose the beginnings of a formal theory of the anchoring relations, and illustrate some of the ideas with examples typical of the real-life use of GIS.

1 Introduction

In this paper we introduce Anchoring, a new approach to handling vague and uncertain location in GIS. We use Anchoring to frame and approach one particular problem: Within the spatial data model of our GIS, how is it possible to reference spatial information whose location cannot be assigned to precise coordinates?

A key motivation behind Anchoring was to develop a framework or model that preserves any imprecision or vagueness, and does not eliminate it through any forced approximation to precise regions. Many existing approaches are based on the faulty assumption that vague or imprecise objects can, indeed must, be associated with regions which represent their spatial extent. No matter whether these regions are 'fuzzy', or the intersection of a set of 'precisifications', we feel that any precise representation of a vague or imprecise spatial extent, though useful at a stage of analysis, fails to capture a key aspect of the region (namely, its vagueness or imprecision), and should therefore be omitted from any spatial data model.

With Anchoring we try to keep the vagueness or uncertainty of an object's location 'alive' by saying only what we can say about it precisely and leaving everything else for subsequent analysis. What can be said precisely about a region whose boundary is imprecise or vague sometimes includes its topological relationships to regions whose boundary is precise. Anchoring uses these relationships to give vague or imprecise regions reference within a spatial data model.

2 The Problem of Indeterminate Location

Frank [3] articulates the problem thus:

A.G. Cohn and D.M. Mark (Eds.): COSIT 2005, LNCS 3693, pp. 1–13, 2005.
© Springer-Verlag Berlin Heidelberg 2005

'Space in today's GIS is always seen as Euclidean space, represented with Cartesian coordinates ... Data that are not related to a point with given position cannot be entered in a GIS. This limits the use of GIS for combining verbal reports about events, such as accidents.' [3]

To illustrate, suppose we wish to record in our GIS information about a car accident, but all we know about the location of the accident is that it is 'somewhere' on the stretch of motorway which lies between two given junctions. The data model of our GIS, quite standardly, uses the coordinatised Euclidean plane; accidents would be represented as geometric points, and stretches of motorway as connected sets of geometric line segments. Our problem is that without assigning the accident to a particular point, specified by a pair of Cartesian coordinates, we cannot represent the accident in the spatial data model. Of course, in reality we expect the emergency services to arrive at the scene of such an accident quickly, and they could then give us a precise description of the accident's location. But what if the data we are entering into our GIS is from historical sources, and the GIS is being used for analysis of this data? Or if the accident was not on a motorway but in a remote area like, say, a national park, and the information is provided through possibly impressionistic verbal reports lacking precise coordinates? In such situations our information might never be precise enough to represent the accident as a geometric point.

This example involves uncertainty, but related problems arise in connection with vagueness—although vagueness and uncertainty are quite distinct concepts, they have enough commonality in the kinds of representational problems they raise to justify our treating them together, at least in some respects. In both cases we see a kind of *indeterminacy*, and it is this aspect which is primarily addressed by our Anchoring theory.

Vagueness arises, amongst other places, in connection with descriptive terms such as 'hill', 'valley', or 'town centre', whose referents, while certainly located in space, cannot be assigned precisely delineated locations, short of an arbitrary and contestable decision. In Figure 1(a), for example, the name 'Greendale' clearly refers to the valley whose presence is evident to the map user from the contours shown; the map rightly makes no attempt to assign a region with precise boundaries as the referent of 'Greendale', and yet it cannot truly be claimed that the map is on *that* account ambiguous or incomplete: after all, we all know what a valley is, and can see from the particular configuration of contours that there is a unique valley associated with the name 'Greendale'. This does not mean that we can always give definite answers to questions such as whether or not Green House is in Greendale: it is part of our common-sense understanding of the notion of a valley that answers to such questions will sometimes be indeterminate. To be entirely certain *which* valley we are talking about, we can always refer to it as the valley of the River Green; and while this fixes the valley beyond doubt it still does nothing to make more precise its exact extent.

Likewise, in Figure 1(b), Green Hill is the hill which has a summit at elevation 220m at the exact coordinates where this is marked, and from our general understanding of what hills are this suffices to establish which hill bears that name, without in any way importing the spurious precision that would come from delineating some precise outer boundary and thereby assigning to the hill some exact spatial extent.

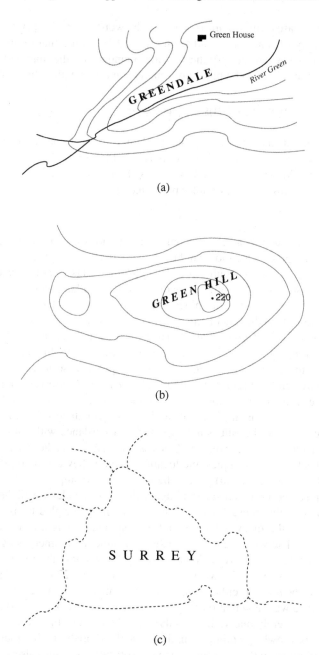

Fig. 1. Placement of names on maps. Unlike 'Surrey', neither 'Greendale' nor 'Green Hill' is associated with a region having a precisely delineated boundary. None the less, all three maps are both clear and informative.

Contrast these cases with Figure 1(c), where the word 'Surrey' is printed to indicate the English county of that name, and from our understanding that counties do have precise boundaries we can identify the relevant boundary on the map and understand the spatial extent of Surrey to be precisely what is contained within that boundary.

3   Outline of Some Existing Approaches

A number of methods have been proposed for handling locational vagueness and uncertainty in the context of information systems in which spatial location can only be specified precisely. We shall review a few of these here, and in particular shall examine how they fare in relation to the examples presented above.

Fuzzy Sets

If we think of our space as a set S of points, each endowed with precise numerical coordinates, then a precise location can be understood as some subset L of that space. For each point $p \in S$, we have either $p \in L$ or $p \notin L$. This is what is meant by saying that L is a *crisp* set, and standard set theory admits no other kind. A crisp set L is completely specified once it is determined, for each $p \in S$, which of $p \in L$ and $p \notin L$ holds. A fuzzy set [8], on the other hand, is a more complex beast: to specify it, we must state, for each $p \in S$, the *degree* to which p may be regarded as belonging to L. This is a number in the range $[0, 1]$, and in principle all real numbers in this range are possible fuzzy membership degrees. Crisp sets arise as the special case in which the number assigned is always either 0 (corresponding to $p \notin L$) or 1 (corresponding to $p \in L$), whereas in a general fuzzy set, the points assigned value 0 are separated from those assigned 1 by a zone consisting of points assigned intermediate values. It is normally assumed, moreover, that the values are assigned in accordance with some continuous function, so that close together points will be assigned close together values.

Referring back to our examples, the location of the valley can be represented as a fuzzy set, as shown in Figure 2(a), where the fuzzy membership values for the valley's location are presented by means of shading, with lighter shades for higher values. A big problem here is that there does not seem to be any principled basis on which to decide exactly how the fuzzy values should be assigned; this is illustrated by Figure 2(b), where a completely different, but equally 'plausible' assignment is used.

In the case of the motorway accident, if all we know about the location of the accident is that it is somewhere between, say, junctions 26 and 27 on the M5, then there does not seem to be *any* set, either crisp or fuzzy, which can represent the location of the accident. Since we know that (at the granularity we are working with) this location is a point, and are merely uncertain as to the exact location of that point, a better way of expressing this knowledge might be in the form of a probability distribution—in this case a flat distribution on the relevant stretch of motorway, and zero elsewhere.

Rough Sets, Egg-Yolk Theory, and Supervaluation

Under this heading we include all those approaches which assign to a vague or uncertain entity not one precise region, but *two*: an 'inner' region containing all points which are uncontestably within the location of the entity of interest, and an 'outer' region

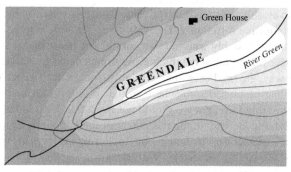

(a) A fuzzy membership function for 'Greendale'

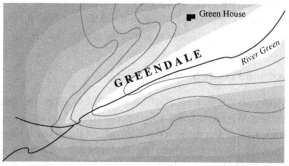

(b) Another fuzzy membership function for 'Greendale'

Fig. 2. Arbitrariness of fuzzy membership functions: How do we choose between the functions shown here, or any of the infinitely many others we could have drawn?

containing all points which are *not* uncontestably *outside* that location. In egg-yolk theory [1], the former is likened to an egg-yolk, the latter to the whole egg; in rough set theory [4], they are the inner and outer approximations respectively.

Supervaluation semantics [2], when applied to spatial location (as opposed to its more usual application area of vague concepts), ends up with a similar picture. If the location of an object is vague, then there may be many different ways of making it precise—these are called 'precisifications' of the location. Supervaluation theory works with propositions, so we must consider the truth or falsity of propositions of the form 'point p is in region R' (i.e., $p \in R$) under different precisifications of R. Propositions which come out true under every precisification are called 'supertrue'—this corresponds to our earlier use of the phrase 'uncontestably true', and therefore the points p for which the proposition '$p \in R$' is supertrue correspond to the egg-yolk; likewise, the whole egg consists of those points for which '$p \notin R$' is *not* supertrue.

Applying this to our hill example, we see that there are many different ways in which we could place the inner and outer boundaries. We are confronted with a similar problem to the fuzzy set case: the theory forces us to make some arbitrary decisions which embody more precise information than we are entitled to assume.

We can, however, salvage something of value from this situation by interpreting the two boundaries somewhat differently. The yolk must certainly only contain points for which $p \in R$ is uncontestably true, but we should not insist that it contain *all* such points, since to determine precisely which those points are is by nature an impossible task (put another way, the predicate 'uncontestably true' is itself vague). Likewise, the outer boundary must be placed somewhere so that it only *excludes* points for which '$p \in R$' is uncontestably *false*, but we should not insist that it excludes all such points. This gives considerable lee-way in the placement of the boundaries, and indeed points the way to the freer approach embodied in our Anchoring theory. This understanding of vague boundaries was well expressed by Wittgenstein:

> The boundaries ... are still only like the walls of the *forecourts*. They are drawn arbitrarily at a point where we can still draw something firm. — Just as if we were to border off a swamp with a wall, where the wall is not *the* boundary of the swamp, it only stands around it on firm ground. It is a sign which shows there is a swamp inside it, but not, that the swamp is exactly the same size as that of the surface bounded by it. [6, §211]

Information Space and Precise Space

By way of introduction to Anchoring, it will be helpful to adopt the following perspective on spatial location. We are talking about geographical features that are located in space. We have a descriptive language for talking about these features, and in particular this allows us to say things about where the features are located in relation to other features, e.g., 'My house is north of the town centre'. If we consider all the things we can say about these features, short of assigning exact absolute locations, we can think of this as constituting an *information space*—it is a space in the sense that it includes information about the locations of features relative to each other, but also in the more abstract sense of being an assignment of properties and relations (spatial or otherwise) to a collection of entities. It is the sort of information that could be captured in a relational database. However a GIS is not just a relational database; the entities it handles are embedded in a space—an 'embedding space' [7]—whose structure constrains the properties and relations that it is possible to assign to those entities. To use a GIS to record information about the objects in the information space, we have to bring those objects into relation with the locations we can define in this embedding space.

We define a *precise space* as any embedding space in which phenomena can be assigned exact positions relative to some numerical coordinate system. The coordinatised Euclidean plane is an example of a precise space. So, too, is the surface of a sphere under the latitude and longitude coordinate system. And aside from mathematical objects, any map, drawing or other representation of the earth's surface can function as a precise space, so long as each point on the representation corresponds to a fixed geographic position according to a fixed scale or projection.

In what we might call the 'classical' GIS data model, the embedding space is always a precise space, and there is only *one* kind of relation between information space and that precise space, namely *exact location*. That is, each element of information space is assigned to a spatial location (point, line, polygon, ...) defined in the precise space. In effect the object is *identified* with the location it is assigned to (see Figure 3).

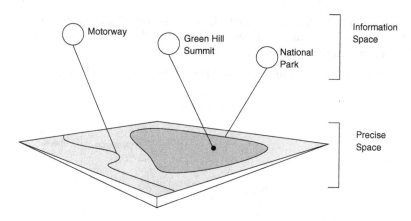

Fig. 3. The 'classical' view of a GIS data model with just one kind of relation (exact location) between objects in information space and locations in precise space

In fuzzy set approaches, it is recognised that for many objects in information space it is not possible to assign an exact location in precise space without falsifying reality. They therefore replace exact location with *fuzzy location*, which assigns to an object O in information space not one exact location but rather, in effect, an infinite set of exact locations, each indexed by a degree of fuzzy membership. On the other hand, rough set and egg yolk approaches assign a *rough location* consisting of just *two* exact locations, the inner and outer approximations.

By presenting things in this way, we intend to highlight the fact that the problem of vague location is *to establish the nature of the relationship between information space and precise space*. Our guiding insight is that rather than constraining this relationship to some particular very precise form (be it exact location, fuzzy location, or rough location), we should take a freer and more relaxed view of the possible relationships between the two spaces, in particular allowing for *more than one* type of relationship between an entity in information space and an exact spatial location.

4   Anchoring Relations

The underlying ontology of Anchoring consists of two levels, representing information space and precise space. The elements of precise space are points and various kinds of sets of points which can function as precise spatial locations; the elements of information space are all the various geographical objects and phenomena of interest. The key idea is that an object in information space may be *anchored* to locations in precise space in various different ways. We leave it open at this stage just how many different ways we will need, but here are some we have found useful so far:

- Object O is *anchored in* region R; that is, whatever may be the nature of O's spatial location, we can certainly say that O is located within R.
- Object O is *anchored over* region R; that is, we can certainly say that R falls within the location of O (in many cases, R will in fact be a point).

– Object O is *anchored outside* region R, i.e., no part of O is located within R.
– Object O is *anchored alongside* region R, meaning that O is so situated that it abuts R (in many cases, R will be a line).

Some of these relations are illustrated, using examples from §5, in Figure 4.

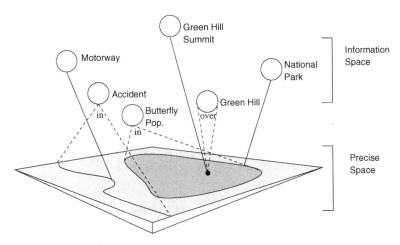

Fig. 4. Broken lines indicate anchoring relations, solid lines indicate exact location

Superficially, it might be thought that all we have done here is to replicate standard relations between regions such as the RCC relations $P(O, R)$, $PI(O, R)$, $DC(O, R)$, $EC(O, R)$. However, this would be to radically misunderstand our intentions: the O term here is *not* a precise spatial location—that is, it is not that we are merely ignorant of its location, but rather that in the case that O is irreducibly vague, it simply does not have a precise location. This prevents us from writing, e.g., $P(O, R)$, since 'P' in RCC denotes a relation between precise locations; but we can still say, in our theory, that O (an object) is anchored in R (a precise region).

Even though objects in information space may be vaguely located, they may bear definite spatial relations to each other. For example, even if the terms 'Central London' and 'Greater London' are not precisely defined, we can certainly say that Central London is part of Greater London, and this places constraints on the possible anchorings: for example, Greater London must be anchored over any precise location that Central London is anchored over, and Central London must be anchored in any precise location that Greater London is anchored in. Such observations suggest the possibility of developing a formal calculus of the anchoring relations. In the next section we sketch the beginning of such a calculus; its further development is work for the future.

An object O that is exactly located at L is anchored *both in and over L*; this provides the Anchoring definition of exact ('classical') location. O still belongs to information space, but now we can reasonably *identify* it with L in the manner of classical GIS. For example, assuming the boundaries of the county of Surrey are precisely delineated, we can use 'Surrey' as the name, not only of a county (i.e., a particular type of administrative unit) but also of an exact spatial location, a subset of precise space. We can go on to say things like 'Box Hill is anchored in Surrey' as a shorthand for 'Box Hill is anchored in the region R such that Surrey is anchored both in and over R'. Bearing this convention in mind, here are some examples of anchoring:

- Brighton is anchored alongside the Sussex coastline.
- The North Atlantic Ocean is anchored outside the Southern Hemisphere.
- Eastern Europe is anchored over both Latvia and Ukraine.
- The Black Forest is part of Southern Germany, and Southern Germany is anchored in Germany, so the Black Forest is anchored in Germany.
- In Figure 1, Greendale is anchored over a stretch of the River Green, and Green Hill is anchored over the summit point at 220m.

4.1 Towards an Axiomatic Treatment of Anchoring Relations

We list here a plausible set of axioms for the part of Anchoring concerned with anchoring in and anchoring over, together with some of their consequences. Our notational conventions are as follows. We write

- bold lower-case letters (e.g., $\mathbf{v}$) to denote elements of information space (these are elements whose locations are described using anchoring relations);
- italic upper-case letters (e.g., P) to denote precise regions in the precise space;
- $\mathbf{v} \rhd P$ to mean that $\mathbf{v}$ is anchored over P.
- $\mathbf{v} \lhd P$ to mean that $\mathbf{v}$ is anchored in P.
- P and O for the RCC relations 'is part of' and 'overlaps'; these are used exclusively between elements of information space, since in the precise space, whose elements are sets of points, we can use standard set-theoretic notations;
- $\mathbf{v} + \mathbf{w}$ and $\mathbf{v} \cdot \mathbf{w}$ to denote the mereological sum and product of $\mathbf{v}$ and $\mathbf{w}$. Again, these are used exclusively over the information space; in precise space we can use the set-theoretical notations $\cup$ and $\cap$.

Using this notation, our proposed axioms are as follows:

1. If any part of $\mathbf{w}$ is anchored over P then $\mathbf{w}$ is anchored over any part of P:

$$\mathbf{v} \rhd P \wedge \mathsf{P}(\mathbf{v}, \mathbf{w}) \wedge Q \subseteq P \rightarrow \mathbf{w} \rhd Q$$

2. If $\mathbf{v}$ is anchored in part of P then any part of $\mathbf{v}$ is anchored in P:

$$\mathbf{v} \lhd Q \wedge \mathsf{P}(\mathbf{w}, \mathbf{v}) \wedge Q \subseteq P \rightarrow \mathbf{w} \lhd P$$

3. If two objects are anchored over the same region, then they overlap, and their common part is also anchored over that region:

$$\mathbf{v} \rhd P \wedge \mathbf{w} \rhd P \rightarrow \mathsf{O}(\mathbf{v}, \mathbf{w}) \wedge \mathbf{v} \cdot \mathbf{w} \rhd P$$

4. The sum of two objects anchored in the same region is also anchored in that region:

$$\mathbf{v} \lhd P \wedge \mathbf{w} \lhd P \rightarrow \mathbf{v} + \mathbf{w} \lhd P$$

5. An object anchored over two regions is anchored over their union:

$$\mathbf{v} \rhd P \wedge \mathbf{v} \rhd Q \rightarrow \mathbf{v} \rhd P \cup Q$$

6. An object anchored in two regions is anchored in their intersection:

$$\mathbf{v} \lhd P \wedge \mathbf{v} \lhd Q \rightarrow \mathbf{v} \lhd P \cap Q$$

7. An object anchored in a region lies within any object anchored over that region:

$$\mathbf{v} \lhd P \wedge \mathbf{w} \rhd P \rightarrow \mathsf{P}(\mathbf{v}, \mathbf{w})$$

8. Any region an object is anchored over is part of any region it is anchored in:

$$\mathbf{v} \lhd P \wedge \mathbf{v} \rhd Q \rightarrow Q \subseteq P$$

Some straightforward consequences of these axioms (together with standard set theory and mereology) are:

9. $\mathbf{v} \rhd P \cup Q \leftrightarrow \mathbf{v} \rhd P \wedge \mathbf{v} \rhd Q$ (From 1 and 5).
10. $\mathbf{v} \lhd P \cap Q \leftrightarrow \mathbf{v} \lhd P \wedge \mathbf{v} \lhd Q$ (From 2 and 6).
11. $\mathbf{v} \rhd P \wedge \mathbf{w} \rhd Q \rightarrow \mathbf{v} + \mathbf{w} \rhd P \cup Q$ (From 1 and 5)
12. $\mathbf{v} \lhd P \wedge \mathbf{w} \lhd Q \wedge \mathsf{O}(v, w) \rightarrow \mathbf{v} \cdot \mathbf{w} \lhd P \cap Q$ (From 2 and 6)

Anchoring is more general than theories based on the 'egg-yolk' idea or rough sets. In such theories, a vague region $\mathbf{v}$ is represented as a pair of precise regions $(P_\mathbf{v}^-, P_\mathbf{v}^+)$, where $P_\mathbf{v}^- \subset P_\mathbf{v}^+$. In anchor notation we would write $\mathbf{v} \rhd P_\mathbf{v}^- \wedge \mathbf{v} \lhd P_\mathbf{v}^+$. But Anchoring does not impose any requirement that there should be, for each vague region $\mathbf{v}$, exactly two precise regions P and Q such that $\mathbf{v} \rhd P \wedge \mathbf{v} \lhd Q$, nor, even if this is the case for a particular $\mathbf{v}$, would it in any way equate $\mathbf{v}$ with the pair (P, Q), instead leaving it open that at a later time, more precise anchoring information concerning $\mathbf{v}$ may be acquired without the identity of $\mathbf{v}$ as a single vague region being disrupted.

5 Illustrative Studies

Although the axioms above may seem straightforward—even trivial—any implementation of Anchoring should respect them, for they will play a role in almost any inference from anchoring information. For example, suppose we know that a certain hut is located on Green Hill, and that the hill is anchored over the hill summit, whose exact location is the region A. If we now learn that the precise location of the hut is a region B, then axiom 5 assures us that the hill is anchored over the union $A \cup B$.[1]

5.1 Uncertain Location: Motorway Accidents

Suppose we want to use an anchoring-enabled GIS to record road accidents. In our example, all we know about a certain accident is that it occurred on the M5 between junctions 26 and 27. To enter this into our system, we activate the Location dialogue, which gives us the option of either entering Exact Location or Anchoring Information.

---

[1] The further, non-monotonic, inference that the hill is anchored over some larger connected region incorporating both A and B (e.g., their convex hull) is not licensed by Anchoring. In future work we propose to explore such non-monotonic extensions to the theory.

We choose the latter, and select the 'Anchored In' option. (Since the location of the accident is effectively a point, anchoring-in implies uncertainty rather than vagueness here.) We specify the stretch of the M5 between junctions 26 and 27. Similarly, we can enter a second accident known to have taken place between junctions 28 and 30.

Later, we want to query our system to find all motorway accidents in Devon during the period of interest. The boundary between Devon and Somerset crosses the M5 between junctions 26 (which is in Somerset) and 27 (in Devon). The system searches the database and identifies our second accident, between junctions 28 and 30, as definitely located in Devon; but it also finds the first one and returns it as 'possible' (pending further information), since it is known to have occurred on a stretch of road part of which is in Devon. Suppose we now add to the database the information that this accident took place west of the point where the A38 crosses over the M5. We now know that this object is anchored in the stretch of motorway between junctions 26 and 27, and also in the stretch of motorway west of the A38 crossing. By axiom 6, the system can infer that the accident is anchored in the stretch of motorway between the A38 crossing and junction 27, and since this stretch is entirely within Devon, the query would this time return that accident as definitely in Devon.

Suppose now that our query is for accidents in the Exeter area (Exeter is situated close to junction 30 of the M5). To define the Exeter area, we again select Anchoring Information rather than Exact Position. Here is a good case for 'egg-yolk' style inner and outer bounds (interpreted in the more flexible Wittgensteinian manner). We say that the Exeter area is anchored over the city of Exeter itself (some suitable exactly-located administrative boundary is presumably available for this in our database), and anchored in a circle of radius 10 miles, say, centred on the centre of Exeter. Then the system will give us a list of definites and possibles, and the possibles can be annotated with the source of uncertainty: some will be possible because their exactly-known location falls within the outer region but not the inner, some will be possible because their own locations are uncertain but they are anchored in a region which overlaps the definite part of the Exeter area, and some will suffer from both types of uncertainty.

It might be objected that we have no business to be handling data as ill-defined as 'an accident on the M5 somewhere between junctions 28 and 30'; but of course we may have no choice. This will often be the case with historical data, where greater precision is not available. But even with more up-to-date information such as recent motorway accidents, we may be at the mercy of information supplied by members of the public, who may be less than precise in their recall of the relevant details. If your GIS insists on only having precise information then you either have to make some of the details up (and no doubt this happens quite often) or you must ignore any information that lacks the required precision (this doubtless happens too). Anchoring provides a way of *not being embarrassed* by the uncertainty or vagueness that inevitably attends much of our information.

5.2   Vague Location: Butterfly Population

Assume that our GIS is used to record information about rare species in a national park. The boundary of the national park is precisely defined, so it can be assigned a precise location. Suppose that a population of Heath Fritillary (*Melitaea athalia*), a rare species of butterfly, lives within the boundary of the national park.[2] By nature, the Heath Fritillary population cannot be assigned a precise location, consisting as it does of many scattered individuals which are constantly moving around. On the other hand, there *are* some definite facts we can record; for example (1) the population is confined within the boundaries of the national park, (2) there are certain point locations within the park where the butterflies have been observed, and (3) there are areas of unsuitable habitat within the park where the butterflies have never been, and are never likely to be observed (see Figure 5). Thus the object P, representing the Heath Fritillary population, is (1) anchored in the precisely delineated region which constitutes the geographical extent of the national park, (2) anchored over each of the individual points where the butterflies have been observed, and (3) anchored outside the regions where the habitat is unsuitable. In the third case, of course, it may be that 'areas of unsuitable habitat' are themselves vague and have to be anchored to some more precisely specifiable regions, in which case it will be these regions which P must be anchored outside.

Given all this information, one could, if necessary, construct a region which amounts to an informed guess as to the location of the population, using the GIS to derive this region if desired. But it would be a mistake to enter this region as the main information concerning the location of the population: it is better to maintain in the database only those pieces of information which are definitely known—in this case the three types of anchor information—and use these to make plausible inferences which may well have to be rescinded if more detailed anchoring information becomes available.

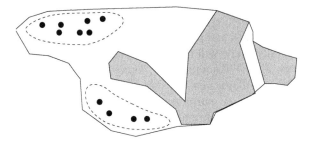

Fig. 5. Butterfly population in a National Park. The black dots represent locations where the butterflies have been observed; the shaded areas represent unsuitable habitat. The broken lines represent a guess as to the region occupied by the butterfly population.

6   Conclusions and Future Work

In this paper we have described a new approach to representing vague or uncertain information concerning spatial location. Locational information about objects in information space is expressed through various anchoring relations which enable us to state exactly what is known regarding the spatial location of an object without forcing us to identify that location with either a region in precise space or any mathematical construct from such regions, such as rough sets or fuzzy sets.

---

[2] In Exmoor, Devon, populations of Heath Fritillary are protected by the Devon Wildlife Trust (http://www.devonwildlifetrust.org), and GIS is used in this work.

We have described the motivation for Anchoring, begun work on a formal theory of anchoring relations, and illustrated these ideas with examples typical of real-life use of GIS. Much remains to be done. First, the formal theory must be extended to cover anchor relations such as anchoring outside and anchoring alongside, and linked more firmly with both the mereotopological structure of information space and the set-theoretic structure of precise space. Second, systematic methods must be developed for applying Anchoring to a wide range of examples of interest to GIS users. Third, we must explore ways of incorporating the theory into existing spatial data models, allowing the kinds of dialogues we have envisaged to be handled correctly and efficiently.

The problem of how to represent vague location has its roots in a much wider problem, namely the integration of field-based and object-based geographical ontologies or data models. Smith and Mark [5] note that

> [t]he completion of such an ontology is a challenging task, involving serious research challenges . . . . But its realization would bring significant benefits, because many serious professional users of geographic information, such as pilots, soldiers, and scientists, hikers, wildfire fighters, and naturalists, must communicate about particular parts of the landscapes as if they were objects, while at the same time drawing on the resources of field-based topographic databases.

The relationship between a hill or a valley and the underlying configuration of contours—as in our Greendale and Green Hill examples—is a good example of this. That valleys, hills, and other geomorphological features do not have precise boundaries reflects their nature as features of the underlying, continuous field of land elevation. Just as people identify hills and valleys on a contour map, all we can ask is that our data models 'point' to areas of precise space in some way associated with these features. This is achieved in our theory by the use of anchoring relations rather than exact location.

References

1. A. G. Cohn and N. M. Gotts. The 'egg-yolk' representation of regions with indeterminate boundaries. In Peter A. Burrough and Andrew U. Frank, editors, *Geographic Objects with Indeterminate Boundaries*, GISDATA 2, pages 171–187. Taylor & Francis, 1996.
2. Kit Fine. Vagueness, truth, and logic. *Synthese*, 30:263–300, 1975.
3. Andrew U. Frank. Spatial ontology: A geographical information point of view. In Oliviero Stock, editor, *Spatial and Temporal Reasoning*. Kluver Academic Publishers, 1997.
4. Z. Pawlak. Rough sets. *International Journal of Computational Information Science*, 11:341–356, 1982.
5. Barry Smith and David M. Mark. Do mountains exist? Towards an ontology of landforms. *Environment and Planning B: Planning and Design*, 30(3):411–237, May 2003.
6. Ludwig Wittgenstein. *Philosophical Remarks*. Basil Blackwell, Oxford, 1975. Edited from his posthumous writings by Rush Rhees and translated into English by Raymond Hargreaves and Roger White.
7. Michael Worboys and Matt Duckham. *GIS: A computing perspective*. CRC Press, Boca Raton, Florida, 2004. Second edition.
8. L. A. Zadeh. Fuzzy sets. *Information and Control*, 8:338–353, 1965.

Gradation and Map Analysis in Area-Class Maps

Barry J. Kronenfeld

Department of Geography, George Mason University,
4400 University Drive, MS 1E2, Fairfax, VA 22030
bkronenf@gmu.edu

Abstract. This paper reports on a preliminary experiment to determine if gradation on area-class maps facilitates complex map analysis tasks. Twenty-nine subjects were shown either a traditional (crisp) map or one of three graded maps created from the same underlying data, and asked to perform analysis tasks involving two different spatial scales and levels of mental map development. Performance on simple lookup tasks was highest using the crisp map, but each of the more complex tasks was performed with greatest accuracy using a map that depicted gradation between regions. The results suggest that portrayal of gradation on area-class maps may facilitate complex analysis tasks involving global map comprehension and memory of map pattern.

1 Introduction

In the late 1980s and early 1990s, the technical debate between the merits of raster vs. vector data models in GIS evolved into a discussion of conceptual models of geographic space that is still ongoing. At the core of this discussion is a binary classification between objects and fields, a distinction that mirrors the atom/plenum view in physics (Couclelis 1992). Objects are generally considered to be independent of one another (Goodchild 1992), have identity and clear boundaries, and be able to be manipulated and moved (Frank 1996). Fields, on the other hand, are spatially continuous and comprehensive schema from which values can be determined at any location within their bounds (Goodchild 1992, Burrough & Frank 1998). The conceptual distinction between objects and fields is easily discerned from most GIS data models: raster grids, contour lines and TINs represent fields, while points, lines and independent polygons implement object conceptualizations.

An exception to the clear linkage between data model and conceptual model is the area-class map, a polygon tessellation in which each polygon is assigned a value in a nominal variable (i.e. assigned to a class). I have found a broad tendency among my students to consider area-class maps as collections of independent objects, and yet their topological dependency identifies them as fields by most technical definitions (e.g. Goodchild 1992). Indeed, some authors classify them as belonging to a separate conceptual model entirely (Plewe 1998, Buttenfield 2000). The difficulty seems to arise from two sources. First, the act of classification attaches a name and thereby an identity to individual polygons. Thus, it is easy to "extract" polygons from an area-class map and treat them as independent regions, whereas, for instance, the individual contour lines of a topological map are rarely considered in isolation. Second, the

A.G. Cohn and D.M. Mark (Eds.): COSIT 2005, LNCS 3693, pp. 14–30, 2005.
© Springer-Verlag Berlin Heidelberg 2005

abrupt nature of boundaries in area-class maps is typical of objects, not fields. This clear discontinuity does not violate any technical rule, but departs from the prototypical field concept of a continuously varying smooth surface.

Given the ambiguity surrounding boundaries on area-class maps, an important question is whether they are in fact a fundamental cognitive element of the region concept, or rather simply an artifact of manual cartography. The latter view holds that boundary delineation is merely an easy drawing task, predicated by the constraints of manual cartographic production but rendered obsolete by the advent of the modern computer. The implication inherent to this view is that human cognition of spatial regions is complex and includes tools for conceptualizing and reasoning about gradation between regions. The former view, in contrast, holds that sharp boundaries are an essential element of human cognition, and suggests that gradation, when it exists in reality, should nevertheless be avoided at least in cartographic representation.

Opinions on this issue are varied, but the question has never been empirically tested. Meanwhile, classification and visualization techniques have progressed to the point where it is possible to produce aesthetically pleasing maps showing gradation between regions (Kronenfeld 2005, Hengl et al. 2002). Such maps may be interesting to look at, but their utility depends on the ability of map users to comprehend and analyze them in different contexts. This in turn depends on the underlying cognitive structure of concepts of geographic regions.

This paper develops one model of how humans might cognize and reason about gradation between regions on an area-class map. The cognitive adequacy of the proposed model is tested via an experiment comparing map comprehension and analysis using maps varying in the degree of visual representation of gradation between regions. The paper proceeds as follows. The next section distinguishes between several concepts related to gradation. In section three, a plausible model of cognition in graded area-class maps is developed, and several experimental hypotheses are formed from this model. Section four describes the experimental design and test instrument. The results are reported in section five, from which the final section draws conclusions and discusses future areas of research.

2   Conceptual Issues

Gradation as used here refers to the gradual change between one set of attribute values and another across a spatial transect. A simple example of gradation is the gradual change in elevation experienced as one proceeds from the bottom to the top of a hill. In this example, there is only one attribute that changes (elevation). In many situations, however, there are numerous attributes that change simultaneously, and it is in such cases that the concept of gradation between *regions* (i.e. areas with similar sets of attributes) becomes useful. For example, in the gradual transition between grassland and desert, numerous variables describing rainfall, vegetation and soil characteristics all change largely in step with one another.

Gradation is related, but not equivalent, to a number of terms used to describe imperfection in geographic information, and so it will be prudent to clarify these terms and their associated concepts here. Duckham et al. (2001) provide a useful typology of imperfection which distinguishes between *error* and *imprecision* at its topmost level: error denotes a difference between observation and reality, while imprecision denotes a lack of specificity in an observation. Their distinction corresponds closely to that made by statisticians between *uncertainty* (the possibility of error) and imprecision.

Both error and imprecision occur in statements about graded regions, but their relationship to gradation often depends on whether one is talking about the regions themselves or the boundaries that delineate them. Statements regarding the error or uncertainty of boundary locations, for example, assume that there is a "correct" location for the boundary, and thus deny the conceptual possibility of gradation. On the other hand, statements regarding the error of assigning a location to a region are not so presumptuous; indeed, it *would* be in error to flatly declare a location to be in one region or the other when in fact it lies in a zone of gradation between regions.

Imprecise statements regarding regions and their boundaries are compatible with, but do not necessarily signify, gradation. The term *vagueness* refers to a special type of imprecision that is most often associated with gradation. Vagueness refers to the lack of clear boundaries between *concepts* (Bennett 2001), i.e. the existence of borderline cases to which it is not clear whether the concept applies (Duckham et al. 2001). Vague statements can, however, be made about sharply delineated regions, as in the statement: "The boundary with Canada lies a few miles to the north." Conversely, statements regarding regions with graded boundaries are often vague but need not be; one might declare, for example, that one has climbed exactly two thirds of the way up a mountain. Fuzzy set theory, in which entities are assigned precise grades of membership in classes, is often criticized for its inability to properly account for vagueness (Haack 1996) and perhaps should be considered a theory of conceptual gradation instead.

According to Varzi (2001), the concept of a vague geographic entity makes no sense, because it cannot be referred to precisely. Using the example of Mount Everest, he asks:

> …just what would this vague object be? How could we
> be so precise as to designate it? *(Varzi 2001, p. 5)*

Identity, however, is a key characteristic of our conceptualization of objects but not fields (Frank 1996). Varzi's comments are therefore most relevant in the context of object-based conceptual models; if regions are considered components of a geographic field, then precise reference to an entity is not necessary. Fields are queried by location first (Burrough & McDonnell 1998), and so region-definitions can be considered attributes to be attached to each location, rather than objects to be delineated.

The distinction between object- and field-based models probably does not fully represent the range of cognitive structures used for spatial reasoning. As noted in the introduction, the application of this distinction to area-class maps is problematic, as regionalizations seem to have characteristics of both objects and fields. As a starting point for determining the role of gradation in cognition of regions, the next section develops a more precise model of how gradation between regions might be conceptualized and used in spatial analysis and reasoning.

3 A Cognitive Model

We will start with a numerical representation of gradation based on fuzzy set theory. The model, which has been widely used, consists of the specification for each location of membership values in a set of classes (Brown 1998, Burrough 1996). Note that these membership values are not intended to portray vagueness, but simply attribute characteristics intermediate between those of class prototypes. The result of such classification in a geographic context is a composite of field layers, each of which represents the distribution of membership in a single class; thus, a location-based query would return a set of membership values in a set of classes. Note that regions are not precisely defined by such a model, but will emerge perceptually if there is sufficient spatial autocorrelation in class membership values.

Interpretation of Grades of Membership

For this method of classification to convey meaningful information, it is necessary to determine what is signified by membership values in classes (Goodchild 2000). This question of meaning is difficult to answer if we conceptualize regions as independent objects, but becomes straightforward in the context of fields. By way of example, the attributes of a location with membership values of $\{0.5, 0.5\}$ in two classes would be inferred to be halfway intermediate to those defined by the classes themselves. Given the height at the base and the top of a slope, this rule could be used to infer the elevation of a location "halfway up the hill" and the answer would seem to match our intuition. Similar examples could be devised for inferring the rainfall and temperature of a location "halfway between" two climate regions, or the vegetation "two thirds of the way between" two biomes.

A formal mathematical description of the above rule can be developed as follows:

$$y_{ij} = \sum_{k=1}^{q} \mu_{ik} c_{kj} \tag{1}$$

where the variables signify the following:

y_{ij}	:	inferred value of attribute j at location i
μ_{ik}	:	membership of location i in class k
c_{kj}	:	prototypical value of attribute j in class k
q	:	number of classes in the classification system

Attributes are thus inferred by linear interpolation between class prototypes. Equivalent formulas have been proposed by several authors independently (McBratney & Moore 1985, p. 184; Zhu 1997, Burrough & McDonnell 1998), suggesting that they might represent a natural inference system.

If the above numerical model is intuitive, then classification using membership values communicates specific information that is not conveyed in traditional area-class maps. The additional information specifies the properties of locations along boundary regions, precisely where error and uncertainty are greatest in crisp maps (Brown 1998). However, the numerical model alone does not explain cognition of spatial *regions*, only the interpretation of individual locations. Further explanation is needed to describe how fields of membership values are modeled via concepts of regions and gradation.

Conceptualization of Gradation Between Regions

Most authors have linked field-based conceptualization of geographic space with experience moving and navigating in complex landscapes (Couclelis 1992, Frank 1996), so a theory of how gradation between regions is actually interpreted should relate to this type of experience. Research in wayfinding has identified landmarks as playing a key role in the development of mental maps, acting as anchor points from which the spatial properties of other locations are inferred (Frank 1996). Couclelis et al. (1987) suggested a "plate tectonics" model of spatial cognition, where such anchor points define regions within which perceived spatial relations are held constant. In this model, perceived changes in relations between regions affect all locations within a region in a coordinated fashion. The researchers reported on the preliminary results of an experiment that supports such a model, but their methods assume regions with discrete boundaries, and do not look for evidence of perceived gradation between regions.

Examining the development of spatial knowledge within individual users, Nyerges (1995) noted the importance of studies that show percepts (perceived images, sounds, etc.) to bypass short-term memory and move directly into long-term memory. Furthermore, these percepts are not filtered when they first enter long-term memory, but only when called back into working memory. This leads to the question of how specific percepts are identified for recall. Consistent with Couclelis et al. (1987), Golledge (1990) believes that at the heart of a mental model must be an *anchor*: "An anchor is a well learned occurrence of something at a location that has relationships to the area (knowledge) surrounding it, i.e. our localized orientation clue" (Nyerges 1995, p. 68). This idea suggests a cognitive structure in which salient, object-like concepts are linked to more complex percepts, which are recalled, through cognitive filters, when the salient concepts are insufficient for a particular task.

These theories suggest a possible model of how information on spatially continuous landscapes is cognitively structured. If the mind registers core regions with known characteristics as anchors, then these could provide links to more complex percepts stored in long-term memory. A "percept" might be a visual mental image or partial image, which would be recalled based on topological relations between anchor points. These percepts, once recalled, could be used to infer characteristics of locations outside of the core regions, or to perform complex analysis tasks.

If the above theory is correct, then we should expect gradation to be more useful for complex tasks than for simple tasks. This is because complex tasks require relational analysis between different map elements, whereas simple tasks require analysis of only one element at a time. When only one map element is analyzed, there is no need for information storage and retrieval, and the simplicity of the relations between map elements and legend information is advantageous. When multiple elements are analyzed, the greater number of classes needed in traditional area-class maps will cause confusion in the storage-retrieval process, as the map user must go back and forth between the map and legend to perform the task. Perceptual information on gradation would cause less confusion because much information could be stored in raw percepts and recalled only when relevant to the task at hand.

On the other hand, if cognitive structures for conceptualizing regions are static and do not include mechanisms for perceiving gradation, then we should expect portrayal of gradation to be an equal or greater hindrance in the performance of complex tasks than in the performance of simple tasks. Perceptual information on gradation would provide little benefit for task analysis, and distract users from more useful map information.

4   Experimental Design

It is a well-known principle in cartography that too much information clutters a map and makes interpretation difficult. This principle can be expressed in terms of the communication model, which developed in the middle of the 20th century and is described in Montello (2002). According to the communication model, maps act as channels that "transmit information from a source (the world) to a recipient (map reader)" (Montello 2002, p. 290). Information is always lost in the transmission from world to map, but errors also occur in the transmission from map to reader. Too much information causes errors in interpretation, and so the manner in which information is encoded should be carefully considered to maximize interpretability.

Considering this cartographic principle, the experiment was designed to compare the effectiveness of maps that contained an equivalent amount of information (i.e. they encoded information from the real world with equal accuracy), but varied in terms of the degree of gradation portrayed. This was achieved by trading off the number of map classes for the degree of gradation. Thus, a map series was created that ranged from a few-class graded area-class map to a many-class crisp map. The map series and methods used to create it are described in detail in Kronenfeld (2005); a general description of the map series is presented here to facilitate description of the experiment.

Map Series

The maps in the series each present an alternate view of the forest regions of New York, New Jersey and Pennsylvania. Data for the maps was derived from USFS inventories (USDA Forest Service 2004) and consisted of twelve layers depicting the distributions of the most common tree genera in the study region. Each layer consisted of a 350x350 raster grid of 2x2km grid cells each containing a numerical *importance value* (IV) representing the local dominance of the tree genus; IVs had a theoretical range of 0% (genus not present) to 100% (all trees belong to genus). Four of the twelve data layers are shown in Figure 1 for illustrative purposes.

From this data, 10 forest region maps were created using automated techniques. Maps with sharp boundaries were classified using k-means clustering, while classification of graded maps used a combination of k-means clustering and PCA-based fuzzy classification (Kronenfeld & Kronenfeld 2004). A 4-class graded map was chosen as one endpoint of the series, as 4 classes was deemed the minimum required to achieve a reasonable depiction of the underlying data. Thirteen classes

Fig. 1. Four of the twelve genus distributions used to create forest region maps. From left to right: birch, cherry, hickory, pine. Darker colors indicate greater abundance, but scale is different on each map.

were required for a crisp map to achieve the same level of encoding accuracy as the 4-class graded map. From the full map series, ranging from the 4-class to the 13-class map, two intermediate maps of 7 and 10 classes were chosen for inclusion in the experiment.

For each of the graded maps, a modified RGB color blending technique was used to paint pixels based on class memberships. Simple linear RGB blending was found to have a desaturating effect on intermediate colors, which has been shown to convey uncertainty (Buttenfield 2000) rather than the desired effect of gradation. For this reason, the linear blending technique was adjusted to enhance saturation for intermediate colors and to assure that color mixtures were perceived as blends of the original colors. Although color schemes exist that create perceptually even gradients, the complexity of implementing of such a scheme was not warranted in the present experiment because precise numerical interpretation of membership values was not required.

Two variables were measured to capture the visual and content characteristics of each map. Map encoding accuracy was measured as the percent variance of the underlying data explained (PVE) by each area-class map. PVE here is the same measure commonly used in principal components analysis. In the context of area-class maps PVE can be interpreted as follows: 100% PVE signifies that all information from the original data is captured by the classified map, while 0% PVE would correspond to a hypothetical 1-class map, in which the single map class describes the average value of each map variable. PVE was held as constant as possible in the map series, so that each map encodes the underlying data with equal accuracy. PVE for the 4-, 7-, 10- and 13-class maps were 85.32%, 85.26%, 85.32% and 85.73% respectively.

Visual fuzziness was measured using a confusion index (CI) that was computed as a function of class membership values. The CI ranged from 0 for pixels assigned to a single forest class, to a maximum theoretical value of 1 for equal blending of the colors of an infinite number of classes. Overall CI for each map was calculated as the average of the CI for each grid pixel. Confusion indices decreased as the number of classes increased, indicating that encoding of information via gradation in the fewer-class maps was replaced by encoding in a greater number of map classes. CI values for the 4-, 7-, 10- and 13-class maps were 0.54, 0.24, 0.13 and 0 respectively.

Fig. 2. A representative region in four maps used in experiment. From 4-class map (left) to 13-class map (right). Darker colors indicate higher membership values.

Figure 2 presents grayscale distributions of one forest class in each of the four maps for illustrative purposes. In the 4-class map (left), significant membership of the class extends through much of the study area, although a core region in the Adirondack Mountains of upstate New York is apparent. As the number of classes is increased, the spatial extent of each individual class decreases, and class boundaries become more distinct. The full-color maps are available from the author upon request.

Legends

The structure of the map legends was designed for quick graphical reference, requiring as little reading of text as possible. Importance values of individual tree genera were presented using proportionally sized circles. A circle was chosen as a symbol so that comparisons could be made easily in both the vertical and horizontal directions. The experiment was designed so that only binary comparison, not precise estimation, of dominance values was required. Colors for the first twelve map classes were selected from the *ColorBrewer* website (Harrower & Brewer 2003). Their color scheme included only twelve colors, so the final color used in the 13-class map was selected by the author to be maximally distinct from the other colors.

Determining which color to assign to which map class was a difficult process because limitations in the expected number of subject participants excluded the possibility of testing several random color configurations. To make comparison between maps as fair as possible, colors were made consistent across maps; that is, the color assigned to a class in any fewer-class map was also assigned to the co-located class on every subsequent map. Beyond this rule, the color scheme was designed to emphasize as much as possible the overall pattern in each map, and to avoid conflicting or ambiguous color combinations.

Analysis Tasks

To test the theory that gradation would be more useful for facilitating complex tasks than simple tasks, two dimensions of map analysis were derived from the cartography-related literature and incorporated into the experiment: type of communication and scale of analysis. Type of communication refers to the position along a sequence of information processing that begins with exploratory analysis and ends in decision-making, and relates to a basic distinction between *reference maps* and *thematic maps* in cartography. According to Petchenik (1979), reference maps are intended to facilitate quick lookup of information, while the purpose of thematic maps is to provide general information on the distribution of phenomena. Thematic maps thus are intended to educate and stimulate exploratory analysis, while reference maps allow quick retrieval of information. MacEachern (1994) distinguishes between four stages of communication: exploration, confirmation, synthesis and presentation.

Table 1. Classification of map analysis tasks for hypothesis testing

exploration vs. presentation	←spatial scale of analysis→	
	local lookup	*global lookup*
	local recall	*(global recall)*

A number of cartographers have also emphasized the spatial scale of analysis by the end user as an important consideration in the production of maps. Bertin (1983, p. 151) distinguished between three levels of reading in graphical images: the overall level, in which the user sees the whole image and compares it to other images; the intermediate level, in which the user sees partial images and compares them with each other; and the elementary reading level, in which single elements are isolated and compared.

The two distinctions above led to a two-dimensional classification of map analysis tasks, summarized in Table 1. First, map analysis tasks may require identification of information patterns at different spatial scales, ranging from the *local* to the *global*; second, analysis may involve identification of information via direct map *lookup*, or via *recall* through cognitive maps developed over time through map reading and exploration.

It was hypothesized that performance of *local lookup* tasks would be quicker and more accurate with the crisp 13-class map, but that *global lookup* and *local recall* tasks would be facilitated by gradation. The hypothesis was somewhat vague, in that it did not necessarily predict that performance on any task would be best using the 4-class map. This is because the visual clarity of the 4-class map was too poor to develop clearly defined core regions as cognitive anchor points. Rather, it was hypothesized that some degree of gradation would result in better performance than no degree of gradation for these tasks. This vagueness brings up the problem of multiple working hypotheses, which will be addressed in the results section.

Test Instrument and Subjects

The test instrument was a Visual Basic program written by the author that led subjects through a series of tasks in five sections, including sections of *local lookup*, *local recall* and *global lookup* tasks. Each subject received a single map version for the duration of the experiment.

Subjects consisted of 29 undergraduate students at the University at Buffalo. Four were recruited from an introductory level cartography-related class, each of whom received a different map version. The remaining 25 subjects were recruited from the general student body.

Table 2. Number of questions of each type given to each subject

task number	task name	task description	number of questions
1	*local lookup*	identify three most abundant genera at location	10
2		determine relative similarity of location pairs	5
3		identify location with most of tree type	8
4	*local recall*	identify three most abundant genera at location	6
5	*global lookup*	match tree type with distribution map	6

Prior to map analysis, subjects were asked for information to determine if prior experiences might affect task performance. Three experience variables were recorded: length of residence in the study area, knowledge of the twelve tree genera, and coursework in cartography-related classes.

Several warm-up tasks were given for the purpose of familiarizing the subjects with the ecoregion map and legend, as well as the option buttons, drop-down menus and other controls used to report their responses. Subjects were then led through five types of tasks. The order and number of questions asked for each type of task are shown in Table 2.

In the *local lookup* task (Figure 3a), subjects were asked to predict the three most common tree genera (referred to simply as "tree types" at a given location, identified by a black pointer line with a white shadow. After each response, subjects were provided feedback in the form of an ordered list showing the actual importance value of each genera.

In the *local recall* task, subjects were again asked to predict the three most common tree types at a given location. This time, however, a blank map was shown and they were required to answer from memory. Subjects were not told in advance that they would need to recall information from memory, so this task relied on the natural formation of cognitive maps, as opposed to directed learning. Feedback was provided in the same way as in the *local lookup* task.

Finally, in the *global lookup* task (Figure 3b), subjects were presented with the actual distribution of one of the 12 tree genera, and told that it represented the abundance of a caterpillar pest. The subjects were required to identify the tree type whose distribution most closely matched that of the caterpillar. Qualitative feedback was given based on the correlation coefficient computed between the distribution of the tree genus actually presented and the one selected by the subject.

To stimulate the formation of a mental map by each user, subjects were asked to perform two other types of analysis tasks between the *local lookup* and *global lookup* tasks. Responses to these tasks are not analyzed here.

Versions

For each of the four map versions, two test versions were created using different locations for each of the *local lookup* and *local recall* questions, to ensure that the selection of locations was not biased toward any specific map. Locations were selected randomly, but were all at least 20 pixels (40km) away from the study area boundary. Additionally, the 12 genera were split into two groups of six, and each subject was shown distributions of the genera from one of these groups for the *global lookup* task.

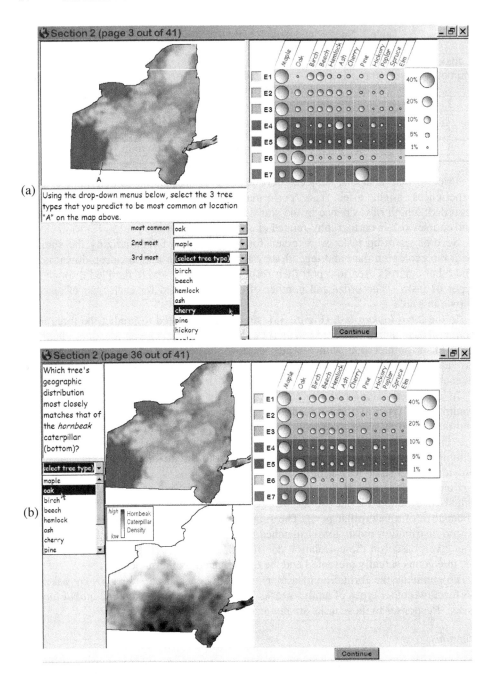

Fig. 3. Screenshots of test instrument. Shown are (a) local lookup task and (b) global lookup task. Local recall task was similar to (a), except map consisted of only a blank outline of the study area and legend was not displayed.

Evaluation

Subjects' answers were compared to the twelve genus distribution layers to evaluate their performance on each task. Because there were good answers as well as best answers to each question, performance on each analysis task was measured on a continuous scale, with a maximum value of 1. To achieve a continuous scoring scale for the *local lookup* and *local recall* tasks, scores were computed as

$$\frac{3 \times IV_1 + 2 \times IV_2 + 1 \times IV_3}{3 \times IV_{max} + 2 \times IV_{2nd} + 1 \times IV_{3rd}} \quad (2)$$

where IV_1, IV_2, and IV_3 signify the importance values of the genera predicted by the subject as 1st, 2nd and 3rd most abundant, and IV_{max}, IV_{2nd}, and IV_{3rd} signify the importance values of the tree types that were actually the most abundant. For the *global lookup* task, the overall correlation coefficient between the tree type whose distribution was displayed and the tree type selected by the subject was used as a measure of task performance.

5   Results

Nine correlation coefficients, between each of the three subject background variables and performance scores on the three analysis tasks, were computed to detect any influence of prior background on the experimental results. Two relationships were found significant or marginally significant: average score on the local recall task was positively correlated with years of residence (r^2=0.09, p=0.03), and average score on the global lookup task was negatively correlated with number of cartography classes taken (r^2=0.04, p=0.09). Graphs of the relationships (Figure 4) showed each correlation to be the result of an individual outlier, and there was otherwise little evidence to suggest that prior background influenced subject performance.

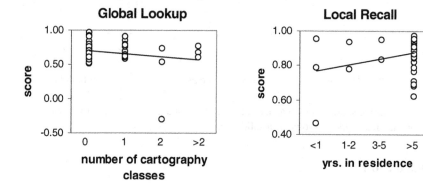

Fig. 4. Statistically significant relationships between prior knowledge and task performance

The t-test was used to evaluate whether performance scores and times for students given the 4, 7 and 10-class maps were statistically different from those of students given the 13-class map. Comparisons were not made among the 4, 7 and 10-class maps because the initial hypotheses were related only to the difference between crisp and graded maps, not the optimal degree of gradation. As expected, performance on the local lookup task was higher for maps that were more crisp (Fig. 5). Only the difference between the 4-class and 13-class maps was statistically significant (p=0.01). This can be attributed to a ceiling effect due to the easy nature of the task for relatively crisp maps; the average score for the seven-class map was already very high (0.96 out of a possible 1).

On the local recall task, performance was more equal across the four maps in the series, but subjects who received the 7-class map scored highest. The difference between scores of subjects given the 7-class and 13-class maps was marginally significant (p=0.09). Interestingly, performance on the 10-class map was poorest overall. On the global lookup task, performance using the 10-class map was significantly higher than for the 13-class map (p=0.03). Average performance was second highest for subjects receiving the 13-class, although the average for the 7-class map was brought down by a single outlier. Excluding this outlier, performance on the global lookup task showed a regular symmetrical pattern around peak performance on the 10-class map.

The time it took subjects to complete each task section was relatively consistent across map versions (Figure 6). Subjects given the 4-class map completed the global lookup task section more quickly than other subjects, but the difference between their times and those of subjects who received the 13-class map were only marginally significant (p=.10).

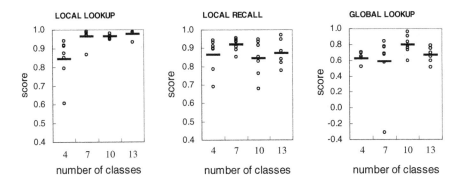

Fig. 5. Performance scores on map analysis tasks. Individual scores marked by circles; average scores for each map version shown as a solid line.

6 Discussion

The concept of *cognitive adequacy* has been used in artificial intelligence research to describe the degree to which a formal representation system corresponds to cognitive mechanisms (Knauff 1999). In the context of area-class maps, inclusion or exclusion of gradation is an important feature of the cognitive adequacy of a representation system. There has been relatively little research on cognitive structures enabling our conceptualization of regions in the context of a field-based conceptual model. It was proposed here that such cognitive structures involve the focal recognition of core regions as anchors, storage of images or other percepts containing information on non-focal regions, and an interpretive linking mechanism to infer information from stored percepts based on relationships to core regions.

The experimental results regarding task performance scores were consistent with this hypothesis. For the equal-accuracy map series, three specific hypotheses – that the crisp map would perform best on the local lookup task, while some degree of gradation would result in better performance for the local recall and global lookup tasks – were each supported. If cognitive structures do not include mechanisms for conceptualizing gradation between regions, then one would expect performance on all three tasks to be best using the crisp 13-class map. This is because information encapsulated in membership value gradients in the fewer-class maps would be more difficult to decode than information encapsulated in the additional regions of the 13-class map.

The experiment presented here had several limitations which suggest avenues for further research. First, the small sample size (n=29) was divided among four different map versions, meaning that only 6-8 subjects received each map version. This reduced the power of the statistical results, as evidenced by the low significance of most comparisons.

Related to the low sample size is the fact that the maps were limited to a single series of equal encoding accuracy. Thus, no comparison was made between performance using crisp vs. graded maps containing the same number of map classes.

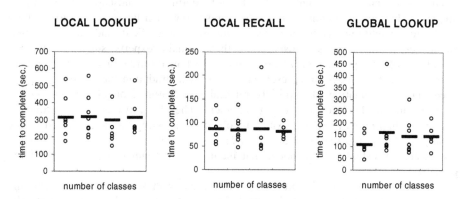

Fig. 6. Performance times on map analysis task sections

The problem is that such comparison by itself would be inconclusive, since differences in performance could be attributed to differences in the overall level of information encoded in each map – the graded map would always encode more information than the crisp map. Constraining the encoding accuracy of the map series limited this effect, under the assumption that, all other things being equal, subjects would perform better (if more slowly) given more accurate representations. However, perceived "clutter" could depend more strongly on the number of map classes than on gradation. Considering the well-known limit of 7±2 items that can be stored in short-term memory, higher performance on the local recall task for the 7-class might be the result of confusion created by the 10- and 13-class maps (although such an explanation is difficult for the higher performance on the global lookup task using the 10-class map). An experiment comparing performance on a crisp vs. graded 7-class map might provide some insight on this issue.

Another consideration regarding interpretation of the experimental results is the fact that it was not specified *a priori* which of the three graded area-class maps would outperform the crisp map on either the local recall or global lookup task. This creates a problem of multiple working hypotheses, which lowers actual statistical significance. On the other hand, of the three specific predictions made, all were supported; this presents a general pattern that is not very likely to have occurred due to random chance alone.

An important technical issue is the effect of legend colors and color blending methods on cognitive perception. The experiment was designed for subjects to make relative comparisons, so that determination of exact membership values from map colors would not be required. This minimized but did not eliminate the effects of differential perceptual qualities of colors. For example, the author noted a perceived dominance of red over weaker yellow in areas where the two colors mixed. Use of a perceptually defined color scheme might help ameliorate this, but it is doubtful that the effect could be fully eliminated due to the tendency for a figure/ground relationship to form between the two colors. Despite this, feedback on the color scheme used in the experiment suggested that overall it was easy to distinguish between classes and identify gradients.

Our understanding of cognitive models of continuous landscapes is still in its infancy. The most common approach to representing such landscapes is to partition them into discrete regions, but the broad recognition of gradients and transition zones among ecologists, soil scientists and other domain experts suggests that the prevalence of this approach may be due more to technical limitations than to cognitive ones. A plausible model was presented here of how gradation might be conceptualized via a combination of explicitly remembered features, or anchor points, and percepts linking these features to one another. The above review of experimental limitations highlights the complexities involved in assessing the cognitive adequacy of any model of the regionalization concept. Overall, however, the experimental results are consistent with a model that includes a mechanism for conceptualizing gradation.

Further research should be directed at corroborating these results and determining the specific mechanism(s) by which gradation between regions is stored cognitively. This was not addressed directly by the present experiment, but two broad possibilities exist. Gradation might be stored in unprocessed images, from which values at specific locations are later recalled. Alternatively, some pre-processing might be performed to create topological relations that include information on the transitional nature of boundaries between regions. If the latter is true, then a spatial data model that replicates these relations might result in better performance on map analysis tasks than the simple raster grid used here.

References

Bennett,B. (2001). What is a forest? On the vagueness of certain geographic concepts. *Topoi*, **20:** 189-201.

Bertin,J. (1983). *Semiology of Graphics*. Madison, Wisconsin: The University of Wisconsin Press.

Brown,D.G. (1998). Classification and boundary vagueness in mapping presettlement forest types. *International Journal of Geographical Information Science*, **12:** 105-129.

Burrough,P.A. (1996). Natural objects with indeterminate boundaries. In: Burrough,P.A. & Frank,A.U. (Eds.), *Geographic Objects with Indeterminate Boundaries*, London: Taylor & Francis.

Burrough,P.A. & McDonnell,R.A. (1998). *Principles of Geographical Information Systems*. New York: Oxford University Press.

Buttenfield, Barbara P. (2000). Mapping Ecological Uncertainty. In: Hansaker, C.T., Goodchild, M.F., Friedl, M.A., and Case, T.J. (Eds.), *Spatial Uncertainty in Ecology*, pp. 116-132. New York: Springer-Verlag.

Couclelis,H., Golledge,R.G., Gale,N. & Tobler,W.R. (1987). Exploring the Anchor-Point Hypothesis of Spatial Cognition. *Journal of Environmental Psychology*, **7:** 99-122.

Couclelis,H. (1992). People manipulate objects (but cultivate fields): beyond the raster-vector debate in GIS. In: Frank,A.U., Campari,I. & Formentini,U. (Eds.), *Theories and methods of spatio-temporal reasoning in geographic space*, pp. 65-77. New York: Springer-Verlag.

Duckham,M., Mason,K., Stell,J.G. & Worboys,M.F. (2001). A formal approach to imperfection in geographic information. *Computers, Environment and Urban Systems*, **25:** 89-103.

Frank,A.U. (1996). The prevalence of objects with sharp boundaries in GIS. In: Burrough,P.A. & Frank,A.U. (Eds.), *Geographic objects with indeterminate boundaries*, London: Taylor & Francis.

Golledge,R.G. (1990). The Conceptual and Empirical Basis of a General Theory of Spatial Knowledge. In: Fischer,M.M., Nijkamp,P. & Papageorgiou,Y.Y. (Eds.), *Spatial Choices and Processes*, pp. 147-168. Amsterdam: Elsevier Science Publishers.

Goodchild,M.F. (1992). Geographical Data Modeling. *Computers & Geosciences*, **18:** 401-408.

Goodchild,M.F. (2000). Introduction: special issue on 'Uncertainty in geographic information systems'. *Fuzzy Sets and Systems*, **113:** 3-5.

Haack,S. (1996). *Deviant Logic, Fuzzy Logic*. Chicago: University of Chicago Press.

Harrower,M.A. & Brewer,C.A. (2003). ColorBrewer.org: An Online Tool for Selecting Color Schemes for Maps. *The Cartographic Journal*, **40:** 27-37.

Hengl,T., Walvoort,D.J.J. & Brown,A. Pixel and colour mixture: GIS techniques for visualisation of fuzziness and uncertainty of natural resource inventories. Hunter, G. and Lowell, K. 2002. Melbourne, Australia, Delft University Press. Accuracy 2002: Proceedings of the 5th International Symposium on Spatial Accuracy Assessment in Natural Resources and Environmental Sciences.

Knauff, M. (1999). The cognitive adequacy of Allen's interval calculus for qualitative spatial representation and reasoning. *Spatial Cognition and Computation*, **1:** 261-290.

Kronenfeld,B.J. & Kronenfeld,N.D. (2004). Minimizing Information Loss in Continuous Representations: A Fuzzy Classification Technique based on Principal Components Analysis.*Conference Proceedings of the joint meeting of The Fifteenth Annual Conference of The International Environmetrics Society and The Sixth Annual Symposium on Spatial Accuracy Assessment in Natural Resources and Environmental Sciences.*, Portland, ME.

Kronenfeld,B.J. (2005). Evaluating Gradation as a Communication Device in Area-Class Maps. *Auto Carto 2005 Proceedings*, CaGIS.

MacEachren,A. (1994). *Some Truth with Maps: A Primer on Symbolization & Design*. Washington D.C.: Association of American Geographers.

McBratney,A.B. & Moore,A.W. (1985). Application of fuzzy sets to climatic classification. *Agricultural and Forest Meteorology*, **35:** 165-185.

Montello,D.R. (2002). Cognitive Map-Design Research in the Twentieth Century: Theoretical and Empirical Approaches. *Cartography and Geographic Information Science*, **29:** 283-304.

Nyerges,T.L. (1995). Cognitive Issues in the Evolution of GIS User Knowledge. In: Nyerges,T.L., Mark,D.M., Laurini,R. & Egenhofer,M.J. (Eds.), *Cognitive Aspects of Human-Computer Interaction for Geographic Information Systems*, pp. 61-74. Dordrecht: Kluwer Academic Publishers.

Petchenik,B.B. (1979). From Place to Space: The Psychological Achievement of Thematic Mapping. *The American Cartographer*, **6:** 5-12.

Plewe,B.S. (1998). Data Modeling, GIS Integration, Vagueness. In: Peuquet,D.J., Smith,B. & Brogaard,B. (Eds.), *The Ontology of Fields: Report of a Specialist Meeting Held under the Auspices of the Varenius Project*, pp. 18-23. Santa Barbara, CA: NCGIA.

USDA Forest Service (2004). What is Foresty Inventory and Analysis? FIA Fact Sheet Series, http://fia.fs.fed.us/library.htm (2 pages).

Varzi,A.C. (2001). Vagueness in Geography. *Philosophy & Geography*, **4:** 49-65.

Zhu,A.-X. (1997). A Similarity Model for Representing Soil Spatial Information. *Geoderma*, **77:** 217-242.

Simulation of Obfuscation and Negotiation for Location Privacy

Matt Duckham[1] and Lars Kulik[2]

[1] Department of Geomatics,
University of Melbourne, Victoria, 3010, Australia
mduckham@unimelb.edu.au
[2] Department of Computer Science and Software Engineering,
National ICT Australia Victoria Laboratory,
University of Melbourne, Victoria, 3010, Australia
lkulik@cs.mu.oz.au

Abstract. Current mobile computing systems can automatically sense and communicate detailed data about a person's location. Location privacy is an urgent research issue because concerns about privacy are seen to be inhibiting the growth of mobile computing. This paper investigates a new technique for safeguarding location privacy, called *obfuscation*, which protects a person's location privacy by degrading the quality of information about that person's location. Obfuscation is based on spatial imperfection and offers an orthogonal approach to conventional techniques for safeguarding information about a person's location. Imprecision and inaccuracy are two types of imperfection that may be used to achieve obfuscation. A set of simulations are used to empirically evaluate different obfuscation strategies based on imprecision and inaccuracy. The results show that obfuscation can enable high quality of service in concert with high levels of privacy.

1 Introduction

Pervasive location-aware systems offer a new class of personalized information based services due to their ability to continuously monitor, communicate, and process information about a person's location with a high degree of spatial and temporal precision and accuracy. Those systems are able to collate large amounts of location information into user profiles that provide a complete history of a user's movements. Although user profiles can be used beneficially to offer highly personalized services to a user, location information is sensitive personal information that needs to be protected.

The protection of location privacy is a crucial factor for facilitating the widespread use of location-aware technologies. Privacy issues are considered to be one of the key research challenges in location-aware computing [11]. Unrestricted access to location information is associated with a range of potential negative effects, including *location-based "spam,"* where businesses could exploit the knowledge of a person's location for unsolicited product marketing; decreased *personal safety*, for example from stalking or assault; and *intrusive inferences*, where a person's political views or individual preferences are inferred from their location (see [7, 13, 9]).

A.G. Cohn and D.M. Mark (Eds.): COSIT 2005, LNCS 3693, pp. 31–48, 2005.
© Springer-Verlag Berlin Heidelberg 2005

1.1   Obfuscation and Automated Negotiation

Our approach to protecting location privacy aims to offer high quality location-based services based on imperfect spatial information. The use of spatial imperfection for privacy is a novel approach suggested in [2], which enables a person to access information relevant to his or her spatial position while safeguarding personal location privacy by revealing the least possible information about that position. We call the process of degrading the quality of information about a person's location, with the aim of protecting that person's location privacy, *obfuscation*.

Individuals using obfuscation should be able to balance their desired level of privacy against their desired quality of location-based service (LBS). In this paper we investigate using automated negotiation in order to achieve a satisfactory balance of the level of privacy and the quality of service. Higher levels of location privacy are likely (although not guaranteed) to entail lower levels of quality for LBS. Achieving the best balance between location privacy and quality of service lies at the heart of successful negotiation strategies. The idea behind automated negotiation is to facilitate practical mechanisms that location-based service providers can implement to attain effective obfuscation based on user preferences, without the need for high levels of explicit user interaction with the obfuscation system.

Obfuscation requires the ability to offer high quality LBSs based on *imperfect* spatial information. This approach is motivated by the initial work on navigation algorithms under spatial imprecision [3], which has developed strategies for providing navigation services to an individual without knowledge of that individual's precise location. Obfuscation allows the identity of a person to be revealed, but that person's location to be hidden. This contrasts with more conventional approaches to location privacy, where a person's identity is hidden but his or her location is revealed (see section 2).

1.2   Imperfect Spatial Information

The use of imperfect spatial information is a key concept in obfuscation. In the literature at least three types of imperfection in spatial information are identified: (1) *imprecision*, which refers to a lack of specificity in information, (2) *inaccuracy*, which is a lack of correspondence between information and reality, and (3) *vagueness*, often characterized by the existence of boundary cases in information [15]. An inaccurate description of an agent's location means that the agent's actual location differs from the conveyed location: the agent is lying about its current location. An imprecise position description could be a region including the actual location (instead of the location itself). A vague description could involve linguistic terms, for example that the agent is "far" from a certain location.

In this article, we compare strategies based on imprecision and inaccuracy to obfuscate an individual's location. We focus on nearest point of interest (POI) queries, which are location-based proximity queries such as "Where is my nearest sushi restaurant?" In particular, we address the question to what extent can a high quality of service be combined with high levels of privacy for nearest POI queries. Important aspects discussed in this paper are the impact of the shape of an obfuscation region and the implications of its initial size for negotiating location privacy.

The paper is structured as follows: Section 2 compares an obfuscation-based approach with current approaches for location privacy. The model for negotiating location privacy is introduced in Section 3. The simulation experiments are explained in Section 4 and their results are given in Section 5. Section 6 concludes the paper and outlines further research.

2   Background

Research into how to safeguard an individual's privacy is becoming an urgent issue in pervasive computing. Most approaches to protecting (location) privacy fall into three categories: *regulation, privacy policies*, and *anonymity*. Each of these approaches plays an important role in providing a complete solution to location privacy, but each approach also has its limitations.

Regulatory approaches to privacy develop rules to govern fair use of personal information, including legislation. Langheinrich [10] gives an overview of the history and current status of privacy legislation and examines international fair information practices. However, regulations often lag behind new technology and ideas, and apply "across the board" making them difficult to tailor to specific contexts that may arise.

Privacy policies stipulate allowed uses of location information. Kaasinen [9] surveys policy-driven approaches to location privacy. Privacy policies rely on trust and, therefore, are vulnerable to inadvertent or malicious disclosure of private information.

Anonymity concerns the dissociation of information about an individual, such as location, from that individual's actual identity. A special type of anonymity is *pseudonymity*, where an individual is anonymous, but maintains a persistent identity (a pseudonym) [12]. Although anonymity techniques are fundamental to privacy protection they have limitations, especially in spatial application domains. A person's identity can often be inferred from his or her location, so anonymity and pseudonymity are vulnerable to data mining [4, 1]. Further, anonymity presents a barrier to authentication and personalization, which are required for a range of applications [8, 10].

Obfuscation offers the potential to extend existing location privacy protection capabilities. First, the aim of obfuscation is to protect information about a person's location, but enable that person's true identity to be revealed (thereby avoiding the difficulties faced by anonymity-based approaches, including problems with authentication and personalization). Second, obfuscation does not rely on any centralized server to broker location-based services or administer privacy policies, making it suitable for highly distributed environments like peer-to-peer systems.

Recent work has extended conventional privacy protection strategies using concepts closely related to obfuscation. Gruteser and Grunwald have investigated an anonymity approach called "spatial cloaking" [6]. Similarly, Snekkenes suggest a privacy policy system based on the "need-to-know principle" [14]. However, the work presented in [2] is the first to directly develop obfuscation as a mechanism for protecting location privacy. Obfuscation is a new direction for privacy research that is explicitly spatial and is *complementary* to conventional privacy protection strategies. In contrast to previous work, obfuscation is not based on, but may be used in combination with, regulation, privacy policies, and anonymity.

3   Obfuscation and Negotiation

The aim of obfuscation is to protect a person's location privacy by degrading the quality of information about that person's location, at the same time as delivering a location-based service of acceptable quality to that person. One way to degrade the quality of information about a person's location is to be imprecise. Instead of providing a single location to a location-based service provider, a person might wish to provide a *set* of locations (an *obfuscation set*, usually denoted O). An orthogonal way to degrade the quality of information about a person's location is to be inaccurate. For inaccuracy, we might generate an obfuscation set O that does not contain that person's true location. Thus, we may identify four possibilities for an individual located at point l and their obfuscation set O:

1. Accurate and precise: $l \in O$ and $|O| = 1$
2. Inaccurate and precise: $l \notin O$ and $|O| = 1$
3. Accurate and imprecise: $l \in O$ and $|O| > 1$
4. Inaccurate and imprecise: $l \notin O$ and $|O| > 1$

Under imprecision, the larger the obfuscation set, the less information is being revealed about the individual's true location, and so the greater the level of privacy that individual is able to enjoy. Under inaccuracy, the greater the distance between elements in the obfuscation set and the individual's true location, the less information is being revealed about the individual's true location, and so the greater the level of privacy that individual is able to enjoy.

3.1   Negotiation

Previous work in [2] has provided the formal basis for using obfuscation based on imprecision in nearest POI queries. The approach uses a negotiation process, summarized in algorithm 1 below. The aim of this negotiation process is to achieve a satisfactory balance of level of privacy and quality of service. In this section, we briefly review the negotiation process, but for more details the reader is directed to [2].

Using a graph-based representation of the geographic environment, the negotiation algorithm first partitions the obfuscation set O into equivalence classes. Elements in each equivalence class has the same POI $p \in P$ as their closest (we assume for simplicity there are no ties: all locations $o \in O$ are closest to one POI $p \in P$). This step can be thought of as building a graph-based equivalent of a Voronoi diagram. Indeed, the term "graph Voronoi diagram" is coined and formally defined in [5].

The algorithm operates using a graph-based representation of geographic space in order to model the constraints to movement that are normally a feature of most location-based services. However, the simplest way to explain the negotiation process is with the analogy of a Voronoi diagram. In Figure 1 the Voronoi diagram has been computed for the locations of a few POIs (white dots), with the shaded circular regions in each sub-figure representing the obfuscation set for the clients actual location (black dot). At each iteration of the negotiation process, there are four possibilities:

Algorithm 1. Negotiation proximity query with obfuscation (after [2])

Data: Obfuscation set O for the agent; graph-based representation of the geographic
  environment G; the set of POIs P

Result: The location $p \in P$ which is the best estimate of the nearest POI given the agent's
  privacy requirements

1.1 Find the relation δ such that for all $o_1, o_2 \in O$, $o_1 \delta o_2$ iff o_1 and o_2 have the same POI
    $p \in P$ as their most proximal;

1.2 Construct the partition O/δ;

1.3 **if** $O \in O/\delta$ **then**

1.4 $\quad$ Return the closest POI for an arbitrary element in O;

1.5 **if** *Agent agrees to identify for its current location l the equivalence class $[l] \in O/\delta$* **then**

1.6 $\quad$ Return the closest POI for an arbitrary element in $[l]$;

1.7 **if** *Agent agrees to identify some new obfuscation set O' such that $O' \subset O$* **then**

1.8 $\quad$ Reiterate algorithm with O' in place of O;

1.9 **else**

1.10 $\quad$ Return some best estimate of the closest POI based on maximizing $|[l']|/|O|$, for
    $\quad$ some arbitrary $l' \in O$;

---

1. If all the locations in obfuscation set are closest to the same POI (within the same proximal polygon), then the location of this is returned as the query result (Figure 1a, Algorithm 1 lines 1.3–1.4).

2. If the agent agrees to identify in which proximal polygon it is actually location, then the POI for this proximal polygon (shown with bold outline in Figure 1b) is returned as the query result (Algorithm 1 lines 1.5–1.6).

3. If the agent agrees to identify some other smaller obfuscation set (shown as dashed line in Figure 1c), then negotiation reiterates with this new obfuscation set (Algorithm 1 lines 1.7–1.8).

4. Otherwise, the POI for the proximal polygon that contains the largest proportion of the obfuscation set is returned (shown with bold outline in Figure 1d) as a best estimate of the closest POI. Note that, as in Figure 1d, this best estimate may not be the optimal answer (Algorithm 1 lines 1.9–1.10).

The analysis in [2] shows that this negotiation process: (1) is well formed, in the sense that it will always terminate; (2) is computationally efficient, in that its underly-

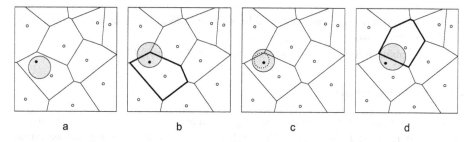

| a | b | c | d |

Fig. 1. Negotiation alternatives, illustrated by a Voronoi diagram

ing algorithm has the same computational complexity as the comparable conventional algorithm for finding the most proximal POI without obfuscation.

The decision as to whether an agent agrees to identify the smaller obfuscation sets required for negotiation branches 2 and 3 above will depend on the balance of level of privacy and quality of service for that agent. Thus, if an agent requires a higher quality of service than can be achieved at the current level of privacy, then it may need to reveal more information about its actual location (a smaller obfuscation set). The goal of the remainder of this paper is to investigate this decision, and some of the other parameters that will affect the balance between quality of service and level of privacy for an agent obfuscating its location.

4   Simulations

The framework set out in the previous section provides the basis for an obfuscation system that enables individuals to access high quality location-based services whilst revealing as little information as possible about their current location. Thus, an obfuscation system aims to achieve a satisfactory balance of level of privacy (LOP) and quality of service (QOS). In general, higher LOPs are expected to lead to lower QOS and vice versa (lower LOPs are expected to lead to higher QOS). There exist a number of different parameters that can be manipulated within an obfuscation system and that complicate the balance of LOP and QOS.

In order to investigate these parameters we developed a simulation environment, programmed in Java. The parameters that can be manipulated within the simulation system include:

- the size, shape, and location of the initial obfuscation set used in the negotiation process;
- the strategies adopted by individual agents during the negotiation process;
- the number and location of points of interest (POIs) available within the spatial environment; and
- the spatial environment itself.

By manipulating these parameters within the simulation system we can empirically investigate the effects of these changes upon the balance of LOP and QOS. The primary research questions this research sets out to answer are:

1. Using obfuscation, is it possible to achieve high QOS at the same time as high LOP for location-based proximity queries?
2. Which obfuscation strategies provide the best balance of QOS and LOP?

4.1   Simulation Strategies

There are several distinct obfuscation strategies that have been tested within the simulation system. Each strategy, described below, has different parameters that may be varied as part of the negotiation process.

- *O-strategy*: As described in in the previous section, during the negotiation an agent can choose to identify in which equivalence class it is currently located to the location-based service provider. Using an O-strategy (*optimized*-strategy) agent chooses to reveal this information at the first negotiation iteration. Thus, with minimal negotiation, an O-strategy agent finds the best possible obfuscation for a particular set of initialization conditions. The size of the initial obfuscation is the only negotiation parameter that can be varied by an O-strategy agent.
- *C-strategy*: The C-strategy (*compact*-strategy) uses a full negotiation process that takes advantage of the spatial structure of the environment. The initial obfuscation for a C-strategy agent is a compact connected "ball" of nearby points. At each iteration a C-strategy agent will discard one or more locations from the edge of the obfuscation region, maintaining a compact connected obfuscation throughout the negotiation process. As for the O-strategy, the size of the initial obfuscation is a parameter that can be varied by a C-strategy agent. Additionally, because a C-strategy agent uses a full negotiation process it may also decide to accept a reduced QOS in return for higher LOPs.
- *L-strategy*: The L-strategy (*lying*-strategy) uses inaccuracy rather than imprecision to obfuscate an agent's location. L-strategy agents provide a single (precise) location, but one that is perturbed from the agent's true location. An L-strategy does not take part in any negotiation process, since it provides a precise location from the outset. However, a L-strategy agent can vary the amount it perturbs its true location, i.e., how much it prepared to lie about its true location.

In addition there are a number of derived or hybrid strategies that agents can adopt. Examples of derived and hybrid strategies investigated include the following:

- *E-strategy*: Like the C-strategy, the E-strategy (*elongated*-strategy) uses a connected region of nearby points and maintains a connected obfuscation throughout the negotiation. However, unlike a C-strategy agent, a E-strategy agent uses a elongated "sausage" rather than a compact "ball" of locations for its obfuscation.
- *R-strategy*: Like the C-strategy, the R-strategy (*random*-strategy) uses a full negotiation process but does not take advantage of the spatial structure of the environment. Instead, R-strategy agents construct an initial obfuscation from random locations within the environment. At each iteration, R-strategy agents randomly remove one or more locations from their obfuscation in order to continue the negotiation process.
- *CR-strategy*: CR-strategy agents initialize their obfuscation as a compact region of connected nearby points (C-strategy), but then randomly remove points from that region during the negotiation process.
- *CRL-strategy*: A CRL-strategy is based on a CR-strategy, but additionally an agent may discard its true location from the obfuscation set, meaning it may provide an obfuscation set that is both imprecise and inaccurate.

4.2   Confidence

In addition to the strategies and environmental parameters, some agents may also specify a threshold value which determines how that agent wishes to balance its LOP against its QOS. This threshold value takes the form of a level of confidence, as a number in the interval $[0.0, 1.0]$. At each iteration of the negotiation process, the negotiation algorithm checks the proportion of the obfuscation set that is closest to each POI. If this proportion is greater or equal to the confidence threshold for any of the candidate POIs the agent terminates the negotiation by requesting the best estimate of the nearest POI. A confidence level of 0.6 means that an agent will accept the best estimate of the closest POI as long as 60% of its obfuscation set is closest to one POI. A confidence level of 1.0 means that an agent will only accept a perfect QOS while a confidence level of 0.0 means that an agent will accept any QOS (see [2] for more details). In effect, the confidence level provides a mechanism to balance QOS and LOP, without needing to explicitly compute QOS (which would require that the agent reveal its true location).

5   Results

5.1   Density of Points of Interest

The density of POIs in the environment is one of the main factors that should determine the balance of LOP and QOS. Higher POI densities are expected to require that an agent must reveal more information about its location (lowering its LOP) in order to achieve the same QOS. To investigate this expectation, 100 simulations were conducted at each of 10 different POI densities. The simulations were conducted using a simple environment of a small regular network of 400 nodes arranged in a grid. For each of the 10 sets of simulations, Figure 2 shows the average LOP, in terms of the number of elements in the final obfuscation $|O|$ (i.e., once negotiation is completed), plotted against the against POI density, measured as the number of POIs used to initialize the set of simulations (for a fixed environment size). The median is preferred to the mean as an average, since the population of results for each simulation were often skewed and contained outliers. For clarity, Figure 2 is presented as a log-log plot, because successive sets of simulations doubled the number of POIs in the environment. The other simulation variables were set at default levels: a confidence level of 1.0 was used for all agents (perfect QOS), and each agent used the entire environment as its initial obfuscation (the largest possible initial obfuscation).

The results show that, as expected, LOP decreases with increasing POI density. At the extreme right of the figure, every node is a POI, so every agent must reveal its precise location in order to find the nearest POI with total confidence. At the extreme left of the figure, the environment contains only one POI, so an agent need not reveal any information about its location in order to find the nearest POI (there is only one).

In between these extremes, the figure shows four response curves for the different obfuscation strategies tested. The first strategy is the O-strategy (optimized strategy), where the agent agrees at the first iteration of negotiation to reveal in which equivalence class it is currently located. As expected this strategy outperforms all other strategies, in the sense that it provides a higher LOP than any other strategy for a particular POI density. The worst strategy is the R-strategy (random), where the obfuscation is composed of points located randomly throughout the environment. The poor performance of the R-strategy is is due to the spatially dispersed nature of its obfuscation. Even for small obfuscations, elements of the obfuscation set O may be scattered across the

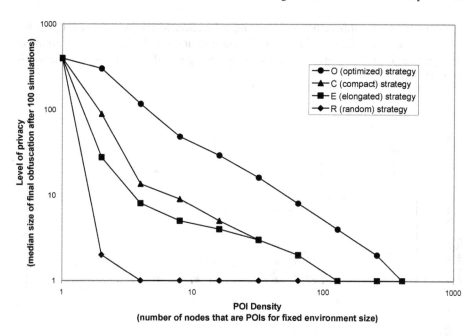

Fig. 2. Effect of POI density upon LOP

environment with no single POI nearest to all of these elements. For similar reasons, the C-strategy, where obfuscations are compact connected subgraphs, outperforms the E-strategy, where obfuscations are elongated connected subgraphs.

A Wilcoxon signed-rank test (a discrete, paired equivalent of the t-test) was used to test the null hypothesis that the observed differences between the results arrived by chance (i.e., sets of results were drawn from the same population). For those data points that are visually distinct on the graph in Figure 2, these tests indicated that the null hypothesis should be rejected at the 5% significance level. In other words, the results indicate that the O-strategy performed as well as or significantly better than the C-strategy, which performed as well as or significantly better than the E-strategy, which performed as well as or significantly better than the R-strategy.

5.2   Initial Obfuscation Size

Another factor that should affect the balance of LOP and QOS is the initial obfuscation size. The size of the initial obfuscation set O constrains the LOP: small obfuscation sets mean lower LOPs, large obfuscation sets allow higher LOPs. To investigate this, 9 sets of 100 simulation runs were performed, changing the size of the initial obfuscation for each set of simulations. Based on the results from Figure 2 the POI density for these simulations was set at a level typical of the mid-range of the simulations (8 POIs in the environment of 400 nodes). The confidence level used in the negotiation process was again 1.0. Figure 3 shows average (median) LOP against initial obfuscation size in terms of $|O|$, the total number of elements in the initial obfuscation set. At the extreme left of the figure, the initial obfuscation size is a single location, leading to the lowest possible LOP. At the extreme right of the figure, the obfuscation size is the entire environment (the agent begins the negotiation process by revealing no information about where it is located).

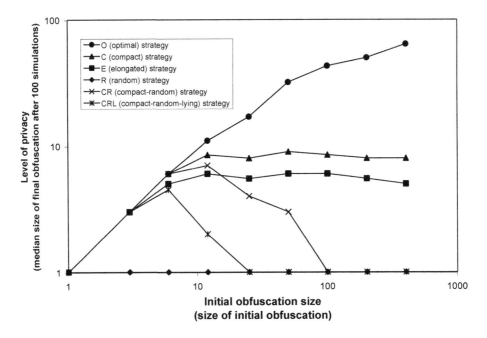

Fig. 3. Effect of initial obfuscation size upon LOP

The figure shows the same ordering of strategies as in the previous section, with the O-strategy outperforming C-strategy, outperforming the E-strategy, outperforming the R-strategy. Again, a Wilcoxon signed-rank test was used to confirm this ordering. Two additional strategies are included in Figure 3, which are hybrids of the C- and R-strategies. The CR-strategy agent selects a compact initial obfuscation in combination with a random negotiation strategy. The CRL-strategy uses the CR-strategy in addition to an agent being able to lie about its true location (i.e., provide an obfuscation that does not include its actual location). These hybrid strategies did not perform well, offering similar performance to the C-strategy only at the lowest initial obfuscation sizes, but degrading rapidly into the R-strategy. In fact, in general these strategies rarely out-performed even the random strategy, providing a strong indication that it is the negotiation process that dominates the effectiveness of the strategy, rather than the initial conditions for the negotiation. This is a helpful result, as is ensures the obfuscation process is not too sensitive to the initial conditions.

The noticeable feature of Figure 3 is that while the LOPs achieved by the C- and E-strategies climb steadily with the O-strategy at lower initial obfuscation sizes, both

C- and E-strategies level off at a maximum LOP at higher initial obfuscation. The maximum level is directly related to the POI density in Figure 2. Repeating these sets of simulations at different POI densities produces similar graphs to Figure 3 which differ in the maxima for the C- and E-strategies. For these sets of related graphs, plotting the maxima for the C-strategy against POI density results in Figure 4.

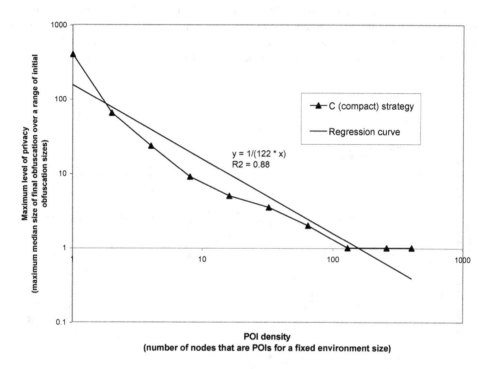

Fig. 4. Maximum median LOP as a function of POI density for a range of initial obfuscation sizes

The import of Figure 4 is that the maximum possible LOP for the C-strategy (and E-strategy) can be directly related to the POI density, independently of the initial obfuscation size. The regression curve in Figure 4 is a simple power function that fits the observed data with a reasonably high product-moment correlation coefficient (r-squared value of 0.88). This is a useful result because knowing that, on average, an initial obfuscation size greater than some threshold will not produce improvements LOP provides a basis upon which to choose an initial obfuscation, one of the primary difficulties facing an obfuscation system.

The reason for this behavior can be explained with the analogy of the Voronoi diagram. For the C- and E-strategies, larger initial obfuscation sets also have a greater overlap with the proximal polygons of several POIs. During negotiation, the obfuscation sets are reduced from the outer boundary. Larger initial obfuscation sets require additional iterations of the negotiation process to ensure that all elements of an agent's obfuscation set are closer to one POI than to any other POIs. Thus, on average the size of the final obfuscation sets are dominated by the density of POIs.

5.3   Confidence

The simulations so far have all used a confidence level of 1.0: the agents need to know with complete confidence which is the nearest POI. While this may be useful in some applications, in many applications users might be prepared to accept suboptimal query results if this also provided higher levels of privacy. Decreasing the level of confidence is expected to increase the LOP, at the expense of decreased QOS. Figure 5 shows the LOPs achieved by the different negotiation strategies across a range of confidence levels. The figure was generated using 11 sets of 100 simulations with a constant mid-range POI density (8 POIs in an environment of 400 nodes) and with the maximum initial obfuscation size.

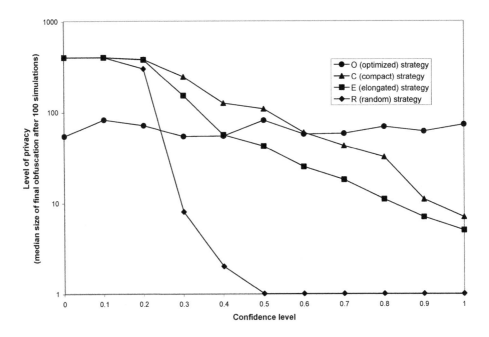

Fig. 5. Effect of confidence level on LOP

As expected, the LOP achieved by the obfuscation system climbs steadily with de-creasing confidence levels. Again, the C-strategy outperforms the E-strategy, which in turn outperforms the R-strategy. Both C- and E-strategies still provide moderate levels of privacy even at the highest confidence levels. However, the O-strategy always ter-minates after one iteration of the negotiation by indicating in which equivalence class it is located. Thus changing the confidence levels has no effect on the performance of the O-strategy. The important point to note in Figure 5 is that at lower levels of confi-dence, the C-, E-, and even R-strategies are capable of providing higher LOPs that the O-strategy. Thus, for agents who do not require a perfect answer to their location-based query may be able to achieve higher LOPs be using a full negotiation strategy, such as the C-strategy.

A similar set of simulations was used to generate Figure 6, which shows the QOS provided to the agent plotted against level of confidence. QOS is measured in terms of the how much further away from the agent's location a is the actual query result provided by the obfuscation service b when compared with than the best possible query result c measured as $d(a,b) - d(a,c)$ where $d(x,y)$ is the network distance between x and y. High values indicate a low QOS, low values indicate a high QOS (hence the scale in Figure 6 is reversed with low values and high QOS at the top of the y-axis). Although, as expected, QOS decreases with level of confidence, it is noticeable that there exists a wide range of confidence levels less than 1.0, which still provide the highest possible QOS. By definition, the O-strategy always delivers a perfect QOS (although discussed above, sometimes at the cost of lower levels of privacy).

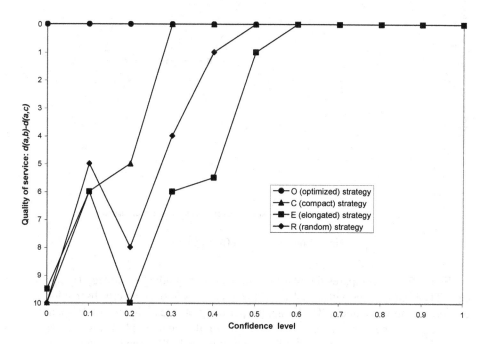

Fig. 6. Effect of confidence level on QOS

5.4 Balancing Quality of Service and Level of Privacy

The previous simulations provide a basic understanding of the effects of varying different obfuscation parameters: different negotiation strategies, POI densities, initial obfuscation sizes, and levels of confidence. We are now in a position to investigate the overall balance of LOP and QOS. Plotting median LOP against median QOS across multiple simulations confirms that it is indeed possible to achieve high QOS and high LOP. Rather than examples of such plots, Figure 7 shows the *lowest* observed QOS against LOP across the entire range of different confidence levels. Thus, Figure 7 represents not the average results, but the worst observed results, in terms of the lowest QOS observed for a particular LOP. Despite representing the worst extremes of the each strategy, the C-strategy in particular still performs well, providing high levels of QOS at all but the highest LOPs. For clarity, the R-strategy performs is not plotted in Figure 7, but as expected performs badly, providing a much lower QOS across almost all LOPs than the other strategies.

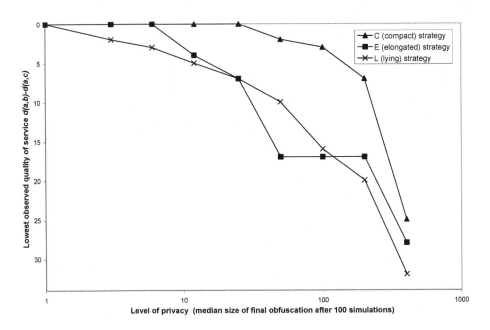

Fig. 7. Lowest observed QOS against LOP

Figure 7 also compares the results with the corresponding L-strategy (lying strategy). The L-strategy provides the LBS with a precise location, but perturbs that location by a predetermined amount. The higher the perturbation, the higher the LOP. To plot the L-strategy and the imprecision-based strategies on the same graph it was necessary to formulate a shared measure of LOP. To achieve this, the LOP for an L-strategy was measured as the size (in terms of the number of elements) of the corresponding C-strategy compact "ball" that has the agent's true location at its center and the perturbed location at its boundary. Based on this relationship, QOS against LOP is also plotted on Figure 7 alongside the C- and E-strategies. Although the L-strategy performs reasonably well, in general the results indicate that the L-strategy does not achieve as high QOS as the C-strategy at the same LOP.

5.5 Environment

As stated previously, the environment for the simulations was a small regular network of 400 nodes arranged in a grid. However, all the experiments were also repeated for a more realistic environment: a generalized map of a portion of central Paris containing a similar number of nodes. The results indicated that, at least within the limited scope of the experiments already described, the use of a less regular environment did not affect the properties of the obfuscation process. However, future work will need to test further types of obfuscation strategy. In particular, the C- and E-strategies tested in this paper use *connected* initial obfuscation sets. A more realistic initial obfuscation set might be all locations within, say, 300m of the agent's actual location. Although in a regular grid environment such an obfuscation will also be connected, in a more realistic, heterogeneous environment, such as central Paris, such an obfuscation will necessarily be connected, which may degrade the performance of the obfuscation process.

5.6 Summary of Results

In summary we draw the following general conclusions from the results of our simulations:

1. Using obfuscation and negotiation, it is possible to achieve high quality location-based services whilst maintaining high levels of location privacy, by revealing only low quality information about location.
2. The final LOP for an agent depends more strongly on the negotiation strategy than on the initial negotiation conditions.
3. Negotiation strategies differ in their suitability for obfuscation. Of the strategies tested, the results suggest that the C-strategy, using a compact region of imprecision and discarding points near the edge of that region during negotiation, outperforms all other non-optimized strategies including a simple L-strategy. Further, at lower confidence levels, the C-strategy can also outperform the O-strategy.
4. The maximum possible LOP for an agent using full negotiation (the C- or E-strategies) across a range of initial negotiation conditions depends strongly on POI density. Within our simulations, the maximum LOP may be related using a simple power law to the POI density. In turn, this provides a maximum initial obfuscation size, above which further increases in initial obfuscation are unlikely to result in concomitant increases in the final LOP.

6 Conclusions and Outlook

Obfuscation enables people to protect their location privacy by revealing the least possible information about their current location when accessing location-based services. Our approach to obfuscation focuses on automated negotiation that enables users to balance the level of location privacy against the quality of location-based service. Since negotiation-based obfuscation is complementary to current approaches to location privacy (regulation, privacy policies, or anonymity), we believe that obfuscation represents an important new direction for location privacy research that has not previously been adequately investigated.

6.1 Balancing QOS and LOP

The simulation results show that a negotiation-based obfuscation strategy for protecting location privacy is able to achieve both high levels of QOS for nearest POI queries in concert with high levels of location privacy. Tailoring the negotiation process to the requirements of a particular user can be achieved using confidence levels. Thus, higher QOS can be guaranteed using high confidence levels; lower confidence levels improves LOP with little or no loss of QOS, at least for higher confidence levels. In summary, these results clearly show that the negotiation process can be used to balance LOP and QOS. These encouraging results warrant further investigation into obfuscation-based strategies, including field tests with real location-based services.

It is expected that in a practical obfuscation system user profiles would be used to govern the automated negotiation process. For example, a simple user profile might contain the minimum LOP a user is prepared to accept (in terms of the minimum obfuscation set size), or the minimum QOS. Current work is investigating the practical aspects of delivering obfuscated location-based services to mobile users.

6.2   Imprecision and Lying

The simulations also indicate that the strategy based on inaccuracy (L-strategy) does not perform as well as the best strategy based on imprecision (C-strategy). This result relies on the establishment of a common measure of LOP for inaccurate and imprecise obfuscation, and must be carefully interpreted. However, it does suggest that the use of imprecision as a core obfuscation strategy warrants further investigation. Imprecision also has the advantage that it does not require the user to explicitly "lie" about his or her location, something that might not be acceptable to some LBS providers. The shape of the obfuscation region has a significant impact for location privacy: compact regions compare favorably with elongated regions. Therefore, strategies that obfuscate a location using regions or blocks should be preferred to strategies that use elongated or linear structures for obfuscation.

6.3   Selecting the Initial Obfuscation Set

Selecting a good initial size for an obfuscation set, which balances location privacy and QOS, is a difficult task. However, our experiments demonstrated that increasing the size of the initial obfuscation set leads to higher levels of privacy only up to a point. These findings suggest that it may be possible to select *a priori* an initial obfuscation region that satisfies the requirements for high LOP and high QOS. This result might be practically useful in a number of ways. For example, a more advanced obfuscation system could enable an obfuscating agent to execute a *prequery*, which determines the average density of POIs within a region. This would only require an agent to reveal minimal information about its current location, but might allow the client to "calibrate" the obfuscation system to select an appropriate initial obfuscation set size that matches the density of POIs.

Other potential mechanisms for setting the initial obfuscation level include: using the entire graph as the initial obfuscation set (computationally practical, as shown in [2]), or selecting "natural" imprecise regions, such "downtown," "Kensington," "Victoria." Such approaches might also be extended to enable obfuscation based on vagueness (e.g., where "downtown" does not have a crisp boundary). Further research on obfuscation for location privacy will address:

- extending obfuscation techniques to other location-based services, in particular navigation services and route queries. Initial work in [3] has already set out the foundations for navigation under imprecision, based on the inherent *instruction equivalence* of navigation instructions.
- extending the static obfuscation model presented in this paper to a truly dynamic model that enables spatiotemporal location-based services.
- counter-strategies for invading location privacy from the perspective of an external agent wanting to undermine a person's location privacy.

Acknowledgments

Dr Duckham is partially supported by an Early Career Researcher Grant and an International Collaborative Research Grant from the University of Melbourne. Dr Kulik is partially supported by an Early Career Researcher Grant from the University of Melbourne. The authors are grateful to Shubham Gupta, who generated the data set of Paris for the simulations, and to the three anonymous reviewers for their helpful comments.

References

1. A.R. Beresford and F. Stajano. Location privacy in pervasive computing. *IEEE Pervasive Computing*, 2(1):46–55, 2003.
2. M. Duckham and L. Kulik. A formal model of obfuscation and negotiation for location privacy. In *Pervasive 2005*, Lecture Notes in Computer Science. Springer, Berlin, 2005.
3. M. Duckham, L. Kulik, and M. F. Worboys. Imprecise navigation. *GeoInformatica*, 7(2):79–94, 2003.
4. S. Duri, M. Gruteser, X. Liu, P. Moskowitz, R. Perez, M. Singh, and J-M. Tang. Framework for security and privacy in automotive telematics. In *Proc. 2nd International Workshop on Mobile Commerce*, pages 25–32. ACM Press, 2002.
5. M. Erwig. The graph Voronoi diagram with applications. *Networks*, 36(3):156–163, 2000.
6. M. Gruteser and D. Grunwald. Anonymous usage of location-based services through spatial and temporal cloaking. In *Proc. MobiSys '03*, pages 31–42, 2003.
7. M. Gruteser and D. Grunwald. A methodological assessment of location privacy risks in wireless hotspot networks. In D. Hutter, G. Müller, and W. Stephan, editors, *Security in Pervasive Computing*, volume 2802 of *Lecture Notes in Computer Science*, pages 10–24. Springer, 2004.
8. J. I. Hong and J. A. Landay. An architecture for privacy-sensitive ubiquitous computing. In *Proc. 2nd International Conference on Mobile Systems, Applications, and Services*, pages 177–189. ACM Press, 2004.
9. E. Kaasinen. User needs for location-aware mobile services. *Personal and Ubiquitous Computing*, 7(1):70–79, 2003.
10. M. Langheinrich. Privacy by design—principles of privacy-aware ubiquitous systems. In G. D. Abowd, B. Brumitt, and S. Shafer, editors, *Ubicomp 2001: Ubiquitous Computing*, volume 2201 of *Lecture Notes in Computer Science*, pages 273–291. Springer, 2001.
11. R.R. Muntz, T. Barclay, J. Dozier, C. Faloutsos, A.M. Maceachren, J.L. Martin, C.M. Pancake, and M Satyanarayanan. *IT Roadmap to a Geospatial Future*. The National Academies Press, Washington, DC, 2003.
12. A. Pfitzmann and M. Köhntopp. Anonymity, unobservability, and pseudonymity—a proposal for terminology. In H. Federrath, editor, *Designing Privacy Enhancing Technologies*, volume 2009 of *Lecture Notes in Computer Science*, pages 1–9. Springer, 2001.

13. B. Schilit, J. Hong, and M. Gruteser. Wireless location privacy protection. *IEEE Computer*, 36(12):135–137, 2003.
14. E. Snekkenes. Concepts for personal location privacy policies. In *Proc. 3rd ACM conference on Electronic Commerce*, pages 48–57. ACM Press, 2001.
15. M. F. Worboys and E. Clementini. Integration of imperfect spatial information. *Journal of Visual Languages and Computing*, 12:61–80, 2001.

Investigating the Need for Eliminatory Constraints in the User Interface of Bicycle Route Planners

Hartwig H. Hochmair[1] and Claus Rinner[2]

[1] St. Cloud State University, Department of Geography,
720 Fourth Avenue South, St. Cloud, MN 56301, USA
hhhochmair@stcloudstate.edu
[2] University of Toronto, Department of Geography,
100 Saint George Street, Toronto ON M5S 3G3, Canada
rinner@geog.utoronto.ca

Abstract. According to choice models in economics, consumer choice can be modeled as a two-stage process, starting with the choice of feasible alternatives, called the screening process, followed by compensatory evaluation of the remaining alternatives. Although spatial decision support systems used in various application areas support the screening process by allowing users to impose constraints on alternatives, this basic functionality is not widely available in current route planners. Based on an Internet survey of potential users, we examine the need for screening functionality in route planners for cyclists. Part 1 of the survey examines the users' demand for context information before stating their route preferences. Part 2 and part 3 investigate the users' demand for constraint functionality with and without context information. The results indicate that eliminatory constraints are essential concepts for the route selection process, and that maps are most effective in presenting context information about route alternatives.

Keywords: Route planner, user interface design, spatial decision support, eliminatory constraints, context-based route selection.

1 Introduction

Route selection problems commonly involve a set of route alternatives from which a choice of an alternative must be made under consideration of several evaluation criteria. With this paper, we address route selection within the framework of multi-attribute decision making (MADM). The MADM framework involves a selection among a limited set of alternatives and has a single, implicitly defined objective (Malczewski 1999). Solving a MADM problem involves the sorting and ranking of alternatives according to an underlying decision rule. A decision rule is a procedure that integrates information on alternatives and the decision maker's preferences to produce an evaluation of the set of alternatives. Two classes of decision rules can be distinguished: compensatory and non-compensatory. The compensatory approach is based on the assumption that the high performance of an alternative achieved in one or more criteria can compensate for the weak performance of the same alternative in

A.G. Cohn and D.M. Mark (Eds.): COSIT 2005, LNCS 3693, pp. 49–66, 2005.
© Springer-Verlag Berlin Heidelberg 2005

other criteria. Contrarily, under the non-compensatory approach a poor performance by an alternative in a criterion cannot be offset by another criterion's good outcome. Evaluation criteria (also called attributes or decision variables) considered in a MADM problem include benefit criteria and cost criteria. Multiplication of these criteria by a constant coefficient represents the objective function which needs to be maximized or minimized in an optimization problem.

Eliminatory constraints impose limitations on the set of decision alternatives. An alternative is feasible if it satisfies all eliminatory constraints. Dichotomizing the set of decision alternatives under consideration into feasible and infeasible is referred to as a screening procedure. Choice behavior can be modeled as combined two-stage process at which the screening process (i.e., the non-compensatory stage) is followed by a compensatory stage (Srinivasan 1988). Non-compensatory constraints eliminate those decision options from further consideration that do not meet thresholds set for evaluation criteria (Malczewski 1999). In this work we analyze the user need for non-compensatory constraints in the user interface and—as a clarification—use the term *eliminatory constraints* as a synonym for non-compensatory constraints.

1.1 Eliminatory Constraints in Spatial Decision Support Systems

A growing number of electronic bicycle route planners that are available as commercial products or as Internet applications provide evidence for the increasing demand for these spatial decision support tools. Existing route planners provide only limited functionality in terms of their decision rule in that they either evaluate a single attribute optimization function (e.g., shortest path), or (rarely) use a compensatory decision rule involving user-defined criteria (e.g., short and scenic route) and importance weights. Non-compensatory techniques using eliminatory constraints are even less frequently found in existing systems.

Online route planners that are primarily designed for bicycle tourists allow for (1) manually selecting from a set of pre-defined routes (EMS 2005), and (2) adding further stops between start and end of the pre-defined route (Ehlers et al. 2002). In the first case, no decision rule is supported. In the second case, a fixed optimization function for a single criterion (i.e., path length) is evaluated. Other online route planners allow for arbitrary selection of start and end nodes by entering addresses or points of interest (MAGWIEN 2005), and provide an additional function for importance weighting of various route selection criteria. Examples include the weighting of slope avoidance (Bikemetro 2005), or fast, scenic, and short routes as realized in the commercial software Rad.RoutenPlaner (Rad.RoutenPlaner 2003). All of these examples have in common that the user cannot explicitly state eliminatory constraints.

In fact, very few route planners provide eliminatory constraint functionality (e.g., BBBike 2005), although non-compensatory decision rules have been implemented in Geographic Information Systems (GIS) for years, e.g. for site selection, route planning, and land-use decision-making (Carver 1991; Jankowski 1995; Eastman 1997). Recent GIS applications have used the Ordered Weighted Averaging method to offer the user a choice between various decision rules including the non-compensatory logical AND and OR operators (Jiang and Eastman 2000; Rinner and Malczewski

2002). Route planning services for wheelchair users provide evidence that eliminatory constraints may in fact be important for route planning tasks. The MAGUS application (Beale et al. 2001; MAGUS 2005) is a GIS-based route planning tool for wheelchair users. It also includes prohibitive obstacles (e.g., steps), which denote eliminatory constraints on a route. Dewey (2001) provides a detailed list of potential barriers for wheelchair users in urban areas. In both applications, the search algorithm selects the set of feasible routes by non-compensatory screening, which ignores value trade-offs (Keeny 1980). In this paper we focus on the user's conceptualization of the best route in terms of compensatory and non-compensatory decision rules. We omit complexity and efficiency considerations of multiple criteria path computation (Horn 1997; Mitchell 2000; Mooney and Winstanley 2003) and assume a given set of alternative routes.

The motivation for the presented work is based on findings of a previous Internet study about route selection criteria for cyclists in unknown urban environments (Hochmair 2004a). The 42 participants were asked to specify those route features (i.e., eliminatory constraints) that should exclude a specific route from the set of potential routes between two locations. The large variety of the 23 eliminatory constraints motivated the following two questions: To which extent should these constraints be taken into account in the user interface design for route planners? Does context information about existing route alternatives affect the user need for constraint functionality?

Table 1. Constraints mentioned for bicycle routes in urban environments

Constraint	freq	Constraint	freq
heavy traffic	31	city center ("avoid" criterion)	3
long detour	12	intersections with heavy traffic	2
steep slope (up/down)	9	routes with public transport	2
no bike lane	7	tunnel	2
boring suburbs or industrial zone	6	many turns at intersections	2
bad surface quality	6	roundabout	1
unsafe districts - high criminal rate	6	narrow streets	1
many traffic lights or other forced stops	4	bad signage	1
illegal street segments	4	many one-way streets in area	1
pedestrian area	4	fee	1
complex or confusing route	3	shopping streets	1
dangerous streets	3	controls by police ("avoid" crit.)	1
stairs	3	construction sites ("avoid" crit.)	1

Table 1 lists the stated constraints found in the 42 filled questionnaires (Hochmair 2004a). The "freq" column shows the number of participants who mentioned the corresponding constraint. The importance of constraints in the user's decision process is also demonstrated by the high number of "avoid" criteria (10) that where mentioned when users were asked for their preferred route characteristics (35 in total). Three of the "avoid" criteria did not overlap with any of the 23 eliminatory constraints and were therefore added to Table 1 to complete the list of constraints.

1.2 Context Information: Information About Choice Alternatives

Work on compensatory decision models provides evidence that the importance weight assigned to a criterion is a function of the range of values for that particular criterion (Srinivasan 1988; Keeny and Raiffa 1993). Thus the user of a decision support system needs to know the range of attribute values from all available choice alternatives (e.g., the length of the shortest and longest alternative) before being able to make a reasonable preference statement. Various research tools, such as HomeFinder (Williamson and Shneiderman 1992), GeoVISTA Studio (Takatsuka and Gahegan 2002), and CommonGIS (Andrienko and Andrienko 2001), include dynamic queries as a standard function. This allows the user to view the range of attributes for all choice alternatives and to filter the set of alternatives by reducing the valid range of attribute values. In contrast, current route planners do not usually offer context information before asking the user for preference statements. Instead, they ask for start and destination, and possibly for route preferences (e.g., Rad.RoutenPlaner 2003; Bikemetro 2005). The user thus decides under uncertainty and cannot assess the consequences of her preference statements. In one part of our research we adopt the findings of Srinivasan (1988) and Keeny and Raiffa (1993) to route planning and examine whether users of route planners feel more comfortable when being provided with context information before making their preference statements, i.e., before assigning weights to route selection criteria. Few route planners (e.g., Demis 2005) allow subsequent modifications of importance weights on initial route suggestions. Through route display on a map, these modifications can be done under more certainty. Generally, in interactive systems such as route planners, the user's model of a system governs her interactions with it (Barfield 1993). If the user lacks information about possible consequences of her selection, the results of taking an action and setting choice preferences may be unexpected and frustrating due to an incorrect user model.

In the case of eliminatory constraints it seems obvious that a function that allows imposing constraints on route alternatives will only be utilized if at least one of the route alternatives is affected by imposing an eliminatory constraint. Otherwise the action taken by the user will have no effect on the decision outcome. We expect that information about existing route alternatives also supports non-compensatory screening, that the user can better assess whether imposing eliminatory constraints on a route is relevant for defining the optimal route and that unnecessary constraint statements in the decision making process can be reduced. The third part of our study therefore investigates whether additional context information reduces the user's demand for eliminatory constraint functionality in user interfaces for route planners.

1.3 Structure of the Paper

The remainder of this paper is structured as follows. Sections 2, 3, and 4 describe the three parts of an Internet survey which investigates the demand for context information and the demand for eliminatory constraint functionality. Each survey question is described through its hypothesis, questionnaire design, and results. Lastly, section 5 integrates the findings with respect to the individual questions and provides an outlook on future work.

2   Question 1: Demand for Context Information

Question 1 addresses the user's comfort with various ways of presenting information about pre-computed route alternatives while making preference statements. Potential users were asked to rank four information designs from best to worst and to evaluate how helpful the particular information is for assigning importance weights to compensatory route selection criteria.

Participants in the study were 32 voluntary Internet users, most of them under-graduate Geography students in St. Cloud and Toronto. Ages ranged from 16 to 53 years (median = 26). Out of the 32 participants, 10 stated to use their bike daily, 13 between one and six days a week, and 9 less than once a week (average = 3.8 days/week, standard deviation = 2.9). 91% of the participants have seen or used route planners for car navigation, whereas fewer participants know route planners for bicycle or pedestrian navigation (38% each). The sample routes used for the study were taken from the suburban area of Minneapolis. Therefore we assume that none of the participants was familiar with the test area.

2.1   Hypothesis

Adopting the findings in Srinivasan (1988), Barfield (1993), and Keeny and Raiffa (1993) for the design of route planners, we hypothesize that the user of such tool feels comfortable with a user interface that visualizes characteristics of existing route alternatives between two selected locations before stating her route preferences. We expect that such design is preferred over a design that omits any type of route information.

2.2   Questionnaire Setup

The questionnaire started with four Web pages, each visualizing a different way of presenting information about feasible route alternatives. Participants had to imagine that they had already selected the start and destination of their trip, so that the system could pre-compute a set of three reasonable routes and show the information about these three routes. Participants could move back and forth to inspect the four Web pages. When done, participants were shown the thumbnails of all four designs on a summary page (Fig. 1), and asked to rate each design on a scale between 4 points (most useful design) and 1 point (worst design). Besides the route information each of the four designs also showed an image of a slider box which simulated a decision situation where the user could use sliders to state the importance weights for four route selection criteria, namely *fast*, *safe*, *simple*, and *attractive* route.

Design a) in Fig. 1 provides no route information but shows only the slider box so that the decision maker has to act under uncertainty. Design b) provides statistical route information visualized as bar charts. Each bar chart depicts the three routes' performances in attribute levels on a range between 0% and 100% for the four criteria *fast, safe, simple*, and *attractive*. Design c) visualizes the pre-computed routes on a street map. Design d) combines the map visualization with the bar charts.

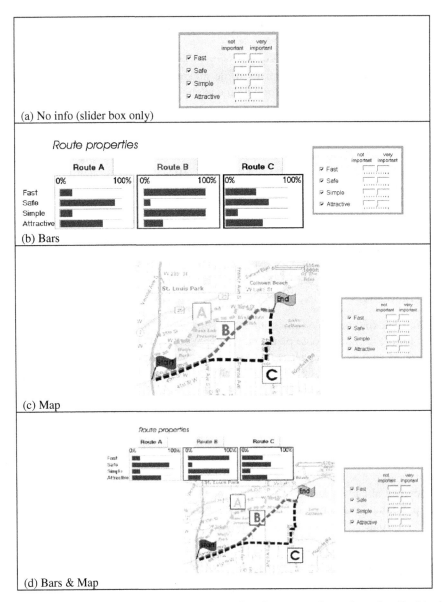

Fig. 1. Various representations of context information about route alternatives visualized in the user interface

2.3 Results

Fig. 2 shows the number of participants that considered a specific information design to be superior to all other designs (ties were excluded). According to the bar chart, and the means and medians of each design, on average, highest preference was assigned to design d), followed by c) and b). Design a) was least preferred.

A Wilcoxon signed-rank test compared medians between any two designs (Table 2). The left three columns compare design a) with b), c) and d). The results show that the medians of preference ratings for all designs that provide any type of information about route alternatives are significantly higher than the median for the design that provides no route information. This confirms our hypothesis: Users of route planners feel more comfortable in stating their preference when information about route alternatives is provided in the user interface.

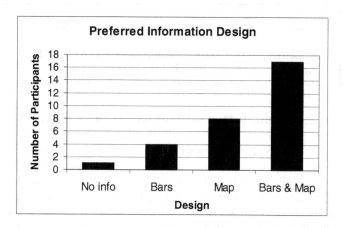

	No info (a)	Bars (b)	Map (c)	Bars & Map (d)
Mean	1.6000	2.4667	2.9000	3.0667
Median	1.0000	2.0000	3.0000	4.0000

Fig. 2. Preferences for various types of route information on a scale from 1 to 4. The bar chart summarizes the favorite design for all 32 participants, with tied favorite designs excluded.

When it comes to comparison between designs that provide information (right three columns in Table 2), no significant difference between the medians can be found at the 95% confidence-level. However, both the magnitude of medians and the distribution of the most favored designs (Fig. 2) suggest that maps are more powerful in supporting the user in stating her preferences than bar charts only, and that the information-richest design combining map and bar chart was the most preferred. The superiority of maps over charts can be explained by several factors (Freksa 1999): Maps derive their basic structure from the spatial structure of the geographic world, thus they intrinsically preserve spatial information. Robinson et al. (1995) describe the objective of map design as evoking in the mind of the map viewer an image of the environment. Specifically, in a map, the "structural relationships of each part to the whole" are important (Robinson et al. 1995, p. 317). In a route planner, this provides relevant context information about routes and their environment. To the contrary, in a statistical representation of route alternatives, the spatial relation is lost, and the information is portrayed by a collection of graphical features that indicate the magnitude of selected spatial and non-spatial attributes. Apparently, the spatial information shown in route maps can trade off the potentially more detailed information in a bar chart of route selection criteria. Finally, Slocum et al. (2004) discuss the use of tables and statistical measures for spatial data. "Raw tables are useful for providing specific information … but they fail to provide an overview of a data set" (Slocum et al. 2004, p. 54). We conclude that tabular data on route alternatives would provide too much detail and too little overview information to the users of a route planner.

Table 2. Significance levels for median comparison of preference values for various information designs

	No info in one of two visualizations			Info in both visualizations		
	No info vs. Bars	No info vs. Map	No info vs. Bars & Map	Bars vs. Map	Bars vs. Bars & Map	Map vs. Bars & Map
Sigma (2-tailed)	.003	.000	.000	.104	.073	.628

3 Question 2: Eliminatory Constraints Without Context

The second question examines the users' demand for a user interface functionality that allows them to impose eliminatory constraints on the route alternatives in addition to a given set of compensatory benefit criteria. We designed two user interfaces by which to test the functionality. The first one offers a set of four decomposable, compensatory higher-level criteria (Fig. 5a), each of which represents a criterion class comprising several lower-level criteria. For example, the criterion class *simple* contains among others the lower-level criteria *safe area*, *lighted at night*, and *bike lane* (which was however not explicitly shown to the participants). The second user interface offers a set of eight compensatory lower-level criteria (Fig. 5b) that were taken from the four criterion classes described before. For each of the two sets, participants had to check those eliminatory constraints from a given list that they considered to substantially complement the functionality of the user interface. No context information about route alternatives was provided for this task.

Participants for question 2 were 34 volunteer Internet users, most of them undergraduate students or university employees. Ages ranged from 19 to 61 years (median = 26). The participants' familiarity with route planners and their frequency of bike use was about the same as for participants of part 1 (section 2). Most participants (30 out of 34) stated that they use their bikes mainly in urban areas (small towns or cities). As the maps presented in question 1 and question 3 show urban areas, and as we try to avoid a potential impact of the type of cycling environment on compared results between question 1, 2, and 3, we excluded data from participants who live in rural settlements or villages. Finally, data from 30 participants have been analyzed for question 2.

3.1 Hypothesis

We claim that the demand for eliminatory constraints depends on the set of compensatory decision criteria that is available in a route planner user interface. The decision maker's objective ("find best route") and the related attributes form a hierarchical structure of evaluation criteria. Basic lower-level attributes can be grouped into more general, higher-level attributes, so called factors (Backhaus et al. 1996). The set of attributes must be complete to cover all relevant aspects of a decision problem (Keeny and Raiffa 1993; Malczewski 1999). Since a higher-level attribute covers more aspects of the decision problem than a single lower-level attribute, fewer higher-level attributes than lower-level attributes will be needed to provide the complete set of attributes for a route selection problem. Eliminatory constraints may be additionally required to constrain the set of feasible routes. Our hypothesis is that additional eliminatory constraints are more frequently selected in conjunction with a set of compensatory lower-level criteria than with a set of compensatory higher-level criteria.

3.2 Questionnaire Setup: The Sets of Higher-Level and Lower-Level Criteria

The two sets of compensatory criteria presented to the participants as part of the user interfaces are based on a previous classification study (Hochmair 2004b). In the study, participants were asked to group 35 bicycle route selection criteria into 3 to 6 classes and to think of appropriate class names. All participants together suggested ten class names (left column in Fig. 3), with various lower-level attributes assigned to selected classes by each user (not shown in Fig. 3).

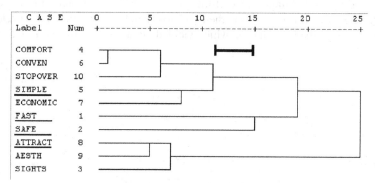

Fig. 3. Dendrogram for hierarchical clustering of suggested classes (classification data from Hochmair 2004b).

Based on classification results we computed a numerical membership value between 0 and 1 for each attribute in each class. For example, a value of 1 for attribute A in class C means that each participant who suggested class C, assigned A to C. We used a hierarchical cluster analysis (furthest neighbor method) to group correlated classes based on the membership values derived from individuals' assignment of attributes to classes. The dendrogram (Fig. 3) reveals that four classes, namely *simple*, *fast*, *safe*, and *attractive* (underlined) are appropriate for use as higher-level classes (bold line). They share only a small number of class members. These four classes were also the most frequently suggested ones in the classification study (Fig. 4). We used this set of four higher-level criteria as part of the first interface presented to the participants for task 2 (Fig. 5a).

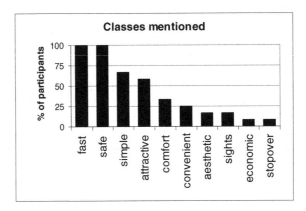

Fig. 4. Percentage of participants that suggested a certain class name in the classification task (after Hochmair 2004b)

In the same classification study, class members (i.e., lower-level attributes) were ranked by a relevance value that sorted lower-level attributes with respect to the class objective and the global importance of the attributes. For the second set of compensatory route selection criteria, we took the two class members with the highest relevance value from each of the four classes, which yielded eight benefit criteria (Fig. 5b).

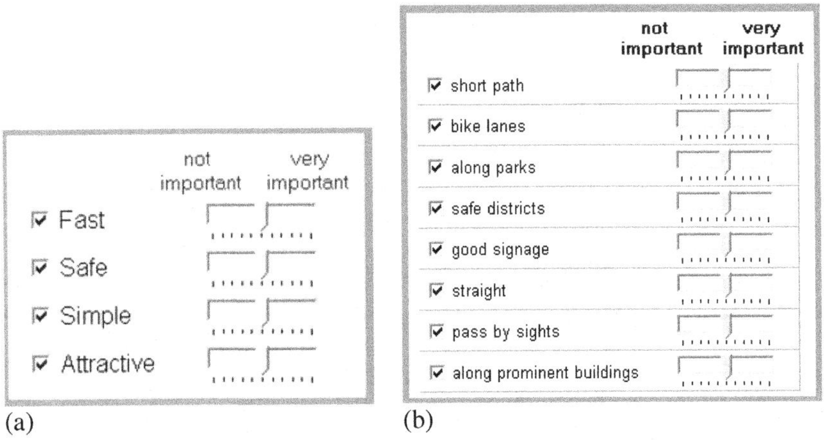

Fig. 5. Sets of higher-level (a) and lower-level (b) compensatory route selection criteria

The images of the slider bars with their set of four or eight compensatory benefit criteria were embedded on the left side of each questionnaire. On the right side, the 26 eliminatory constraints (taken from Table 1) were presented as checkboxes (see also Fig. 7). Participants were asked to check those eliminatory constraints that they considered as relevant for defining the best route in accordance with the given set of compensatory criteria presented on the left. Both questionnaires were shown in random order to each participant to avoid potential impacts arising from the sequence of presentation. Using checkboxes seems to be an appropriate method for mapping the

eliminatory constraint functionality to the interface, as the use of checkboxes for binary choices is suggested in related literature on human computer interaction (Koyani
et al. 2003; Shneiderman and Plaisant 2004).

3.3 Results

On average each participant checked 6.67 eliminatory constraints (median(H)=7) in
combination with the set of four higher-level compensatory criteria (Fig. 5a) and 8.40
eliminatory constraints (median(L)=8) in combination with the set of eight lower-
level compensatory criteria (Fig. 5b). The number of selected eliminatory constraints
across participants is normally distributed for both conditions (One-Sample Kolmo-
gorov-Smirnov Test), but a Paired Samples T-Test did not show significantly different means between the two conditions. This could be due to the large standard deviations in relation to the difference of means. We infer that the participants of the study
had an inherent inclination to selecting more or fewer eliminatory constraints independently of the two conditions, and suggest examining the number of selected eliminatory constraints one-by-one instead.

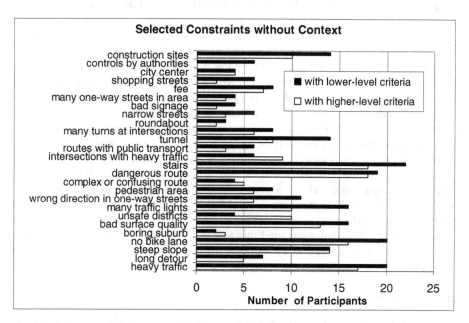

Fig. 6. Selected eliminatory constraints in combination with two different sets of compensatory
route selection criteria. Median with higher-level criteria (6) is significantly smaller than median with lower-level criteria (7.5).

A Wilcoxon Signed-Rank Test for Paired Data showed a significant difference
(p<0.01) between the medians of the number of times each eliminatory constraint was
selected in conjunction with the two sets of compensatory benefit criteria (me-
dian(H)=6; median(L)=7.5), thus supporting the hypothesis stated in section 3.1. Fig.
6 visualizes the results for all eliminatory constraints. Filled bars (i.e., the upper bar in

each paired group) show participants' responses for the given set of lower-level compensatory criteria, whereas white bars show responses in connection with the given set of four higher-level compensatory criteria.

This result leads to the conclusion that it is not the number but the semantic comprehensiveness of compensatory route selection criteria that supports easy-to-use route planners while preserving high functionality. As demonstrated, a more complex structure of eight lower-level criteria cannot ascribe the user's objective of defining a best route equally well as a set of only four higher-level criteria. This statement refers to the demand for further eliminatory constraints.

Although the results of the survey suggest an importance ranking for the inclusion of eliminatory constraint functionality (Fig. 6), a particular criterion should only appear in a route planner user interface if it is relevant for the set of currently available route alternatives (e.g., "avoid stairs", only if at least one route includes stairs). Otherwise, the user would be asked to make preference statements although these would not affect the decision result, as it is the case with several existing route planners.

4   Question 3: Context-Dependent Eliminatory Constraints

The third part of the study examines the potential impact of context information on the user's demand for eliminatory constraint functionality. The results of question 1 (section 2) have shown that users tend to prefer being provided with context information about route alternatives before stating their preferences. Context information may decrease the user's demand for eliminatory constraint functionality, as context information allows the user to assess under a higher degree of certainty which eliminatory constraints may be needed to define the optimal route. In consequence, "blind" checking of arbitrary eliminatory constraints, which have no effect on the decision outcome (recall the "avoid stairs" example from above), should be unnecessary.

This part of the survey will also examine, whether the findings from section 3, namely a decreased demand for eliminatory constraint functionality in connection with higher-level compensatory criteria, also hold if the user is provided with context information during the preference statement process.

The 21 participants that completed this task are among the 32 Internet users participating in question 1 (section 2) and ages for this group ranged from 16-53 years (median = 26).

4.1   Hypothesis

This part of the study analyses the demand for eliminatory constraints with respect to two independent variables, namely the level of generality of compensatory criteria (*high, low*), and the type of route information provided (*bar chart*, or *map*, or *bar chart combined with map*). The dependent variable is the number of eliminatory constraints selected for any of those combinations. We hypothesize that the decreased need for eliminatory constraints in combination with higher-level criteria found above can also be observed when context information about existing route alternatives is provided. Furthermore, we hypothesize that context information reduces the demand for eliminatory constraints when compared to decision situations without context information. With reference to the last part of the hypothesis, we will compare the findings from this question with those from question 2 (section 3.3).

4.2 Questionnaire Setup

The different values of the two independent variables result in a total of six possible combinations for the user interface design (Table 3).

Table 3. The six user interface designs used for the third study

Level of generality for compensatory criteria	Information Design		
	Bars	Map	Bars & Map
high (4 sliders)	Bars(4)	Map(4)	Bars & Map(4)
low (8 sliders)	Bars(8)	Map(8)	Bars & Map(8)

The questionnaire was realized in the form of six separate Web pages. Each page contained three elements, two of them being varied according to Table 3. The three elements are: an image showing the set of compensatory route selection criteria on the left side (see also Fig. 5a and b), an image visualizing route information at the top of the page (information design as used in Fig. 1b, c, d), and the list of 26 eliminatory constraints presented as checkboxes. The design variant in Fig. 7 shows one of the six combinations, namely four higher-level compensatory criteria together with attribute information as bar charts (design *Bars(4)* in Table 3).

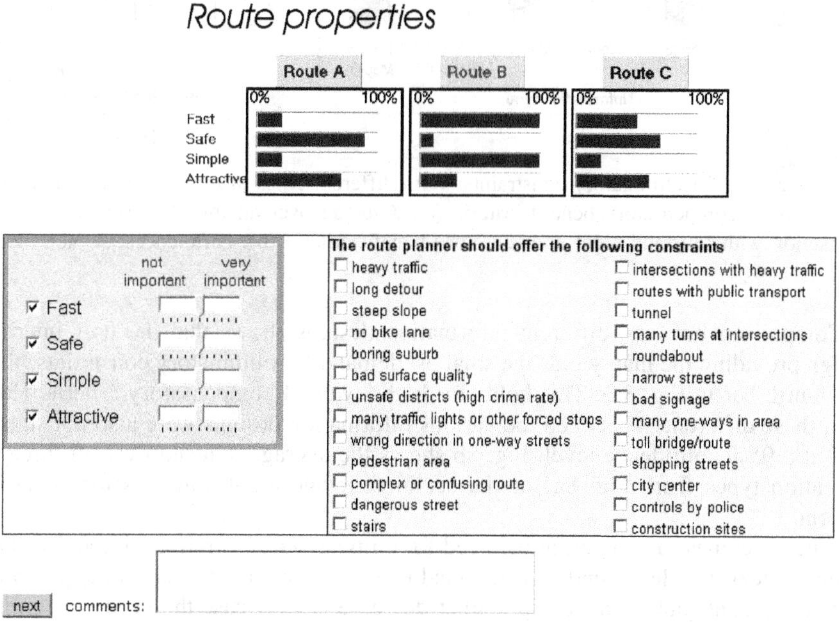

Fig. 7. One of the six user interface designs used for question 3

Participants were asked to check those eliminatory constraints that they considered as relevant for defining the best route in accordance with the given set of compensatory criteria on the left and the route information at the top. All six pages were shown in random order to each participant. Moving back to previous pages was not possible.

4.3 Results

Fig. 8a shows the number of eliminatory constraints that participants selected on average for each of the six user interface designs. Dotted bars refer to a design with four higher-level compensatory criteria, whereas bars with horizontal pattern refer to eight lower-level compensatory criteria.

It can be seen that for all three information designs (bars, map, bars & map) the number of selected eliminatory constraints is slightly higher for eight than for four compensatory route selection criteria in correspondence with results from question 2. Thus the predicted trend for the demand for eliminatory constraints in combination with higher-level and lower-level compensatory criteria was also observed in conjunction with context information. However, a Wilcoxon signed rank test reveals that the differences in each group are not significant at a 95% confidence level.

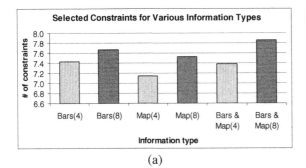 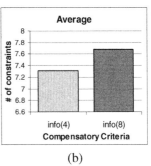

(a)   (b)

Fig. 8. Demand for eliminatory constraints: With different types of route information and different sets of compensatory benefit criteria (a). Average over all three information types in connection with four and eight compensatory benefit criteria (b). Differences are not significant.

Comparison between different information designs shows that the user interface design providing the map yields the smallest demand for eliminatory constraints (third and fourth bar in Fig. 8a). This holds for both levels of compensatory criteria. However, these differences between the various information designs were also not significant at a 95% confidence level. Fig. 8b shows the average demand over all three information types from Fig. 8a for the set of four and eight compensatory decision criteria.

When it comes to comparing the need for eliminatory constraints with and without given context, no clear trend can be found (Fig. 9). This means that in the given designs additional route context did neither decrease nor increase the need for eliminatory constraints significantly. The number of selected constraints without context information (first and third bar in Fig. 9) are derived from the findings in section 3.3, while the second and fourth bar are taken from Fig. 8b.

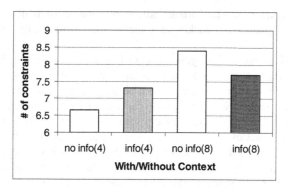

Fig. 9. Demand for eliminatory constraints with and without context information in combination with four and eight compensatory benefit criteria

A possible interpretation of this result is that the user interface was too overloaded so that the participants could not take into account the additional context information when thinking about utilizing additional eliminatory constraints. This in turn means that the route planner would need to restrict the availability of eliminatory constraints to relevant constraints only, which are those that affect the set of feasible routes. Only then would the user be able to grasp all choice options as well as the context information provided. Another interpretation of the results is that the provided context information was just not detailed enough for the participants to exclude some of the previously selected eliminatory constraints. As a general conclusion from the results of the third question we assume that, as long as the user does not know which undesired features, such as stairs, are actually part of any route alternative at hand, the user's willingness to check such constraint is low, as she cannot assess consequences of taking such action.

5   Conclusions and Future Work

This paper presents results from a survey that focused on three questions about the need for context information and eliminatory constraints in user interfaces for bicycle route planners. It could be shown that users feel more comfortable in their preference statements on compensatory route selection criteria if information about existing route alternatives is provided, preferably as a route map in combination with attribute bar charts (question 1). The results of question 2 revealed that the demand on eliminatory constraints depends on the generality of the compensatory benefit criteria provided in the user interface. For both sets of benefit criteria tested, more than six avoid criteria were selected on average. This demonstrates the need for including eliminatory constraints in route planning tools in addition to benefit criteria. The last part of the survey (question 3) showed a tendency of maps to reduce the need for eliminatory constraints to a greater extent than the two alternative information designs. However, no evidence could be found that context information generally affects the user's demand for eliminatory constraint functionality. But the same way context information is helpful when stating preference for compensatory route selection criteria, information about "avoid" characteristics that are actually part of the set of available routes will be a good help for the decision maker in selecting the most useful eliminatory constraints.

These conclusions are yet too general to yield guidelines for user interface design for practical use. However, some tendencies can be derived: Maps should be included in the user interface of route planners once it comes to preference statements, as maps contain spatial information which is hard to represent by means of textual or statistical representation. In addition, eliminatory constraint functionality is useful as it allows the user to define the set of feasible routes. The choice of selection functions provided in the user interface should be made with care (by the tool itself), and adapted to the set of available route alternatives at hand. Dummy functions, which have no effect on the outcome of the route planner, should be hidden or disabled. One goal of future work is to examine what the user's most intuitive sequence of preference statements is when defining the optimal route, and how the user interface can support this sequence. Our assumption is that the user's cognitive abilities are overstrained when being asked to make all preference statements at once, which could for example be avoided through a step-wise refinement process of preference statements. Such process includes the use of a hierarchy of compensatory decision criteria, and dynamic selection functionality for given attribute values. A route planner with such functionality would serve as an exploratory, visual decision analysis tool. Another open question is the impact of user task and time restriction on the preferred representation of context information. Limited time for route choice may for example lead to preference for a user interface that shows maps only and omits route statistics, thus changes the result of the first study.

Acknowledgements

The authors thank the Transregional Collaborative Spatial Cognition Research Center (SFB/TR8) for providing the server infrastructure used for the Internet survey. This research has been partially supported by the Natural Sciences and Engineering Research Council of Canada (NSERC) with a grant to the second author. Chris Sidlar and Ben Spigel provided useful feedback on the manuscript.

References

Andrienko, N. and Andrienko, G. (2001). Intelligent support for geographic data analysis and decision making in the Web. *Journal of Geographic Information and Decision Analysis*, 5 (2), 115-128.

Backhaus, K., Erichson, B., Plinke, W., and Weiber, R. (1996). *Multivariate Analysemethoden*. Berlin: Springer.

Barfield, L. (1993). *The User Interface - Concepts & Design*. Wokingham, England: Addison-Wesley Publishing Company.

BBBike (2005). Online Radroutenplaner für Berlin. Retrieved 02/28/2005 from *http://www.radzeit.de/*

Beale, L., Picton, P., Matthews, H., and Field, K. S. (2001). *MAGUS: a GIS application for wheelchair users in urban environments*. Twenty-first Annual ESRI International User Conference, San Diego.

Bikemetro (2005). Online Bicycle Route Planner for California. Retrieved 02/28/2005 from *http://www.bikemetro.com/route/routehome.asp*

Carver, S. J. (1991). Integrating multi-criteria evaluation with geographical information systems. *International Journal of Geographical Information Systems*, 5 (3), 321-339.

Demis (2005). Online Bicycle Route Planner. Retrieved 02/28/2005 from *http://www3.demis.nl/fietsplanner/*

Dewey, C. (2001). Navigationsdienste für Rollstuhlfahrer - Berücksichtigung besonderer körperlicher Anforderungen bei der computergestützten Wegesuche (Diploma Thesis, Fachbereich Geowissenschaften, Westfälische Wilhelms-Universität, Münster).

Eastman, J. R. (1997). *IDRISI for Windows, Version 2.0: tutorial exercises*. Worcester, MA: Graduate School of Geography, Clark University.

Ehlers, M., Jung, S., and Stroemer, K. (2002). *Design and Implementation of a GIS Based Bicycle Routing System for the World Wide Web (WWW)*. Spatial Data Handling 2002, Ottawa.

EMS (2005). Radroutenplaner Emsland. Retrieved 06/26/2005 from *http://www.emsland-route.de/*

Freksa, C. (1999). Spatial Aspects of Task-Specific Wayfinding Maps. A representation-theoretic perspective. In J. S. Gero and B. Tversky (Eds.), *Visual and Spatial Reasoning in Design* (pp. 15-32). Sydney: Key Centre of Design Computing and Cognition.

Hochmair, H. H. (2004a). Decision support for bicycle route planning in urban environments. In F. Toppen and P. Prastacos (Eds.), *Proceedings of the 7th AGILE Conference on Geographic Information Science* (pp. 697-706). Heraklion, Greece: Crete University Press.

Hochmair, H. H. (2004b). Towards a classification of route selection criteria for route planning tools. In P. Fisher (Ed.), *Developments in Spatial Data Handling* (pp. 481-492). Berlin: Springer.

Horn, J. (1997). Multicriteria Decision Making. In T. Bäck, D. B. Fogel and Z. Michalewicz (Eds.), *Handbook of Evolutionary Computation* (pp. 1-15). Bristol (UK): Institute of Physics Publishing.

Jankowski, P. (1995). Integrating geographical information systems and multiple criteria decision making methods. *International Journal of Geographical Information Systems*, 9 (3), 251-273.

Jiang, H. and Eastman, J. R. (2000). Application of Fuzzy Measures in Multi-Criteria Evaluation in GIS. *International Journal of Geographical Information Science*, 14 (2), 173-184.

Keeny, R. L. (1980). *Siting energy facilities*. San Diego, CA: Academic Press.

Keeny, R. L. and Raiffa, H. (1993). *Decision Making with Multiple Objectives: Preferences and Value Tradeoffs*. Cambridge, UK: Cambridge University Press.

Koyani, S. J., Bailey, R. W., and Nall, J. R. (2003). *Research-based Web Design & Usability Guidelines*, Dept of Health & Human Services, National Cancer Institute, Washington, DC. Available from *http://www.usability.gov*

MAGUS (2005). MAGUS project page. Retrieved 02/28/2005 from *http://www.magus-online.org.uk/index.htm*

MAGWIEN (2005). Magistrat Wien: Routensuche für Radfahrer. Retrieved 08/04/2004 from *http://service.wien.gv.at/wien-grafik/cgi-bin/wg?app=13&tmpl=wo*

Malczewski, J. (1999). *GIS and Multicriteria Decision Analysis*. New York: John Wiley.

Mitchell, J. S. B. (2000). Geometric shortest paths and network optimization. In *Notebook of Computational Geometry* (pp. 633-701). Amsterdam: Elsevier.

Mooney, P. and Winstanley, A. (2003). The JPathFinder Multicriteria Path Planning Toolkit. In M. Gould, R. Laurini and S. Coulondre (Eds.), *AGILE 2003* (pp. 283-291). Lausanne: Presses polytechniques et universitaires romandes.

Rad.RoutenPlaner (2003). Software CD ROM. Retrieved 02/28/2005 from *http://www.tvg-software.de*

Rinner, C. and Malczewski, J. (2002). Web-enabled spatial decision analysis using Ordered Weighted Averaging (OWA). *Geographical Systems, 4* (4), 385 - 403.

Robinson, A. H., Morrison, J. L., Muehrcke, P. C., Kimmerling, A. J., and Guptil, S. C. (1995). *Elements of Cartography*. Hoboken, NJ: Wiley.

Shneiderman, B. and Plaisant, C. (2004). *Designing the User Interface: Strategies for Effective Human-Computer Interaction*. Boston: Pearson Education, Inc.

Slocum, T. A., McMaster, R. B., Kessler, F. C., and Howard, H. H. (2004). *Thematic Cartography and Geographic Visualization*. Upper Saddle River, NJ: Prentice Hall.

Srinivasan, V. (1988). A Conjunctive-Compensatory Approach to the Self-Explication of Multiattributed Preferences. *Decision Sciences, 19*, 295-305.

Takatsuka, M. and Gahegan, M. (2002). GeoVISTA Studio: A Codeless Visual Programming Environment for Geoscientific Data Analysis and Visualization. *The Journal of Computers & Geosciences, 28* (10), 1131-1144.

Williamson, C. and Shneiderman, B. (1992). The dynamic HomeFinder: Evaluating dynamic queries in a real-estate information exploration system. In *Proc. ACM SIGIR '92 Conference* (pp. 338-346). New York: ACM.

Path Memory in Real-World and Virtual Settings

Adam Hutcheson and Gary L. Allen

Department of Psychology, University of South Carolina,
Columbia, SC 29208 USA
hutchesa@mailbox.sc.edu, allen-gary@sc.edu

Abstract. Using a technique common in the orientation- specificity literature, we set up an experimental study in an outdoor setting and used it as the basis for a parallel study involving a virtual version of that setting. After viewing the path, participants made a series of directional and scene recognition judgments involving locations along that path after either being moved directly to the testing site or being moved there via a circuitous route. At the site, participants were situated in alignment or counter-alignment with their previous viewing position. Similarities between performance in real-world and virtual setting included participants' use of stable landmarks in both environments and improvement in performance through repeated testing. In addition, previously published patterns of performance that signify specific means of achieving orientation-free performance in this task were not replicated, presumably because the time-space dimensions of the setting. We concluded that in light of the similarities and lack of strong differences in results, virtual environments can be viable options for researchers wishing to eliminate confounding and nuisance variables that may be present when doing spatial tests in large-scale spaces.

Keywords: spatial memory, orientation specificity, spatial alignment effects.

1 Spatial Memory in Real-World and Virtual Environments

Outdoor tests of spatial abilities have been used in many instances for many different purposes. Whether one is interested in distance estimation, route learning, or object location, tasks placed in outdoor settings are easy to create and carry out. The placement of spatial tasks in real-world outdoor settings brings with it the characteristic of rich spatial information that is inherent in the environment, itself. When we walk outside, we can gather spatial information from nearby landmarks, the sounds that surround us, and even celestial markers such as the sun and stars.

These factors, while sometimes necessary in everyday navigation, create a problem when trying to control every aspect of spatial information that a participant in a research study receives in a task. The absence of complete control that exists when doing research on spatial abilities in a real-world setting has lead many researchers to the conclusion that there should be a way to test the same abilities but with more complete control. The solution that has risen with the creation and advances in computer technology is the Virtual Environment (VE).

The two main types of VE presentation are desktop systems and immersive-display systems [4, 6]. Desktop systems are those which display the VE on a fixed computer

A.G. Cohn and D.M. Mark (Eds.): COSIT 2005, LNCS 3693, pp. 67–82, 2005.
© Springer-Verlag Berlin Heidelberg 2005

screen. An immersive-display system includes the use of a head-mounted display that has two small screens show the image as one three-dimensional image and is worn by the participant like a pair of eyeglasses. Regardless of which system that is used to display the VE, the interfaces used for movement may involve the keyboard, a joystick, the computer mouse, or a specialized treadmill device.

1.1 Parallel Real-World and Virtual Environments

One powerful purpose of a VE is to accurately simulate a real-world environment with complete control over all environmental characteristics. As this is a relatively new area of research in the spatial domain, there are few studies that have been done to show the similarities in participant behavior between real-world and virtual environments.

An early study concerned with comparing behavior in these two types of environments was done by Witmer, Bailey, and Knerr [14]. It looked at the effects of VEs on route learning. They were interested in participants' ability to traverse a route in a real building after rehearsing the route by one of three methods: VE rehearsal, building rehearsal, and symbolic rehearsal. The VE group was given a computer-generated version of the building to practice with before being asked to move through the real building. The building rehearsal group walked in the real building and was asked to identify significant locations along their route. The symbolic rehearsal group verbally rehearsed the directions of the route aloud. Even though the VE group showed a higher number of errors when reproducing the route, they did show a more rapid rate of improvement across repeated trials than the other two groups. Their conclusion was that virtual environments could be used almost as effectively as a real environment.

In a study by Oman, Shebilske, Richards, Tubré, Beall, and Natapoff [5], three-dimensional spatial memory was tested in real and virtual settings. Their primary focus was on how humans can orient to their position even without the presence of gravity to create a single axis of body rotation and movement. By this, they were referring specifically to the ability of astronauts to orient to their position within their spacecraft which makes full use of all three spatial dimensions. Their purpose was to see if participants could remember the placement of 6 objects located around them independent of their body's position. Participants were placed into two groups: a physical display group and a virtual display group. Participants in the physical display group sat inside a booth with a computer screen attached to each of the six sides of the booth. These computer screens displayed the objects with which they were tested. The virtual display group used a head-mounted display unit for the same purpose as the computer screens in the physical display group. As the participant moves their head around, all objects come into view. Participants in both groups were asked to study the array of objects and then answer questions about the array by determining the "correctness" of the viewpoint that was presented to them. They found that the percentage of correct responses and the reaction times of the virtual display group were similar to, and even slightly better than, the physical display group. They conclude that, given a real situation that needed the use of all three-dimensions (such as in the weightlessness of space), similar results would be obtained.

Péruch and Wilson [6] were interested in how active and passive transport would affect the transfer of spatial information from a VE to a real-world environment. By passive transport, they were referring to situations such as the experience of a passenger of a car during travel. Active transport, then, would refer to the driver who is actively exploring the environment. Participants were placed into conditions that allowed either active or passive transport along a route on a college campus. Also, the effect of virtual and real environments was examined by alternating how the participants learned and later reproduced the route. Half of the participants learned the route on a computer-generated version of the campus and then tested along the real route. The other half learned the route in the real environment and was later tested using the VE. The results of this study showed that neither type of transport nor type of learning phase affected the participants' ability to learn the route. They concluded that testing in virtual and real environments will point to similar outcomes.

In a very recent study by Waller [13], participants' ability to navigate in a large-scale virtual environment was assessed and correlated with self-reports of sense of direction. The VE was presented as a desktop display showing the exterior of a "building". The participant views a "walk" around this building exterior and is then asked to make judgments as to the overall shape of the building. They were also asked to point to an unseen landmark located near the virtual building to assess their sense of direction within the virtual domain. To test sense of direction in the real-world, participants were asked to point to unseen landmarks located outside of the real building in which they were taking the test. Results from this study indicated that mental representations of the familiar real-world environment were correlated almost as highly with representations of the virtual environment as it did with the self-reports of sense of direction. It was concluded that VEs can be used as a great tool in testing spatial memory for large-scale spaces.

1.2 Orientation Factors in Spatial Memory

As Montello et al. [4] pointed out, orientation factors (especially the issue of orientation-specificity) are theoretically very salient in theoretical considerations of spatial cognition and, by extension, in evaluations of VE's. Orientation-specificity is the idea that a person stores the memory of a path or an object array in the same direction of which it was learned. If someone does not store the memory in an orientation-specific manner, it is considered orientation-free. The standard path used in orientation-specificity studies was first developed in experiments by Presson and Hazelrigg [7]. This "Presson path" was U-shaped, with a leg on one side that was longer than the other. Participants were placed into one of three conditions: viewing a map of the path, looking at the path from a single vantage point, or actually walking on the path. Participants were asked to either point to the path locations in the same alignment as in the learning phase or were asked to point to the locations in a contra-aligned manner. Alignment effects were used to get a measure of orientation specificity. If a participant is tested in the same orientation in which they learned the path, they should not show high latencies and errors in pointing. These participants would not be showing alignment effects. A significantly larger latency in pointing coupled with higher errors in pointing exhibited by the participants in the contra-aligned (opposite orientation) condition would indicate an alignment effect. They found significant alignment effects for the map learning condition but not for the walking or looking conditions.

A follow-up study by Presson, DeLange, and Hazelrigg [8] suggested that the size of the spatial array itself could determine whether performance is orientation dependent or independent. Small arrays, either maps or small paths, were more likely to yield orientation-specific performance while larger arrays yielded orientation-free performance.

Another study showing the orientation-specificity of spatial representations was Roskos-Ewoldsen, McNamara, Shelton, and Carr [10], who used a room-sized path like the one described above in a Presson and Hazelrigg [7] study. The participants were allowed to view a path from a single vantage point. A blindfold was then placed over their eyes to prevent them from seeing the path during testing. The experimenter wheeled them around in a cloverleaf formation that intersected the path in order to properly disorient them. One important result from this study is that when participants were tested at an off-path location, performance tended to be orientation-specific.

Sholl and Nolin [12] studied the contention that large arrays are necessary but not sufficient to produce orientation-free performance. They used "Presson Paths" of varying sizes (with path legs varying in length from 2 ft to 13 ft). They also varied the angle of the gaze by having some participants sit while studying the path and others stood up. Alignment effects were present in the small path conditions but not in the large path conditions. They concluded that large path layouts are necessary for orientation-free performance. They said that the angle of gaze during the learning phase must be close to eye level of the participant. Also necessary was the use of on-path testing. If any of these requirements were violated, orientation-specific performance was observed.

Orientation-free memory of the type found by previous research could be accomplished by different processes. In particular, Sholl and Bartels [11] hypothesized two distinct means of maintaining orientation-free performance. The first, known as the updating hypothesis, is the idea that a person constantly retrieves information about their location relative to their body from working memory. According to this hypothesis, studies using passive transport in a wheelchair did not disorient the participant when orientation-free performance was found. The second hypothesis asserts that during passive transport, the participant creates virtual-views of the path from vantage points that they never actually experienced. This requires information about the path to be retrieved from long-term memory to form mental images. This hypothesis explains why orientation-free performance can still be observed even when the participant is properly disoriented.

Sholl and Bartels [11] completed three experiments to look at these two hypotheses in a standard path memory task. For the first experiment, participants viewed the path, were disoriented by passive transport in a cloverleaf trajectory, and then taken to an off-path test site. Participants were told, "You are at location [x] with location [y] in [front/back] of you. Point to location [z]". The second experiment involved either passively transporting the participant to an on-path test site in a cloverleaf trajectory or directly from the study site. The pointing instructions were the same as in the first experiment. The third experiment was identical to the second except that participants were not told their locations before they were asked to make pointing decisions. They concluded that the updating hypothesis explains how women form orientation-free representations because they show strong effects for alignment and the trajectory taken to the test site. They also conclude that the virtual-views hypothesis sufficiently explains how men form orientation-free representations because they do not show strong effects for alignment or the trajectory taken to the test site.

1.3 Looking for Orientation-Free Memory in Real and Virtual Environments

Research findings pointing to conditions yielding orientation-free representations suggest the need for additional experiment to replicate and extend such results. Specifically, it would be useful to replicate the pattern of results leading to the conclusion that men use virtual views and women use updating to form orientation-free representations of a path. Accordingly, we set up our first study as a replication and extension of Experiment 2 in the Sholl and Bartels [11] study. Just as in that study, our hypothesis was that orientation-free performance would be observed when participants were tested at an on-path location. Along with having the participants point to various locations on the path, they were also asked to identify pictures from novel views. Also, this study took place on an outdoor path rather than an indoor room-sized path to test generalizability of the Sholl and Bartels [11] findings.

Our predictions were as follows. If men are using virtual-views to form orientation-free representations, then they would show large alignment effects after being wheeled directly to test site but not after disorientation process. If women simply update their location based on their body movements, then they should show large alignment effects after proper disorienting but not if wheeled directly to test site. These predictions are summarized in Table 1.

An added feature of this study was the repeated testing of recognition memory for scenes from the path. Giving the Picture Verification task twice allowed us to see if working memory was playing a role in orientation-free performance as suggested by the updating hypothesis. If this were true, participants using updating would show higher error in picture verification in the second presentation which is immediately preceded by a short interference task.

Table 1. Expected pattern of results based on Sholl and Bartels [11]

	Men		Women	
	0-Turn	3-Turn	0-Turn	3-Turn
Pointing Error	Align < Contra-align	Align > Contra-align	Align < Contra-align	Align < Contra-align
Pointing Latencies	Align < Contra-align	Align > Contra-align	Align < Contra-align	Align < Contra-align

Our plan was to follow this initial study conducted in a real-world environment with another experiment using the same spatial relations portrayed in a desk-top VE.

Previous studies have suggested significant differences between learning in real and virtual environments. For example, Richardson, Montello, & Hegarty [9] found more accurate orientation when participants were tested when aligned with the original viewpoint than when contra-aligned with the perspective seen at the beginning of exploration. Similarly, a number of studies have shown orientation-specificity as reflected in the speed and accuracy with which viewed scenes versus novel scenes were verified after learning layout from VEs [1], [2]. Accordingly, we anticipated the possibility of little evidence of orientation-free memory in the experimental results from the VE study.

2   Orientation Effects in Memory After in Situ Viewing

2.1   Method

A path very similar to the ones used in previous studies was used in this experiment. The path was created by placing orange construction cones on a sidewalk. Colored flags with numbers (1-4) were used to identify the construction cones. Because we used a preexisting sidewalk, we had to find one that most closely matched the dimensions of the path used in the previous studies. As can be seen in Figure 1, the distance between locations 2 and 3 are proportionally larger than those previous paths. The distances between the path locations were as follows: from 1 to 2 was 33 feet, from 2 to 3 was 210 feet, and from 3 to 4 was 85 feet.

Participants were volunteers from the participant pool at the University of South Carolina, who receive research participant credit for their psychology classes. They fulfilled a course requirement for their participation. A total of 32 participants (16 men, 16 women) were needed. Each participant completed the practice and test trials on a Dell Latitude CPx laptop computer with a thirteen inch color monitor. The participant sat approximately sixteen inches away from the screen. All computer trials were administered using the Superlab software. An "Arrow Dial" was placed to the right of the computer to be used during the Point to the Location trials. It was constructed from a mouse pad with an arrow on the top which could be rotated around a circle of tick-marks like those on a clock. A digital stopwatch was also used to time the participant in the "Point to the Location" trials.

A plywood desk was attached to the wheelchair to provide a stable place to put the computer. Extra strength Velcro was used to keep the computer from sliding around or falling off of the desk. The desk was firmly tied to the armrests of the wheelchair with thin nylon cord in four locations.

Upon reaching the study site, the participant was tapped on the shoulder as an instruction to remove the blindfold. From this vantage point, all four path locations were easily visible. He or she was given approximately 30 seconds to study the path. At that time, replaced the blindfold and was transported to the test site.

Participants were either taken directly to the test site or were pushed along a three-turn route very similar to the ones used in the previous orientation-specificity studies. Also, they were either aligned or contra-aligned when placed at the test site. These various routes and alignment conditions are shown in Figure 1. The closet-like booth was then placed around them. They were tapped on the shoulder to indicate that they could remove the blindfold and begin the test.

The participants were asked to complete three tasks when they reached the test site. During the "Picture Verification" task, a direction was presented on the screen that said, "Imagine that you are standing at location [x] facing [y]. Is this what you would see?". After pressing the space bar, a picture was presented that showed the location that they should be facing (from the directions given) but they were either taken from the correct standing location or from one of the other two locations. Each picture was presented two times to allow both a "Yes" response and a "No" response for a total of 24 responses. This task was presented at both the beginning of the test and at the end to determine if there was any decay of the information.

The "Point to Location" task included questions very similar to those in the Picture Verification task except the participant was asked to point to a third location using the Arrow Dial to the right of the computer. For instance, the direction would say, "You are standing at location [x] with location [y] in front of you. Point to location [z]." The responses on the Arrow Dial and the time from presentation of the direction to the completion of response were recorded. There were a total of 12 responses for this task.

The third task was a simple interference task used to potentially take up space in working memory. This task asked the participant to help a cartoon car arrive at a garage by using the same "Yes" and "No" responses. The interference task was placed in the middle of the test.

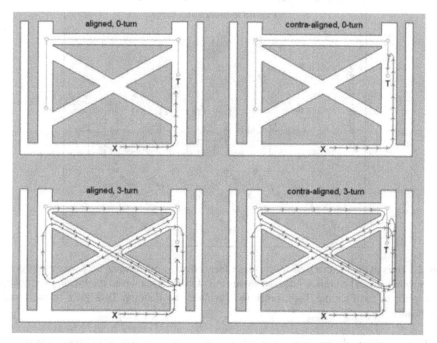

Fig. 1. Circles on the indicated path show locations of the four landmarks; lines with arrows show the path of travel from the study site to the test site for the zero-turn (top two panels) and three-turn (bottom two panels) conditions and for the aligned (left two panels) and misaligned testing conditions. The "X" represents the location of the path study site. The "T" represents the site used for testing.

2.2 Results

2.2.1 Picture Verification Task

A 2 (sex of participant) x 2 (order of tasks) x 2 (alignment condition) x 2 (number of turns during transport) x 2 (repeated tests) ANOVA with four participants per cell was performed on the number of incorrect responses in the Picture Verification Task. This analysis yielded a main effect for sex of participant (across first and second presentations of pictures), $F(1, 16) = 6.55$, $p < .05$, $\eta^2 = .29$, as well as significant interactions involving sex of participant X alignment, $F(1, 16) = 13.82$, $p < .01$, $\eta^2 = .46$, alignment X order, $F(1, 16) = 6.06$, $p < .05$, $\eta^2 = .29$, and sex X alignment X order, $F(1, 16) = 6.60$, $p < .05$, $\eta^2 = .27$. No other significant main effects or interactions were found including the repeated testing of picture verification.

Post-hoc comparisons ($p < .05$) showed no order effect for men; they consistently performed with lower error when aligned than when contra-aligned regardless of which task came first. In contrast, task order affected women's errors. When Picture Identification came first, no alignment effect was observed, and level of accuracy was similar to that of men who were contra-aligned. When the pointing task came first, however, picture identification was adversely affected with alignment but not with misalignment (see Figure 2).

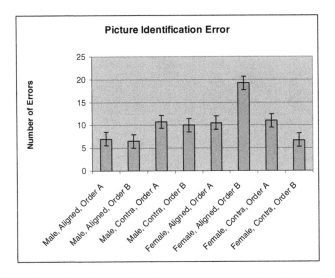

Fig. 2. Errors in the Picture Verification Task for men and women when they were aligned with the viewing site and contra-aligned during testing. Order (A) involved the Picture Verification Task prior to the Pointing Task; order (B) involved the Pointing Task prior to the Picture Verification Task.

A set of planned pair-wise comparisons ($p < .05$) was conducted based on the sex X alignment X turn interaction found in the Sholl and Bartels [11] study and there were significant differences between two sets of groups: males who were aligned, 0-turn differed from females who were aligned, 3-turn and males who were aligned, 3-turn differed from females who were aligned, 3-turn.

A 2 (sex of participant) x 2 (order of tasks) x 2 (alignment condition) x 2 (number of turns during transport) x 2 (repeated tests) ANOVA with four participants per cell was performed on picture verification latency. The analysis yielded only one main effect, which was of repeated testing, $F(1, 16) = 75.91$, $p < .01$, $\eta^2 = .83$. No other main effects or interactions were found to be significant. This finding indicates that participants were faster in the second presentation of the pictures than in the first.

2.2.2 Pointing Task

A 2 (sex of participant) x 2 (order of tasks) x 2 (alignment condition) x 2 (number of turns during transport) ANOVA with four participants per cell was performed on absolute pointing error. This analysis yielded significant interactions for sex X alignment, $F(1, 16) = 7.90$, $p < .01$, $\eta^2 = .33$, and sex X order, $F(1, 16) = 10.95$, $p < .01$, $\eta^2 = .41$. There were no significant main effects and none of the other interactions was significant. Post-hoc comparisons $(p < .05)$ showed that pointing error was less for aligned men than for contra-aligned men, but it was greater for aligned women than for contra-aligned women (see Figure 3).

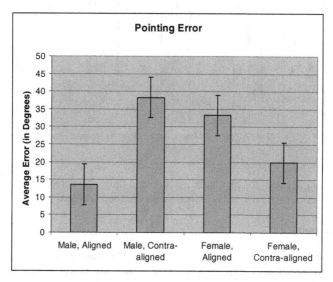

Fig. 3. Error was less for men when aligned than when contra-aligned but greater for women when aligned than when contra-aligned

A set of planned pair-wise comparisons $(p < .05)$ was conducted based on the sex X alignment X turn interaction found in the Sholl and Bartels [11] study, and there were no significant differences between means.

A 2 (sex of participant) x 2 (order of tasks) x 2 (alignment condition) x 2 (number of turns during transport) ANOVA with four participants per cell was performed on pointing latency. There was a significant main effect of sex of participant, $F(1, 16) = 13.51$, $p < .01$, $\eta^2 = .46$, and a significant sex X alignment interaction, $F(1, 16) = 5.50$, $p < .05$, $\eta^2 = .26$. There were no other significant main effects or interactions.

The main effect of sex showed that males performed significantly faster than females. Post-hoc comparisons showed that men's latencies were shorter than women's when testing was aligned with viewing, but there was no difference when testing was contra-aligned with viewing (see Figure 4).

A set of planned pair-wise comparisons ($p < .05$) was conducted based on the sex X alignment X turn interaction found in the Sholl and Bartels [11] study and found that only the males who were aligned, 0-turn differed significantly from the females who were aligned, 3-turn.

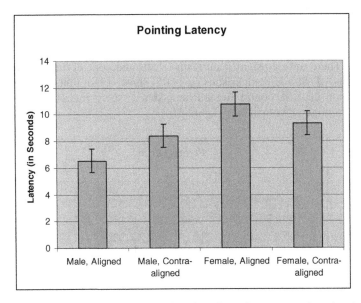

Fig. 4. Pointing latencies for men were quicker than those for women when the viewing sites and testing sites were aligned but not when they were contra-aligned

2.3 Discussion

Some of our findings were easily interpreted. The main effect of sex in the picture verification task showed that men made significantly fewer errors than did women overall. This was consistent with the male superiority shown in Sholl and Bartels's [11] results with pointing responses. Also, it was not surprising was that participants made more errors during the first presentation of the picture task than in the second presentation. Evidently, they benefited from having extra exposure to the stimuli and possibly from completion of the intervening pointing task. The intervening task designed to disrupt working memory had no apparent detrimental effect on performance.

Much more difficult to interpret were results involving women's performance with misaligned testing conditions. Men who were aligned with the study perspective during testing generally showed better recognition, less pointing error, and quicker pointing responses than did women who were aligned. They also showed more accurate pointing than did other men who were contra-aligned, an evident sign of orientation-specificity in the spatial representation. The story was different with women. They showed superior pointing performance when they were contra-aligned and superior picture verification when contra-aligned, given that they had done the pointing task preceding the picture verification task. As Montello, Hegarty, Richardson, & Waller [4] pointed out, if the circuitous route disoriented participants, then a finding of no difference between aligned and contra-aligned testing would be expect. However, no known theory-based explanation would predict superior performance in a contra-aligned position. Although we have no definitive interpretation for this difference, we speculate based on some informal feedback that women used information from the test photographs themselves in an attempt to maintain orientation. Information in the photos may have provided stable landmarks that could be used in orienting. Previous studies on sex differences have shown that women tend to use landmarks more than do men in wayfinding tasks [3]. Perhaps the men in this study made less use of contextual features in the environment and focused more exclusively on the task path and targets.

The results obtained from this study did not parallel those from Sholl and Bartels's [11] Experiment 2. Whether this lack of correspondence was due simply to the space-time scale involved with our larger setting or to some other procedural difference cannot be determined from the results. Nevertheless, our results could not be used to differentiate between types of orientation-free representations.

3 Orientation Effects in Memory After Virtual Viewing

3.1 Method

The path that the participants viewed was a panoramic picture taken of the original path used in Experiment 1. This picture was taken on the same day that the pictures used for the Picture Verification Task were taken. Also, the panoramic picture that was used for the "study site" in Experiment 2 was taken from the same location that served as the study site for Experiment 1.

Participants completed this experiment on a Dell Latitude CPx computer and sat approximately sixteen inches away from the monitor. To recreate the transportation from the study site to the test site, we placed a blind fold over the participants' eyes and told them to imagine being pushed in the wheelchair in this environment. Just as in Experiment 1, the "Picture Verification" Task and "Point to the Locations" Task were given using the Superlab software.

During the procedure, the "study site" was presented on the screen, and participants were instructed to study the path for approximately 30 seconds. When they had committed the path to memory, the imagined transport to the test site began. The experimenter actually rotated the participant's chair in accordance to the actual transport conditions. The most important aspect of the procedure was that the time that elapsed between the study and test phases was the same as in Experiment 1: 1 minute for 0-turn condition and 4 minutes for 3-turn condition. After transport ended at a desk placed at the test site, the Superlab program began and the participant completed the same tasks as in Experiment 1.

3.2 Results

3.2.1 Picture Verification Task

A 2 (sex of participant) x 2 (order of tasks) x 2 (alignment condition) x 2 (number of turns during transport) x 2 (repeated tests) ANOVA with four participants per cell was performed on the number of incorrect responses out of 48 possible. Results showed a main effect for repeated testing, $F(1, 32) = 7.85$, $p < .05$, $\eta^2 = .20$, and a significant three-way interaction involving sex X order X repeated testing, $F(1, 32) = 4.26$, $p < .05$, $\eta^2 = .12$. No other significant main effects or interactions were found. The means involved in the main effect for repeated testing showed that participants made fewer errors after the second presentation than after the first presentation.   Post-hoc comparisons $(p < .05)$ performed on the means involved in the three-way interaction further showed that men's error decreased from the first to the second administration of the Picture Verification Task regardless of task order but that women's error decreased from the first to the second administration of the task only when the Pointing Task was administered first (see Figure 5).

A set of planned pair-wise comparisons $(p < .05)$ was conducted based on the sex X alignment X turn interaction found in the Sholl and Bartels [11] study, and these revealed no significant differences between means.

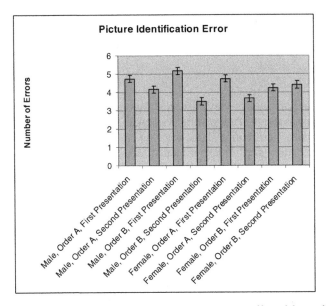

Fig. 5. Men's errors in the Picture Verification Task were unaffected by task order, while women's errors improved if the Pointing Task were administered first but not if the Picture Verification Task were administered first

A 2 (sex of participant) x 2 (order of tasks) x 2 (alignment condition) x 2 (number of turns during transport) x 2 (repeated tests) ANOVA with four participants per cell was performed on picture verification latency. The analysis yielded a significant main effect of repeated testing, $F(1, 32) = 138.63$, $p < .01$, $\eta^2 = .81$. This shows that the participants were significantly faster for the second presentation (2.65 seconds) than for the first presentation (4.44). This analysis also yielded significant interactions for alignment X repeated tests, $F(1, 32) = 4.36$, $p < .05$, $\eta^2 = .12$, and alignment X turn X order, $F(1, 32) = 4.99$, $p < .05$, $\eta^2 = .14$. These interactions were subsumed by the interaction of alignment X turn X order x repeated tests, $F(1, 32) = 5.38$, $p < .05$, $\eta^2 = .14$. No other interactions or main effects were significant.

The simplest way to explicate the complicated four-way interaction is to look at the shortest and longest latencies for the groups. The shortest latency was found during the second picture verification trial for the group that was contra-aligned, had 0-turns, and did the Pointing Task first (2.05 seconds). The longest latency was found during the first picture verification trial for the group that was aligned, had 0-turns, and did the Pointing Task first (6.46 seconds). This was the only significant mean difference in the pair-wise comparison of means. In no instance was it the case that the second picture verification trial showed a longer latency than the first. Alignment and repeated testing drove this interaction. Post-hoc pair-wise comparisons ($p < .05$) were performed on the alignment X turn X order interaction. The only significant difference that was found was between the group that was aligned, 0-turn, and performed the pointing task first (5.02 seconds) and the group that was contra-aligned, 0-turn, and performed the point task first (2.67 seconds).

A set of planned pair-wise comparisons ($p < .05$) was conducted based on the sex X alignment X turn interaction found in the Sholl and Bartels [11] study, and these revealed no significant differences between means.

3.2.2 Pointing Task

A 2 (sex of participant) x 2 (order of tasks) x 2 (alignment condition) x 2 (number of turns during transport) ANOVA with four participants per cell was performed on absolute pointing error. A main effect of alignment was found for pointing error, $F(1, 32) = 4.36$, $p < .05$, $\eta^2 = .12$. The aligned group has significantly less error (27.08°) than the contra-aligned group (53.3°). No other main effects or interactions were found.

A set of planned pair-wise comparisons ($p < .05$) was conducted based on the sex X alignment X turn interaction found in the Sholl and Bartels [11] study, and there were no significant differences between means.

A 2 (sex of participant) x 2 (order of tasks) x 2 (alignment condition) x 2 (number of turns during transport) ANOVA with four participants per cell was performed on pointing latency. A significant interaction was found for alignment X turn X order, $F(1, 32) = 5.96$, $p < .05$, $\eta^2 = .16$. Post-hoc pair-wise comparisons ($p < .05$) showed the source of the interaction was the greater latencies for the group that was contra-aligned, had 0-turns, and did Picture Verification first (12.8 seconds). This mean was greater than the comparable mean for any other group.

A set of planned pair-wise comparisons ($p < .05$) was conducted based on the sex X alignment X turn interaction found in the Sholl and Bartels [11] study, and there were no significant differences between means.

3.3 Discussion

Just as in Experiment 1, error in identifying pictures was significantly higher for the first presentation of pictures than for the second. Again, it appeared that the participants benefited from just seeing the task photographs a second time, a result borne out in the latency data, as well. There was a minor exception to this generalization about improved performance. Women who saw the pictures after first doing the pointing task showed no improvement in performance with repeated testing. Thus, for women, doing the pointing task between the two administrations of the picture verification task rather than before them made a difference. The lack of difference for the women who first did the pointing task, however, did not really support the idea that the working memory task intervening between the two picture verification sets had a severely detrimental influence.

For pointing error (measured in degrees), there was a main effect of alignment. The participants who are in the aligned condition had significantly less error than the participants in the contra-aligned condition. This is consistent with the view that participants should show more error in the contra-aligned condition because they have to "flip" their mental representation of the path in order to point to the locations. The latency data was consistent with this interpretation to an extent. The aligned, 0-turn group showed the shortest latency while the contra-aligned, 0-turn group showed the longest. These findings support the idea of orientation-specificity in the underlying spatial representation.

With respect to the patterns of behavior differentiating between virtual views and spatial updating as sources of orientation-free representations, the results of this study were similar to those of Experiment 1. The pattern of behavior reflecting orientation-free memory for men and women in Sholl & Bartels [11] study was not found. Again, we cannot say if the chief factor for this outcome was the larger time-space setting involved in the procedure, but it seems a reasonable account.

4 Conclusion

Our purpose in conducting these studies was to examine similarities and contrasts in orientation effects seen in memory for spatial layouts under regular viewing conditions and under desk-top VE conditions. Our primary vehicle was to look for patterns of behavior that are consistent with previously published means of attaining orientation-free representations, specifically, virtual views and updating proposed by Sholl and Bartels [11]. Unfortunately, results from neither of our studies revealed evidence of these strategies or means for storing such information. We speculate that the most likely reason for this outcome is the scale of time-space involved. Sholl and Bartels's [11] studies involved a pathway laid out in an indoor laboratory space. Our studies presented the spatial information in a considerably larger outdoor context. Accordingly, the memory demands in terms of encoding the distance and direction relations involved, inhibiting attention to environmental information outside the task, and maintaining the information from viewing site to test site were more substantial in our study. At this point, we will begin a more systematic study of these potential differences to narrow down the possibilities.

More generally, our results suggest some generalizations that cut across real-world and virtual experience. Obviously, despite some subtleties, alignment makes a difference, and thus we have some notion that consistency of orientation is influential and beneficial under actual and virtual circumstances [2, 4]. In linking these studies, however, care must be taken with interpretation. The "acid test" for the view-dependent form of orientation-dependent spatial representation reported in Christou and Bülthoff [2] involves faster recognition of scenes actually viewed within the VE rather than for veridical scenes not seen. Similarly, in Richardson et al. [9] is was alignment with original point of view prior to exploration. Our inferences regarding orientation-dependent performance is more similar to that of Richardson et al [9] in the sense that they are based on quicker and more accurate performance with aligned than with contra-aligned testing. When we compare our method with that of Christou and Bülthoff [2], we conclude that almost most, if not all, of the testing views in our study were novel in terms of egocentric viewing experience. Yet alignment had its influence.

It is not clear that alignment effects in our study were more pervasive with virtual experience than in situ as could well have been anticipated because of the limited field of view. In addition, our results make it very clear that cognitive context can have a substantial impact on orientation performance as assessed by pointing tasks and picture recognition tasks. In particular, results from both studies showed a degree of sensitivity to task order. In general, recognition memory tended to be more valid if an initial recognition memory test preceded, rather than followed, a test of directional knowledge requiring pointing.

As far as differences between actual and virtual experience were concerned, they appear to be rather subtle. The most apparent of these effects involved sex-related differences, especially a novel finding involving women's performance in the contra-aligned condition after actually viewing the environment. No such outcome was observed in the study of virtual experience. Subsequent replication is needed to validate the finding in actual spaces, but if it does appear to be consistent, we suggest looking at attentional differences between men and women in such tasks. We believe that attention to landmark information in the task setting that is not central to the task may differentiate men and women to some extent when orientation tasks are presented in real-world outdoor settings.

References

1. Albert, W.S., Rensink, R.A., Beusmans, J.M.: Learning Relative Directions Between Landmarks in a Desktop Virtual Environment. Spatial Cognition and Computation, Vol. 1. (2000) 131-144
2. Christou, C.G., Bülthoff, H.H.: View Dependence in Scene Recognition After Active Learning. Memory & Cognition, Vol. 27. (1999) 996-1007
3. Lawton, C.A.: Gender Differences in Way-Finding Strategies: Relationship to Spatial Ability and Spatial Anxiety. Sex Roles, Vol. 30. (1994) 765-779
4. Montello, D.R., Hegarty, M., Richardson, A.E., Waller, D.: Spatial Memory in Real Environments, Virtual Environments, and Maps. In: Allen, G. (ed): Human Spatial Memory: Remembering Where. Lawrence Erlbaum Associates, Mahwah, NJ (2004) 251-285
5. Oman, C.M., Shebilske, W.L., Richards, J.T., Tubré, T.C., Beall, A.C., Natapoff, A.: Three Dimensional Spatial Memory and Learning in Real and Virtual Environments. Spatial Cognition and Computation, Vol. 2. (2000) 355-372

6. Péruch, P., Wilson, P.N.: Active Versus Passive Learning in Real and Virtual Environments. Spatial. Cognitive Processes, Vol. 5. (2004) 218-227
7. Presson, C.C., Hazelrigg, M.D.: Building Spatial Representations Through Primary and Secondary Learning. Journal of Experimental Psychology: Learning, Memory, and Cognition, Vol. 4. (1984) 716-722
8. Presson, C.C., DeLange, N., Hazelrigg, M.D: Orientation Specificity in Spatial Memory: What Makes a Path Different From a Map of the Path? Journal of Experimental Psychology: Learning, Memory, and Cognition, Vol. 15. (1989) 887-897
9. Richardson, A.E, Montello, D.R., Hegarty, M.: Spatial Knowledge Acquisition From Maps and From Navigation in Real and Virtual Environments. Memory & Cognition, Vol. 27. (1999) 741-750
10. Roskos-Ewoldsen, B., McNamara, T.P., Shelton. A.L., Carr, W.: Mental Representations of Large and Small Spatial Layouts Are Orientation Dependent. Journal of Experimental Psychology: Learning, Memory, and Cognition, Vol. 24. (1998) 215-226
11. Sholl, M.J., Bartels, G.P.: The Role of Self-To-Object Updating in Orientation-Free Performance on Spatial-Memory Tasks. Journal of Experimental Psychology: Learning, Memory, and Cognition, Vol. 28. (2002) 422-436
12. Sholl, M.J., Nolin, T.L.: Orientation Specificity in Representations of Place. Journal of Experimental Psychology, Learning, Memory, and Cognition, Vol. 23. (1997) 1494-1507
13. Waller, D.: The WALKABOUT: Using Virtual Environments to Assess Large-Scale Spatial Abilities. Computers in Human Behavior, Vol. 21. (2005) 243-253
14. Witmer, B.G, Bailey, J.H., Knerr, B.W.: Virtual Spaces and Real World Places: Transfer of Route Knowledge. International Journal of Human-Computer Studies, Vol. 45. (1996) 413-428

Shortest Path Search from a Physical Perspective

Takeshi Shirabe

Institute for Geoinformation and Cartography,
Technical University of Vienna, 1040 Vienna, Austria
shirabe@geoinfo.tuwien.ac.at

Abstract. Shortest path problems are simple yet rich in applications in many areas including spatial information theory and geographic information science. Though they explicitly or implicitly involve some object that travels in a given network, existing algorithms and applications tend to focus only on geometric properties of the network (e.g. arc length and connectivity), and overlook physical properties of the object (e.g. velocity and acceleration) which may not be constant during a trip. This paper introduces a physical perspective to shortest path search by enforcing laws of motion, to enhance the estimation of travel times and the search for optimal paths with respect to physical quantities rather than socio-economic values.

1 Introduction

Network is a widely used geometric model for spatial information. In general a network represents connectivity (with arcs) and degrees of separation (with numerical weights) between things (with nodes). Complexity can be increased in an open-ended manner by assigning additional attributes to arcs and nodes and/or by introducing artificial arcs and nodes.

As with the structures of networks, the kinds of questions and problems concerning networks are many. Of all, shortest path problems are among the most discussed network problems in the literature and practice. A shortest path problem, in its purest form, searches for a sequence of arcs connecting two specified nodes of a network that minimizes the total arc weight. Different and/or additional requirements make many possible variations. Fastest paths, least (monetary) cost paths, and "simplest paths" [5, 16] are such examples. While these problems are valuable on their own, it is also important to note that they frequently appear as a subproblem of larger and more complex network analyses such as location-allocation, districting, and traffic assignment.

A shortest path problem can be seen as a purely geometric problem if a given network has fixed topology and metrics. This is the case when the concept of time is left out and arc weights are assumed not to change with time. If the problem is considered in a more realistic setting, however, the weight of each arc may need to be expressed as a function of time to reflect changing network conditions (e.g. traffic volume). Such "dynamic" shortest path problems have recently gained increasing attention [2, 4, 8, 10, 13, 14, 17].

The temporal dimension certainly adds more realism to shortest path problems. It, however, introduces another, possibly more intricate, issue. That is, to model truly

A.G. Cohn and D.M. Mark (Eds.): COSIT 2005, LNCS 3693, pp. 83–95, 2005.
© Springer-Verlag Berlin Heidelberg 2005

dynamic network phenomena, not (just) networks but objects passing through them (e.g. people and vehicles) should be treated in a time-dependent manner. This implies that the motion of an individual object should be explicitly described and taken into account in search of paths. Such an approach supports an emerging spatial analysis paradigm, "time geography" [9] and "people-based representation" [12], that focuses on individuals' spatial behaviors rather than their aggregates.

The individual's microscopic motion is not easy to capture with preset arc attributes. Consider a street network model as an example. To avoid having a vehicle "jump" from node to node, one may assume that it cruises at a uniform speed within each arc and changes its speed instantaneously at each node (if no other information is available). Then is it fair to say that a path with one large turn and a path with many small turns take the same amount of time and effort to travel, just because their total lengths are equal? This particular problem might be partially resolved by imposing some extra weight on each successive arc pair as a turn penalty; then all arc (and arc pair) weights are fixed values or functions. In theory, however, if a vehicle takes different routes, it might enter even the same arc at different speeds and thus traverse that arc in different times because it can *accelerate*. The problem then comes down to that such fixed weights would contradict physical laws of motion.

This does not suggest that existing geometric (rather than physical) approaches be abandoned. On the contrary, it is usually reasonable to assume that the time taken by an object to reach an intended speed is negligible compared with other factors. This assumption significantly reduces computational complexity. Nevertheless, there are cases where explicit physical considerations make nontrivial impact on shortest path search. They include cycling in a hilly winding topography, driving an emergency vehicle in a city, and flying with a limited amount of fuel. These require stricter assumptions of moving objects, which allow for acceleration, force, energy, and other physical quantities.

The present paper does not intend to exhaust all physical (or even Newtonian) effects, but aims to illustrate their relevance to network analyses by dealing with one simple problem. It is a fastest path problem for a single object subject to external forces in a static network. For simplicity, we assume that the object is a zero-dimensional particle as assumed in most elementary physics.

The rest of the paper is organized as follows. Section 2 reviews concepts of networks. Section 3 discusses both social and physical factors that might constrain motion in networks. Section 4 models a shortest path problem with physical constraints. Section 5 discusses possible extensions. Section 6 concludes the paper.

2 Networks

We begin with introducing basic concepts of networks, and then review two special kinds of networks useful for later discussions: one with turn penalties and the other with time-dependent arc weights.

2.1 Elements of a Network

A network $G=(N,A)$ is a mathematical structure that consists of a set N of nodes (or vertices) and a set A of pairs of distinct nodes called arcs (or edges). These elements are typically associated with real numbers, often referred to as weights. Arc (i,j) is said to be directed if it is distinguished from arc (j,i); undirected, otherwise. A network with directed (resp. undirected) arcs is called a directed (resp. undirected) network. Throughout this paper, networks are assumed to be directed. This does not lose generality since any undirected network can be converted to a directed network by replacing each undirected arc with two directed arcs of opposite directions.

There are a variety of spatial phenomena whose underlying structures are effectively modeled by networks. Examples include traffic, water, and communication flows. The arrangement of arcs and nodes, together with their weights, determines where and how things can move. In a typical street network, for example, a node represents an intersection, and an arc represents a street segment between two intersections. The weight of an arc may indicate the time it takes to traverse that arc, while the weight of a node may indicate the time it takes to cross that node.

2.2 Networks with Turn Penalties

Unlike those weights associated with single arcs, weights relating to multiple arcs generally require more careful treatment. One such example is found when turning from one arc to another is penalized with some extra weight, e.g. due to traffic lights. Since classical shortest path algorithms (such as Dijkstra's) do not assume these turn penalties, many alternative procedures have been developed [e.g. 3, 6, 11, 16, 18]. One of the oldest yet still effective approaches is to convert a network with turn penalties into one with no such penalties. This can be done in the following manner. First, modify a given network G by adding an artificial node (say node 0) and an artificial arc with no weight from node 0 to a chosen source node. Then, construct a new network $G*$ by making a node for each arc of G and an arc for each turnable pair of arcs of G. The weight of each arc of $G*$ is set to the associated turn penalty plus the weight of the second arc of the corresponding arc pair of G. The derived network is called "pseudo-network" [3] or "pseudo-dual graph" [16].

Fig. 1 illustrates an example. The network on the left taxes some penalty (reported in the middle table) on each turn, in addition to arc weights. Given node 1 as the source node, the corresponding pseudo-network (on the right) summarizes information on both the arc weights and turn penalties of the original network, without assist of the turn penalty table.

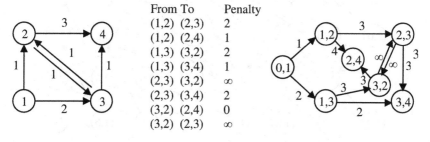

From	To	Penalty
(1,2)	(2,3)	2
(1,2)	(2,4)	1
(1,3)	(3,2)	2
(1,3)	(3,4)	1
(2,3)	(3,2)	∞
(2,3)	(3,4)	2
(3,2)	(2,4)	0
(3,2)	(2,3)	∞

Fig. 1. A network with turn penalties and its pseudo-network

2.3 Time-Dependent Networks

Arc weights may change over time. Networks with such time-varying weights are often referred to as time-dependent networks or dynamic networks, and associated problems tend to be more difficult to analyze than static counterparts. If weights are assumed to change only at a limited set of discrete times, however, a dynamic network can be transformed into a "time-expanded network" [1] which has a static form. Given a dynamic network G and a set T of times to be considered, a time-expanded network G^{**} can be constructed in the following manner. First, make a node (i,s) for each node i of G and each time s of T. Then, make an arc $((i,s),(j,t))$ if the following condition is met: If arc (i,j) of G is entered at time s, it will be exited at time t. The weight of each arc $((i,s),(j,t))$ of G^{**} is often given as a function of s. If it represents the travel time, it equals $t-s$.

Fig. 2 shows an example. The network on the top is time-dependent, as the two numbers beside each arc indicate its travel time when being entered before time 2 (inclusive) and that for after time 3 (inclusive), respectively. In the time-expanded form (on the bottom), each node points to a node of the original network and tells when it is entered (which is limited to 0 to 7 in this case, but can be extended as long as needed), and arc weights as travel times are implied by their corresponding time intervals. The present dynamic network does not allow waiting at any node and then the time-expanded network does not posses the "first-in-first-out (FIFO) property" [15] which could be exploited for efficient shortest path algorithms. It, however, contains no directed cycles so that the search for a shortest path—with respect to time—is still trivial. For instance, if one starts at node 1 at time 2, one can reach node 4 as early as at time 4 by taking path 1-2-4 in the original network (which corresponds to path $(1,2)$-$(2,3)$-$(4,4)$ in the expanded network).

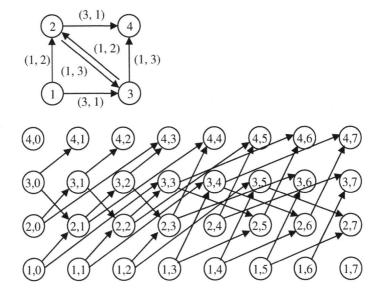

Fig. 2. Time-dependent network and its time-expanded form

3   Constraints on Motion in Networks

When an object travels in a network, its motion is limited in many ways. Though different constraints are employed in different circumstances, many of them fall in two general classes: social and physical.

3.1   Social Constraints

In most real networks, objects are not free to move anywhere and anyway they *could*. To see it, consider a street network again. One-way streets must not be entered from the wrong direction. Left turns are often prohibited at busy intersections. There are speed limits on almost all roads. And red lights say stop. Aware or not, there are many more rules the driver is supposed to observe on the streets. These constraints are social in the sense that network travelers may be capable to go against/beyond them but agree not to do so. Many of the existing network analysis tools are good at implementing social constraints like those traffic regulations.

3.2   Physical Constraints

While there are artificial or social conventions that restrain objects from certain behaviors, there are natural or physical principles that dictate how objects move (or do not move). Even a reckless driver brakes before turning a corner—or skids off the road. Hill climbing is slow and takes a good deal of energy. And of course a car stops when it runs out of gas. As such, physical constraints are more universal—though they need adaptations for different applications [7]—since no object with mass can escape their influence.

If we limit our scope to the geographic scale, these constraints are largely governed by Newton's laws of motion. The first of his laws states that an object remains at rest or in uniform motion in a straight line if no external force acts on it. According to the second law, an unbalanced force causes an object to accelerate in the direction of the force and by the magnitude of the force divided by the object's mass. Letting a scalar m denote the mass of the object, a vector $\mathbf{F}$ the force exerted to the object, and a vector $\mathbf{x}$ the location of the object as a function of time t, the following equation encapsulates the second law.

$$\frac{d^2\mathbf{x}}{dt^2} = \frac{\mathbf{F}}{m} \tag{1}$$

Finally Newton's third law of motion says that if an object A exerts a force on an object B then the object B exerts a force of the same magnitude but in the opposite direction on the object A.

Currently few applications of shortest path problems explicitly enforce these laws. Still it seems possible to approximate how an object may progress in a network over time. For example, an object's average speed within an arc is given by the ratio between the length and travel time assigned to that arc, and its moving direction is implied by the direction of that arc. While these two physical quantities (or collectively referred to as velocity) help keep track of an object following a known path, they are seldom utilized for finding an optimal path. Shortest paths with respect to time, however, should be affected at least by how fast and in which direction the object is about to move at the origin.

Acceleration has been even less relevant to shortest path problems. In most cases, intra-arc motions are ignored or assumed to be uniform. In practice, however, an object needs to adjust its velocity for a certain movement, which, in turn, affects the travel time of each arc. This is easy to see from a model of a car race where there are many different lap times for one possible (thus shortest) path.

Further less discussed are forces. It is often taken for granted that a network traveler just moves at the pace indicated by each arc weight, without being concerned with what actually makes it move. As described earlier, an object accelerates only if an unbalanced force is applied. Even while an object maintains a constant velocity, it usually needs some force to beat friction and gravity. Thus it is important to beware what forces might act on an object and how they might affect its motion in a network.

4 Modeling Shortest Path Problems with Physical Constraints

We have seen that some physical constraints are understood as consequences of the forces acting on an object. They are here taken into account in addressing a shortest path problem. The problem is considered in a small grid network illustrated in Fig. 3, where a small particle-like object is to travel from node 1 to node 6.

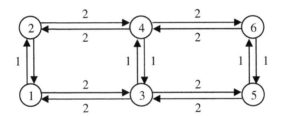

Fig. 3. A grid network. Associated with each arc is its fixed length.

We limit our scope to a relatively simple (yet illustrative) situation by making the following assumptions on the network and the object.

- The mass of the object is constant.
- The object is initially at rest.
- The object is driven only by pushes (parallel to arcs), frictional forces, and gravitational forces (perpendicular to the plane the network is embedded in).
- The object may be pushed in any direction, but with a limited magnitude.
- Given a normal force between the object and any arc, the magnitude of the resulting frictional force is constant (regardless of the object's motion) and proportional to the normal force.

The following notation is used throughout this section.
g: gravitational coefficient
μ: coefficient of friction between the object and any arc
m: mass of the object
f: magnitude of the maximum push applicable to the object

4.1 Shortest Paths to Fastest Paths

In the given network it is trivial to find paths with minimum length. If the length of a path is defined as the sum of the lengths of the arcs in the path, any path from node 1 to node 6 consisting of one vertical arc and two horizontal arcs (e.g. path 1-3-4-6) is a shortest path with a length of 5. If a uniform penalty, say 1, is assigned to each turn, there are two shortest paths (i.e. path 1-2-4-6 and path 1-3-5-6) with a length of 6.

Then which path takes the least amount of time to pass through? If the travel time of each arc were proportional to its length, a "fastest" path would be one of those "shortest" paths mentioned above. Under the influence of forces (which cause acceleration), however, the travel time of each arc is not an intrinsic property of it, but depends on how the object moves within it. This differentiates fastest paths from shortest paths. For example, set the variables g, μ, m, and f to 10, 0.1, 1, and 3, respectively. Then, according to our assumptions and Newton's laws, the fastest path is 1-2-4-6 with a travel time of $2+\sqrt{3/2}$ ($\approx$3.22), followed by path 1-3-5-6 with a travel time of $1+\sqrt{6}$ ($\approx$3.45), and the third fastest path is 1-3-4-6 with a travel time of $\sqrt{2}+\sqrt{3}+\sqrt{3/2}$ ($\approx$4.37). While the winding shape of third fastest path apparently causes delays, the two fastest paths have the same geometric structure yet their travel times are different. This is intuitively explained as follows. The fastest path permits the object to continue to accelerate down the long stretch 2-4-6, but the second fastest path does not allow it to take the same advantage of the equally long stretch 1-3-5 because a complete stop is necessary to make a 90-degree turn at node 5.

As such, it is, in principle, possible to find a fastest path by enumerating all paths and computing their travel times. This is, however, practically prohibited since there could be an exceedingly large number of possible paths. Thus we will seek to design an approximation procedure in subsequent subsections. To do so, we first evaluate how arc travel time is affected by acceleration.

4.2 Arc Travel Time as a Function of Entering Velocity and Exiting Velocity

In theory, how long it takes the object to travel an arc depends on its initial velocity and its acceleration. And certainly the object can minimize the travel time by moving with its full acceleration all the way.

To formulate this, assume that the object enters an arc (i,j) at time 0 and let

x_{ij}: length of arc (i,j)

t_j: time at which the object reaches node j

v_i: speed of the object at node i

v_j: speed of the object at node j

a^+: magnitude of the maximum acceleration of the object in the direction of arc (i,j)

The following equations then describe: if the object starts with the speed v_i and moves with the constant acceleration a^+ (both in the direction of arc (i,j)), it will reach the speed v_j and have traveled as far as x_{ij} at time t_j.

$$v_j = v_i + a^+ t_j \tag{2}$$

$$x_{ij} = v_i t_j + \frac{1}{2} a^+ t_j^2 \tag{3}$$

By solving these for v_j and t_j, we obtain

$$v_j = \sqrt{v_i^2 + 2a^+ x_{ij}} \tag{4}$$

$$t_j = \frac{-v_i + \sqrt{v_i^2 + 2a^+ x_{ij}}}{a^+} \tag{5}$$

The formula (5) gives the (minimum) travel time of arc (i,j) when the objects enters the arc at a specified speed in the direction of the arc.

Here notice that unless v_j is zero, the direction of motion of the object at j is that of arc (i,j). In other words, to make a 90-degree or U-turn at node j, the object needs to slow down to stop at node j. There may be a more general case where the object is required to achieve a certain speed on the arrival at node j to make particular motions.

This additional constraint prevents the object from using the maximum acceleration throughout. Instead, the object must change from acceleration to deceleration (i.e. acceleration in the opposite direction) at an intermediate point p of arc (i,j) to achieve the required speed at node j.

To describe this formally, add the following variables to those already defined.

x_{ip}: distance between node i and point p

t_p : time at which the object reaches point p (assuming it leaves the node i at time 0)

v_p: speed of the object at point p

a^- : magnitude of the maximum acceleration of the object in the opposite direction of arc (i,j)

Then, similarly to the previous case

$$v_p = v_i + a^+ t_p \tag{6}$$

$$x_{ip} = v_i t_p + \frac{1}{2} a^+ t_p^2 \tag{7}$$

$$v_j = v_p - a^- (t_j - t_p) \tag{8}$$

$$x_{ij} - x_{ip} = v_p (t_j - t_p) - \frac{1}{2} a^- (t_j - t_p)^2 \tag{9}$$

For this system of equations to be consistent, the acceleration in both directions must be great enough (i.e. $a^+ \geq \dfrac{v_j^2 - v_i^2}{2x_{ij}}$ and $a^- \geq \dfrac{v_i^2 - v_j^2}{2x_{ij}}$) to achieve the required speed at node j. If this condition is satisfied, the system is solved for x_{ip}, v_p, t_p, and t_j as

$$x_{ip} = \frac{v_j^2 - v_i^2 + 2a^- x_{ij}}{2(a^+ + a^-)} \tag{10}$$

$$v_p = \sqrt{\frac{a^+ v_j^{\,2} + a^- v_i^{\,2} + 2a^+ a^- x_{ij}}{a^+ + a^-}}$$ (11)

$$t_p = \frac{\sqrt{(a^+ + a^-)(a^+ v_j^{\,2} + a^- v_i^{\,2} + 2a^+ a^- x_{ij})} - (a^+ + a^-)v_i}{a^+(a^+ + a^-)}$$ (12)

$$t_j = \frac{\sqrt{(a^+ + a^-)(a^+ v_j^{\,2} + a^- v_i^{\,2} + 2a^+ a^- x_{ij})} - a^+ v_j - a^- v_i}{a^+ a^-}$$ (13)

The formula (13) gives the (minimum) travel time of arc (i,j) when the objects enters and exits the arc at respective specified speeds in the direction of the arc.

Finally we compute a^+ and a^-. Assuming that no arc is inclined, only pushes and frictional forces contribute to the object's acceleration. Therefore, according to Newton's second law

$$a^+ = (f - \mu mg)/m$$ (14)

$$a^- = (f + \mu mg)/m$$ (15)

4.3 Network Expansion with Speeds

Arc travel times have been expressed as a continuous function of entering and existing velocities. Other than enumeration of all paths, there seems no easy procedure for finding an exact fastest path in a network with such a property. So we will attempt to discretize this function to create an ordinary network that approximates the motion-dependent network. To do so, we first construct a pseudo-network to resolve directions and then "expand" it with a limited number of allowable speeds at each node. The procedure is as follows.

Let N and A denote the sets of nodes and arcs of a given network G, respectively, and S be a set of at-node speeds to be considered, and assume that each arc (i,j) in A has been assigned a length $x_{ij} > 0$. Then construct a network G^{**} in the following procedure.

1. Construct a pseudo-network $G^* = (N^*, A^*)$ of a given network G.
2. Make a node in G^{**} for each node $(i, j) \in N^*$ and each speed $s_j \in S$. Let (i,j,s_j) denote this node.
3. Make an arc in G^{**} for each pair of nodes (h,i,s_i) and (i,j,s_j) and denote it by $((h,i,s_i), (i,j,s_j))$ if the following condition holds: If the object under consideration enters arc (i,j) of G at speed s_i of S in the direction of the arc, it can exit the arc at speed s_j. To verify this, use the formulas (13), (14), and (15). If true, record the minimum travel time obtained as the weight of arc $((h,i,s_i), (i,j,s_j))$.

We refer to the resulting network as a "speed-expanded network," which is similar to the time-expanded network except that speeds have replaced times.

For example, consider the network of Fig. 3 and let S be $\{0, 1, 2, 3, 4\}$. Then we have a speed-expanded network with 75 nodes—as the original network G contains 14 arcs (plus one artificial arc). Assuming that the variables g, μ, m, and f remain 10, 0.1, 1, and 3, respectively, Fig. 4 shows a subset of the corresponding pseudo-network G^* that involves node (2,4) (on the left) and a subset of its speed-expanded form G^{**} that involves node (2,4,0) (on the right). The pseudo-network tells, for example, that one may turn from arc (2,4) to arc (4,6) in G and the length of the latter arc is 2 (no turn penalty). On the other hand, the speed-expanded network tells that if one leaves arc (2,4) in G at speed 0, one can traverse arc (4,6) in G in a time of 1.73, 1.53, or 1.44 depending on the speed at the end of that arc. These suggest that weights sensitive to turns and speeds cannot be handled by a pseudo-network alone.

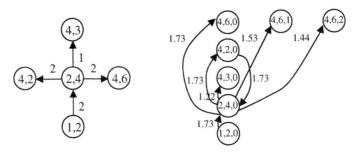

Fig. 4. Subsets of a pseudo-network and its speed-expanded form

The grid network (Fig. 3) forces the object to make a complete stop to make a turn, and the speed-expanded network is built accordingly. It is, however, important to note that one may give an allowable speed limit or range to each turn—e.g. depending on its turning angle—in order to model more realistic intersections. If it were the case here, the speed-expanded network would contain such arcs that emanate from a node with non-zero speed.

As with the case of the time-expanded network, existing shortest path algorithms are applicable to the speed-expanded network. The number of nodes in the expanded network equals the number of nodes in the pre-expanded network multiplied by the number of at-node speeds $|S|$. Note that $|S|$ is generally independent of—thus does not grow with—the original network size. For the present problem, path (0,1,0)-(1,2,0)-(2,4,2)-(4,6,3) has been found to be a shortest path. This corresponds to the exact fastest path 1-2-4-6 obtained earlier, but its travel time is overestimated as 3.4. This implies that our approximation procedure may overlook exact solutions. Precision and accuracy can be increased by adding finer elements to S.

5 Extensions

The approach presented above has addressed a simple fastest path problem. This section briefly discusses how it is adapted to other cases.

5.1 Fastest Path Problems in Elevated Networks

We have so far assumed that no arc is inclined. This assumption is relaxed here. A network with inclined arcs can be constructed from one with no inclined arcs by assigning each node a height value and letting each arc be on a line connecting two (elevated) nodes. Though real networks may be on more undulated topography, it is not the case here.

In this setting, the formula (13) is still useful to compute the travel time of each arc, and thus a speed-expanded network is similarly created. The only modification needed is on the acceleration, which is determined by the forces applied. Consider an object moving on an inclined arc. Then, if the combination of a push and a gravitational force acted on the object is strong enough to beat a frictional force (i.e. $f + mg \sin \theta > \mu mg \cos \theta$) (c.f. Fig. 5), the formula (14) is modified to

$$a^+ = (f + mg \sin \theta - \mu mg \cos \theta) / m \qquad (16)$$

If the combination of a push and a frictional force acted on an object is strong enough to beat a gravitational force (i.e. $f + \mu mg \cos \theta > mg \sin \theta$), the formula (15), is modified to

$$a^- = (f - mg \sin \theta + \mu mg \cos \theta) / m \qquad (17)$$

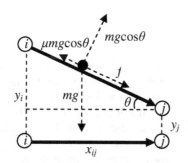

Fig. 5. A given arc (lower bold arrow) with length x_{ij} connecting two nodes (lower circles) has been transformed to an inclined arc (upper bold arrow) connecting two elevated nodes (upper circles) with heights y_i and y_j. An object (black dot) on the inclined arc is subject to the four forces (dashed arrows) including a push, a gravitational force, a normal force, and a frictional force with f, mg, $mg\cos\theta$, and $\mu mg\cos\theta$ as their respective magnitudes. The arc's incline θ is given by. $\arctan \dfrac{y_i - y_j}{x_{ij}}$.

5.2 Minimum Work Paths

Incorporation of forces in a shortest path analysis enables one to estimate the amount of work required for an object to travel to the destination. The speed-expanded network is useful for this purpose, too, except that the weight of each arc is not the minimum travel time but the minimum work that takes the object to traverse that arc (which is given by the product of f and $x_{ij}/\cos\theta$ in Fig. 5). In this case, the object takes as much advantage of gravitational and frictional forces as it can to save work.

The result may be too idealistic to match the object's actual energy consumption. It is because not all generated energy (e.g. from fuel combustion) contribute to motion. Still it serves as an indicator of the relative economy of one path over another.

5.3 General Dynamic Networks

The shortest path problem discussed in this paper has added a new dimension to networks, that is, a set of possible velocities at each node. Their directions are addressed by a pseudo-network, and their magnitudes are addressed by a speed-expanded network. In the latter network, associated with each node are speeds, but can be their derived quantities such as squared speeds or levels of kinetic energy, and still similar results will be obtained (with different degrees of approximation). These happen to be physical quantities, but can be replaced by other kinds of *states* (social, economic, mental, cognitive, etc.) the object may posses in other contexts. In this sense, both the time-expanded network and the speed-expanded network are merely special cases of a more general "dynamic network"—dynamic, i.e. varying with the state of the object. Furthermore, in principle, there can be more than one variable describing one state. Therefore, in a further more general dynamic network, the weight of each element (i.e. node, arc, or their grouping) changes according to one or more state variables.

6 Conclusions

A preliminary work has been presented for incorporating physical aspects into network analyses (shortest path problems in particular). In this scheme, an object may change its velocity within each arc, and it does so only when acted on by external forces which include gravitational and frictional forces. These physical phenomena are nothing new, but this paper has shown that their incorporation into network models complements traditional approaches.

The present scheme is by no means complete, as it misses a number of factors that might significantly affect the ways objects move in real networks. Two of them seem to deserve more immediate attention. First, social characteristics of networks have been completely omitted but should be equally important. On the streets (rather than on racing circuits), for example, vehicles rarely have chances to perform their full potential, partly due to strict traffic regulations. Second, this paper restricts its scope to a single object. While the basic concepts presented here are applicable to multiple objects, too, their interactions will increase complexity. More realistic network models as such should be explored in future research.

The physical perspective has introduced computational challenges, too. The spatial phenomena under consideration are essentially continuous in terms of space and time and thus require a higher degree of mathematical sophistication including ordinal differential calculus. The resulting shortest path problems are no longer linear and can be NP-hard for which there are no known algorithms that take only polynomial time with respect to the network size. We have employed a dynamic programming approach for approximate solution, but more effective and efficient algorithms are expected to be investigated.

Acknowledgements

The author would like to thank the anonymous reviewers for their valuable comments.

References

1. Ahuja, R. K., Magnanti, T. L., and Orlin, J. B. (1993) Network flows: theory, algorithms, and applications, Englewood Cliffs, NJ: Prentice Hall.
2. Ahuja, R.K., Orlin, J.B., Pallottino, S., Scutella, M. (2002) Minimum time and minimum cost path problems in street networks with traffic lights. Transportation Science 36, 326-336.
3. Cadwell, T. (1961) On finding minimum routes in a network with turn penalties. Communications of the ACM 4(2): 107-108.
4. Chabini, I. (1998) Discrete dynamic shortest path problems in transportation applications: Complexity and algorithms with optimal run time. Transportation Research Record 1645, 170-175.
5. Duckham, M. and Kulik, L. (2003) "Simplest" Paths: Automated route selection for navigation. In Kuhn, W., Worboys, M. and Timpf, S. (eds) Lecture Notes in Computer Science 2825, Springer, 169-185.
6. Easa, S. M. (1985) Shortest Route Algorithm with Movement Prohibitions. Transportation Research 19B, 197-208.
7. Frank, A. U. (2001) Tier of ontology and consistency constraints. International Journal of Geographical Information Science 15, 667-678.
8. Halpern, J. L. (1977) Shortest Route with Time-Dependent Length of Edges and Limited Delay Possibilities in Nodes. Zeitschrift fur Operations Research 21, 117-124.
9. Hägerstrand, T. (1970) What about people in regional science? Papers of the Regional Science Association 14, 7-21.
10. Kaufman, D. E., and Smith, R. L. (1993) Fastest Paths in Time-Dependent Networks for IVHS Application. IVHS Journal 1, 1-11.
11. Kirby, R. F. and Potts, R. B. (1969) The minimum route problem for networks with turn penalties and prohibitions. Transport Research. 3, 397-408.
12. Miller, H.J. (1991) Modelling accessibility using space-time prism concepts within geographical information systems. International Journal of Geographical Systems 5, 287-301.
13. Orda, A., and Rom, R. (1990) Shortest-Path and Minimum-Delay Algorithms in Networks With Time-Dependent Edge-Lengths. Journal of the Association for Computing Machinery. IVHS Journal 37, 603-625.
14. Orda, A. and Rom, R. (1991) Minimum weight paths in time-dependent network. Networks 21(3), 295-320.
15. Pallottino, S. and Scutellà, M.G. (1998) Shortest path algorithms in transportation models: classical and innovative aspects. In Marcotte P. and Nguyen S. (eds) Equilibrium and Advanced Transportation Modelling, 245-281.
16. Winter, S. (2002) Modeling Costs of Turns in Route Planning. Geoinformatica 6, 345-361.
17. Ziliaskopoulos, A. and Mahmassani, H. S. (1993) Time-Dependent Shortest Path Algorithm for Real-Time Intelligent Vehicle Highway System Applications. Transportation Research Record 1408. 94-100.
18. Ziliaskopoulos, A.K. and Mahmassani, H.S (1996) A Note on Least Time Path Computation Considering Delays and Prohibition for Intersection Movements. Transport Research 3, 359-367.

Operationalising 'Sense of Place' as a Cognitive Operator for Semantics in Place-Based Ontologies

Pragya Agarwal

School of Computing, University of Leeds, Leeds, UK
pragya@comp.leeds.ac.uk

Abstract. Meanings of geographic concepts have to be grounded in real world and in human commonsense to formulate semantically-enriched geographic ontologies. Cognitive concepts, such as the notion of geographic *place*, have been shown to be inherently vague. This vagueness results primarily due to a lack of understanding of the modifiers linking the cognitive semantics with the real world, and because of the ambiguous ontological and semantic distinctions with similar spatial concepts. In this paper, an experimental framework is developed to demonstrate that the notion of 'sense of place' can be operationalised as a cognitive operator to ground the meanings of *place* in the real world as well as in human behaviour, and for defining the ontological distinctions with other spatial and location identifiers, such as *neighbourhood*. The results from human subject experiments are used as a basis for extracting the key parameters associated with 'sense of place'.

1 Introduction

Recent research in semantics and ontologies has shown that meanings for terms and concepts in the ontology have to be grounded in the real world (Kuhn 2003). Most geographic concepts are cognitively formed, resulting in ambiguity because of the inherent vagueness in human cognition, perception and natural language description. The meanings for these concepts have to be, therefore, grounded in human behaviour in the real world to develop context-oriented dynamic ontologies. Studies in location-based information systems have shown that the definition and assignation of place labels is problematic and confusing (Espinoza *et al.* 2001). This is true for real as well as virtual and conceptual spaces. The significance of *place* definition in development of geographic models and ontologies, and the problems in resolving the semantic ambiguity inherent in *place* has previously been discussed in Agarwal (2004a, 2004b).

This paper addresses the problem of defining identifiers for *place* to ground place-based ontologies, and demonstrates that this can be resolved by operationalising the notion of 'sense of place' that gives a place its meaning and identity. 'Sense of place', in existing literature (Tuan 1977, Datel and Dingemans 1984, Johnston *et al.* 2000), has, however, been proposed as an abstract notion and has not been operationalised in place-based models. A computational theory for 'sense of place' will enable identification of minimal parameters necessary for simulating 'sense of place' in place-based models, in virtual environments, in using place indexing for creation of

A.G. Cohn and D.M. Mark (Eds.): COSIT 2005, LNCS 3693, pp. 96–114, 2005.
© Springer-Verlag Berlin Heidelberg 2005

query-based systems and even in development of computer-based automated narrative generators that use place as the basis for relating human activity to space. These indicators will be useful in development of context-aware computing and in linking human activity to physical spaces, where context-aware computing is used to include: (1) the presentation of contextualized information and services; (2) context based automatic execution of services; and (3) the attaching of contextual information for later retrieval (Dey *et al.*, 2001). Furthermore, with the concept of place learning in cognitive maps becoming more explicit, there are considerable implications on how place is treated and conceptualised in place-based searches and navigational models. Such results have implications on the design of place-based location-based services (LBS) that are closer to human commonsense and reasoning.

'Sense of place' is shown in this paper to be the individuating characteristic that makes a space specific and imparts the meaning in a place, and is the key consideration in place-knowledge construction. The emphasis is on 'geographic' place and the work is carried out in large-scale urban environments. The experiments are designed to identify the primitives and relations that are influential in determining a 'cognitive sense of place', introduced as SOP_c in this paper. In this way, 'sense of place' can be used to define differences between place and neighbourhood by operationalising it in terms of reasoning with real world objects. The experimental framework also indicates that cognitive contexts can be captured in an ontology by identifying the core parameters and rules associated with spatio-temporal reasoning through human subject experiments. It is expected that this study will indicate the way forward for better representation of real world features in geographic models by a clearer assignation of semantics and relations in a spatial ontology that is grounded in human reasoning in the real world.

The rest of the paper is structured as follows. Section 2 outlines the theoretical framework, and the research questions. The experimental framework, along with the major hypotheses for the experiments is described in section 3. The experimental setup is described in section 4 and the major results are discussed in section 5. Section 6 presents some concluding remarks and outlines future directions.

2   Theoretical Framework

Place learning as a significant factor in navigational tasks has demanded much attention in cognitive science, psychological and biological literature. *Place* is associated with ascertaining a 'memory of a location' (McNamara 1991, Golledge 1992:202), and with the distance and direction estimates for locations in the environment (Chown *et al.* 1995). Spatial learning and knowledge acquisition tasks, such as wayfinding, instrumental in developing a cognitive schema in the environment, are associated with place learning (Gale *et al.* 1990, Cornell *et al.* 1994, Jacobs 2003). Places are considered as a collection of objects and landmarks (Lynch 1960, Presson 1987, Sorrows and Hirtle 1999), and distance estimates are crucial to the formation of a cognitive sense of place (Downs and Stea 1973, Golledge 1992). *Place* recognition has also been associated with *neighbourhood* recognition and estimation, and shown to be linked to an array of landmarks, their locations and geometric arrangements (Arleo and Gerstner 2000), with distance estimates significant in *place* learning for human navigation (Steck and Mallot 2000). The notion of *place* is related to the notion of 'proximic *region*' as a 'condensed *space*' (Hudson 1998), and that of 'social *neighbourhood*s' (Galster 1986, Forrest and Kearns 2001). Norberg-Schulz (1980) outlined that the cognitive organisation of space in the way that people orient themselves and define experiences in space are critical for determining a model of *place*.

Many studies have stressed on the importance of 'sense of place' in explaining and understanding *place*. Within geographic discourses, the notion of 'sense of place' has been proposed as a concept that, in general terms, means an 'awareness of one's natural environment' but also is crucially linked (especially in phenomenological conceptualisations) to determine the meanings and formation of *place*. Experiments in virtual environments have started to focus on creating places with 'sense of place' by including people's experiences and meanings in virtual landscapes (McCall *et al.* 2004), where presence or 'being there' is employed as one of the measures for 'sense of place' and the emphasis is on the immersive, embodied nature of 'sense of place'. While 'sense of place' is proposed as a localised individual notion in theories by Tuan (1977, 1990) and Relph (1976), it also emerges as a global notion in Massey (1993), where 'sense of place' is not regarded as nostalgic but progressive, with *places* continuously evolving as 'articulated moments in networks of social relations and understanding' (Massey 1993:66). The notion of 'sense of place' has also found its way in management strategies in resource management and for participatory planning (see Cantrill 1998, Williams and Stewart 1998). In humanistic theories, 'sense of place' originated as a study of the human experience and imagination as it is affected by the geographic locations (Johnston *et al.* 2000), but is also pervading the larger body of work where 'sense of place' is being examined as resulting from interconnected social, environmental and behavioural processes (Shamai 1991). 'Sense of place' is linked to the formation of deep emotional connections with specific location, and the feelings of attachment that results from familiarity, memory and association (Relph 1976, Withers 1996). 'Sense of place' has been studied as an 'orientation mechanism'- to know where one was (Starrs 1994), and is linked to the feeling of closeness with the environment (Cuba and Hummons 1993, Feld and Basso 1996).

Although the large body of literature in 'sense of place' has been in the non-positivistic traditions, a few have attempted to frame and operationalise it from a behavioural perspective. Researchers have resisted the attempt to assign a precise definition for 'sense of place'. Relph (1976:4) says that 'sense of place' 'is not just a formal concept awaiting precise definition, and clarification cannot be achieved by imposing precise but arbitrary definitions', and Lewis (1979:40) said that 'it is quite useless to try measuring it'. Nevertheless, there have been attempts to explain and measure the 'sense of place', with the proposed grounding in human behaviour. Lewis (1979:28) said that for 'sense of place', 'it is often easier to see its results in human behaviour than to define it in precise terms'. Relph (1976:20) stated that 'by taking *place* as a multifaceted phenomenon of experience and examining the various properties of *place*, such as location, landscape, and personal involvement, some assessment can be made of the degree to which these are essential to our experience and 'sense of place'. Peterson and Saarinen (1986:164) highlight the importance of 'local symbols' in the environment for formation of 'sense of place' and thereby conceptualisation of a *place*, and Datel and Dingemans (1984:135) said that the 'sense of place' is formed through a 'bundle of meanings, symbols and qualities that a person of a group associates (consciously or unconsciously) with a locality or *region*'. Few empirical studies have been carried out to define this concept and to convey its applicability in understanding and explaining *place*. In most empirical studies (Proshansky *et al.* 1983, Pellow 1992, Riley 1992), ranking procedures were adopted to scale the level of 'sense of place' with two (Goldlust and Richmond 1977), three (Shamai 1991) and four levels (Shamai and Kellerman 1985). This is based on the theory that there are different levels of interactions with the environment and consequently the assessment of *place* and the level of 'sense of place' are scaled accordingly (Canter 1977, 1997). Relph (1976) also defined different degrees of sensing a *place*, and Shafer (1984) specified different levels of intensity of feelings towards *place*s. These studies are limited in that they only define the level of the 'sense of place', and the issue of what a 'sense of place' means is left undetermined, Hence, the problems with terminological and cognitive heterogeneity underlie any results obtained from these studies. Moreover, most of these studies are focused on specific *place*s and on their description and understanding and the generic applicability of any such definitions for understanding of *place* is not explained.

2.1 Research Questions

Both urban and philosophical research indicates that time cannot be separated from space in forming a 'sense of place' (Parkes and Thrift 1978, Brentano 1988), and 'space and time coordinates are elements of language inter-related in describing an environment' (Capra 1975:173). Studies have shown that the cognitive ordering of time is not necessarily linear (Prince 1978) and several parallel time-paths exist, dividing and ordering spatial knowledge within cognitive models of the urban environment. Various studies have examined the nature of temporal knowledge in human behaviour for environmental learning (Giraudo and Péruch 1988, Clayton and Habibi 1991, McNamara 1992). None has, however, looked specifically at the nature of temporal knowledge and relations formed in the process of place formation. The notion of triggered response has shown that places can be conceptualised as a series of temporally arranged views (Schölkopf and Mallot 1995, Mallot and Gillner 2000), and, hence, temporal sequencing of information occurs through the process of place learning. If place is linked critically to temporality and time is associated with the formation of place associations, then, is the temporal evaluation of an environment place-dependent? Do objects inside a place appear to be closer to each other in time?

As GIScientists, we know that 'everything is related to everything else, but near things are more related than distant things' (Tobler 1970: 236). It is accepted that nearness relations in geographic space are context-dependent (Denofsky 1977, Worboys 2001, Bittner 2002). The notion of distance decay defined as 'the attenuation of a pattern or process with distance' (Johnston *et al.* 2000) represents statistically the concept of time-space convergence and compression, resulting in an observed decrease in social and spatial interaction with distance. Places were also defined as 'proximal spaces' (Harrison and Dourish 1996), and linked to notions of condensed space and bounded space. The boundary effects of place have been qualified both theoretically and in the social sense for the concentration of activities and communities (Massey 1993, Pirinen 2002). However, the clustering effect of place in the behavioural and cognitive determination of qualitative relations in space and time are not explained through previous studies. Therefore, does formation of place decay with distance, and is distance a determinant in defining place and a cognitive 'sense of place'? In addition, are boundary effects of a place significant factor in the formation of qualitative proximity relations in the environment? In the next section, the experimental framework is discussed, leading to the development of certain key hypotheses that form the basis of the experimental work.

3 Experimental Framework

The cognitive theories in place-learning, and the phenomenological notion of 'sense of place', form two disparate theories for explaining *place* and its meanings. The experimental framework in this paper brings these two theoretical frameworks together; one that is instrumental in defining meanings of place and the other for establishing its parameters. This is done by defining 'sense of place' as not only an experiential notion, but also explaining it in terms of spatial (and temporal) knowledge and learning in the environment. This is similar to 'unwelt', a concept proposed by the German naturalist, Jacob van Uxekell, where the physical world is ordered by a set of organised experiences, 'unwelten' that depends on individual sensory and cognitive apparatus. In this way, the representation systems of individuals in ordering spatial and temporal discontinuities in urban environments can be used to determine the ontological commitments in forming a 'sense of place', and the notion of a multi-dimensional 'cognitive sense of place', termed SOPc, can form the basis for an integrated foundation for explaining *place*. 'Sense of place', from a cognitive perspective, is used here to imply a 'conceptualisation of the extent and the determination of the objects, and their relations, based upon the ontological model of landmark-neighbourhood for places'. In the experimental framework described in this paper, 'sense of place' is considered as a mechanism through which individual conceptualisations of *place* are grounded in a collective notion, defining the meanings of *place* and its links in the real world. 'Sense of place', is shown to be a representative of ontological commitments in the real world for the formation of a *place*, and investigated as a grounding mechanism through which *place* ontologies can be linked back to human cognitive system as well as to the real world objects.

Current research in GIScience proposes four-dimensional models where space and time are integrated, rather than separate entities (Galton 2003). It is also shown from previous experiments that spatial relations are often extended to define temporal relations (Clayton and Habibi 1991), and that distance is often defined in terms of time (Herman *et al.* 1983). Research in geographic time (Montello and Golledge 1998) is still looking for answers to questions such as, 'are scales in space and time logically linked?' The application of proximities in space is applied in this experimental framework to the notion of 'nearness in time'. This method provides an opportunity to test whether spatial expressions and knowledge are linked logically and are transferable to temporal estimates within human reasoning. If objects that are completely inside a place appear closer than objects lying outside it, this will have implications on how place is represented in geographic modelling. By applying the notion of 'nearness' in time (as well as in space), the basis of temporal estimates in linear or cyclic time is avoided, and a qualitative basis for temporal relations is enabled. Distance estimates are aimed at assessing whether the conceptualisation of place is accompanied by a heightened sense of time within the place, and whether the formation of place leads to a clustering of objects in space as well as in time. Previous studies on place have shown that it is conceptualised in different context scales, from 'self' to 'community', and at different spatial scales, from 'local' to 'global'. At a local level, spatio-temporal knowledge acquisition is dependent on the direct interaction of the individual with the environment. The information within the cognitive set and the spatio-temporal relations defined within it are constantly updated at this local scale through direct and indirect interaction with the environment, through the actions of movement, memory comparisons and associations, and through hierarchical ordering of components within the environment. In this process, spatial divisions of association are assigned and categories of place defined, and, therefore, cognitive dimensions play a significant role in place conceptualisation at a local scale. For the purpose of the experiment developed in this research, the conceptualisation of *place* is considered at an immediate local level, and the global conceptualisation is not considered explicitly. It is expected that some rules will be transferable from the local scale to the global scale of *place* conceptualisation, but this will require further testing.

Based on the questions raised in the previous section, the major hypotheses for the experimental framework can be outlined as such:

H1: The conceptualisation of place affects the spatial and temporal reasoning and knowledge in the environment.

1. The distance estimates in the environment are place-dependent.
2. Place is proximal space and boundaries are an integral aspect of place formation. Objects are estimated to be closer when they lie inside a place than when they lie outside it.
3. The 'sense of place' is linked to a sense of space as well as a sense of time, and is dependent on the degree of familiarity.

4 Experimental Setup

The experiments described in this paper were carried out in Nottingham, UK, and were largely concentrated around the University of Nottingham campus. Similar to Agarwal (2004a, 2004b), the experiment design was developed after a comprehensive review of previous behavioural and cognitive studies (Battig and Montague 1968, Lloyd *et al.* 1996, Mark *et al.* 1999, Montello *et al.* 2001), and questionnaire design literature (Courtenay 1978, Fowler 1989). A sample size of 50 was selected and it was ensured that adequate scientific standards are followed. A pilot study was undertaken as a pre-test in 2002 to develop viable methods of data collection, based on a sound theoretical background and the requirements of the study. The analysis of responses at this stage allowed identification and verification of suitable parameters, analysis and inference methods. Although it has been shown by psychological studies that there is a difference in cognitive processing between sexes, inter-group analysis was not considered relevant within the remits of this study. The length of residence was excluded from the final experiment design since the analysis in the pilot study showed that it did not yield conclusive results regarding the meaning and conceptualisation of place. Although previous theoretical and empirical studies have shown that cross-cultural factors can have an influence on the perception of place (Duncan and Ley 1993, Twigger-Ross and Uzzell 1996), this did not appear to be significant in this experiment. The total number of males $N_1=28$, and the number of females $N_2=22$. The subjects were predominantly white, with 92% White and 8% British Asians. All subjects were native English speakers. Random assignment was carried out for selecting the sample population. In addition, the experimental framework required a concentrated locational reference for the subjects to avoid locational biases, and to minimise the effects that any extraneous factors specific to certain locations might have on the results. From the 50 subjects, the number of respondents living on the University Campus $N_1=28$, and the number of respondents living in an area just outside the campus, on the west side of the University $N_2=22$. The response time was constrained to 60 seconds to avoid deliberation.

4.1 Stage One

The volunteers were given a list of 29 locations (termed 'objects' hereafter) that were compiled from the list of significant 'buildings' and 'places' from tourist brochures, council websites and a range of different travel guides. The experiment required the respondents to grade these on a category scale of 1 to 5 (1=very good, 2=good, 3=reasonable, 4=poor, 5=very poor) for factors such as significance in the city, historicity, attractiveness, personal attachment and level of familiarity. These were the control factors that were identified from theoretical literature as influencing the cognitive 'landmarkedness' of an object in the environment across a range of environment, self and activity contexts. The responses were then collected, scaled on the basis of the 'level of familiarity', used as the primary factor in landmark and place association, and then aggregated to assign a score to each of the 29 objects used in the questionnaire. It was found that 15 returned incomplete questionnaires, and were excluded from the analysis. The 35 complete and legible responses were used for the identification of landmarks, and only these 35 respondents were then used as 'subjects' in the second stage of this experiment. From the 29 control objects, ten that had the maximum score from the 35 responses were extracted as a subset denoting 'landmarks' in the cognitive set of the subjects and used in the second stage.

4.2 Stage Two

This stage of the experiment was carried out once the responses from the first stage had been compiled and analysed. This second stage was aimed at the identification of the cognitive association of place and was developed based on the landmarks identified from the first stage. The total number of respondents was N =35, that returned completed questionnaires from stage one of this experiment. The total number of objects, acting as spatial referents (or landmarks) for the cognitive set of each individual are O=10. For N=35 subjects, number of males $N_1=24$, number of females $N_2=11$. Since place was operationalised at a local level within this experimental setup, the notion of neighbourhood was employed to assess whether any clear distinctions were formed between place and neighbourhood, when both were formed at the local level of association and cognitive conceptualisation. The factors for assessment in this questionnaire were degree of familiarity with the landmark (very familiar/not so familiar or high/low), inside the place that the subject lives in (yes/no), inside the neighbourhood that the subject lives in (yes/no), and the distance assessment to the landmark (very near/near/far/very far) from home (geographic location), both spatially and temporally. Within the questionnaire, it was made clear to the subjects that the temporal distance measure did not imply travel time, but instead the comparative age (i.e. the position on the temporal scale for the 'age' or the life trajectory). In addition, since all objects extracted from the list in the first stage had associated historicity, these were close to each other in temporal distance (in historical time-line), which meant that these objects were more or less equidistant from the locations of the respondents. The respondents were concentrated within one area of the city and hence resided in an area with a similar temporal profile. Although the landmarks were chosen from the first stage of the study to be 'most familiar', and the subjects were familiar to a certain degree with all ten objects, the measure in this stage of the experimental design was used to specify the level of familiarity for each object relative to the others in the group of ten employed for this study.

Firstly, the subjects (N=35) were asked whether the landmark was in 'the neighbourhood that you live in' and 'in the place that you live in'. The subjects were then instructed to state the distance relations from 'where they lived' to each landmark. A scale of qualitative distance relations of categories very near, near, far and very far was used instead of a graded scale that is normally employed for such assessment (Herman *et al.* 1983). It had been seen from the pilot study that the respondents found it easier to interpret and use a qualitative scale, as opposed to a numerical scale. This is because the qualitative scale was closer to the intuitive distance estimated and relations that were formed within the cognitive schema, and employed linguistic expressions that were used in developing a conceptual structures. Because a pair-based relationship was required for the landmarks and subjects, the stimulus was provided in the form of an inter-relationship measure, and it was expected that the individual biases and scales in determining what 'near' and 'far' meant was avoided. It was shown in previous studies (Worboys 2001) that the rule of symmetricity does not apply to distances in geographic space. However, within the limits of this study, only the distances in one direction were queried. The results from the pilot study had shown that the subjects found it difficult to assess directions in the other direction, such as from the landmark to where they lived, and 88% of the respondents stated that it meant the same thing as assessing the distance in the other direction. Because the aim was to assess whether the objects seemed closer within the cognitive schema of the individuals, and an egocentric stimuli was being employed, the distance assessment in one direction was considered adequate.

5 Results

The data that resulted from the experiment are shown in tables 1, 2 and 3. Table 1 shows the counts for *distance estimates in space* and in *time* for the perceived location of objects *in place* (yes/no). Table 2 shows the counts for perceived location of objects *inside or outside the neighbourhood*. Table 3 shows the relation of counts between the estimated location of objects *in place* within the conceptual schema of the environment, and the corresponding perceived location for the object in relation to the conceptual *neighbourhood*. Here, space_dist=distance in space, time_dist=distance in time, and deg_famil=degree of familiarity.

Table 1. Relative counts for distance estimates in space and time to the estimated location of objects

	in place		space_dist	counts1	time_dist	counts2
yes	1	very near	1	84	1	34
	1	near	2	157	2	146
	1	far	3	49	3	108
	1	very far	4	7	4	9
no	2	very near	1	1	1	1
	2	near	2	3	2	5
	2	far	3	27	3	32
	2	very far	4	22	4	15

Table 2. Relative counts for distance estimates in space and time relative to neighbourhood conceptualisation

	in neighbourhood		space_dist	counts1	time_dist	counts2
yes	1	very near	1	84	1	34
	1	near	2	123	2	122
	1	far	3	4	3	53
	1	very far	4	1	4	3
no	2	very near	1	1	1	1
	2	near	2	37	2	29
	2	far	3	72	3	87
	2	very far	4	28	4	21

Table 3. Relative counts for estimated location in place to estimated location in neighbourhood and corresponding degree of familiarity.

	in place		in neigh	counts1		deg_famil	counts2
yes	1	yes	1	210	high	1	236
	1	no	2	87	low	2	61
no	2	yes	1	2	high	1	13
	2	no	2	51	low	2	40

The Kendall's coefficient of concordance showed an extremely high goodness of fit for agreement across the subjects for the different variables in the model, with Kendall's W=0.953, Chi-square=133.375, p<0.01, and df =4. Reliability testing gave a high Cronbach's alpha, =0.6687 (p<0.01, df =34, F=544.0455). Factor Analysis showed no difference in latent dimensionalities of the variables (p<0.01). Normality conditions were not satisfied for all the variables, so suitable non-parametric tests were applied for statistical testing. These tests showed that *distance in space* and *distance in time* had the same distribution, p=0.000. A similar distribution was seen for *location in place*, and for *estimated distance in space* and *in time*, p=0.000. An independent-samples test done on the 10 objects for the 35 subjects indicated that there was a strong probability of the results being from the same population, and there was no significant variation in the population for the general trends exhibited in the model, p=0.000. High positive covariance and correlation was seen among all the five variables of in_place, in_neighbourhood, space_distance, time_distance and degree_familiarity. The four variables explain the qualitative estimates of *distance in space*, with $R^2 = 0.541$ and p<0.01. The results suggest that *distance in time* (VIF=1.623, beta= -0.292), *in place* (VIF=1.205, beta=0.193), *in neighbourhood* (VIF=1.712, beta=0.771), and *degree of familiarity* (VIF=1.229, beta=0.089) were all significant factors in the regression model which showed a good fit, $R^2=0.541$, F (4, 30) =8.830, p=0.000, as presented in table 4.

Table 4. Regression Model for dependent variable=distance in space, $R^2=0.541$, p<0.01

Model		Sum of Squares	df	Mean Square	F	Sig.
	Regression	.877	4	.219	8.830	.000
	Residual	.745	30	.025		
	Total	1.623	34			

The strongest impact upon the estimated distance in space (*space_dist*) is from the estimated location of the object in the neighbourhood (*in_neighbourhood*), Spearman's rho, r=0.577, p=0.000. For the *degree of familiarity* and *distance in space*, r=0.329, p=0.05; for *distance in time* and *location in neighbourhood*, r=0.697, p=0.000; and for the location *in place* and *degree of familiarity*, r=0.380, r=0.024<0.05. On testing the relation of the other variables with the estimated *distance in time*, the results showed that the variables *distance in space* (VIF=1.953, beta=-0.352), *in neighbourhood* (VIF=1.809, beta=0.814), *in place* (VIF=1.258, beta=0.125) and *degree of familiarity* (VIF=1.245, beta=0.020) adequately described the model, $R^2=0.447$, F (4, 30) =6.067, p= 0.001. The regression model is shown in table 5.

Table 5. Regression Model for dependent variable=distance in time, $R^2=0.447$, p<0.01

Model		Sum of Squares	df	Mean Square	F	Sig.
	Regression	1.032	4	.258	6.067	.001
	Residual	1.276	30	.043		
	Total	2.307	34			

The strength of relationship between the location *in place* and the estimated distance *in space* is shown to be high (chi-square=142.602, df =3, p=0.000), with a positive strength of association. This signifies that the *distance in space* was estimated to be closer if the object was *in place* (r=0.529, p=0.000), showing a possible significance of place-boundaries on qualitative spatial distances in the cognitive schema. The difference between expected and observed frequencies was high showing that the trends in the model were not random. Figure 1(a) shows the variation in cognitive *distance in space* with estimated cognitive location *in place*. The dependence of the temporal knowledge to the cognitive notions of *place* was also estimated through significance testing. Results showed that although there was a significant association (chi-square = 68.933, p=0.000), with a positive association (r=0.389, p=0.000), the association between temporal knowledge and the conceptualisation of *place* was not as strong as that with spatial knowledge. Figure 1(b) shows the relation between *estimated distances in time* and the *location in place*.

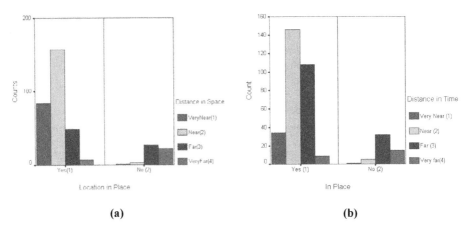

(a)   (b)

Fig. 1. Relationship between location in place and (a) distance estimates in space, (b) distance estimates in time

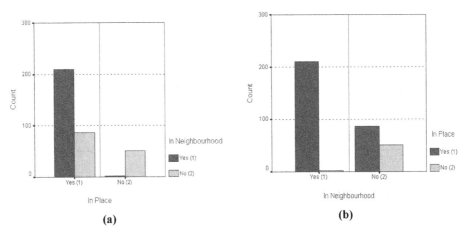

(a)   (b)

Fig. 2. Relationship between conceptualisations of place and neighbourhood, where, (a) objects not in place are not in the neighbourhood, (b) objects in neighbourhood are in place

A significant co-dependence between the conceptualisations of *place* and *neighbourhood* is demonstrated in the results (chi-square=84.367, p=0.000). Figure 2 (a) and (b) shows the significance of this relationship (r=0.491, p=0.000), with the results indicating that there is a high chance of the object being conceptualised *in neighbourhood* if it is conceptualised as located *in place*. When the objects were conceptualised to be *in place*, there is an equal chance of them being conceptualised to be situated in the *neighbourhood*. If the object is not conceptualised to be located *in place*, there is a significantly low chance of it being located in the cognitive *neighbourhood*, too. The results imply that *neighbourhood* is conceptualised as a subset of *place*.

The *degree of familiarity* was hypothesised to be a determining factor in the formation of a *cognitive* 'sense of place', and significance tests showed a strong association between the conceptualisation of the object to be *in the place* and have a high *degree of familiarity* (chi-square=65.918, p=0.000, r=0.435). Objects located inside the cognitive notions of *place* have high familiarity associations. The *degree of familiarity* was also strong for objects that were conceptualised *in the neighbourhood* (chi-square=349.000, p=0.000, r=0.428). The results suggest that *place* is closer in the familiarity set of the individuals than *neighbourhood*, although only marginally. From these results, *neighbourhood* is indicated as a lower-order concept to *place*. It is also shown that *place* is a closer concept in the cognitive set of individuals than *neighbourhood*, with the strength of attachment stronger to a *place* than to a *neighbourhood*. The effect of *place* being a more proximic concept in the familiarity set has to be verified by assessing the effect of *neighbourhood* conceptualisation on spatial and temporal relations formed in the environment. The relation between the estimated location of objects with relation to conceptualised *place* and *neighbourhood* and the degree of familiarity to the objects is shown in figure 3(a) and (b). Statistical tests validated the indications from the previous results that *place* is a more proximic concept within the cognitive and familiarity-set of the individuals than neighbourhood. The distance estimates in *space* and *time* were assessed for the corresponding location with relations to the conceptual *neighbourhood*.

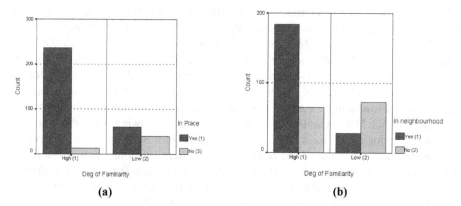

(a)   **(b)**

Fig. 3. Objects with a (a) high degree of familiarity are conceptualised to be located in place, and (b) a low degree of familiarity have a lower probability of location in neighbourhood

The perceived location of the object in *the neighbourhood* has a significant influence on the *estimated distance in space* (r= 0.726, p= 0.000), with objects in the *neighbourhood* having a significant chance of appearing to be *very near*. There is a high probability that the two variables are independent (chi-square =184.022, p=0.000). Figure 4(a) shows this relationship between neighbourhood conceptualisation and spatial knowledge. Tests for temporal distance estimation as a factor of perceived location *in neighbourhood* show a significant chance that temporal knowledge is neighbourhood-dependent (r=0.528, p=0.000, chi-square=97.476, p=0.000). Objects inside *neighbourhood* are conceptualised as being close in *time*. The relationship between location *in neighbourhood* and the relative estimation of temporal distance is shown in figure 4(b).

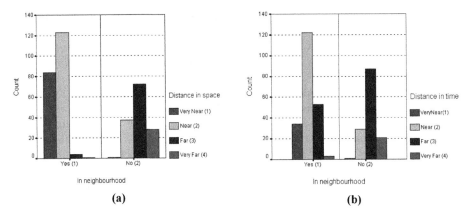

(a)   **(b)**

Fig. 4. Relationship between conceptualised neighbourhood and (a) distance estimates in space, (b) corresponding temporal knowledge in the environment

Finally, the results from the experiment were interpreted in a unified manner. This was done to define the relationship between the conceptualisation of *place* and the other variables, and to determine the relations between them, together with the appropriate weights. For this purpose, regression analysis was performed with '*in place*' as the dependent variable. The results show that *in neighbourhood* (VIF=2.154, beta=0.140, p=0.028), *distance in space* (VIF=2.595, beta=0.331, p=0.000), and *degree of familiarity* (VIF=1.429, beta=0.167, p=0.001) are all significant factors in the regression model (R^2=0.362, df= (4,345), p<0.01). However, the variable *distance in time* shows low significance for the probability that it is related to the dependent variable (VIF=1.519, beta=0.082, p=0.122), thereby showing that the temporal knowledge is not a significant factor in the formation of a 'sense of place'. Therefore, the variable was taken out of the equation and the regression model was run again. The results suggest that the three variables, *distance in space* (beta=0.361, p=0.000), *in neighbourhood* (beta=0.160, p=0.010) and *degree of familiarity* (beta=0.170, p=0.001) are all significant factors in the model (r=0.528, R^2=0.357, p<0.01) (see table 6).

Table 6. ANOVA model for dependent variable (*in place*)

ANOVA

	Sum of Squares	df	Mean Square	F	Sig.
Regression	16.059	3	5.353	64.054	.000
Residual	28.915	346	.084		
Total	44.974	349			

The results suggest that qualitative estimation of distances in space is affected by the conceptualisation of the extent of *place* and *neighbourhood*, along with the degree of familiarity with the objects in the environment. Qualitative notions of spatial and temporal distances are *place* and *neighbourhood*-dependent. Distance estimates in *space* are place-dependent, and objects are perceived to be closer if they are in a *place* than when outside it. Spatial distance estimates are more sensitive to these factors than temporal estimates. Although objects have a strong probability of being perceived to be nearer if they are in the individual's *place*, this association is even stronger for a *neighbourhood* conceptualisation. The cognitive conceptualisation of *place* and *neighbourhood* extent affects the estimation of distances in the environment. Extend and boundaries have a significant ontological status in *place* conceptualisation and in cognitive knowledge of the environment. H1.2 is, therefore, accepted. It is seen that objects that are not close together in Euclidean space are perceived to be closer if they are perceived to be inside the conceptualised *place* and *neighbourhood*. However, it is difficult to establish an equivalent rule for temporal distances from the results obtained in this experiment. *Place* and *neighbourhood* have a high degree of co-dependence, and there is a high probability of *neighbourhood* being conceptualised as a sub-set of *place* within the cognitive schema. However, *place* has a higher degree of familiarity associated with it in the cognitive set of the individuals than that of *neighbourhood*. The degree of familiarity is not necessarily dependent on the location of the objects in the *neighbourhood* or *place*, although it plays a significant role in determining the extent of both *place* and *neighbourhood*, with a higher influence on *place* than on *neighbourhood*. H1.3 is not conclusively supported, as the results of the experiment do not show sense of time as an equally significant factor in the formation of a 'sense of place' as sense of space. Neither is there adequate indication that the conceptualisation of *place* and *neighbourhood* has a strong impact on the temporal reasoning and formation of temporal relations in the environment. Temporal reasoning did not establish any conclusive relationships with the degree of familiarity in the environment. However, it is conclusively shown that distance estimates and spatial reasoning is place-dependent, thereby validating H1.1. A larger sample size may be instrumental in verifying this further. Overall, hypothesis H1 is verified, with the results suggesting inter-dependency between spatial and temporal distance estimates, and the conceptualisation of *place* in an environment.

From these results, the 'cognitive sense of place' (SOP_C) can be defined by a unified determination of the effect of all the variables employed in the experimental design, and the dimensions that became apparent. SOP_C can be expressed as **<environmental knowledge: spatial, familiarity, neighbourhood: boundaries, k>** Here, SOP_C is expressed as a relationship between environmental knowledge (with spatial knowledge as the significant variable), familiarity in the environment, conceptualisation for neighbourhood that is operationalised through its extent and determined by boundaries as a significant variable. This equation will require further development using mechanisms that are more formal. This equation also needs to be refined to explain factor K that is used here to represent any symbolically important aspect which has been omitted within the constraints of this experiment. However, this schematisation provides the foundation for further experiments, and a framework of better understanding of individual differences in relation to the formation of a 'sense of place'. This understanding is important for defining ontological commitments and for characterising cognitive theories for *place* in the real world.

6 Conclusions

The experimental framework and results presented in this paper provide directions for formalising the seemingly abstract concept of 'sense of place' to operationalise it as a grounding mechanism in place-based models and ontologies. The experiments have probed the elusive concept of 'sense of place' and demonstrated that the parameters for SOP can be developed, based on experimental work with human subjects, which can be used to define distinctions between *place* and *neighbourhood*. Experimental results in this paper indicate a strong correlation between the formation of 'sense of place' with spatial reasoning, and a 'cognitive sense of place' can be operationalised as a factor of spatial knowledge, degree of familiarity and conceptualisation of boundaries, both for *neighbourhood* and for *place.* Abstract notions such as familiarity are made explicit within this setup and included in the equation. The results are significant for understanding the cognitive dimensions of *place*, and for formulating cognitive theories of *place* and the impact that it has on spatio-temporal reasoning and behaviour in geographic environments. It is shown that object-neighbourhood ontology does not completely explain the nature of *place* formation, and more dimensions must be considered for formulating any definitive theories of *place*.

The methods developed from this paper can be tested in non-urban environments by applying these to indigenous communities. This will also enable the assessment of the generalisability of the results reported here. Future work will also extend on this paper to refine this model for 'sense of place' by employing more variables within the experimental framework and by using more robust mathematical and logic-oriented formalisation procedures. Such formalisations will be useful in development of context-aware computing for location-based services, and also enable the simulation of 'sense of place' in place-based models, in virtual environments, and in development of computer-based automated narrative generators that use place as the basis for relating human activity to space.

Acknowledgements

I am grateful to Tony Cohn and Paul Aplin for commenting on the initial drafts of this paper, and to Dan Montello for his constructive comments on the experimental framework described in this paper. Thanks are also due to the three anonymous reviewers whose comments helped improve the quality of this paper and to the students and faculty at The University of Nottingham who took part in the experiments.

References

Agarwal, P., 2004a, 'Contested nature of 'place': knowledge mapping for resolving ontological distinctions between geographical concepts', In Egenhofer, M., Freksa, C., Harvey, M. (Eds.) Third International Conference, GIScience 2004, Lecture Notes in Computer Science, Springer-Verlag, Berlin, LNCS 3234, pp. 1-21.

Agarwal, P., 2004b, Topological operators for ontological distinctions: disambiguating the geographic concepts of place, region and neighbourhood, Spatial Cognition and Computation, Vol. 5, Issue 1, pp 69-88.

Arleo, A. and Gerstner, W., 2000, Spatial cognition and neuro-mimetic navigation: A model of hippocampal place cell activity. Biological Cybernetics, 83: 287-299

Battig, W. F. and Montague, W. E., 1968, Category Norms for Verbal Items in 56 categories: A Replication and Extension of the Connecticut Norms. Journal of Experimental Psychology Monograph, 80 (No. 3, Part 2): 1-46.

Bittner, T., 2002, Reasoning about Qualitative Spatio-Temporal Relations at Multiple Levels of Granularity. In: F. vanHarmelen (Editor), Proceedings of the 15th European Conference on Artificial Intelligence – ECAI 2002. IOS Press, Amsterdam.

Brentano, F., 1988, Philosophical Investigations on Space, Time and the Continuum. CroomHelm, London.

Canter, D., 1977, The Psychology of Place. Architectural Press, London.

Canter, D., 1997, The Facets of place. In: G. T. Moore and R. W. Marans (Editors), Advances in Environment, Behavior, and Design. Plenum, New York, pp. 109-148.

Cantrill, J.G. 1998. The environmental self and a sense of place: communication foundations for regional ecosystem management. Journal of Applied Communication Research 26:301-318.

Capra, F., 1975, The Tao of Physics, Shambhala, New York.

Chown, E., Kaplan, S. and Kortenkamp, D., 1995, Prototypes, location and associative networks (PLAN): Towards a unified theory of cognitive mapping. Cognitive Science, 19: 1-51.

Clayton, K., and Habibi, A., 1991, Contribution of temporal contiguity to the spatial priming effect. Journal of Experimental Psychology: Learning, Memory, and Cognition, 17: 263-271.

Cornell, E. H., Heth, C. D. and Alberts, D. M., 1994, Place recognition and way finding by children and adults. Memory & Cognition, 22(6): 633-643.

Courtenay, G., 1978, Questionnaire Construction, In Hoinville, G., and Jowell, R (Editors), Survey Research Practice: studies attempting to exploit the quantitative nature, Heinemann Educational Books: London.

Cuba, L., and Hummon D. M., 1993, A place to call home: identification with dwelling, community, and region. Sociological Quarterly 34 (1):111-131.

Datel, R.E. and Dingemans, D.J., 1984, Environmental perception, historical preservation, and sense of place, In: T.F. Saarinen, D. Seamon and J.L. Sell (Editors) Environmental Perception and Behaviour: An Inventory and Prospect, Chicago: University of Chicago Research Paper 209, pp. 131-144

Denofsky, N., 1976, How near is near? AI Memo 344. MIT AI Lab, Cambridge MA.

Dey A., G. Abowd and D. Salber, 2001, A conceptual framework and a toolkit for supporting the rapid prototyping of context-aware applications, Special issue on context-aware computing, Human–Computer Interaction (HCI) Journal, vol. 16, no. 2–4, pp. 97–166.

Downs, R. M. and Stea, D., 1973, Image and Environment: cognitive mapping and spatial behaviour. Aldine Publishing Coy.

Duncan, J., and D. Ley, Eds. 1993. Place/culture/representation. London, England: Routledge.

Entrikin, J.N., 1991, Betweenness of place: towards a geography of modernity.London: Macmillan.

Espinoza, F., Persson, P., Sandin, A., Nyström, H., Cacciatore. E. and Bylund, M., 2001, GeoNotes: Social and Navigational Aspects of Location-Based Information Systems, in Abowd, Brumitt and Shafer (eds.) Proceedings of Ubicomp 2001, pp 2-17.

Feld, S., and K.H. Basso., 1996. Senses of place. Santa Fe, N.M.: School of American Research Press.

Forrest, R. and Kearns, A., 1999, Joined-up Places? Social Cohesion and Neighbourhood Regeneration, York: York Publishing Services for the Joseph Rowntree Foundation.

Fowler, F.Jr., 1989, The Significance of Unclear Questions, In Cannell, C., Oksenberg, L., Kalton, G., Bischoping, K. and Fowler, F.Jr. (Editors), New techniques for Request for Pretesting Survey Questions, Research Report, Survey Research Center, The University of Michigan.

Gale, N., Golledge, R. G., Pellegrino, J. W. and Doherty, S., 1990, The acquisition and integration of route knowledge in an unfamiliar neighbourhood. 10: 3-25.

Galster, G., 1986, 'What is neighbourhood? An externality space approach.' International Journal of Urban and Regional Research, 10: 242-261

Galton, A., 2003, Desiderata for a Spatio-Temporal Geo-ontology. In: W. Kuhn, M. Worboys and S. Timpf (Editors), Spatial Information Theory: Foundations of Geographic Information Science, Proceedings of International Conference COSIT'03. Springer-Verlag, Kartause, Ittingen, pp. 1-13.

Giraudo, M.-D., and Péruch, P., 1988, Spatio-temporal representation of urban space. Journal of Environmental Psychology, 8:9-17.

Goldlust, J. and Richmond, A., 1977, Factors associated with commitment to identification with Canada, In: W. Isajiw (Editor), Identities, Peter Martin Associates, Toronto. pp. 132-153

Golledge, R. G., 1992, Place recognition and wayfinding: Making sense of space. Geoforum, 23(2): 199-214.

Harrison, S. and Dourish, P., 1996, Re-place-ing space: The roles of place and space in collaborative systems, Proceedings of Computer Supported Collaborative Work. ACM, Cambridge, MA, pp. 67-76.

Herman, J. F., Norton, L. M., and Roth, S. F., 1983, Children and adults' distance estimations in a large-scale environment: Effects of time and clutter. Journal of Experimental Child Psychology, 36: 453-470.

Hirtle, S., 2001, Dividing up space: Creating Cognitive structures from Unstructured Space, Meeting on Fundamental Questions in GIScience, Manchester, UK.

Hudson, A. C., 1998, Placing trust, trusting place: on the social construction of offshore financial centres. Political Geography 17(8): 915-937

Jacobs, L. F., 2003, The Evolution of the Cognitive Map. Brain Behaviour and Evolution, 62: 128–139.

Johnston, R. J., Gregory, D., Pratt, G., and Watts, M. (Editors) 2000, The Dictionary of Human Geography, 4th Edition, Blackwell Publishers Ltd., Oxford, UK, pp. 332

Kuhn, W., 2003, Semantic Reference Systems. International Journal of Geographical Information Science, 17(5): 405-409.

Lewis, P., 1979, Defining sense of place, In: W. P. Prenshaw and J. O. McKee (Editors). Sense of Place: Mississippi, University of Mississippi, Jackson, MI. pp. 24-46

Lloyd, R., Patton, D. and Cammack, R., 1996, Basic-Level Geographic Categories. Professional Geographer, 48: 181–194.

Lynch, K., 1960, The Image of the City. MIT Press, Cambridge, MA.

Mallot, H.-A. and Gillner, S., 2000, Route navigation without place recognition. What is recognized in recognition triggered responses? Perception 29: 43-55

Mark, D. M., Smith, B. and Tversky, B., 1999, Ontology and Geographic Objects: An Empirical Study of Cognitive Categorisation. In: C. Freksa and D. M. Mark (Editors), Spatial Information Theory: A Theoretical Basis for GIS. Springer-Verlag, Berlin, pp. 283-298.

Massey, D., 1993, Power-geometry and a progressive sense of place. In: Bird, Jon; Curtis, Barry; Putnam, Tim; Robertson, George (Editors) Mapping the futures: Local cultures, global change, London and New York: Routledge, pp. 59-69.

McCall, R., O'Neill, S. and Carroll, F., 2004, Measuring Presence in Virtual Environments. In , Conference on Human-Factors in Computing: CHI'2004 . Vienna, Austria: Association of Computing Machinery (demo paper).

McNamara, T. P., 1991, Memory's view of space, In: The Psychology of Learning and Motivation (Vol. 27, pp. 147-186), G. H. Bower (Editor). Academic Press, San Diego, CA.

McNamara, T. P., 1992, Spatial representation. Geoforum, 23(2): 139-150.

Montello, D. R. and Golledge, R., 1998, Scale and Detail in the Cognition of Geographic Information: Report of a Specialist Meeting held under the auspices of the Varenius Project, May 14-16, UCSB.

Montello, D. R., F.Goodchild, M., Gottsegen, J. and Fohl, P., 2001, Things'll Be Great When You're Downtown: Behavioral Methods for Determining Referents of Vague Spatial Queries. In: M. Cristani and B. Bennett (Editors), Workshop on Spatial Vagueness, Uncertainty, and Granularity, CD proceedings, Leeds, UK.

Norberg-Schulz, Christian., 1980, Genius Loci. Towards a Phenomenology of Architecture. London.

Parkes, D. and Thrift, N., 1978, Putting Time in its Place. In: T. Calstein, D. Parkes and N. Thrift (Editors), Making Sense of Time. Edward Arnold, London, pp. 119-129.

Pellow, D., 1992, Attachment to the African compound. In: I. Altman and S. M. Low (Editors) Place attachment, New York: Plenum Press, pp.187-210.

Peterson, G. and Saarinen, T., 1986, Local symbols and sense of place, Journal of Geography, 164-168.

Pirinen, M., 2002, In Place, Out of Place: Meaning of Place and Question of Belonging for Adult-Third Culture Kids. Masters Thesis, University of Joensuu, Finland.

Presson, C. C., 1987, The developments of landmarks in spatial memory: The role of differential experience. Journal of Experimental Child Psychology, 44: 317-334.

Prince, H., 1978, Time and Historical Geography, In T. Calstein, Parkes, D. and Thrift, N. (Editors), Making Sense of Time, Vol. 1, pp 17-37, London, Edward Arnold

Proshansky, H. M., Fabian, A. K., and Kaminoff, R, 1983, Place-identity: Physical world socialization of the self, Journal of Environmental Psychology 3: 57-83. Also in Groat, Linda, (Editor) Giving Places Meaning. London: Academic Press, 1995

Relph, E., 1976, Place and Placelessness. Pion Limited, London.

Riley, R. B., 1992, Attachment to the ordinary landscape. In: I. Altman and S. M. Low (Editors) Place attachment, New York: Plenum Press, pp. 13-35

Schölkopf, B. and Mallot, H. A., 1995, View-based cognitive mapping and path planning. Adaptive Behavior (3): 311-348.

Seamon, D., 1984, Phenomenologies of Place and Environment, Phenomenology and Pedagogy, 2:130-135.

Shafer, B. C., 1984, Debated problems in the study of nationalism, Canadian Review of Nationalism, 11: 1-19.

Shamai, S., 1991, Sense of place: An empirical measurement, Geoforum, 22: 347-358.

Shamai, S. and Kellerman, A., 1985, Conceptual and experimental aspects of regional awareness: an Israeli case study, Tijdschrift voor Economische en Social Geografie, 76, 88-89.

Sorrows, M. E. and Hirtle, S. C., 1999, The Nature of Landmarks for Real and Electronic Spaces. In: C. Freksa and D. M. Mark (Editors), Spatial Information Theory: Cognitive and Computational Foundations of Geographic Information Science, International Conference COSIT '99. Lecture Notes in Computer Science 1661 Springer, Stade, Germany, August 25-29, pp. 37-50

Starrs, P.F., 1994, The importance of places, or, a sense of where you are. Spectrum: The Journal of State Governments 67 (3): 5-17

Steck, S.D. and H.A. Mallot, 2000, The role of global and local landmarks in virtual environment navigation. Presence, Teleoperators 9:69-83

Tobler, W. R., 1970, A Computer Model Simulating Urban Growth in the Detroit Region. Economic Geography, 46: 234-240.

Tuan, Yi-Fu, 1977, Space and Place: The Perspective of Experience, University of Minnesota Press, Minneapolis.

Tuan, Yi-Fu, 1990, Topophilia: A study of Environmental Perception, Attitudes and Values, New Jersey, Prentice Hall Inc, Englewood Cliffs.

Twigger-Ross, C.I., and D.L. Uzzell, 1996, Place and identity processes. Journal of Environmental Psychology 16:205-220.

Williams, D.R., and Stewart S.I., 1998, Sense of place: an elusive concept that is finding a home in ecosystem management. Journal of Forestry 96 (5): 18-23.

Withers, C. W. J., 1996, Place, memory and monument: memorizing the past in contemporary Highland Scotland, Ecumene, 3:325-344.

Worboys, M. F., 2001, Nearness relations in environmental space. International Journal of Geographic Information Science, 15(7): 633-653.

Data-Driven Matching of Geospatial Schemas

Steffen Volz

University of Stuttgart, Institute for Photogrammetry, Geschwister-Scholl-Str. 24D,
70174 Stuttgart, Germany
steffen.volz@ifp.uni-stuttgart.de

Abstract. The emergence of spatial data infrastructures offering geospatial information from heterogeneous sources involves the need to achieve an integration of databases from different data providers within a common platform. Generally, database integration consists of two steps: schema integration and object
integration. Concerning schema integration, a crucial part is the identification of
semantically corresponding elements in different schemas. This process is referred to as schema matching. In this paper, we present a data-driven approach
for the matching of geospatial schemas. It is based on the idea of exploiting the
instance-level relations between multiple representations (of one and the same
real world object) that have been captured on the basis of different schemas. As
a result, correlation measures are derived describing the degree of correspondence between the object classes of different schemas.

1 Introduction

In recent years, tendencies to provide global, generic geospatial information platforms
that are accessible by different kinds of GIS applications can be observed (e.g. [11]).
Building up such platforms is basically identical with the task of integrating existing
local, heterogeneous databases into one system that allows access to the associated
data sources. Mostly, such an infrastructure is realized as a federated database system
which acts as a single, homogeneous database to the global applications, facilitating a
common view on the underlying data [6].

Database integration for the development of federated systems generally involves
two steps: schema integration and object (or data) integration. Schema integration
comprises the transformation of existing local schemas into a global schema for the
data to be integrated. According to [14], this is identical with the problem of integrating independently developed ontologies into a single ontology in an artificial intelligence setting. Object integration can be defined as the solution of conflicts between
inconsistent and contradictory data instances which are representing the same real
world entity.

This work aims at contributing to the research in the domain of schema integration,
especially focusing on the detection of semantic correspondences between elements of
different geospatial schemas, which is generally referred to as schema matching [7].

Within this paper, a data-driven approach for the matching of spatial schemas will
be presented. It has been implemented for street data of the Geographic Data Files
(GDF) [5] schema and the schema of the German Authoritative Topographic Carto-

A.G. Cohn and D.M. Mark (Eds.): COSIT 2005, LNCS 3693, pp. 115–132, 2005.
© Springer-Verlag Berlin Heidelberg 2005

graphic Information System (ATKIS) [21]. GDF is an international standard that has especially been developed for car navigation applications and thus is used to describe and transfer street networks and street related data whereas ATKIS contains data of different topographic units like settlement, vegetation, traffic, etc. Both data models capture street objects as linear features in approximately the same scale (1:25 000).

1.1   Multiple Conceptual Schemas and Multiple Representations in GIS

In the initial stage of the development of computer applications, scenarios are set up in order to derive the use cases for which the planned software has to serve. If, for example, a car navigation application has to be implemented, it has to be able to handle street data on which routing algorithms can operate, it has to be able to deal with restrictions (e.g. forbidden maneuvers), etc. But some aspects of reality, like e.g. the vegetation, are rather unimportant for navigation purposes and can be neglected. So the car navigation application has a certain view of the world, focusing on what is important for it, whereas, for example, a land planning application probably sees the world quite differently. This means that each application or at least each type of application has a different view on reality depending on its use cases. It is the task of humans to convey exactly those views to applications which are necessary for them to fulfill their specific purposes in an optimal way. Generally, this process is called conceptual schema modeling.

Now since identical entity types of the real world (like streets, buildings, etc.) are observed from different application perspectives, there are *multiple conceptual schemas* about the same real world phenomena, potentially bearing contradictions.

On the basis of these multiple conceptual schemas, different institutions and companies are capturing their data. Consequently, one and the same real world object, like for example the Fifth Avenue in New York City, is available in *multiple representations* in different geospatial databases. These multiple representations are often inconsistent and thus severe problems can occur if they have to be processed together, i.e. if they have to be combined during data analysis or if updates have to be forwarded from one representation to a corresponding one.

1.2   Approach of This Work

Generally, the approach of this work is based on the idea to build up explicit relations between multiple representations, i.e. between corresponding digital instances of different geospatial databases which have been captured according to different conceptual schemas. These relations have to describe, how consistent/inconsistent two corresponding representations are with respect to geometry, topology or thematic properties. Within this work, they are called Multi-Representational Relations or MRep Relations in short.

Thus, a formal structure has to be built up for MRep Relations at first. Afterwards, MRep Relations must be generated for real data which are available in multiple representations. In our case, we investigate street data from ATKIS and GDF which are using differing conceptual schemas to describe street objects. By analyzing the MRep Relations and especially the affiliations of the instances constituting corresponding representations to object classes in their source schemas (i.e. ATKIS and GDF), schema similarities can be detected (see Fig. 1).

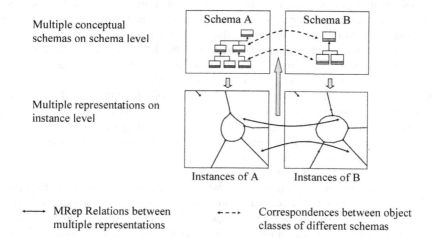

Multiple conceptual
schemas on schema level

Multiple representations on
instance level

Fig. 1. The basic assumption of our approach is that we can derive correspondences between different conceptual schemas by analyzing the MRep Relations which have been created for the underlying data sets (after [17]).

1.3 Outline of This Article

The paper is organized in 5 sections. In the following section 2 related work concerning the identification of corresponding elements in different databases is presented. Section 3 describes the basis of the proposed approach which consists of modeling and building explicit links between multiple representations. In section 4, the data-driven schema matching and its results are explained in detail. Finally, section 5 concludes this paper and gives and outlook on future issues.

2   Related Work

The problem of identifying corresponding elements in different, inconsistent databases has been investigated from different perspectives. For this work, especially the research in the fields of geospatial sciences and in the database domain is of importance.

Regarding GIS, many approaches to overcome semantic heterogeneities have been made using ontologies or semantic mappers (translators). Moreover, the issue of how to match and how to merge spatial data instances stemming from different sources has been addressed. In the database domain, basically similar problems were encountered. Research activities here rather focused on schema integration and schema matching as well as on the detection of duplicate objects in different databases, however without regarding the spatial component.

2.1   Research in the GIS Domain

One of the most important topics regarding research in GIS is concerned with the question of how interoperability between spatial information systems can be achieved. Many approaches deal with the semantic integration of different concepts about real world objects using ontologies. For [16], semantic integration is similar to a communication process since two partners who want to communicate have to have the same understanding about the things or objects they are talking about. For this reason, they must use a common vocabulary. However, in reality, different GIS communities are using different vocabularies, but the differences between these vocabularies are not clear. For this reason, we can use ontologies to provide an explanation for "the intended meaning of a formal vocabulary" [9]. To describe the meaning of such a formal vocabulary consisting of entities, classes, properties and functions related to a certain view of the world, ontologies use a specific vocabulary themselves (like RDF, UML, OWL, etc.), with which concepts and relations between them (like part-of, is-a, is-similar-to, etc.) can be expressed. Ontologies can be "a simple taxonomy, a lexicon or a thesaurus, or even a fully axiomatized theory" [8].

In [8], an architecture for Ontology-Driven GIS (ODGIS) is proposed. Here, ontologies are translated into classes which are stored on an ontology-server that can be browsed by users. Thus, for each entity type of the real world like "lake", its roles (e.g. geographical region, recreation area), its attributes (e.g. acidity, water volume), its functions (e.g. fishing, swimming) or its parts (e.g. beach, cove) can be displayed. If the user selects a certain ontological description and a certain geographical region, mediators are started and look for data instances (e.g. all lake objects) in the affiliated databases that comply with the search criteria. The authors consider their approach to be flexible and for this reason opposed to standardization efforts.

As [10] see it, two cases have to be considered concerning the integration and interoperability of spatial databases:

1. Conceptual schemas are based on the same ontology: merely synonyms and homonyms have to be detected to perform an integration.
2. Conceptual schemas are based on different ontologies (of different communities): a shared ontology has to be found by analyzing the similarities and relations between the source ontologies.

For the latter case, the authors are presenting a formalism for the representation of ontologies by means of Description Logic (DL). Each user community can define its perception of an object using the specified DL and then different ontologies can be related and merged using a reasoning system.

Another example on how to integrate different semantics of spatial data is provided by [3]. The approach consists of two components, the Semantic Wrapper and the Semantic Mapper. Objects of different spatial databases are wrapped by the Semantic Wrapper and have to conform to a predefined interface so that they can be recognized by the Semantic Mapper. This interface is specific for a certain application domain like transportation, topography, etc. On the level of the Semantic Mapper, the semantics of two objects can be compared and the schematic and semantic differences between them can be resolved.

A similar concept is described in [13]. The author proposes semantic reference systems (in addition to spatial reference systems) that provide a framework allowing standardized semantic transformations (in addition to coordinate transformations of geometric data) between thematically differing spatial data sources As a result, an ontology mapping can be created. In [20], integration is basically seen as a product operation of different spatio-thematic layers and an algebraic model for the construction of these products is set up. As a prerequisite, the different thematic classes of the spatio-thematic layers have to be related by human activity. A prototype is presented that allows integrating different land cover data sources on the basis of the developed algebra.

Besides the efforts dealing with the integration of different semantic concepts, important research has been carried out in the field of matching corresponding object instances or multiple representations, respectively, which are stored in different geospatial databases. For example, [1] has implemented an approach for matching street network nodes of two different GDF datasets which have been acquired by different companies (NAVTEQ and TeleAtlas). The algorithm developed here is based on the idea of describing intersections of streets, i.e. nodes of a street network, by an explicitly defined code. The code consists of point coordinates and the number, abbreviations and names of incident streets. For each intersection, such a code is created. By comparing the codes of the intersections within the different GDF data sets and by assigning the intersections with the most similar codes to each other, references can be derived.

A fundamental, line-based matching approach for street network data of ATKIS and GDF has been presented by [18]. In a first step, the algorithm finds all potential correspondences of topologically connected line elements in two source data sets by performing a buffer operation. The matching candidates are stored in a list. This list is ambiguous and typically contains a large amount of $n{:}m$ matching pairs. Then, unlikely matching pairs are identified and eliminated using relational parameters like topological information and feature-based parameters like line angles. The result is a smaller but still ambiguous list with potential matching pairs. These matching pairs are evaluated with a merit function in order to compute a unique combination of matching pairs which represents the solution of the matching problem. This is a combinatorial problem which is solved with an A* algorithm.

The problem of spatial data conflation (or merging) is for example being tackled by [4]. The merging process is defined here as "feature deconfliction", where all parts of a matched feature pair are unified into a single "better" feature. The conflation algorithm has to decide, which properties are preserved in the resulting instance. In their approach, the authors are also taking into account the data quality information of the corresponding instances.

2.2 Research in the Database Domain

In this section, three research fields of the database domain that are of relevance for our work are presented: schema matching and schema integration as well as duplicate detection for data cleansing purposes.

As mentioned before, schema matching can be defined as the process of detecting semantically similar elements of two schemas. Its application domains are manifold, covering e.g. data warehouses or web-oriented data integration, etc. but most importantly schema integration. There are different types of techniques with respect to schema matching. It can either be merely schema-based, only taking into account schema-level information like names, descriptions, data types, relationship types (part-of, is-a, etc.), constraints and schema structure whereas other approaches are also investigating instance data. Furthermore, only individual schema elements can be assigned to each other or combinations of elements (schema structures) can be matched [14].

If schema matching is performed manually, it is a time consuming process and therefore it should be automated as much as possible. For this purpose, [7] have built up a library of different match algorithms (called "matchers"). For example, they use

- *Simple matchers*: String matching techniques like n-gram (strings are compared according to sequences of n characters), edit distance (computation of the number of edit operations necessary to transform one string into another), etc. and
- *Hybrid matchers*: Matching operations which are using a fixed combination of matchers to improve the match result, e.g. a combination of n-gram and edit distance or a combination of name and data type matching, etc.

They further exploit results of previous matches since they assume that many schemas are identical or at least very similar. Suppose we have already derived match results between schemas S1 and S2 as well as between schemas S2 and S3. If we know that S1 and S2 are very similar, than we can easily reuse the available match results to obtain a match result between S1 and S3, assuming a transitive nature of the similarity relation between the elements. Finally, all results of the different matchers (i.e. simple and hybrid matchers as well as the results of previous matches) are combined to compute a total similarity. The authors have evaluated their approach on real world schemas and achieved promising results.

Schema integration is the process of building an integrated schema from existing, independently developed local schemas, thus providing a global conceptual schema for all data available within a federated database system [6]. Different kinds of conflicts have to be resolved during the integration phase, e.g. description conflicts (like conflicts because of different ranges of values or different numerical accuracies for identical attributes), conflicts due to different data models (e.g. object-oriented versus relational concepts) or also structural conflicts (like different representations of class properties, on the one hand as attributes of a class or on the other hand as references to subclasses). In [6], different techniques to accomplish schema integration have been briefly discussed. Most methods use so-called assertions or inter-schema correspondences, respectively, as a basis for schema integration. Such correspondences can be derived using schema matching. On the basis of assertions, integration rules specify how corresponding schema elements have to be integrated.

With regard to data cleansing, the identification of instances in different databases that represent the same real world object is important to guarantee a high data quality. However, this is a difficult task. Especially the detection of corresponding elements in semi-structured data sets like nested XML data involves problems. They are dealt with in [19]. The authors apply a threshold-based similarity function which relies on the edit distance measure in order to identify corresponding string objects. To reduce the number of edit distance computations which are very expensive, different filter methods have been implemented. Thus, corresponding instances and structures can be identified efficiently.

3 Modeling and Building MRep Relations

According to [15], "geographic data set integration (or map integration) is the process of establishing relationships between corresponding object instances in different, autonomously produced, geographic data sets of a certain region." Following this notion, it is our goal to set up a model in order to be able to explicitly describe the relations between multiple representations in GIS, which we call Multi-Representational Relations or MRep Relations. Consequently, in the following sections first the structure of MRep Relations is defined. Then, on the basis of the developed model, we build up MRep Relations between street data sets of ATKIS and GDF.

3.1 The Structure of MRep Relations

Basically, multiple representations could only be linked using bidirectional references. However, in our research we want to collect more information about the relations between corresponding representations. For this reason, we have to explicitly define a formal specification of possible MRep Relations. The information stored in MRep Relations can then be exploited for deriving semantic correspondences between object classes of different schemas.

First of all, the identifiers of those instances which make up a representation in data set A as well as the identifiers of the instances constituting the corresponding representation in data set B have to be specified and so the cardinality of an MRep Relation can be implicitly deduced. It can either be *1:1*, *1:n*, *n:1* or *n:m*.

Furthermore, the degree of similarity between two corresponding representations has to be determined and stored in an MRep Relation. Since geospatial objects do have a geometric, a topological and a thematic component, similarities between representations can be assessed on these three levels. However, within this work, only geometric and topological similarity indicators were generated. Since we used linear street data, appropriate geometric similarity measures based on the comparison of angle and length differences and also distance values like the average line distance and the Hausdorff distance were introduced. Topological similarity was detected using adjacency relations of corresponding street representations. Finally, the different partial similarity measures were aggregated into a total similarity value. This was done using a simple weighted sum approach where the absolute values of the individual similarity measures were first divided into 7 classes from 0 (lowest similarity) to 6 (highest similarity) to obtain so-called evaluation values. Then, each evaluation value was weighted with a factor. The weight factors were specified on the basis of the operator's expertise regarding the influence of the different partial similarity values on the total similarity. The total similarity value was normalized onto an interval ranging from 0 to 100 units, with 100 representing the highest similarity. The aggregation of similarity measures is a difficult problem that can be further optimized within our approach, e.g. using machine learning techniques (see e.g. [2]).

Besides the similarity values, MRep Relations have to provide further information. For schema matching purposes, especially the affiliations of the instances of a representation to certain object classes of the source schema has to be known. There are more attributes belonging to MRep Relations like comparisons of data quality parameters but they can be neglected here since they are out of the scope of this paper.

MRep Relations are stored and exchanged using an XML-based format called *MultiRepresentational Relation Language* (MRRL) which was specified within this work. The following extract of MRRL shows the basic structure of an MRep Relation (see Fig. 2).

```
<mreprelation>
   <attributes>
      <general_atts>
         <source_ids>atkis.349;</source_ids>
         <target_ids>gdf.826;gdf.827; </target_ids>
         <cardinality>1 : 2</cardinality>
         <total_similarity>90,56</total_similarity>
      </general_atts>
      <semantic_atts>
         <source_classes>Lane;</source_classes>
         <target_classes>Intersection;Road;</target_classes>
      </semantic_atts>
      <geometric_atts>
         <length_difference>13.84</length_difference>
         <avg_line_distance>8.83</avg_line_distance>
         <hausdorff_distance>11.45</hausdorff_distance>
            ...
      </geometric_atts>
      <topologic_atts>...</topologic_atts>...
   </attributes>
</mreprelation>
```

Fig. 2. Part of an MRRL instance file displaying some structural elements of an MRep Relation

3.2 Generating MRep Relations for Real Data

For two structurally similar test areas in the inner city of Stuttgart/Germany, each about 4 square kilometers in size, street data sets of ATKIS and GDF were available in multiple representations. The ATKIS object instances either were associated with the classes "Strasse" ("Street"), "Fahrbahn" ("Lane") or "Weg" ("Way"), whereas the object instances of GDF consisted of three types of Road Elements: Road Elements which belonged to a complex feature of the object type "Road" (ROAD), Road Elements which belonged to a complex feature of the object type "Intersection" (ISEC), or Road Elements which did not belong to a complex class (RDEL).

In order to reduce the global geometric deviation between the data sets to be matched, a rubber-sheeting transformation was applied at first. For the acquisition of MRep Relations, a semi-automatic software, the so-called Relation Builder Toolbox, has been developed (see Fig. 3). It has been implemented within the framework of the publicly available Java-based GIS environment JUMP [12]. Using this tool, a human operator can manually select the instances of corresponding representations by analyzing mainly geometric and topological, however also a few thematic criteria (like road category, etc.; street names were not available for ATKIS data and thus could not be used). For the matching, the operator is provided with a set of rules that are necessary in order to achieve replicable results amongst different operators. One of these rules says that 1:1 matching pairs (MRep Relations) should be generated whenever possible, even when two structures consisting of multiple road segments are unambiguously the same and could be assigned to each other as a whole (see Fig. 4.1, case 1). If no equivalent segments can be clearly identified, the representations have to be extended as long as a match is recognizable, even though one representation might be topologically split (see Fig. 4.1, case 2). MRep Relations also have to be built in cases where the geometric deviation is rather strong but the topological situation speaks for equivalence (see Fig. 4.1, case 3). The rules for the operator also comprise specifications on how to handle different matching cases: Fig. 4.2 first shows the normal case where *1 or n* road segments of data set *a* are matched to *1 or m* road segments of data set *b*. However, in the test data sets other matching types occurred: there is also a mixed case in which a node exists in one representation that has 3 incident edges, and additionally a separate case, where 2 topologically separate representations of data set *a* have to be assigned to only one corresponding representation in data set *b*. In the latter cases, an average line geometry has to be calculated from the source representations so that the geometric similarity measures can be derived.

It has to be pointed out that the rules for building MRep Relations can only hardly be formalized and rather have to be understood as guidelines since there are often ambiguous cases which can only be solved by the intellectual skills of the operator. If a clear formalization was possible, the development of automatic matching methods would be more successful.

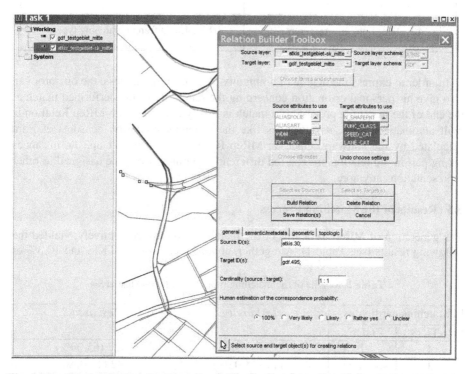

Fig. 3. The Relation Builder Toolbox allows selecting corresponding representations and automatically calculates MRep Relations between them

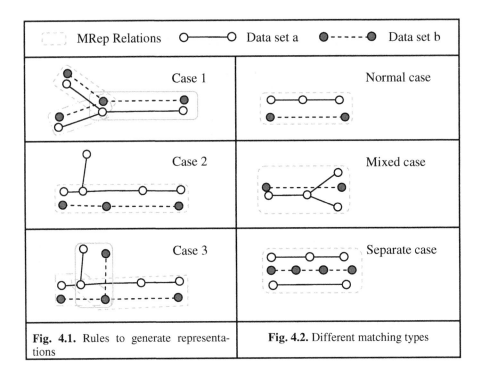

Fig. 4.1. Rules to generate representations

Fig. 4.2. Different matching types

In order to express the degree of ambiguity of a matching process, the operator can also give his personal estimation concerning the reliability of the performed match at the end of the matching process. If a match is finally confirmed, an MRep Relation is built automatically, i.e. its attributes like the different similarity measures etc. are calculated by the software. Once an MRep Relation has been set up, the instances taking part in the relation are colored differently and they cannot be assigned to other representations anymore.

3.3 Results of the Matching Process

The generation of MRep Relations or the matching process, respectively, yielded the following results (see Table 1): Altogether, about 11.1% of the ATKIS and 10.3% of

Table 1. Results of the matching process for the two test areas

Matching cardinality		*Test area 1*	*Test area 2*
ATKIS	GDF		
1	1	512 (69.28%)	427 (65.09%)
1	n	111 (15.02%)	92 (14.02%)
n	1	62 (8.39%)	72 (10.98%)
n	m	54 (7.31%)	65 (9.91%)
Total number of MRep Relations		739	656

the GDF instances in test area 1 and about 19.6% of the ATKIS and 17.2% of the GDF instances in test area 2 did not have a counterpart in the other data set. Consequently, for these instances no MRep Relations could be built. All other instances could take part in an MRep Relation. There were 739 MRep Relations for test area 1 and 656 for test area 2. The statistics about the cardinalities of the matches are depicted in Table 1: about two thirds had a cardinality of 1:1, approximately 25% were 1:n, and the rest were n:m.

4 Deriving Schema Correspondences

The basic assumption of the schema matching approach is that if instances of one data set belonging to a certain object class in schema a are mainly assigned to instances of another data set belonging to a certain object class in schema b, we do have clear evidence for a semantic correlation of the respective object classes of the different schemas (see Fig. 5).

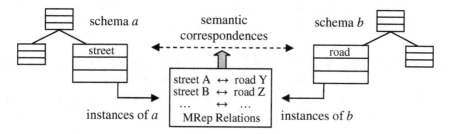

Fig. 5. Basic idea of the schema matching approach: by analyzing the affiliations of corresponding instances to object classes in their source schemas, semantic correlations between the respective object classes can be deduced

In the following sections, the approach is explained in detail. First, the different types of class correspondences which can be deduced from MRep Relations are illustrated. Then, the process of calculating correlation measures between object classes of different schemas based on the analysis of MRep Relations is explained. Finally, the results of the data-driven schema matching approach are presented and discussed.

4.1 Categories for Class Correspondences Within MRep Relations

Since a digital representation of a real world object can consist of one or more individual instances, it is possible that it is not only composed of instances belonging to only one object class but of instances of different classes. For this reason, we have to differentiate three categories of possible class correspondences within MRep Relations (see Fig. 6).

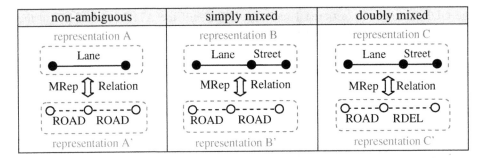

Fig. 6. The three categories for possible class correspondences within MRep Relations

If we have two corresponding representations each only composed of instances belonging to exactly one class, the conclusion regarding the matching of object classes is *non-ambiguous*. However, if the instances of one representation originate from different object classes of their source schema whereas the other representation is homogeneous, this case implies ambiguities and is called *simply mixed* according to our terminology. The last category *doubly mixed* describes the situation where both representations are constituted by instances which stem from heterogeneous object classes.

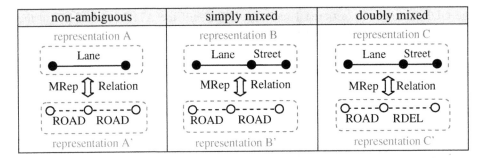

Fig. 7. Class combinations within MRep Relations for test area 2

Fig. 7 shows all class combinations of the different correspondence categories which appeared in test area 2 along with the cardinalities of the MRep Relations. For example, it can be seen (see area included by the dashed line) that within 429 MRep Relations, Street (Strasse) objects of ATKIS could be unambiguously assigned to GDF Road Elements. 333 of the 429 MRep Relations had a cardinality of 1:1, 42 of 1:n (1 Street : n Road Elements), etc.

4.2 Calculating Correlation Measures

Having determined the class combinations within MRep Relations of the two test areas, the correlation measures between the object classes of ATKIS and GDF can be derived. The calculation process consists of two steps:

1. First, all MRep Relations in which instances of one class take part have to be considered. The total number of these MRep Relations (e.g. 636 for test area 1 regarding instances of the GDF Road Element class, see Table 2 below) is taken as 100%. Then, the proportion of each class combination type with respect to this total number is determined. For example, the combination Road Element→Way (Weg) occurs 67 times in test area 1. This corresponds to 10.53% of all 636 class combinations of Road Element.

2. In the second step, all class correspondences of the mixed types have to be recalculated in order to derive only unambiguous correlation values for the object classes. Here, a simple approximation was used assuming an equal distribution within the different class combinations. Thus, for instance, the class combinations between {Road Element→Street/Way} which could be found 31 times in both test areas altogether (see Table 2) were counted in equal shares (15,5 each) for the unambiguous class combinations {Road Element → Street} and {Road Element →Way}. This method lead to the results shown in Table 3 for the Road Element class of GDF.

Table 2. Results of the first step in calculating correlation measures between object classes regarding the Road Element class of GDF

Correspondences of Road Element	Test area 1	Test area 2	Sum
Road Element → Street	526 (82.7%)	429 (75.26%)	955 (79.19%)
Road Element → Way	67 (10.53%)	107 (18.77%)	174 (14.43%)
Road Element → Lane	15 (2.36%)	5 (0.88%)	20 (1.66%)
Road Element → Lane/Street	1 (0.16%)	-	1 (0.08%)
Road Element → Street/Way	18 (2.83%)	13 (2.28%)	31 (2.57%)
Road Element/Intersection→Street	4 (0.63%)	5 (0.88%)	9 (0.75%)
Road Element/Intersection→Lane	1 (0.16%)	-	1 (0.08%)
Road Element/Road → Street	-	4 (0.7%)	4 (0.33%)
Road Element/Road → Lane	3 (0.47%)	4 (0.7%)	7 (0.58%)
Road Element/Road→Lane/Street	1 (0.16%)	3 (0.53%)	4 (0.33%)
Total number	636	570	1206

Table 3. Results of the second step: unambiguous correlation measures for the Road Element object class of GDF

Correlations of Road Element	Total values
Road Element → Street	978.5 (81.98%)
Road Element →Way	189.5 (15.88%)
Road Element → Lane	25.5 (2.14%)
Total number	1193.5

4.3 Final Schema Matching Results

If both steps for the calculation of semantic correlation measures described in the previous section have been carried out analogously for all object classes under consideration, we receive the final result of the data-driven schema matching (see Fig. 8).

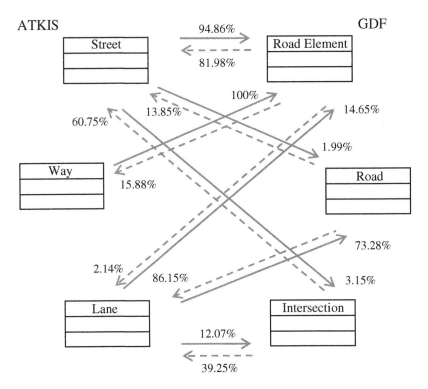

Fig. 8. Correlations of ATKIS classes (——➤) and GDF classes (- -➤) as derived by the data-driven schema matching (all MRep Relations considered)

As it can be seen, some significant inter-schema correspondences could be detected. For example, all ATKIS Way representations were only assigned to representations of Road Elements. Applying a set-based terminology here, we can say that the Way object class is semantically included in the Road Element class (Road Element ⊇ Way). Furthermore, no Road or Intersection representations have been matched with Way objects and vice versa. Thus, these object classes can be considered as being disjoint (Intersection ≠ Way, Road ≠ Way). For all other correlations semantic overlaps can be observed (e.g. Road ∩ Lane, Road Element ∩ Street) with the degree of overlap varying from very high (like Street→Road Element) to very low (like Street→Road). Obviously, no equivalence relations of the type $Class_{ATKIS} \equiv Class_{GDF}$ could be detected between the investigated object classes. For human operators, the results of the schema matching approach seem to be reasonable, i.e. knowing the semantics of the schemas or the object classes investigated, respectively, the correspondences found can basically be confirmed.

In a second variant of the data-driven schema matching approach, only those MRep Relations were analyzed which showed a total similarity value larger than 80, i.e. only the matchings on the instance level which were highly reliable have been incorporated in the calculation process. Thereby it should be investigated if the correlations on the schema level are more distinct if the correspondences between multiple representations are clearer. Actually, this assumption could be validated, as the results of Fig. 9 show.

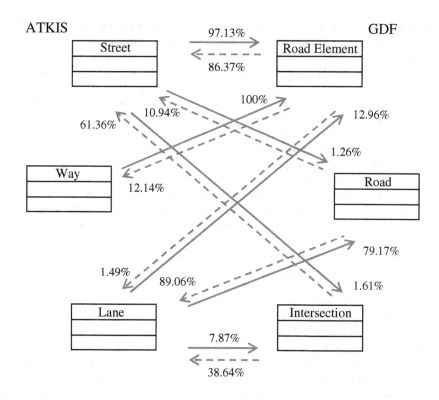

Fig. 9. Correlations of ATKIS classes (⟶) and GDF classes (- -⟶) as derived by the data-driven schema matching (only MRep Relations with a total similarity > 80 considered)

5 Conclusion and Outlook

The goal of this research was to find semantic correspondences of elements belonging to different spatial schemas, namely object classes from GDF and ATKIS. For this purpose, explicit relations, so-called MRep Relations, were designed in order to connect corresponding representations on the instance level. They facilitate the description of the consistencies and inconsistencies that exist between multiple representations of geospatial objects. However, MRep Relations can up to now only be built for points or linear objects and they are only applicable to features of approximately the same scale. It is an open question if solutions for the multi-scale problem can also be achieved by the MRep Relation approach, but generally geometric and topological differences become the stronger the more the scales differ and thus the detection of correspondences might be even more difficult.

For two test areas, MRep Relations have been derived for street data sets of ATKIS and GDF. By analyzing the affiliation of the instances of corresponding representations to object classes of their source schema, values reflecting the correlation between object classes on the schema level could finally be deduced. One advantage of the proposed approach is that the schema matching can be carried out fully automatically if MRep Relations exist. Another merit results from the fact that no knowledge about the schemas to be matched is required within the process since it is only data-driven.

In the near future, the proposed approach shall also be applied to detect correspondences between different attributes and between schema element combinations, i.e. schema structures. To achieve further progress, we are aiming at a combination of multiple schema matchers as it was proposed by [7], leading to a refinement of the matching process. It is also planned to derive MRep Relations in a more automatic way using matching algorithms similar to the ones applied by [18] in order to reduce human interaction during the process.

The results of the data-driven schema matching shall particularly enhance schema integration procedures. For this reason, methods have to be found that exploit the information delivered by the proposed schema matching process for the generation of a global schema. Regarding object oriented data models, an approach called upward inheritance is well suited and can be applied on top of the data-driven schema matching method. It takes inter-schema correspondences (or assertions) between object classes from different schemas – as they are produced by the presented approach – as an input. All object classes of the schemas to be integrated including their subclass relations are then transferred into the global schema. If there is an equivalence relation between two classes of the different schemas, they are merged into one class within the global schema. If one is included in the other, they are represented as superclass (including class) and subclass (included class) in the federated schema. Finally, when they are disjoint or when they overlap a new, common superclass is introduced in the global schema from which the disjoint or overlapping classes are derived (see [5]). Besides their application for schema integration, the results of the schema matching approach can – according to [14] – also be used for schema mapping, i.e. for the process of transforming data from a source schema into a target schema.

Furthermore, MRep Relations are not merely restricted to schema matching applications, but they can also be exploited to perform an analysis of data sets containing multiple representations. First attempts have been made in this direction, trying to achieve a shortest path analysis on street networks in multiple representations. The basic scenario here is that we have two network graphs *a* and *b* which are partially overlapping. We now want to find the shortest path from a starting point in graph *a* (that has *no* corresponding representation in graph *b*) to a target point in graph *b* (that has again *no* counterpart in graph *a*). If we take into account the MRep Relations which have been created for the representations in the overlapping area of the graphs, a change from one graph to another can be realized.

Additionally, MRep Relations can be seen as a basis for the update of corresponding representations within a federated database system. Assuming that an object which is taking part in one or more MRep Relation(s) is changed, using backward references from the changed object to the MRep Relation(s) that belong(s) to it, all corresponding representations available within the federated system can be detected and updated. Using the information stored in an MRep Relation, hints about the algorithm that has to be applied in order to perform the update correctly can be deduced, e.g. if we have a high geometric similarity of two corresponding representations, the algorithm could just assign the new geometry of the altered object as the new geometry of the object to be updated. Similarly, the content of an MRep Relation can also be analyzed for the conflation (merging) of corresponding, multiple representations into one single, "superior" representation, e.g. if the deviation of the position accuracy of two representations is low, the geometry of the resulting representation could just be the "mean" (average) geometry.

Acknowledgements

The research presented here is supported within a Center of Excellence called "Spatial World Models for Mobile Context-Aware Applications" under grant SFB 627 by the German Research Council. The test data have kindly been provided by the NAVTEQ company and the state survey office of the federal state of Baden-Wuerttemberg.

References

1. Bofinger, J.M.: Analyse und Implementierung eines Verfahrens zur Referenzierung geographischer Objekte, Diploma Thesis at the Institute for Photogrammetry, University of Stuttgart, unpublished, (2001), 76 p.
2. Bilenko, M., Mooney, R.J.: Employing Trainable String Similarity Metrics for Information Integration. In: Proceedings of the IJCAI-2003 Workshop on Information Integration on the Web, Mexico, (2003), pp. 67-72.
3. Bishr, Y. A., Pundt, H. Rüther, C.: Proceeding on the road of semantic interoperability - design of a Semantic Mapper based on a case study from transportation. In: Včkovski, A., Brassel, K.E., Schek, H.-J. (eds.): Proceedings of the 2nd International Conference on Interoperating Geographic Information Systems, Zurich, Lecture Notes in Computer Science, Heidelberg, Berlin, (1999), pp. 203-215.
4. Cobb, M., Chung, M., Miller, V., Foley, H., Petry, F., Shaw, K.: A rule-based approach for the conflation of attributed vector data, GeoInformatica 2(1), (1998), pp. 7-35.
5. Comité Européen de Normalisation (CEN): Geographic Data Files Version 3.0, TC 287, (1995), GDF for Road Traffic and Transport Telematics.
6. Conrad, S.: Schemaintegration – Integrationskonflikte, Lösungsansätze, aktuelle Herausforderungen, Informatik – Forschung & Entwicklung, 17, (2002), pp. 101-111.
7. Do, H.H., Rahm, E.: COMA – A system for flexible combination of schema matching approaches. In: Proceedings of the 28th Intl. Conference on Very Large Databases (VLDB) Hongkong, http://www.vldb.org/conf/2002/S17P03.pdf, (2002), 12 p.
8. Fonseca, F., Egenhofer, M., Agouris, P. and Câmara, G.: Using Ontologies for Integrated Geographic Information Systems, *Transactions in GIS* 6(3), (2002), pp. 231-257.
9. Guarino, N.: Formal ontology and information systems. In: Guarino, N. (edt.): Proceedings of the 1st International Conference on Formal Ontologies in Information Systems (FOIS), Trento, Italy, (1998), pp. 3-15.
10. Hakimpour, F., Timpf, S.: Using ontologies for resolution of semantic heterogeneity in GIS. In: Proc. of the 4th AGILE Conference on Geographic Information Science, Brno. http://www.ifi.unizh.ch/dbtg/Projects/MIGI/publication/agile2001.pdf, (2001), 12 pages.
11. Hohl, F., Kubach, U., Leonhardi, A., Rothermel, K., Schwehm, M.: Nexus – an open global infrastructure for spatial aware applications. In: Proceedings of the 5th International Conference on Mobile Computing and Networking, Seattle, USA, (1999), pp. 249-255.
12. JUMP, 2005. http://www.jump-project.org/

13. Kuhn, W.: Semantic Reference Systems. Int. J. Geographical Information Science 17(5), (2003), pp. 405–409.
14. Rahm, E., Bernstein, P.A.: A survey of approaches to automatic schema matching. VLDB Journal, Vol. 10, No. 4, (2001), pp. 334-350.
15. Uitermark, H.: The integration of geographic databases. Realising geodata interoperability through the hypermap metaphor and a mediator architecture. In: Rumor, M., McMillan, R., Ottens, H.F. (eds.): Proceedings of the 2nd Joint European Conference & Exhibition on Geographical Information (JEC-GI) '96, Vol. I, Barcelona, (1996), pp. 92-95.
16. Uitermark, H., Vogels, A., van Oosterom, P.: Semantic and geometric aspects of integrating road networks. In: Včkovski, A., Brassel, K.E., Schek, H.-J. (eds.): Proceedings of the 2nd International Conference on Interoperating Geographic Information Systems, Zurich, Lecture Notes in Computer Science, Springer-Verlag, Heidelberg, Berlin, (1999), pp. 177-188.
17. Volz, S., Walter, V.: Linking different geospatial databases by explicit relations. In: Proceedings of the XXth ISPRS Congress, Comm. IV, Istanbul, Turkey, pp. 152-157. (2004), pp. 445–473.
18. Walter, V., Fritsch, D.: Matching spatial data sets: a statistical approach. Int. J. Geographical Information Science 13(5), (1999), pp. 445–473.
19. Weis, M., Naumann, F.: Detecting duplicate objects in XML documents. In: Proceedings of the SIGMOD International Workshop on Information Quality in Information Systems (IQIS) '04, Paris, (2004), pp. 10-19.
20. Worboys, M.F., Duckham, M.: Integrating spatio-thematic information. Geographic Information Science – second international conference GIScience 2002. In: Egenhofer, M., Mark, D. (eds.), Lecture Notes in Computer Science 2478, (2002), pp. 346-361.
21. Working Committee of the Surveying Authorities of the States of the Federal Republic of Germany: Authoritative Topographic Cartographic Information System (ATKIS), (1988), http://www.atkis.de

The Role of Spatial Relations in Automating the Semantic Annotation of Geodata

Eva Klien and Michael Lutz

Institute for Geoinformatics (IfGI),
University of Münster,
Robert-Koch-Str. 26-28, 48149 Münster, Germany
{klien, m.lutz}@uni-muenster.de

Abstract. How can the usability of distributed and heterogeneous geographic data sets be enhanced? Semantic interoperability is a prerequisite for effectively finding and accessing relevant data in different application contexts. By using geospatial domain ontologies and semantic annotations of geodata based on these ontologies semantic interoperability can be achieved. However, since no automated methods for the semantic annotation of geodata exist this remains a laborious task, which data providers are neither willing nor capable to perform. In this paper we propose a method for automating the annotation process based on spatial relations. At the domain level, spatial relations play an important role for defining and identifying geospatial concepts. At the data level, spatial relations may be expressed through spatial processing methods, as we can calculate relations like topology, direction or distance between two spatial entities. We show how this potential can be exploited for automating the semantic annotation of geodata. The approach is illustrated by introducing a case study for annotating data containing representations of floodplains.

1 Introduction

Distributed and heterogeneous geographic data sets have a great potential for applications ranging from environmental planning to emergency management or e-commerce. However, even though syntactical standards for Spatial Data Infrastructures (SDIs) already enable the retrieval and multiple exploitation of geodata (cf. [1]), still many problems impede efficient usability. Being able to assess semantic interoperability is a precondition for effectively finding and accessing relevant data in different application contexts. One of the shortcomings of current SDIs is the missing support for this assessment.

An important means for achieving semantic interoperability are *ontologies*, which capture consensual knowledge and formalize this knowledge in a machine-interpretable way [2]. In SDIs, ontologies can be employed for making the semantics of the information content of geospatial web services explicit. In [3, 4] we have shown how ontologies can be used to realise semantic matchmaking during service discovery and retrieval. The backbone of our approach is an infrastructure of geospatial domain ontologies and semantic annotations of the geodata. Domain ontologies represent the basic concepts and relations to which all members of an information

A.G. Cohn and D.M. Mark (Eds.): COSIT 2005, LNCS 3693, pp. 133–148, 2005.
© Springer-Verlag Berlin Heidelberg 2005

community commit. They provide the foundation on which the geodata is semantically annotated. The common commitment ensures semantic interoperability [5].

So far, no automated method for the semantic annotation of geodata exists. Manual annotation is difficult, time consuming, and expensive and data providers who are no ontology engineering specialists will be neither willing nor capable to perform it. We propose a method for automating the annotation process that relies on the specific characteristics of geographic information.

Spatial relations between entities are characteristic for geographic information and they are often as important as the entities themselves [6]. In geospatial domain ontologies, taxonomic and non-taxonomic relations are used to define concepts of the physical world and to differentiate between them. At this level, spatial relations play an important role for defining and identifying spatial concepts, but when reasoning about these concepts, spatial relations are not treated differently from other non-taxonomic relations. At the data level though, the spatial relations may be expressed through spatial processing methods, as we can calculate e.g. the topology, direction or distance between two spatial entities. In this paper we illustrate how this potential can be exploited for the semi-automatic semantic annotation of geodata.

In this paper we focus on spatial *relations*. In our future work, the approach will be extended to include spatial *attributes* and non-spatial characteristics. The approach is illustrated by introducing an example ("annotating data containing floodplains") and consists of the following steps:

- Extract all concept definitions from the geospatial domain ontology that contain spatial relations.
- Translate spatial relations (e.g. *adjacent to*) into a corresponding spatial analysis method, which is implemented as a sequence of GIS operations.
- Apply these spatial analyses on the geodataset to be annotated.
- Identify sets of spatial entities that share a characteristic set of relations and can then be referenced to the corresponding geospatial concept.

The remainder of the paper is structured as follows. We first introduce a motivating example to clarify in what context we refer to semantic annotation and why we think a method for automating the process is needed (section 2). In section 3, spatial relations are discussed with respect to their role in the semantic annotation process. In section 4, we illustrate the general idea by conducting a walk-through the annotation process and continue with explaining the proposed method in detail. Section 5 provides an overview on related work. Finally, we discuss the approach and identify some future work (section 6).

2 Motivation: Ontology-Based Discovery and Retrieval of GI

Ontologies can be applied for making the semantics of the information content of geospatial web services explicit in order to enhance geographic information (GI) discovery and retrieval in geospatial web service environments [4].

In the following we introduce an example to illustrate our approach for solving semantic heterogeneity problems in SDIs and to describe the semantic matchmaking mechanism which underlies the ontology-based discovery and retrieval.

2.1 Example "Floodplain"

Floodplains are crucial elements for the task of flood management. They serve as a natural water retention area after a river broke its banks during a flooding event. If sufficient area along the river banks has the function of a floodplain, some of the river's water load will "naturally" be absorbed and the flooding event will be less critical for populated areas that lie further downstream.

Floodplains can be looked at from several different perspectives: "To define a floodplain depends somewhat on the goals in mind. As a topographic category it is quite flat and lies adjacent to a stream; geomorphologically, it is a landform composed primarily of unconsolidated depositional material derived from sediments being transported by the related stream; hydrologically, it is best defined as a landform subject to periodic flooding by a parent stream. A combination of these [characteristics] perhaps comprises the essential criteria for defining the floodplain"[7].

We will use this definition for formalizing the concept of a *floodplain* in our geospatial domain ontology. It is important to note, that the relation *lies adjacent to* is interpreted in the sense of "near or close to but not necessarily touching" (WordNet 2.0[1]).

2.2 Semantic Heterogeneity Problems

In order to avoid future flooding disasters, the planning department of a city council has decided to identify potential areas in the district that may be re-designated as floodplains. The task of John, the planner in charge, is a) first to find data that contains the relevant information (*discovery*) and b) to access this data and retrieve the information (*retrieval*).

In current standards-based catalogues users can formulate queries using keywords and/or spatial filters. The metadata fields that can be included in the query depend on the metadata schema used (e.g. ISO 19115) and on the query functionality of the service that is used for accessing the metadata. Even though natural language processing techniques (e.g. [8]) can increase the semantic relevance of search results with respect to the search request, keyword-based techniques are inherently restricted by the ambiguities of natural language. As a result, keyword-based search can have low recall if different terminology is used and/or low precision if terms are homonymous or because of their limited possibilities to express complex queries [9].

For example, if John uses "floodplains" as a keyword he may fail to find existing Web Feature Services (WFS) that offer information on floodplains, because their metadata description uses a different terminology. Furthermore, he might also discover data sources that are annotated with this keyword but not appropriate for answering his purposes, e.g. a service providing areas that are officially appointed and protected for having the function of a floodplain according to national legislation. Another obstacle often encountered is missing metadata entries. In that case a successful search will not be possible at all.

Once John has discovered a dataset and wants to access it via its WFS interface, he faces yet another major difficulty. The *DescribeFeatureType* request [10] returns the application schema for the feature type, which is essential for formulating a query filter. John now runs into trouble if the property names are not intuitively interpret-

---

[1] http://wordnet.princeton.edu/cgi-bin/webwn/

able or if the feature type "floodplain" is not explicitly stored in the schema. In our example, it might be sufficient to offer John a natural language description for each property. However, our work is aiming at automating the process of discovery and retrieval and this makes a machine-interpretable description of the properties indispensable.

2.3 Semantic Matchmaking

Fig. 1 illustrates the matchmaking which underlies ontology-based discovery and retrieval. The geospatial ontology contains the basic terms of a domain (e.g. geomorphology). It is assumed that all actors within a domain share a common understanding of the concepts and relations provided at the domain level [5]. The information sources, i.e. the geodata are annotated based on the concepts and relations provided in the geospatial ontologies. In our example the information source is a geodataset that contains polygons with land use attributes. John, the user of geospatial web services, is looking for information sources that will answer his question. His query for "lowlands adjacent to a river that are subject to flooding" is formulated based on a geospatial ontology.

The semantic annotations of the geodata available are created in the same way as John's query and stored in a catalogue. Thus, John's query concept becomes machine-comparable to all geodata descriptions in this catalogue.

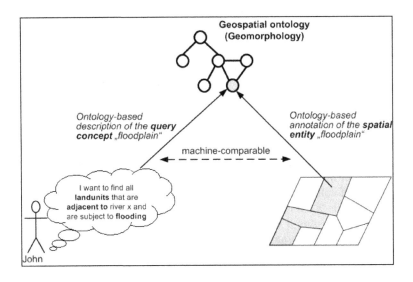

Fig. 1. Ontology-based discovery of geospatial data

As shown in [3] the integration of the matchmaking capability into SDIs overcomes some of the semantic heterogeneity problems in service discovery and thus leads to increased recall and precision.

However, in order for this approach to become widely accepted in the GI community it is essential to provide methods and tools that support the user in creating semantic descriptions. So far, no automated method for the semantic annotation of geodata exists. It remains a laborious task and data providers who are no ontology engineering specialists will neither be willing nor capable to perform it.

3 Spatial Relations

In the geospatial domain, relations among spatial entities are often as important as the entities themselves [6]. For example, for a farmer it is crucial to know that a planned plantation is on lowland *adjacent to* a river. This implies that the plantation will probably be covered by rising water once in a while and the farmer is well advised to choose plants that may cope with these conditions. This makes the representation and processing of spatial relations crucial in geographical applications.

Spatial relations have been classified in all possible various ways. We refer to a classification provided in [11], where spatial relations are classified according to their characteristic behavior in space.. *Topological relations* refer to properties like connectivity, adjacency and intersection among geospatial entities. They stay invariant under consistent topological transformation, such as rotation, translation, and scaling. *Direction relations* deal with order in space (e.g. north, east, south, and west). They are based on the existence of a vector space and, therefore, are subject to change under rotation, while invariant under translation and scaling of the reference frame. The third major type of spatial relations is *distance relation*. They refer to the geographical distances among geospatial objects (e.g. *A is close to B, X is very far from Y*). They reflect the concept of metric, thus change under scaling but stay invariant under translation and rotation.

Inherent to all these relations is the vagueness and imprecision in natural language expressions. Moreover, terminology and semantics of the relations varies across application domains [12]. Consider for example the spatial relation *adjacent* taken from our floodplain example. According to WordNet 2.0, *adjacent* has three slightly different senses:

(1) nearest in space or position; immediately adjoining without intervening space;
(2) having a common boundary or edge; touching;
(3) near or close to but not necessarily touching;

For the interpretation of a floodplain as defined in section 2.1, the first two senses are not applicable. Floodplains do not have to touch a river directly as long as the intervening space does not prevent the rising water from flooding the area.

Attempts to capture the semantics of spatial relations have been undertaken from both, the cognitive [13-15] and the mathematical viewpoint [16-19]. Still, there have only been few attempts to link the formal models of spatial relations developed for GIS with people's intuitive understanding of spatial relations as expressed in natural language [11]. Nevertheless, well-defined primitive operators like the Egenhofer operators for topological relations [20] may be used as the backbone for the definition of terms used in GIS and spatial query languages (cf. [21]).

Spatial relations are usually not explicitly stored together with geographic objects but have to be inferred from the objects' geometry [12]. By extracting them, hidden information in geospatial data becomes explicit. Depending on the application domain, some spatial relations may be more significant than others for identifying relevant implicit information. For the enterprise of using spatial relations for identifying concept characteristics in datasets, it will eventually be necessary to decide on a core set of relations for the geospatial domain of discourse.

In the next section we describe how we want to exploit the potential of spatial relations in order to identify characteristic concept information for the semi-automated annotation of geodata.

4   A Method for Automating Semantic Annotation

In this section we present a method for automating the semantic annotation of geographic datasets within a specific application domain. We first introduce the general idea of using spatial analysis methods that are associated with spatial relations in ontologies to derive annotations for datasets (section 4.1). We then use the case study of annotating data containing floodplains to illustrate the different steps that eventually lead to the semi-automated creation of semantic annotations (section 4.2). In the remaining subsections each of the building blocks of the suggested methodology is presented in detail. These building blocks are:

- a geospatial ontology that defines spatial concepts based on their spatial relations and attributes (section 4.3);
- a method for associating the characteristic spatial relations in this ontology with spatial analysis methods (section 4.4); and
- a reference dataset that is needed to calculate relationships between its well-known reference entities and the unknown ones in the dataset to be annotated (section 4.5).

In this paper we concentrate on the role of spatial *relations* to introduce the fundamental idea of our approach. We are aware, that in order to arrive at reasonable results, the approach will have to be extended to include other (non-spatial) relations and (spatial and non-spatial) *attributes*, e.g. geometry, shape or extent, in the analysis. This will be part of our future work.

4.1   Using Spatial Analyses for Creating Annotations

When reasoning about concept definitions that are (partly) based on spatial relations (e.g. to infer subsumption relationships between them), the spatial relations are not treated differently from other non-taxonomic relations. They simply represent implicit domain knowledge about what it means to be an instance of that concept. This is illustrated in Table 1, where the non-taxonomic relations a) "adjacentTo" and b) "owner" produce the same behaviour for subsumption reasoning.

When dealing with concrete datasets, however, this implicit knowledge can be compared with the inferred characteristics from the objects' geometry, and the results of this comparison can be used for annotation. This requires that each type of spatial relation that has been identified on the domain level is associated with a spatial analysis method (see Fig. 2 for an example). This method provides a formal definition of the semantics of the spatial relation. Note, that this definition is particular for the chosen domain, because the interpretation of the relation can differ significantly depending on the viewpoint (e.g. *adjacent* in section 3). A more detailed description on how spatial relations are associated with spatial analysis methods is given in section 4.4.

Table 1. Two DL inferences, using spatial (a) and non-spatial relations (b)

(a)	Boathouse ≡ House ⊓ ∃ adjacentTo.Waterbody	
	RiverBoathouse ≡ House ⊓ ∃ adjacentTo.River	⇒ RiverBoathouse ⊑ Boathouse
	River ⊑ Waterbody	
(b)	Palace ≡ House ⊓ ∃ owner.Nobleman	
	RoyalPalace ≡ House ⊓ ∃ owner.King	⇒ RoyalPalace ⊑ Palace
	King ⊑ Nobleman	

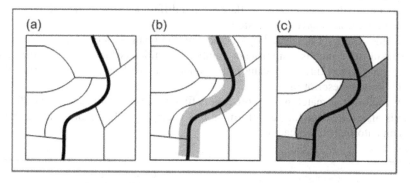

Fig. 2. Using a spatial analysis method associated with the spatial relation adjacent to (a river) to annotate the dataset shown in (a): In (b) a buffer is generated, in (c) the features intersecting the buffer are selected as being "adjacent to the river"

4.2 Walk Through for the "Floodplain" Example

Jane is working at a company that produces thematic datasets for all kind of geographic issues. The company owns a large database of geographic information and wants to make this commercially available for more customers via a geospatial web services environment. The semantic annotation for a specific domain view will consist of the following steps. Before the automated process starts, Jane has to select the domain of discourse. In our example, Jane wants to annotate her data for the geomorphology domain. The annotation procedure then consists of the following steps (Fig. 3):

1. All concept definitions that contain spatial relations are identified in the geomorphology ontology. From each of these concept definitions, the characteristic spatial relations are extracted. This extraction (and the subsequent analysis) is "controlled", in the sense that the system will analyse the dataset by looking explicitly for the concepts defined in the ontology (rather than performing an "uncontrolled"

search for arbitrary patterns in the dataset). The process can be depicted as a decision tree, e.g. the system identifies a land unit L as a floodplain if L fulfils the following criteria:

- L is adjacent to a river
- L is flat
- L is at most 2 m higher than the adjacent river

2. For each spatial relation, the corresponding spatial analysis method will be extracted. For example, *adjacent* is implemented as a sequence of GIS operations (section 4.4).
3. The GIS operations are applied to the geodataset to be annotated (*AnnoDS*). In order to be able to calculate the relation R between two entities x and y (e.g. "x is adjacent to some river") a reference dataset (*RefDS*) with the well-known geometry of y (e.g. all rivers) is required.
4. The spatial entities that meet the characteristic spatial relations of floodplains are stored as the result of this analysis step.
5. Steps 2-4 are repeated for other characteristics that define the analysed concept. In our example, this means that the flatness and difference in altitude compared to the adjacent river also have to be tested.
6. The final result set is created by intersecting the result sets of all analysis steps. If this result set contains a significant number of entities (which is greater than a certain user-defined threshold value), the geodata set will be annotated with "floodplains" in the description.
7. The result of the matchmaking process is finally presented to Jane for verification. She is also asked for further information if necessary. The ontological description is then automatically created.

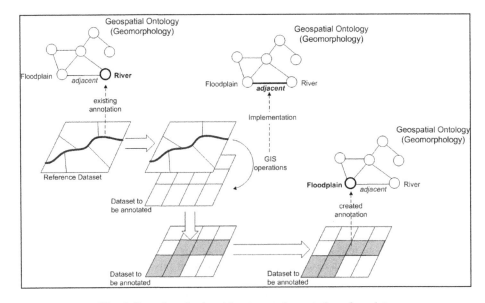

Fig. 3. Procedure for (semi-) automated annotation of geodata

4.3 Defining Geospatial Concepts Based on Characteristic Spatial Relations

What is special about geospatial concepts on the domain level? They describe geographic entities, i.e. entities that are associated with a geometry and a location relative to the earth. In consequence, geospatial concepts stand in high complex relationships to underlying physical reality and these relations may serve to define concepts and distinguish between sub-concepts on the domain level. For example, *floodplain* might be seen as a subconcept of *meadow*. One of the characteristics that distinguish *floodplains* from a *meadow* relies in the spatial relation of lying *adjacent* to a river. Such characteristic relations have to be identified by a domain expert during the ontology modelling process. For extracting the spatial characteristics of floodplains, we adopt the definition of a floodplain introduced in section 2.1. In the following, we illustrate how these can be used for formally defining the *floodplain* concept (1).

$$\forall x \, (\text{Floodplain}(x) \Leftrightarrow \text{Landunit}(x) \wedge \text{HasSlope}(x, \text{Flat}) \wedge \exists y \, [\text{River}(y) \wedge$$
$$\text{Adjacent}(x, y) \wedge \text{lessThan}(\text{Difference}(\text{Altitude}(x), \text{Altitude}(y)), 2)]) \tag{1}$$

Formula (1) states that all floodplains are landunits that are flat and adjacent to a river, and whose altitude does not differ by more than 2 meters from that of the adjacent river (and that all such landunits are floodplains). For the presented method it is important that all relations and attributes used in this definition can be inferred from the dataset using some kind of spatial analysis method. Therefore, only the defining *spatial* features of floodplains but not their *non-spatial characteristics* (e.g. being subject to periodic flooding, being composed of sediments) are taken into account at this stage. However, the non-spatial characteristics are an important part of the concept definitions as well. They will be subject to different kinds of analyses in future work in order to enhance the annotation results.

Some representational difficulties arise from the definition given in (1). The floodplain's "lowness" can only be described with respect to the adjacent river. Such concept interdependencies are often crucial for describing the semantics of a concept. For example, we distinguish between different spatial entities due to physical processes that lead to observable distinctions in the landscape (we can observe that some areas adjacent to a river are flooded and some are not – depending on their altitude compared to the river). For our approach, this means that the representation language chosen must be expressive enough for describing these kinds of concept interdependencies.

Formal concept definitions like (1) constitute the ontological knowledge at the domain level. Each relation could now be implemented with an analysis method that may be applied on the geodata. In the scope of this paper and to illustrate the idea we concentrate on the spatial relation *adjacent*. Nevertheless, an exhaustive classification algorithm for the concept *floodplain* would require the implementation of all relations used in its definition.

4.4 Associating Spatial Analysis Methods with Spatial Relations

The association of relations in the ontology with spatial analysis methods can be done in different ways. An analysis method can be associated as a "black box" containing the spatial relation it implements. This has the advantage that the description of the method and thus the association with the spatial relation in the ontology is simple. However, this also means that the implementation of the spatial relation is not transparent to a service provider like Jane. It can be assumed that there are always a number of different possible implementations for one spatial relation, especially if one takes into account more "fuzzy" relations like *side by side* instead of only precisely defined ones like *meet* [20].

We therefore reject the possibility to represent the implementation as a "black box" in favour of representing the analysis methods based on primitive operations defined in the ISO 19100 series of standards and specifications of the Open Geospatial Consortium (OGC). This strategy provides flexibility for adjusting the semantics of spatial relations in different application domains by changing the underlying implementation (e.g. *adjacent* could also be implemented using other primitives). Moreover, this implementation, and thus the semantic interpretation of a spatial relation remain transparent to the user.

Table 2. Example for representing the implementation of a spatial analysis method (RefDS represents the reference dataset, and AnnoDS the dataset to be annotated)

Algorithm for implementing "adjacent to some X"	Reference Standards and Specifications
select all features from *RefDS* where featureType = X	Web Feature Service
create empty set *A*	
for each selected feature *f*	
A.add(f.geometry.buffer(d : Distance))	ISO 19107, ISO 19109
create empty set *B*	
for each feature *g* in *A*	
for each feature *h* in *AnnoDS*	
if (*g*.geometry.intersects(*h*.geometry))	ISO 19107, ISO 19109
B.add(*h*)	
return *B*	

Table 2 shows an informal representation of an algorithm containing the spatial analysis steps involved in implementing *adjacent*, which is based on the following specifications and standards:

- The *Web Feature Service (WFS) Implementation Specification* [10] defines the *GetFeature* operation for selecting features of a particular feature type.
- In *ISO 19109 "Rules for Application Schema"* [22] states that the geometric characteristics of a feature are described by one or more *spatial attributes* whose values are given by a geometric object (GM_Object) or a topological object (TP_Object). We introduce the *geometry* attribute to refer to the geometric representation of a feature.

- *ISO 19107 "Spatial Schema"* [21] defines the notions of GM_Object and TP_Object and a number of operations that can be applied to them. In our example, we use the *buffer* operation, which returns a buffer polygon, and the *intersects* operation, which returns a boolean value to indicate whether two geometries intersect.

4.5 Annotating a Reference Dataset

We have argued that spatial relations are especially useful for extracting implicit information from geodatasets and that their characteristics make them a perfect candidate for the extraction of implicit information from geodata. However, if a concept definition is based on a spatial relation to another geographic feature *of a particular type,* it is necessary to first identify these related features. For example, for calculating a relation like "adjacent to a *river"* the river features in the dataset (or in a different dataset covering the same spatial extent) have to be known. This can be achieved by providing *reference datasets* that have already been annotated. That is, the *river features* in the reference dataset would already be associated with the *river concept* of the domain ontology. One interesting question in this context is to what extent the provision of reference datasets could be substituted by a recursive process (e.g. for calculating the floodplain's characteristic spatial relation "is adjacent to river", the system would first have to identify rivers in a dataset by calculating *their* spatial characteristics and so on).

We propose to introduce a reference dataset for each geospatial domain ontology. The role of a reference dataset in our example can be fulfilled by the national topographic map, e.g. ATKIS (Amtliches Topographisch-Kartographisches Informationssystem)[2] in Germany. Other reference datasets needed may include a Digital Terrain Model (DTM) for calculating e.g. slope and altitude of unknown spatial entities.

5   Related Work

In this section, we relate the presented approach for semi-automatic annotation of geodata to existing work in the area of spatial data mining and to other approaches for automatic annotation.

5.1 Spatial Data Mining

Spatial data mining is the process of discovering interesting and previously unknown, but potentially useful patterns from spatial databases [23-25]. We incorporate a similar strategy for automatically extracting relevant information from a geospatial database. But, instead of "mining" the dataset for potentially interesting patterns we define the spatial constraints a priori in the geospatial concepts of our domain ontology. These spatial constraints are then implemented into a supervised analysis process that aims to identify a specific concept (and not some previously unknown, potentially useful pattern). Evidently our approach requires far less complex techniques than those applied in Spatial Data Mining.

---

[2] http://www.adv-online.de

5.2 Automatic Annotation

With the emergence of the Semantic Web, the creation of semantic metadata by annotating documents has become a major concern in the community [26]. Several approaches are concerned with automating the process of semantically annotating information for the Semantic Web [27, 28].

Some research in this area has focused on the idea that spatial characteristics play a central role for effectively supporting information retrieval and annotation. In [29], Manov *et al.* demonstrate how their annotation platform can be extended by using spatial knowledge in conjunction with information extraction. In their approach, integrated gazetteers (like the Alexandria Digital Library Gazetteer) provide the additional spatial knowledge. At the MINDSWAP group, Hiramatsu and Reitsma [30] have worked on a geographic ontology to circulate geographically referenced information on the Semantic Web. Their idea is to associate georeferenced data (instead of using some gazetteer) to any other non-spatial information related to the geographic feature, i.e. they want to make use of the inherent characteristic spatial relations in order to add semantics to hypertext coded information. Similar work is done in the SPIRIT project, where knowledge stored in geographical data is made usable in an internet search engine [31]. While these approaches aim at annotating hypertext or enable spatially enhanced internet search, our work provides a method for the semantic annotation of geodata in order to enable its ontology-based discovery and retrieval through web services.

Also related to our work is the automatic extraction of "classical" metadata (like ISO 19115) from geographic data [32]. We believe that the method presented in this paper will not only be useful for semantically annotating but also for populating missing entries in the standard metadata documents for geographic information.

6 Discussion and Future Work

We propose a method for automating the semantic annotation of geodata based on spatial relations and suggest to apply spatial analysis methods in order to extract information on spatial relations useful for annotation. Compared to knowledge extraction techniques like string-based attribute analysis, the calculation of spatial relations remains independent from the textual description of geographic features and their properties. This has the advantage that semantic heterogeneity problems inherent in the processing of natural language descriptions are avoided. In this paper we have concentrated on the role of spatial relations. Taking into account the different types of spatial relations, topological relations seem to be especially useful as they stay invariant under transformations. This is a valuable characteristic since we have to deal with heterogeneous data sources in a variety of formats, scales, and projections.

A crucial issue not yet decided on is the choice of representation language for the ontological knowledge. As has been illustrated in section 4.3, the representation language must be expressive enough for describing concept interdependencies, i.e. a floodplain's "lowness" can only be described with respect to the adjacent river. In Description logic (DL), which has been used in our previous work on ontology-based GI discovery and retrieval [3, 4], "it is impossible to describe classes whose instances are related to another anonymous individual via different property paths" [33]. Thus, current DL-based ontology languages like OWL [34] are not applicable. First-order-logic (FOL) provides the expressivity needed in our approach. However, while reasoning in DL is decidable and therefore guaranteed to terminate, proving entailment in FOL is only semi-decidable [35]. Therefore, the final decision on a representation language will have to take into account the tradeoff between expressivity of the languages and the complexity of their reasoning problems.

In section 4.4 we have described our strategy on how to define a spatial analysis method that implements a single spatial relation by combining well-defined primitive operators. The question of how to explicitly associate a spatial analysis method to a spatial relation remains. A possible approach is outlined in [36], suggesting to integrate the invocation of executable programs into static ontological knowledge. Likewise, the strategy on how to generate and formalise the analysis algorithm for the entire concept definition remains an open issue. For this task a workflow description is needed, that is applicable for representing a decision tree as outlined in section 4.2. For this, we will examine current workflow description languages like BPEL (Business Process Execution Language) and PSL (Process Specification Language).

Apart from the benefits for automating the process of annotating geodata, our approach might also contribute to enhance retrieval capabilities in geospatial web service environments. For example: A user wants to retrieve "all motorbike roads", with motorbike roads being a concept in the domain ontology. The geometric characteristic of a motorbike road, i.e. its high twisting grade, is associated with a spatial analysis method. Thus, the system can apply this spatial analysis method for the on-the-fly retrieval of motorbike roads from a street network. There is no need for finding a dataset that explicitly stores motorbike roads.

Another application area for which the approach might be beneficial is the automation of metadata population for standard metadata like ISO 19115. In the ISO case, fields that might be filled with information extracted by the semantic annotation process are the following: descriptiveKeywords, topicCategory, geographicBox, geographicDescription [37].

In our future work we will extend our approach by providing implementation strategies for spatial attributes, like geometry, extent, and shape. Spatial attributes probably have an equally high potential for identifying characteristic information in geodata as spatial relations. For example, the analysis of the "straightness" of a water course might identify an entity as channel rather than river. However, the applicability of such an analysis highly depends on the resolution of the geodata. If all watercourse geometries are generalized in straight lines, the straightness attribute is of no value for information extraction. These dependencies on representation, resolution, and scale of the data have to be taken into account.

At this stage, we assume that the analysis of spatial characteristics will be the core methodology for identifying characteristic concept information in geodata. However, the analysis of spatial characteristics will probably not suffice for describing the semantics at the conceptual level in many cases (or not be applicable at all). Consequently, besides taking spatial attributes into account, we will also refine the presented approach by combining spatial analyses with the analysis of non-spatial attributes. This combination will eventually lead to reasonable results in the annotation process.

The vagueness of the specification of spatial characteristics is also a crucial issue. Consider the formalisation of the floodplain concept we provided in section 4.3: How many meters away from the water body still counts as adjacent? What degree of flatness still counts as flat? How is the difference of altitude between floodplain and adjacent river determined? In our future work, we will have to consider vagueness of spatial relations when specifying the associated analysis methods.

The performance of the spatial query techniques have to be evaluated and, if necessary, optimized. Methods for spatial query optimisation have been discussed in [12, 38].

We are aware that a fully automated process is out of scope. Therefore, we plan to develop a user interface that guides the data provider through the annotation process.

Acknowledgements

We would like to thank Werner Kuhn and Florian Probst for their valuable input at various stages of this work. Our thanks also go to the anonymous referees for providing valuable comments that helped to improve the content of the paper. The work presented in this paper has been supported by the German Federal Ministry for Education and Research as part of the GEOTECHNOLOGIEN program (grant number 03F0369A) and can be referenced as publication no. GEOTECH-142.

References

1. OGC: OpenGIS Reference Model. Open GIS Consortium (2003)
2. Studer, R., Benjamins, V.R., Fensel, D.: Knowledge Engineering: Principles and Methods. Data and Knowledge Engineering. 25(1-2) (1998): 161-197
3. Klien, E., Lutz, M., Einspanier, U., Hübner, S.: An Architecture for Ontology-Based Discovery and Retrieval of Geographic Information. Presented at 7th Conference on Geographic Information Science (AGILE 2004). Heraklion, Greece. (2004)
4. Lutz, M., Klien, E.: Ontology-Based Retrieval of Geographic Information. International Journal of Geographical Information Science (IJGIS), forthcoming
5. Wache, H., Vögele, T., Visser, U., Stuckenschmidt, H., Schuster, G., Neumann, H., Hübner, S.: Ontology-Based Integration of Information — A Survey of Existing Approaches. Presented at IJCAI-01 Workshop: Ontologies and Information Sharing. Seattle, WA. (2001)
6. Papadias, D., Kavouras, M.: Acquiring, Representing and Processing Spatial Relations. Presented at Sixth International Symposium on Spatial Data Handling. Edinburgh, Scotland. (1994)
7. Schmudde, T.H.: Floodplains. In: Fairbridge, R.W. (ed.): The Encyclopedia of Geomorphology. New York. (1968) 359-362
8. Richardson, R., Smeaton, A.F.: Using WordNet in a Knowledge-based Approach to Information Retrieval (Technical Report CA-0395). Dublin City University: Dublin, Ireland (1995)
9. Bernstein, A., Klein, M.: Towards High-Precision Service Retrieval. Presented at The Semantic Web - First International Semantic Web Conference (ISWC 2002). Sardinia, Italy. (2002)
10. OGC: Web Feature Service Implementation Specification. Open GIS Consortium (2002)

11. Shariff, A., Egenhofer, M., Mark, D.: Natural-Language Spatial Relations between Linear and Areal Objects: the Topology and Metric of English-Language Terms. International Journal of Geographical Information Science. 12(3) (1998): 215-246

12. Clementini, E., Sharma, J., Egenhofer, M.: Modeling Topological Spatial Relations: Strategies for Query Processing. Computers and Graphics. 18(6) (1994): 815 - 822

13. Herskovitz, A.: Language and Spatial Cognition. In: Joshi, A. (ed.): Studies in Natural Language Processing. Cambridge University Press: Cambridge. (1986)

14. Mark, D., Svorou, S., Zubin, D.: Spatial terms and spatial concepts: geographic, cognitive, and linguistic perspectives. Presented at International Geographic Information Systems (IGIS) Symposium. Arlington, VA, USA. (1987)

15. Talmy, L.: How Language Structures Space. In: Pick, H. ,Acredols, L. (eds.): Spatial Orientation. Theory, Research and Application. Plenum: New York. (1983): 225-282

16. Egenhofer, M., Herring, J.R.: A Mathematical Framework for the Definition of Topological Relationships. Presented at the 4th International Symposium on Spatial Data Handling. Zurich, Switzerland. (1990)

17. Frank, A.U.: Qualitative Spatial Reasoning about Distances and Directions in Geographic Space. Journal of Visual Languages and Computing. 3 (1992): 343-371

18. Cohn, A.G.: A Hierarchical Representation of Qualitative Shape Based on Connection and Convexity. In: Frank, A.U., Kuhn, W. (eds.): Spatial Information Theory-A Theoretical Basis for GIS. Springer, Berlin-Heidelberg-New York. (1995): 311-326

19. Papadias, D., Sellis, T.: The Semantics of Relations in 2D Space Using Representative Points: Spatial Indexes. In: Frank, A.U. ,Campari, I. (eds.): Spatial Information Theory - Theoretical Basis for GIS. Springer Verlag, Heidelberg-Berlin. (1993): 234-247

20. Egenhofer, M.: Reasoning about Binary Topological Relations. Presented at Advances in Spatial Databases, 2nd International Symposium. Zurich. (1991)

21. ISO: ISO 19107 - Spatial Schema. ISO TC 211 (2002)

22. ISO/TC-211: Text for DIS 19109 Geographic information - Rules for application schema Vs. 2.0. Draft Version. International Organization for Standardization. (2001)

23. Shekhar, S., Zhang, P., Huang, Y., Vatsavai, R.: Trends in Spatial Data Mining In: Kargupta, H., et al. (eds.): Data Mining: Next Generation Challenges and Future Directions. AAAI Press. (2004): 357 - 380

24. Koperski, K., Han, J., Adhikary, J.: Mining Knowledge in Geographical Data. Communications of the ACM. 26(1) (1998): 65 - 74

25. Roddick, J., Lees, B.G.: Paradigms for spatial and spatio-temporal data mining. In: Miller, H. ,Han, J. (eds.): Geographic Data Mining and Knowledge Discovery (2001)

26. Handschuh, S., Staab, S. (eds): Annotation for the Semantic Web. Frontiers in Artificial Intelligence and Applications. Vol. 96. IOS Press: Amsterdam, The Netherlands. (2003)

27. Handschuh, S., Staab, S., Ciravegna, F.: S-CREAM -- Semi Automatic Creation of Metadata. Presented at the Semantic Authoring, Annotation and Markup Workshop, 15th European Conference on Artificial Intelligence (ECAI02). Lyon, France. (2002)

28. Dingli, A., Ciravegna, F., Wilks, Y.: Automatic Semantic Annotation Using Unsupervised Information Extraction and Integration. Presented at the Knowledge Markup and Semantic Annotation Workshop at the Second International Conference on Knowledge Capture (K-CAP 2003). Sanibel, Florida, USA. (2003)

29. Manov, D., Kiryakov, A., Popov, B., Bontcheva, K., Maynard, D., Cunningham, H.: Experiments with geographic knowledge for information extraction. Presented at the NAACL-HLT 2003, Workshop on the Analysis of Geographic References. Edmonton, Alberta, Canada. (2003)

30. Hiramatsu, K., Reitsma, F.: GeoReferencing the Semantic Web: ontology based markup of geographically referenced information. Presented at the Joint EuroSDR/EuroGeographics workshop on Ontologies and Schema Translation Services. Paris, France. (2004)

31. Heinzle, F., Sester, M.: Derivation of Implicit Information from Spatial Data Sets with Data Mining. Presented at the XXth Congress of the International Society for Photogrammetry and Remote Sensing (ISPRS). Istanbul, Turkey. (2004)

32. Manso, M.A., Nogueras-Iso, J., Bernabe, M.A., Zarazaga-Soria, F.J.: Automatic Metadata Extraction from Geographic Information. Presented at the 7th Conference on Geographic Information Science (AGILE 2004). Heraklion, Greece. (2004)

33. Grosof, B., Horrocks, I., Volz, R., Decker, S.: Description Logic Programs: Combining Logic Programs with Description Logic. Presented at 12th Intl. Conf. on the World Wide Web (WWW-2003). Budapest, Hungary. (2003)

34. W3C: OWL Web Ontology Language Overview. (2004)

35. Russell, S., Norvig, P.: Artificial Intelligence - A Modern Approach. (2002)

36. Borchert, R.: How can a knowledge base run executables on the frame level? Presented at the International Protege Workshop. Manchester, England. (2003)

37. ISO/TC-211: ISO 19115:2003. Geographic information - Metadata. International Organization for Standardization. (2003)

38. Papadias, D., Theodorodis, Y.: Spatial relations, minimum bounding rectangles, and spatial data structures. International Journal of Geographical Information Systems. **11**(2) (1997): 111-138

Anatomical Information Science

Barry Smith[1,2], Jose L.V. Mejino Jr.[3],
Stefan Schulz[4], Anand Kumar[2], and Cornelius Rosse[3]

[1] Department of Philosophy, University at Buffalo, Buffalo, NY 14260, USA
[2] Institute for Formal Ontology and Medical Information Science,
Saarland University, Saarbrücken, Germany
[3] University of Washington, Seattle, USA
[4] Freiburg University Hospital, Freiburg, Germany
phismith@buffalo.edu
{mejino, rosse}@u.washington.edu
akumar@ifomis.uni-saarland.de
stschulz@uni-freiburg.de

Abstract. The Foundational Model of Anatomy (FMA) is a map of the human body. Like maps of other sorts – including the map-like representations we find in familiar anatomical atlases – it is a representation of a certain portion of spatial reality as it exists at a certain (idealized) instant of time. But unlike other maps, the FMA comes in the form of a sophisticated ontology of its object-domain, comprising some 1.5 million statements of anatomical relations among some 70,000 anatomical universals or kinds. It is further distinguished from other maps in that it represents not some specific portion of spatial reality (say: Leeds in 1996), but rather the generalized or idealized spatial reality associated with a generalized or idealized human being at some generalized or idealized instant of time. It will be our concern in what follows to outline the approach to ontology that is represented by the FMA and to argue that it can serve as the basis for a new type of anatomical information science. We also draw some implications for our understanding of spatial reasoning and spatial ontologies in general.

1 The Foundational Model of Anatomy

The Foundational Model of Anatomy (FMA) is a computer-based representation of the entities and relations which together form the phenotypic structure of the human organism [1,2]. It provides a qualitative spatial reference system for the human body that is designed to be understandable to human beings and also to be navigable by computers. It is intended as a general-purpose resource, which can be used by any biomedical application that requires anatomical information, from radiology (in supporting automatic image analysis) to pharmacokinetics (in representing the pathways of drugs as they are absorbed by, distributed through, metabolized in and excreted from the body).

The FMA began its life as a classification of anatomical entities called the University of Washington Digital Anatomist Vocabulary. In recent years it has grown from a list of terms linked by *is_a* and *part_of* relations to a sophisticated spatial-

A.G. Cohn and D.M. Mark (Eds.): COSIT 2005, LNCS 3693, pp. 149–164, 2005.
© Springer-Verlag Berlin Heidelberg 2005

structural ontology of the human organism at all biologically salient levels of granularity, comprehending some 1.5 million statements of ontological relations among some 70,000 anatomical universals. The acronym 'FMA' is currently used in the biomedical informatics community both for this ontology and also for its representation in computerized form within the Protégé 2000 frame-based ontology editing environment [2,3].

We shall argue in what follows that the FMA provides a starting-point for a new type of anatomical information science, representing a new application domain with potentially valuable implications also for other branches of Spatial Information Theory.

2 Types of Relations

The FMA relates exclusively to continuant entities (i.e. to entities, such as molecules, cells, lungs, which endure through time while undergoing changes of various sorts) [4]. The Structural Informatics Group at the University of Washington, which developed and maintains the FMA, has itself initiated work on two complementary ventures, called PRO and PathRO – for 'Physiology' and 'Pathology Reference Ontology', respectively [5] – which deal with those occurrent *processes* in which the anatomical entities at different levels of granularity participate. Here, however, we shall concern ourselves exclusively with continuant entities, which exist at the level of particulars or tokens (having determinate spatial locations at each specific point in time) as instantiations of certain corresponding universals or types (kinds, classes).

We can distinguish a number of distinct types of relations between continuant universals which are employed in the construction of an ontology like the FMA [3,6]:

1) *is_a* relations, linking one universal to another (more general) universal in a subsumption hierarchy; examples: *liver is_a organ*, *lacrimal lake is_a anatomical cavity*
2) static physical relations between continuant universals; examples: *lobe of liver part_of liver, nuclear membrane adjacent_to cytoplasm*
3) relations between universals instantiated at different stages in the development of an organism; examples: *zygote derives_from ovum*, *adult transformation_of child.*

Cross-cutting all of these are distinctions between:

a) instance-level relations (such as the parthood relation between your left thumb and your left hand), which obtain between instances of anatomical universals within the canonical organization of the human body;
b) relations involving also non-canonical anatomical instances including instances of pathological anatomical universals such as *wounded knee located_in leg, amputation stump connected_to arm*;
c) relations involving entities (*implants, food*, etc.) imported into the human body such as *shrapnel contained_in pleural cavity*;
d) relations involving entities (*biopsied samples, excreta*, etc.) exported from the human body.

The FMA itself focuses on relations of types 1) and 2) under heading a). In what follows we expand our scope to include also relations of other types, drawing on recent work, summarized in [7,8], involving not only the FMA's developers but also representatives of other influential research groups in biomedical ontology.

3 Canonical Anatomy

The term 'anatomy' is used to refer both to anatomical *science* and to that anatomical *structure* which this science describes, a certain ordered aggregate of material objects and physical spaces filled with substances (such as blood) which together constitute a biological organism [2]. In the case of the FMA the structure in question is what is called the 'canonical' structure of the adult human body, whereby the idea of canonicity (first proposed for the FMA in [9]) has no analogue in geospatial science. For where geospatial maps deal in every case with specific instances (with specific portions of the surface of the earth), the FMA deals not with the *instances*, the individual human beings, whose bodily organization has been investigated over the centuries with the aid of surgical dissection, radiological imaging and other techniques, but rather with a certain ('canonical') idealization thereof (actually with two idealizations, corresponding to the male and female adult human beings, respectively). The FMA, that is to say, is a collection of generalizations pertaining to the structure of the normal human body, generalizations which are deduced from qualitative observations and which have been refined and sanctioned by successive generations of anatomists and presented in textbooks and atlases of structural anatomy. One needs to take such an idealization as target in a venture like the FMA since the effort to do justice to anatomical structure in all its variants and instantiations would, in the absence of such an idealized reference frame, give rise to an endeavor of unmanageable complexity.

4 Boundaries

A further apparent distinction between the geospatial and anatomical domains from an ontological perspective turns on the fact that, where anatomy embraces within its purview primarily three-dimensional entities such as cells, organs and whole organisms, geospatial ontologies are focused on the (broadly) two-dimensional entities that form the surface of the earth. Applications of geospatial reasoning have thus far been correspondingly concerned for example with the movement of objects across this surface and with associated questions of land-use, soil-type, forest-coverage, and so forth. Closer inspection reveals, however, that both anatomical and geographic information sciences must deal with entities in all spatial dimensions. Thus the FMA deals not only with material objects but also with both fiat and bona fide boundaries of two, one and zero dimensions [10,11], and a geospatial ontology like that which underlies the Spatial Data Transfer Standard (SDTS) [12] comprehends not only two- (one- and zero-) but also three-dimensional universals (called 'entity types') such as *fumarole*, *grave*, *mount* and *trough*.

While the FMA deals primarily with material objects and their boundaries, it also deals with portions of body substances (e.g. of water, urine, or menstrual fluid) and with the body spaces (cavities, conduits) which these occupy [13,14]. GIScience deals similarly not only with material geographic objects such as mountains and forests, but also with non-material geographic objects (such as valleys and craters) having some of the features of containers or conduits. In an extended sense it deals also with the substances (above all portions of fresh and salt water) which occupy these.

Considerable progress has been made on the geospatial side, not only in the standardization of geospatial terminology, but also in the development of formal theories and tools for both quantitative and qualitative spatial reasoning [15,16,17,18], theories and tools which have since been applied in the biological domain [19,20]. In the geospatial case, the tools in question are applied primarily in reasoning about the fixed spatial regions on the surface of the earth with which spatial objects can be associated. Analogous tools for region-based reasoning are more difficult to develop and apply in the domain of anatomy because of the elasticity of the human body as contrasted with the earth as base reference object [4]. (It is an essential feature of the heart, for example, that it is constantly in the process of becoming spatially deformed.) On the other side, however, geospatial ontology is less advanced than anatomy in that it has nothing like the formal sophistication in its treatment of ontological relations, and nothing comparable to the coverage, in terms of systematicity and number of universals treated, that is manifested by the FMA. Thus the SDTS comprehends only some 200 entity types.

5 The Proper Treatment of Relations in Ontologies

We can conceive ontologies for present purposes as *controlled, structured vocabularies* designed to support the integration of data and information deriving from heterogeneous sources. An ontology like that of the FMA is structured through assertions of the form '*A relation B*' (where '*A*' and '*B*' are terms in the FMA vocabulary and '*relation*' stands in for '*part_of*' or some similar expression). Such assertions express general statements about the corresponding universals, which correspond to the sorts of statements found in scientific textbooks. To link such ontologies to reality, however, we need to take account not only of the universals described in scientific theories but also of the corresponding instances or tokens which we find about us in reality, and this means that we need to deal not only with the universal–universal relations commonly treated of in work in ontology, but also with instance–universal and instance–instance relations [7].

Thus for example the thesis according to which *lobe of liver part_of liver* – which expresses a universal-universal relation – gets its reference to reality in virtue of the fact that it is in part a thesis about instances, to the effect that:

> every canonical instance of the universal *lobe of liver* is a part (in the standard mereological, sense of 'part') of some instance of the universal *liver*.

Note the all–some structure of this assertion, which is copied also in parallel universal-universal assertions involving other spatial relations such as adjacency, attachment and continuity [5,21].

As already noted, one important distinction between geospatial and anatomical spatial reasoning turns on the different roles that universals and instances play in their respective map-like representations. On the geospatial side, universals are captured in the *legends* that we find in the corners of maps, legends which can themselves be seen as forming miniature ontologies in their own right. On the anatomical side, in contrast, universals play a central role, not least in virtue of the tremendous increases in our knowledge (for example under the auspices of the Human Genome Project) about the ways in which universals on the coarse-anatomical levels are connected to universals at level of finer grains, down to molecules. Thus where the question of which universals need to be distinguished by an anatomy ontology was once resolved by visual inspection, scientific anatomy rests increasingly on empirical research in the domain of genetics. The parts of the body demarcated on the basis of phenomenology will in the future be acknowledged as genuine anatomical structures only after it has been demonstrated that there are structural genes whose coordinated expression in the development of organisms of the corresponding types brought forth the relevant instances. Hence the FMA in its full version must contain a place also for developmental transformations.

6 Anatomical Entities

Four upper-level universals of the FMA are: *anatomical structure*, *anatomical substance*, *anatomical space*, and *anatomical boundary*.

Anatomical structures are *material* entities such as organisms, organs, cells, and biological macromolecules, which have their own inherent three-dimensional shapes.

Body substances are portions of blood, water, urine or cerebrospinal fluid, entities which inherit their three-dimensional shapes from whatever are the relevant containers. The portion of blood in your right ventricular cavity at some specific time has a shape which it inherits from the surrounding ventricle.

Body spaces are *immaterial* anatomical entities (cavities, orifices, conduits), whose shape, again, is inherited from the relevant surrounding anatomical structure. They are distinguished from spatial regions in that they are *parts of organisms*, which means that they move from one spatial region to another with the movements of their hosts.

Anatomical boundaries are distinguished from anatomical entities in the other three classes by the fact that they are of lower dimension, and stand in a relation of boundary dependence [22] upon some relevant anatomical structure or landmark [6].

7 Anatomical Structures

Mereotopologically speaking, anatomical structures are marked by the fact that they are maximally self-connected, which means that they have their own complete three-dimensional connected physical or bona fide boundaries. Virtually all anatomical structures, however, are connected to neighboring anatomical structures via conduits which link the anatomical spaces within them. To take account of this fact we require a reading of 'maximally self-connected' that allows for corresponding portions of fiat boundaries – we can think of these as punctures in their external surfaces whose areas are quite small in relation to the corresponding total boundary. ('Fiat boundary', here, signifies a boundary in a continuant entity which corresponds to no physical discontinuity or bona fide boundary in the entity itself, but rather to a delineation which is drawn by human beings. [10])

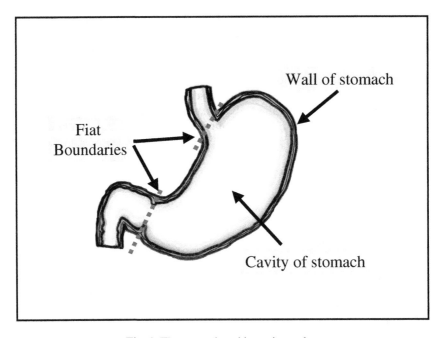

Fig. 1. The stomach and its major outlets

Which small portions of fiat boundaries we can ignore in specifying anatomical structures is not a trivial matter. Consider the small portions of fiat boundaries we need to allow in delimiting an anatomical structure such as a stomach or kidney (see Figs. 1 and 2). The stomach, we might think, would be an unproblematic fiat entity, because it is merely a segment of a certain tubular continuum which includes also for example the esophagus and the small and large intestines. In some cases we can find bona fide *landmarks*, specific changes within the mucosal and muscular layers which form the walls of the relevant cavities, for the drawing of the fiat boundaries which extend laterally across the relevant cavity. However, the cases of the stomach and esophagus and of the kidney and ureter show that not all anatomical structures, at those places on their surface where they are demarcated by fiat boundaries, are demarcated by fiat boundaries which are *landmarked* in this sense.

In the case of the kidney we have an anatomical structure that is separated from its surroundings largely by bona fide boundaries and only by small sections of fiat boundary, but in such a way that the fiat boundary in question is located in a non-fiat entity (the urinary tract) (Figure 2).

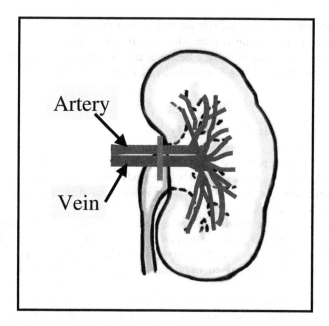

Fig. 2. The kidney with vertical bar designating a fiat boundary in relation to the arterial and venous systems

8 Fiat Boundaries and Partitions

Fiat boundaries come in two types: those which demarcate physical anatomical entities (for example *the plane of the esophagogastric junction*, which demarcates the esophagus from the stomach) and those which demarcate anatomical spaces (for example *the plane of the pelvic inlet*, which demarcates the abdominal from the pelvic cavity). Our talk of *'planes,'* here, draws attention to the fact that most anatomical fiat boundaries have geometrically regular shapes, just as is often the case in the geospatial realm (the borders of Colorado or Manitoba).

In addition to anatomical structures in the technical sense of the FMA, we can recognize also:

1. fiat *parts* of anatomical structures (for example the fundus of the stomach), which are not complete;
2. fiat *aggregates* of anatomical structures (for example the aggregate of the upper and lower limbs), which are not connected themselves.

The recognition of fiat entities of these sorts allows us to do justice to the fact that one and the same anatomical structure can be partitioned in different ways [23]. The stomach can be decomposed in one context into its fundus, body and pyloric antrum and in another into its wall and cavity. The FMA sees the former as a fiat partition into *regional parts*, the latter as what it calls a 'compositional partition' into *constitutional parts* [24]. While constitutional parts are genetically determined, regional parts – for example the loin or the epigatrium – are defined in part by arbitrary coordinates. Even if we remain with bona fide parts, however, then we need to acknowledge cross-cutting demarcations at different levels of granularity. Thus for example the bona fide boundary between the gray and the white matter of the brain is cross-cut by the bona fide boundaries of the neurons which pass between them.

9   Connectedness and Continuity

The body's component parts are intimately interconnected. Indeed, if we leave aside the cells floating free in blood and other body substances, then practically all anatomical structures are connected to other anatomical structures through different kinds of continuities or junctions.

The FMA analyzes the relation of connectedness in terms of three different kinds of relations: *continuous_with*, *attached_to* (e.g. of muscle to bone) and *synapsed_with* (of nerve to nerve and nerve to muscle – a special type of attachment relation, not here further discussed, obtaining at the level of granularity of axons and dendrites).

Two continuants are *continuous* on the instance level if and only if they share a fiat boundary. A continuant is *self-connected* if and only if any division of the entity into two parts yields parts which are continuous. The relation of continuity on the instance level is of course always symmetric. On the class level, however, this is not the case. To see why not, consider the relation between the lymph node and the lymphatic vessel. Each lymph node is continuous with some lymphatic vessel, but there are lymphatic vessels (e.g. lymphatic trunks such as the thoracic duct) which do not stand in continuous connection to any lymph nodes. We thus have *lymph node continuous_with lymphatic vessel* (because for every instance of the former there is some instance of the latter with which the former is continuous), but not *lymphatic vessel continuous_with lymph node.*

To understand the relation *attached_to*, consider the junction depicted macroscopically in Figure 3, which shows a bone and a muscle, the latter consisting of a tendon and a muscle belly, and (on a finer-grained level) of collagen fibers, muscle fibers and bone matrix. The bone itself is well delimited: it ends where the bone matrix ends. The same applies to the muscle fibers which, due to their contractile elements, are clearly demarcated from the tendon. But collagen fibers cross all of these boundaries. One fiber might overlap with the muscle fascia and the tendon, another with the tendon and the bone.

Attachment, too, is symmetrical on the instance level. On the class level, however, it is not in general a symmetrical connectivity relation. To see why not, consider the universals *placenta* and *uterus*. Every instance of the former is indeed attached to some instance of the latter; not, however, conversely.

While the corresponding instances have their own bona fide boundaries, the distal tendon comes into intimate contact with that circumscribed area of the bone where extensions of its collagen fiber bundles of the tendon (the Sharpey's fibers in Figure 3) penetrate the bone and intermingle with collagen fibers in the bone's own matrix. The tendon may thus be separated from the bone only by severing Sharpey's fibers.

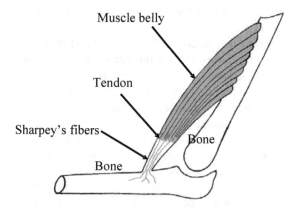

Fig. 3. An anatomical junction

10   Location and Containment

In addition to the relations of instantiation (between an instance and a universal) and parthood (between one continuant instance and another continuant instance), the FMA contains also a treatment of location. To understand location formally, we associate with the human body a collection of regions (*relative places* in Donnelly's terms [25]) and define a function which assigns to each anatomical entity *c* and time *t* the corresponding region r(*c, t*) which *c* exactly occupies at *t*. We can then define the relation of location for anatomical instances as follows: [26]

   c **located_in** *d* **at** *t* =def. r(*c, t*) **part_of** r(*d, t*) **at** *t*

where relations picked out in **bold** obtain at the instance level. On the level of universals we have:

   C located_in D =def for all *c, t*, if *c* **instance_of** *C* **at** *t* then there is some *d* such that *d* **instance_of** *D* **at** *t* and *c* **located_in** *d* **at** *t*.

Trivially, by this definition, all parts of anatomical structures are located in the corresponding wholes. But there is a second type of location relation distinguished in the FMA, which is that of *containment*. This holds between material anatomical entities (body substances and anatomical structures) and the anatomical spaces in which they are contained. For example: the right lung is contained in the right half of the thoracic cavity; a Ca^{++} ion is contained in an intracellular space in a heart muscle cell.

   If we move away from canonical anatomy, we encounter cases where, for example, a lobe of a liver is removed from a donor and transplanted into a hepatectomized patient. There is then an instance of the universal *lobe of liver* which is for a certain time not a part of any instance of *liver*. However, even when such non-canonical cases are taken into account, it still has to be true, at least under the conditions dictated by present technology, that every instance of *lobe of liver* stands in an instance-level parthood relation to some instance of *liver* at some time (more precisely still: every instance of *lobe of liver* stands at the beginning of its existence in such a relation to some instance of *liver*).

Two entities may thus be related in terms of parthood only during a certain phase of their simultaneous existence. It is for this reason that parthood relations between continuants on the instance level must be indexed by times [7]. This is not specific to living systems. For example, a screw can be part of an engine and it can then be substituted by another, replacement screw. In contrast to artifacts, however, biological objects are engaged in a constant exchange of matter with their environment, so that many parthood relationships at finer levels of granularity are short-lived. Moreover, the dynamic phenomena of matter exchange [27] indicate that there must be relations intermediate between parthood and containment, realized for example when you take a bite out of an apple (and the relevant portion of apple moves from being part of the apple to being contained in you oral cavity). Consider what happens when food is degraded in course of digestion into sugars, amino acids and fatty acids. Those portions of such substances in the lumen of the stomach are then *contained* therein, while those that have traversed the epithelium are successively *parts* of epithelial cells and then of blood or lymph.

11 Criteria of Parthood

An account of the processes in question must accordingly allow for the existence of transitions between containment and parthood. How, given the above, are we to distinguish genuine parthood from the relation of being merely spatially included within (i.e. from the relation **located_in** as defined above)? Is an embryo part of, or merely located in, a uterus? [28] Is a bolus of food part of, or merely located in, a digestive tract? Is an oxygen molecule part of, or merely located in, a lung? We here offer four kinds of criteria which may be of assistance in answering such questions (for further detail see [29]).

1. *Genetics*: The parts of the body should be of the same genetic origin as the body itself. Thus the embryo, on this criterion, is not a part of the body of the mother. This criterion faces problems for example in application to oxygen or nitrogen molecules in the body (since these do not have a genetic origin) or to the mitochondria found in nearly every cell of the body (which have their own DNA).
2. *Sortality*: If continuant c is part of continuant d, then c and d must be of the right sorts to make this possible (they must instantiate appropriate universals). Thus if d is an organism, then it is ruled out that a should be an artifact (e.g., a heart pacemaker, a bullet), or a second whole organism (a symbiont, parasite, prey, embryo or fetus).
3. *Life Cycle*: Unless this is already ruled out by sortal constraints, we can infer from the fact that c is located in d during the whole of the life cycle of d, that c is part of d (the right ventricle of the heart is for this reason part of the heart).
4. *Function*: We can infer parthood from location, finally, where an object c, located in a second object d, has a function whose exercise or performance is essential to d's survival or to the maintenance of d's proper functioning. The

functioning of the heart or of the brain is essential for the survival of the whole human body in this sense, where a given volume of urine is not essential to the survival of the bladder, and hence the urine is not counted as part of the bladder. This criterion faces problems for example in application to hair (is hair essential to the proper functioning of the human body?) or to the kidney and of those other organs supplied to the body in pairs [30].

12 Holes and Parts

A further family of problems connected with the location relation in anatomy turns on the fact that the boundaries of the objects with which we have to deal may themselves be difficult to specify. Many anatomical objects are sponge-like; they are replete with vessels, capillaries, cavities, holes and ducts of various sorts. This is true of all the body apart from the cornea and lens. A clear delimitation of an anatomical object, often including reference to a plurality of distinct levels of granularity, is therefore essential for making any assertion about location.

Is a small object such as a calculus located in a duct inside a gland also located in the gland itself, or, in the case when the lumen of the duct communicates directly with the exterior, located only in the exterior surrounding space? The answer to this question depends on whether we admit spaces as parts of anatomical objects in cases such as this.

The range of problematic borderline cases connected with surface structures is depicted in Figure 4. Of the white and gray volumes falling below the (rough) line of demarcation of the surface of the body in question, which are parts of the exterior of this body and which are parts of the body itself?

Analogous puzzles arise also in connection with spatial discontinuities. Accessory spleens such as are illustrated in stylized fashion in Figure 5 can be found in more than 10% of the population. This phenomenon can be accounted for in two ways, either by admitting one discontinuous object with three parts, or (with the FMA – which is generally averse to the admission of discontinous anatomical entities) by admitting three distinct objects which collaborate in the exercise of a certain function.

In sum, the specification of anatomical part, location and connection relations, and also of the degree of spatial overlap between anatomical structures, is often problematic because the relevant spatial extensions are difficult to delimit and because the relevant anatomical entities continuously lose and gain parts and continuously exchange matter with their environment. It will likely never be the case that we can formulate a criterion for parthood that can be guaranteed to yield a determinate answer to the question *is x a part of y?* in every single case. From this some might be tempted to conclude that the notion of parthood, at least in the biological domain, has an ineliminable element of indeterminacy. We, however, prefer to see the indeterminacy as lying rather in our partial knowledge of the relations between the corresponding entities in reality.

Fig. 4. Problematic surface structures

Fig. 5. Normal spleen (left); accessory spleens (right)

13 Non-canonical Anatomy

The FMA is a representation of canonical anatomy, which means that its individual variables range over those adult male and female human beings who satisfy the generalizations which appear in textbooks of structural anatomy and which conform to a pattern repeatedly observed by many generations of anatomists and surgeons over several centuries. By appealing to the device of specifying different ranges of variables, we can modify the scope of the FMA to represent generalizations belonging to the different branches of anatomy, for example to canonical human beings at various stages of embryological development, and even to organisms of other species. It can allow us also to represent the generalizations governing the anatomical variants yielded by the presence of, for example, coronary arteries or bronchopulmonary segments, which deviate from canonical anatomical patterns of organization in various well-understood ways. We here conclude with some brief and speculative remarks on the proper treatment of pathological structures within an ontology like the FMA.

While the universal *colon* as it appears in the FMA comprehends only instances of normal colons (however the term 'normal' is to be defined), if the FMA ontology is to serve as a basis also for *non-canonical* anatomy then it must have the facility to extend this range of instances to include also abnormal colons. The resulting framework of *pathological anatomy* might then include assertions such as

> *abnormal colon is_a colon*;

> *colon carcinoma pathological structure part_of abnormal colon.*

Moreover (recalling our 'all–some' reading of the class-level relation *part_of* above), we might have:

> *colon carcinoma pathological structure part_of colon.*

Every colon carcinoma is part of some colon, even though not every colon has some colon carcinoma as part.

We might have

> *abnormal colon transformation_of colon*

> *colon with carcinoma is_a abnormal colon*

> *colon with carcinoma transformation_of normal colon*

where *C transformation of D* is defined as obtaining whenever *C* and *D* are continuant universals which are such that every instance of *C* was also an instance of *D* at some earlier time in its existence [7].

14   Towards Anatomical Information Science

Leaving aside a number of abstract domain-independent treatments of spatial structures and relations, the primary focus of the discipline called 'Spatial Information Theory' has thus far been in the area of geospatial information. We believe however that Spatial Information Theory ought also to encompass the theory of spatial properties and relations in other domains, not least because – as we hope to have shown in the foregoing – the latter can introduce important phenomena of a kind not thus far considered in the literature.

Because of the special role of the canonical (idealized) human body, and because of the complementary special role of variant and pathological anatomic structures in anatomical information science, many features of the type of spatial information science realized in the FMA will be unfamiliar to those working on spatial representation and reasoning in the geographical domain. In cartographic terms, canonical anatomy would correspond to a map of an idealized portion of geographic reality (an idealized city, say, or an idealized lake or continent). Corresponding to actual maps (of actual cities or actual continents) is *instantiated anatomy*, which comprehends anatomical data about actual human beings of a type that might be recorded in a clinical record or captured in a radiographic image [9]. Instantiated anatomy deals with individual, living, human subjects, but in a way that relies on the categories or kinds depicted in canonical anatomy. Practically all of geography is instantiated geography and geospatial information science is in consequence characterized by the existence of a large mass of spatially referenced instance data and of powerful systems for reasoning with this data, combined with a treatment of the corresponding universals which is relatively impoverished from the theoretical point of view.

There is also a *normative* dimension of the discipline of canonical anatomy, which has no direct counterpart in the geospatial domain. For while there are healthy and unhealthy cities, it is not the case that all healthy cities have a more or less identical groundplan. Moreover, the geospatial domain has no counterpart of the contemporary evidence-based discipline of medicine, and thus no counterpart of its central organizing discipline of canonical anatomical science (and thus no scientific interest in, for example, maps of ideal cities). But for this reason, too, there is no counterpart of pathological anatomy in the domain of geospatial science, which is to say: no science of the determinate ways in which geospatial entities such as cities or lakes depart from some normative ('normal') case.

Thus the SDTS contains within its list of attributes no terms for what we might think of as disorders of its entity types, as contrasted with the 900,000 or so terms included in SNOMED-CT, the systematized nomenclature of clinical terms maintained by the College of American Pathologists [31], a large fraction of which refers to disorders. There is no counterpart, either, of the ways in which human anatomy can be related to the anatomy of other species as a basis for the detection of what may be medically relevant homologies [32].

The existence of the FMA means that anatomical information science rests on an impressive tool for the treatment of anatomical universals, even though both the associated instance data and the tools for reasoning with such instance data are still impoverished. Some progress is being made on the side of instance-level anatomical information science. Again, however, the problems of elasticity, movement, and growth of bodily organs present considerable obstacles to the development of corresponding tools for instance-based spatial reasoning [33]. With the development of genomics-based individualized medicine, and with associated increases in the sophistication of electronic health records and medical image analysis, we believe that the imbalance between class- and instance-based anatomical data will in the coming years be gradually resolved. Corresponding tools for representation and reasoning with anatomical instance data will thus increasingly be needed [34], and we can anticipate that the FMA will play an important role in their development by being used in tandem with some of the reasoning tools developed in recent years in the spatial domain.

Acknowledgments. This paper was written under the auspices of the Wolfgang Paul Program of the Alexander von Humboldt Foundation, the European Union Network of Excellence on Semantic Interoperability and Data Mining in Biomedicine and the Volkswagen Foundation under the auspices of the project "Forms of Life."

References

1. http://sig.biostr.washington.edu/projects/fm/
2. Rosse, C. and Mejino, J. L. V. Jr. A Reference Ontology for Bioinformatics: The Foundational Model of Anatomy. *J Biomed Informatics*, 2003; 36:478–500.

3. Noy, N. F., Mejino, J. L. V. Jr., Musen, M. A. and Rosse, C. Pushing the Envelope: Challenges in Frame-Based Representation of Human Anatomy. *Data and Knowledge Engineering* 48 (2004) 335–359.

4. Grenon, P. and Smith, B. SNAP and SPAN: Towards Dynamic Spatial Ontology, *Spatial Cognition and Computation*, 4: 1 (March 2004), 69–103.

5. Cook, D. L., Mejino, J. L. V. Jr. and Rosse, C. Evolution of a Foundational Model of Physiology: Symbolic Representation for Functional Bioinformatics. *Proceedings MedInfo* 2004: 336–340.

6. Mejino, J. L. V. and Rosse C. Symbolic modeling of structural relationships in the Foundational Model of Anatomy, *Proceedings of KR-MED* 2004 (First International Workshop on Formal Biomedical Knowledge Representation), 48–62.

7. Smith, B., Ceusters, W., Klagges, B., Köhler, J., Kumar, A., Lomax, J., Mungall, C., Neuhaus, F., Rector, A., Rosse, C. Relations in Biomedical Ontologies. *Genome Biology*, 2005, 6 (5), R46.

8. Donnelly, M., Bittner, T., and Rosse, C. A Formal Theory for Spatial Representation and Reasoning in Biomedical Ontologies. *IFOMIS Reprints* 2005.

9. Rosse, C., Mejino, J. L., Modayur, B. R., Jakobovits, R., Hinshaw, K. P., Brinkley, J. F. Motivation and Organizational Principles for Anatomical Knowledge Representation: The Digital Anatomist Symbolic Knowledge Base. *Journal of the American Medical Informatics Association* 1998; 5: 17–40.

10. Smith, B. On Drawing Lines on a Map, in Andrew U. Frank and Werner Kuhn (eds.), *Spatial Information Theory. A Theoretical Basis for GIS* (Lecture Notes in Computer Science 988), Berlin/Heidelberg/New York: Springer, 1995, 475–484.

11. Smith, B. and Varzi, A. C. Fiat and Bona Fide Boundaries, *Philosophy and Phenomenological Research*, 60: 2, March 2000, 401–420.

12. US Geological Survey Spatial Data Transfer Standard (SDTS) Information Site (http://mcmcweb.er.usgs.gov/sdts/).

13. Casati, R. and Varzi, A. C. *Holes and Other Superficialities*, Cambridge, MA: MIT Press 1994.

14. Donnelly, M. On Parts and Holes: The Spatial Structure of the Human Body, in: M. Fieschi, E. Coiera, and Y. J. Li (eds.) *Proceedings Medinfo* 2004, 351–356.

15. Cohn, A. G., Bennett, B., Gooday, J. and Gotts, N. Qualitative Spatial Representation and Reasoning with the Region Connection Calculus, *Geoinformatica* 1, 1997, 1–44.

16. Smith, B. Mereotopology: A Theory of Parts and Boundaries, *Data and Knowledge Engineering* 1996; 20: 287–303.

17. Donnelly, M. Layered Mereotopology. *Proc IJCAI* 2003, 1269–1274.

18. Casati, R. and Varzi, A. C. *Parts and Places: The Structures of Spatial Representation*, MIT Press, Cambridge, 1999.

19. Bittner, T. Axioms for Parthood and Containment Relations in Bio-Ontologies. *Proceedings of KR-MED* 2004 (First International Workshop on Formal Biomedical Knowledge Representation), 2004; 4–11.

20. Schulz, S. and Hahn, U. Mereotopological Reasoning about Parts and (W)holes in Bio-Ontologies, *Proceedings of FOIS*, New York: ACM Press, 2001: 198–209.

21. Smith, B. and Rosse, C. The Role of Foundational Relations in the Alignment of Biomedical Ontologies, *Proceedings of Medinfo* 2004; 444–448.

22. Smith, B. On Substances, Accidents and Universals. In Defence of a Constituent Ontology, *Philosophical Papers*, 27 (1997), 105–127.

23. Bittner, T. and Smith, B. A Theory of Granular Partitions, *Foundations of Geographic Information Science*, London: Taylor & Francis, 2003.

24. Mejino, J. L. V., Agoncillo, A.V., Rickard K. L. and Rosse C. Representing Complexity in Part-Whole Relationships within the Foundational Model of Anatomy, *Proceedings of AMIA Symp.* 2003; 450–454.
25. Donnelly, M. Relative Places. In A. Varzi and L. Vieu (eds.) Formal Ontology in Information Systems, *Proceedings of FOIS 2004*, Amsterdam: IOS Press, 249–260.
26. Donnelly, M. A Formal Theory for Reasoning about Parthood, Connection, and Location. *Artificial Intelligence* 160 (2004) 145–172.
27. Cohn, A. G. Formalizing Bio-Spatial Knowledge, *Proceedings of FOIS* 2001, ACM Press, New York, 2001, 198–209.
28. Smith, B. and Brogaard, B. Sixteen Days, *The Journal of Medicine and Philosophy*, 28 (2003), 45–78.
29. Schulz, S., Daumke, P., Smith, B., Hahn, U. How to Distinguish Parthood from Location in Bioontologies, *Proceedings of AMIA Symposium*, 2005, in press.
30. Johansson, I., Smith, B., Munn, K., Tsikolia, N., Elsner, K., Ernst, D. and Siebert, D. Functional Anatomy: A Taxonomic Proposal, *Acta Biotheoretica*, in press.
31. http://www.snomed.org
32. Travillian, R. S., Rosse, C. and Shapiro, L. G. An Approach to the Anatomical Correlation of Species through the Foundational Model of Anatomy. *Proceedings of AMIA Symposium*, 2003; 669–673.
33. Pilgram, R., Fritscher, K. D., Fletcher, P. T., and Schubert, R. Shape Modelling of the Multiobject Organ Heart, *IASTED: International Conference on Biomedical Enigineering – BioMED 2004*, Acta Press, 2004, 157–160.
34. Ceusters, W. and Smith, B. Tracking Referents in Electronic Healthcare Records, *Proceedings of Medical Informatics Europe*, 2005, in press.

Matching Names and Definitions
of Topological Operators

Catharina Riedemann

Institute for Geoinformatics, Robert-Koch-Str. 26-28,
48149 Münster, Germany
`riedemann@uni-muenster.de`

Abstract. In previous empirical work humans did not recognize the definition
of most topological operators for regions by their names in two geospatial information systems (GIS). This work differentiates the not corresponding definitions and comprises more topological terms in order to find better
term/definition matches. The main hypothesis is that the majority of acceptable
matches – defined as matches selected by the majority of subjects - are others
than those used in GIS. In an explorative questionnaire study, 34 native German
speaking subjects matched German topological terms to graphical depictions of
topological relations. The acceptability of matches was tested with the approximate binomial test (null hypothesis: 50 percent of the subjects or less select a match). Ten matches are acceptable (significance level 0.10 or higher);
only one of them appears in the GIS. This supports the main hypothesis and indicates how to revise topological operators in GIS.

1 Introduction

Topological operators, which analyze neighborhood relations between spatial objects,
are an essential part of geospatial information systems (GIS) [1]. Observations in student classes and personal experience revealed that users have problems identifying and
remembering what a particular operator exactly does. In the final examination of a
class the suitable ArcGIS[1] operator for selecting all streets that overlap a water protection area had to be chosen. The students opted for eight of nine available operators; less
than 40 percent chose the intended one. A possible reason is that the user interface
does not clearly convey the concepts implemented in the software. There are textual
and graphical operator depictions, the choice of which so far rests on experience and
intuition. Treating them separately helps locating problems. A previous empirical study
examined for two GIS products if people recognize the concepts of topological operators by their names [15]. Subjects matched topological terms to graphical depictions of
topological relations. The graphics they selected for a term were taken as an indicator
of the concept they linked with the term. Only the significant minority of subjects
matched the majority of GIS names with their GIS definitions.

The study comprised more topological terms than those appearing in the two GIS.
These are analyzed here in search of alternative names. The perspective is different
from that in [15]: instead of merely verifying if a subject's name/definition match

---

[1] ArcGIS 8.3 of Environmental System Research Institute (ESRI).

A.G. Cohn and D.M. Mark (Eds.): COSIT 2005, LNCS 3693, pp. 165–181, 2005.
© Springer-Verlag Berlin Heidelberg 2005

corresponds to a GIS name/definition match it is of interest now with which definition a term is matched most often. Therefore, the range of definitions must be extended beyond those used in the two GIS as well.

The hypothesis is that the majority of acceptable matches are others than those employed in GIS. Acceptable matches are defined as matches selected by the majority of people. For the individual matches the approach is explorative. While [15] provides information about GIS terms matching GIS definitions, for non-GIS matches there are no sound a priori hypotheses. Consequently, for all term/definition matches it is just assumed that they are selected by the majority of people. Significance is expected only for a very limited number of matches. They establish alternative names better conveying to users what the operators exactly do. Furthermore, they supply material for theoretical considerations why certain terms better match to certain definitions than others and prepare the ground for justified a priori hypotheses. As the author is a native speaker of German, German terms are studied in order to understand subtleties.

This work is restricted to relations between two regions. On the one hand, they are the simplest case in the formal model employed here for their unambiguous description. On the other hand, they may be regarded as ontological primitives [2]. This need not imply that humans do not have concepts of lines or points; it rather states that there is no doubt about regions.

The remainder of this article is structured as follows. After looking at further previous research, the method of this work is described. Then the results are presented and discussed, before concluding with open issues.

2   Related Work

First, the formal analysis of topological relations is briefly introduced in order to provide an unambiguous language for denoting them. Afterwards, topological operators of GIS are presented with the help of this language, before finally turning to the issue of naming entities at user interfaces.

2.1   Formal Analysis of Topological Relations

The two major formal theories of topological relations, the region connection calculus [13] and the intersection model [4], result in identical sets of relations between two regions. As the intersection model is more common in the GIS domain, it is used here.

The basic idea of the intersection model is to regard an object as consisting of an interior and a boundary. All four combinations of interiors and boundaries of two objects are looked at. Each combination can be in one of two states: the object parts can intersect or not. Of the 16 different configurations of empty and non-empty intersections only eight correspond to spatial configurations of regions for our case of two regions in two-dimensional space (see Fig. 1, (d) and (e) comprise two converse relations each). Egenhofer and Herring have presented graphical and verbal examples of these relations, making them easier to grasp (Fig. 1). The important point here is that these are examples, which means that there are other graphical as well as verbal variants that meet the specification of, for example, situation (b) as well. Think of the two

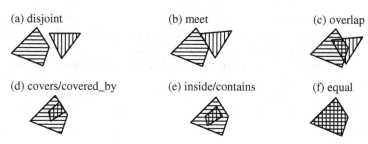

Fig. 1. Graphical and verbal examples of relations between two regions (figure adopted from Egenhofer and Herring [4])

objects meeting only at one point instead of along a line, and think of the term "touch" instead of "meet".

The intersection model has been extended to lines and points, which required introducing the exterior as an additional object part and lead to the nine intersection model [3]. This did not change anything for regions; the nine intersection model for them results in the same eight relations. Nevertheless, the extension is worth mentioning in this context. For a region and a line, 19 relations exist, for two lines even 33. While Egenhofer and Herring kept on drawing graphical examples, they refrained from giving names this time. Obviously, natural language does not provide enough handy terms for all situations. Although this problem does not seem to be of great relevance for the manageable amount of eight relations between to regions, it is an open question how handy the terms chosen for them really are. Maybe the apparent simplicity is deceptive.

2.2 Topological Operators in GIS

The two systems involved in the study, ArcGIS and GeoMedia[2], do not make available the precise definitions of their topological operators to users. Hence, test drawings containing examples of all four intersection relations had to be created to run the operators against them [15]. Table 1 shows the results; the operator definitions are expressed by marking with crosses the four intersection relations they identify. The operators are either atomic, i.e. they consist of a single relation, or they are a combination of several relations. Combinations are logical disjunctions, i.e. it suffices that a spatial situation represents one of the relations in a combination in order to belong to this combination[3]. In the following, the term "combination" will be used no matter how many relations are combined, i.e. there can also be atomic combinations with only one relation. A combination is an unambiguous definition of an operator. Combinations are represented by the numbers of the relations they combine, as they appear in Table 1. "c123", for example, denotes a combination of the relations "disjoint", "meet", and "overlap", and "c6" denotes the atomic combination "equal". Note that combination numbers have the prefix "c".

Although both systems have eight operators this does not imply identical sets. Six combinations occur in both GIS; whereas four combinations are not agreed upon

---

[2] GeoMedia Professional 5.1 of Intergraph Corporation.
[3] A spatial configuration of two regions cannot represent more than one relation.

across system boundaries. The combinations c8 respectively c568 are the inverse of the combinations c7 respectively c467 and can be regarded as not original. The number of matching original combinations thus amounts merely to four.

There is not only a difference in the combinations of operators, but also a difference in naming. Not taking into account the words "completely" and "entirely" with combinations c7 and c8, there are two identical combinations named differently (c6 and c2345678). It is an open question if this matters to users and, if yes, which term fits better.

Table 1. Definitions of topological operators in ArcGIS (A) and GeoMedia (G); the outlined circle is the subject, the filled circle is the object (combination c7, e.g., reads for ArcGIS: the outlined circle completely contains the filled circle)

Nine intersection relations	1	2	3	4	5	6	7	8
Operator combinations and names [4]								
c2 — A: - / G: meet		x						
c3 — A: are crossed by the outline of / G: -			x					
c6 — A: are identical to / G: are spatially equal						x		
c7 — A: completely contain / G: entirely contain							x	
c8 — A: are completely within / G: are entirely contained by								x
c467 — A: contain / G: contain				x		x	x	
c568 — A: are contained by / G: are contained by					x	x		x
c2456 — A: touch the boundary of / G: -		x		x	x	x		
c345678 — A: - / G: overlap			x	x	x	x	x	x
c2345678 — A: intersect / G: touch		x	x	x	x	x	x	x

2.3 Naming Topological Operators

Names in software have been studied in various application domains, for example text editing [6, 8, 9]. It was found that only 10-20 percent of unfamiliar users will hit the term chosen by a system designer when they have to enter a term to invoke a command [6]. The hit rate could be doubled by eliciting terms with real users and choosing the one applied most often as "best possible name". The risk of entering a term unknown to the system is today avoided by selecting commands out of given menus instead of invoking them by entering terms. Still, misunderstanding terms and thus selecting wrong commands cannot be prevented this way.

---

[4] The German names can be looked up in Table 2.

Human factors started getting wider attention in the field of GIS in the early 1990ies, which also included the vocabulary problem. Frank noted that "changing from a textual user interface to a menu driven one, or a graphical one, does not alone substantially improve the usability of a system if the command set is still using baroque terminology" [5, p. 13].

Human concepts of spatial relations were studied empirically. A series of tests done by Mark and Egenhofer examined the topological relations between a road and a park. The basic goal was to check the cognitive validity of the nine intersection model. In a grouping task subjects were given a number of graphics representing topological relations and asked to form groups of them [11]. In an agreement task subjects were given pairs of graphics and sentences and asked to which degree they agree that the sentence describes the topological relation depicted [10, 11]. In a drawing task subjects were given a sentence describing a topological relation and asked to draw an according situation [12]. The nine intersection model turned out to be a critical part of predicates in spatial language, perhaps the most important one. The agreement task showed that humans do not only group relations, but that these groups overlap. As the grouping task does not allow overlapping groups, it is not suitable for clarifying the meaning of topological terms. The constraint to put each graphics into exactly one group most likely produces unnatural groups [11]. In contrast to this, the agreement task is suitable for clarifying concepts. The drawing task again is problematic in this aspect. People draw what comes most easily to their mind. These are rather prototypes of the topological relations than less typical examples. Thus, it is not astonishing that the most frequently drawn spatial relations correspond to the relations that were most frequently selected as group prototypes in the grouping task [12]. The grouping task was applied by other authors to relations between two regions [7, 14]. These are the only studies the author is aware of that comprise larger samples of German subjects. Their natural language group descriptions could have been a source for deriving the lacking a priori hypotheses mentioned in the introduction, if they were published in the articles.

3   Method

The agreement task described in section 2.3 was adopted. Instead of the original rating scale with five options to determine a degree with which a subject agrees that a sentence describes a depicted situation, only the answers "yes" respectively "no" were allowed. One reason is that current GIS do not support fuzzy concepts. For a comparison between the operators appearing in GIS and the concepts revealed in this study it is favorable to demand the same crisp decision. Furthermore, a scale with an odd number of options allows making no decision, which was not desired.

48 German natural language sentences were compiled describing the topological relation between two regions A and B. The user interfaces of ArcGIS and GeoMedia were exploited, Mark's and Egenhofer's list of 64 road/park sentences of [12] was translated and adapted to the region/region situation, and five people were asked to produce as many terms for spatial relations as they could. The latter lists were reduced to pure topological terms by the author. The application neutral mathematical notion of two regions called "A" and "B" was used. Subjects could follow this mathematical notion or imagine any application context. A specific context was not relevant, because the focus was on "general purpose" GIS and not at specific applications. All

A ist in **B**   A schneidet **B**

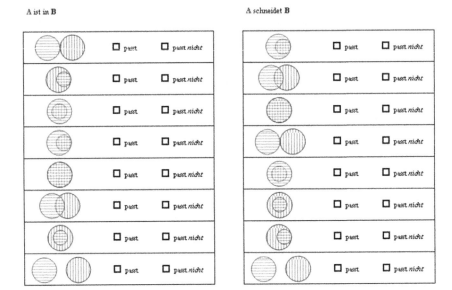

Fig. 2. Page of the questionnaire (translations in brackets not in the original)

sentences were formulated according to the schema "A *relation* B", e.g. "A *berührt* B" ("A *touches* B") and arranged in random order in a questionnaire. The same graphical examples of the eight nine-intersection relations were added to each sentence, in own random order for each sentence. Subjects had to mark for each graphical example if it fits to the sentence or not. Thus, matches between terms and combinations were established. To facilitate recognition of the regions, they were hatched and colored differently, and these colors were also used for the identifiers "A" respectively "B" in the sentences. Fig. 2 shows a page of the questionnaire.

The instruction was (originally in German):

"In the following you find 48 different statements about the relation between two regions A and B. Each statement is accompanied by eight graphical depictions. Each graphical depiction shows the regions A and B in a different spatial arrangement. Region A is always the region hatched horizontally in red. Region B is always the region hatched vertically in blue. Please mark for each arrangement with a cross, if it fits to the statement or not (**this means to set one cross per arrangement**). **At least one arrangement must be marked as fitting.** You can also mark several or all arrangements as fitting. Please answer spontaneously and do not think too long about it. There is no right or wrong answer. Please fill in the questionnaire on your own and ask nobody except the experimenter for help."

The questionnaire was on paper and could be completed in any place. It took about half an hour to answer all questions.

The population about which at best statements can be made are all native German speakers. 34 native German speaking subjects (21 male, 13 female; average age of 34, standard deviation 11) participated in the test. They were recruited among the people working at the Institute for Geoinformatics and among the author's friends and did it

Table 2. Terms in the study[5]

Sentence #	Sentence in questionnaire	English translation[6,7]
	ArcGIS	
1	A wird gekreuzt durch den Umriss von B	are crossed by the outline of
2	A benutzt ein Linien-Segment gemeinsam mit B	share a line segment with
3	A ist identisch zu B	are identical to
4	A ist enthalten von B	are contained by
5	A berührt die Umrandung von B	touch the boundary of
6	A überschneidet B	intersect
	GeoMedia	
7	A trifft auf B	meet
8	A ist räumlich identisch mit B	are spatially equal
9	A ist enthalten in B	are contained by
10	A überlappt B	overlap
11	A berührt B	touch
	ArcGIS (A) and GeoMedia (G)	
12	A enthält vollständig B	completely contain (A) entirely contain (G)
13	A ist vollständig enthalten in B	are completely within (A) are entirely contained by (G)
14	A enthält B	contain (A, G)
	Non-GIS	
15	A ist getrennt von B	A is separated from B
16	A ist an B	A is at B
17	A grenzt an B	A borders on B
18	A tangiert B	A is tangent to B
19	A stößt an B	A abuts on B
20	A schneidet B	A intersects B
21	A fällt zusammen mit B	A coincides with B
22	A ist außerhalb von B	A is outside of B
23	A ist innerhalb von B	A is inside B
24	A ist vollständig in B	A is completely in B
25	A ist vollständig innerhalb von B	A is completely inside B
26	A ist in B	A is in B
27	A ist verbunden mit B	A is connected with B
28	A interagiert mit B	A interacts with B
29	A ist zusammen mit B	A is together with B
30	A ist in Kontakt mit B	A is in contact with B

---

[5] Throughout the text the English terms are used for the sake of readability for non-German speakers. But it must be kept in mind that evaluation is based on the German terms and not on their English translations. Characteristics of German terms might not hold for the corresponding English terms and vice versa (see also footnote 6).

[6] Most likely the English translation does not precisely convey the meaning of the original German sentences and the differences between them.

[7] For the sentences derived from GIS terms, the terms of the English product version are shown.

gratuitous and without any other agreements. Geospatial technology is envisioned to become part of everyday software like office and Internet applications, so in future everybody is (potentially) exposed to GIS functionality. The Internet scenario also means being exposed to a variety of manufacturers and products and not accepting long learning phases. Furthermore, users will have different levels of GIS experience, and the software must accommodate all. One way is to see users as human beings first with commonalities in spatial experience and language and to look for these commonalities. Special groups may demand specific terminology, but as stated above, the focus here is on "general purpose" GIS. Due to these considerations half of the subjects had GIS experience, the other half not. GIS experience is taken as an indicator for general spatial awareness and training. It need not mean that subjects with GIS experience had a specifically well understanding of topological operators. As the GIS experienced subjects came from the geoinformatics department, they probably know more than one system and most likely the two included in the study.

Of the 48 sentences only 30 are included in the analysis reported here (see Table 2). 12 were discarded, because their spatial terms are not purely topological, but contain metrics, e.g. "surround" (requiring that one region embraces the other region to a certain minimal extent, which is a metric distinction of the pure topological "touch" or maybe "disjoint"). Unfortunately, this was only noticed after subjects had completed the questionnaire. Another six sentences were excluded afterwards, because their level of detail is different from the general level. An example of the general level is "A is at B"; "A is at the boundary of B" is an example of a more detailed sentence. In the former case the relation refers to the regions as a whole, while in the latter the relation speaks about a part of a region, the boundary. Including parts of regions in the sentences bears the tendency to rather speak in terms of the intersection model than expressing human concepts. The only sentences remaining regardless of this aspect are those stemming from the GIS, because these terms are being tested.

The match between a term and a combination is defined as acceptable, if it is selected by the majority of subjects. Accordingly, it is not acceptable, if 50 percent of the subjects or less select it. A match is better than another, if more subjects select it.

The frequency of each match was determined and those selected by the majority in the sample were separately tested according to the aforementioned hypotheses[8]:

H_0: $\Pi_{match} <= 0.5$
H_1: $\Pi_{match} > 0.5$.

These are hypotheses comparing a proportion against a reference value. The matches are independent Bernoulli variables (the tested match means success, any other match means failure regarding this match), each match having a specific success probability. Thus, the binomial test can be used. As the sample size is greater than 30, normal distribution of the success probabilities can be assumed, which allows using the approximate binomial test.

The main hypothesis that the majority of acceptable matches are others than those used in GIS cannot be tested with the binomial test, because across the matches different success probabilities occur.

---

[8] As mentioned in the introduction, this is explorative testing and must not be mistaken as actual hypothesis testing.

4 Results and Discussion

Table 3 shows the results. The terms are represented in the columns; the combinations are represented in the rows. Only the 59 (of 255 possible) combinations are included that were selected at least once by a subject. The cells contain the number, i.e. the absolute frequency, of subjects matching the term and the combination meeting in this cell by selection in the questionnaire. Empty cells mean that this term and combination were never associated. The terms are arranged in the same order as they appear in Table 2. The combinations are arranged such that also first the GIS and then the non-GIS combinations appear. Inside these categories the order of the numbers by which they are represented (see section 2.2) is preserved. Terms and combinations used in the two GIS are shaded. Furthermore, the terms are underlined, and the combinations are printed in bold face. This corresponds to the formatting in the cells: the maximum value within a column is underlined; the maximum value within a row is printed in bold face. Consequently, a value representing the maximum of its column as well as of its row appears underlined and in bold face. The stars mark values that successfully passed the significance test and denote the level of significance: one star means ten percent, two stars five percent, and three stars one percent error probability. Thick cell boundaries mark the best significant matches for a combination. The last row contains the invalid answers.

Ten significant matches are contained in Table 3 and only one of them is a match implemented in one of the two GIS products: term 8 ("are spatially equal") denoting combination c6 in GeoMedia. Thus, the hypothesis is supported that the majority of acceptable matches are non-GIS matches. The following specific aspects emanate from the result table.

1. A term is specific for a combination if it has a clear maximum with this combination (in the following such a term is called a "specific term"). This means the majority of subjects must have selected this one combination for the term. The majority is statistically significant if the number of subjects amounts to 21 or more. Only ten out of 30 terms, a third, are specific.
2. The highest number of subjects selecting a term with one of the combinations is 30, which corresponds to 88 percent of the sample. In a statistical test with an error probability of five percent this becomes significant for 76 percent. This is a good, but singular result. The rest of the significant terms show frequencies in the lower twenties. If there is only the term at the user interface, it must be doubted that this is enough to make an operator reliably recognizable. But usually it is accompanied by a graphical depiction and the user can try out operations. If the graphics, however, are poor and testing is always necessary, these additional options can also be counterproductive. Together with the previous issue this reveals that most of the natural language terms are rather ambiguous than clear.
3. Specific GIS terms are 8, 12, and 13, which makes only about 20 percent of all GIS terms. There are more specific non-GIS terms: 15, 16, 21, 24, 28, 29, and 30.

4. The GIS term 8 ("are spatially equal") empirically turns out as specific for exactly the combination that is implemented in GeoMedia (c6). This is not true for terms 12 and 13 (complete containment and inverse in both systems). They are implemented as combinations c7 and c8; while subjects select them with combinations c467 and c568 (see Fig. 3). This shift from atomic to compound combinations suggests that complete containment is not understood in terms of unrelated boundaries, but in terms of one interior completely being part of the other without considering the boundaries.

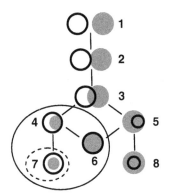

Fig. 3. Complete containment implemented in ArcGIS and GeoMedia (dashed outline) it (continuous outline)

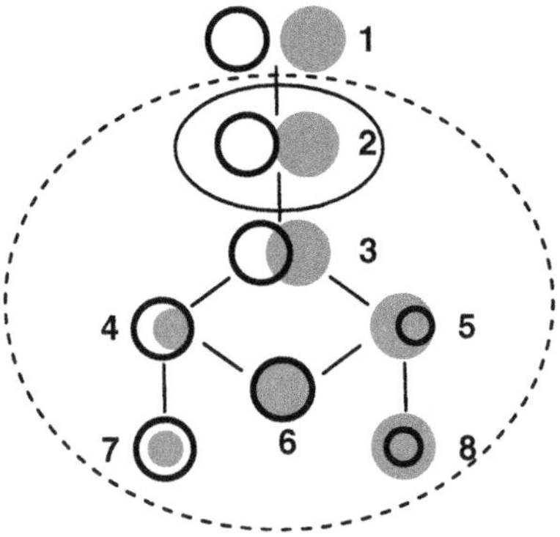

Fig. 4. "Touch" in GeoMedia (dashed outline) and alternative interpretation (continuous outline)

5. The significant terms are distributed among six combinations. For three of these combinations two respectively three significant terms exist, in two cases with differing significance levels. Terms like these with a similar frequency of selection for one combination can be regarded as synonyms for this combination (at least from the purely topological point of view). For combination c6, there are the synonyms "are spatially equal" ("ist räumlich identisch mit") and "coincides with" ("fällt zusammen mit"); for combination c568 the wording is more similar: "are completely within" ("ist vollständig enthalten in") and "is completely in" ("ist vollständig in"). "Interacts with" ("interagiert mit"), "is together with" ("ist zusammen mit"), and "is in contact with" ("ist in Kontakt mit") are synonyms for combination c2345678. In the current design of the two GIS only one term appears for each combination at the user interface. The possibility of presenting synonyms should be considered. This could help to clarify the meaning of otherwise misinterpreted terms. An example is the GeoMedia "touch" ("berührt") operator, which by name could be understood as representing combination c2. Instead, combination c2345678 is referred to (see Fig. 4). Synonyms like "interact" or "not disjoint" would help to specify the implemented operation.

Table 3. Results (table continued on next page)

	Term #[9] Combination	1	2	3	4	5	6	7	8	9	10	11	12	13	14	15	16	17	18	19	20	21	22	23	24	25	26	27	28	29	30
ArcGIS	c3	19	2				8					9									20										2
ArcGIS	c2456		13		6							1					1	2	2												
GeoMedia	c2				1			13				10					21*	20	15	20								2			5
GeoMedia	c345678		3		6	15		2		9	15	1			7							5		2	3	1	1	5		4	1
Both GIS	c6		3	15	1			1	22**	1			2	1	1							24**	1	1	1					1	
Both GIS	c7												3		1																
Both GIS	c8			1					1					3									2		4	7	1		1		
Both GIS	c467												23**		18																
Both GIS	c568				15					14				23**			1							15	23**	15	15			1	
Both GIS	c2345678				1	1	9			1		8					4		3		1	3				15	23**			25*	21*
Non-GIS	c1			1												30**								12					1		
Non-GIS	c12		1													3								18							
Non-GIS	c14																							1							
Non-GIS	c23							3				2					1	3				1	1			3	1				1
Non-GIS	c25				1			1				1					1		1	2						1					
Non-GIS	c26				1																										
Non-GIS	c34	1																													
Non-GIS	c36	2				4				1												1				1					
Non-GIS	c37	1																													
Non-GIS	c47					2				1			4		4																
Non-GIS	c58			5						5			1		4										8	3	8	10			
Non-GIS	c68								1					1											1	1					
Non-GIS	c123		2					1																							
Non-GIS	c126		1					2																							
Non-GIS	c178															1															
Non-GIS	c234					1																									
Non-GIS	c236								1			1										1				1	1			4	1
Non-GIS	c245		1			2		2				1					3	5	10	9											1
Non-GIS	c256					1													1												
Non-GIS	c268																								1						
Non-GIS	c347	1				1				1					1																
Non-GIS	c358			1																1											
Non-GIS	c456		1																												
Non-GIS	c567																								1						

---

[9] The numbers are identical to those used in Table 2, so the terms can be looked up there.

Term #[9] Combination	ArcGIS						GeoMedia					Both GIS			Non-GIS															
	1	2	3	4	5	6	7	8	9	10	11	12	13	14	15	16	17	18	19	20	21	22	23	24	25	26	27	28	29	30
c678								2																						
c1236		14						7																						
c2345	1	1			1						1					2	1													
c2347																	1													
c2356											1							1												
c3456	1																													
c3467						1			3			1		2					1										1	
c3568	1			2		1		2															2			2				
c5678														2																
c12347																						1								
c12457																						1								
c23456	5	8			18		1			1	7					1	2	1	1	3	1					8	1			4
c23458																	1													
c23568							1												1											
c34567	1									1																				
c34578		1								1																				
c34678										1																				
c35678																							2							
c45678																						2	1						1	
c234567	1			1																										
c234568				1																										
c245678																													1	
c1234568				1																										
c12345678		1											1																	
invalid			1	1				1																		2	1			1

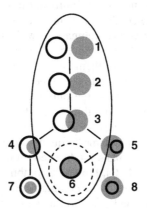

Fig. 5. Equality combination in ArcGIS (dashed outline), which a number of subjects interpreted in this way, and an alternative view of nearly as many subjects (continuous outline)

6. If one term appears with several combinations in equally high frequency, this is a homonym, which in contrast to synonyms bears the danger of misunderstanding and selecting the wrong operator. This can only happen with insignificant terms. There is one term in this study, "are identical to" (3) that shows nearly equal frequencies with combinations c6 and c1236. Fig. 5 reveals that the relations of the latter combination are all depicted with regions A and B having equal size (and the same shape is used for all regions). The regions were deliberately given equal size and shape in order to avoid a bias due to metric characteristics. Unfortunately, it seems that equal size and shape superimpose topological equality here. Supposedly, most of the subjects selecting combination c1236 would have chosen c6 if the regions in relations 1, 2, and 3 were at least sized differently. This would constitute another synonym for combination c6.

7. The size and shape issue discussed under number 6 does not appear with the equality operator of GeoMedia. This gives rise to the assumption that the operator name in GeoMedia is more appropriate. After all, equally sized and shaped but topologically not equal regions can occur in reality. The ArcGIS term is "are identical to"; GeoMedia uses "are spatially equal". Maybe there is also a difference between "identical" and "equal", but presumably the addition "spatially" is the important point. It indicates that equality shall be understood in terms of the space occupied: if two objects share the same space, they are equal. This includes that these objects have the same size and shape. The other way round, two objects of equal size and shape need not share the same space. Consequently, it is advantageous to precisely name the equality concept, particularly because even more notions of equality exist, for example equal identifiers.

8. The significant terms only cover five of the ten combinations occurring in GIS. Maybe the study did not comprise enough terms, so applicable ones exist, but were not there. As different sources were used, this is not very likely, at least not for all five remaining combinations. Another explanation is that some combina-

tions do not have specific terms. But if they are used by humans they must be denotable somehow. Maybe no single significant term applies to them. Finally, the reason could be that these combinations do not correspond to human concepts, at any rate not in the same way as combinations with significant terms.

9. 20 out of the 30 terms in the study are not specific for GIS combinations. Still, the highest frequency of most of them is with combinations implemented in GIS. There are merely three exceptions to this. One is "is separated from" (15), associated with combination c1. This is the only significant case of the three and the one with the highest frequency of all matches. The others are "is outside of" (22), which with a frequency of 18 is associated with combination c12, and "touch the boundary of" (5), showing the same frequency with combination c23456. These cases hint at the fact that humans might need more combinations than those implemented in GIS. In GeoMedia, however, combination c1 is realized by a "not" operator, which can be applied to the complementary combination c2345678.

10. The combinations c7, c47, and c467 (variants of A being contained in B) are obviously selected less often than their inverse combinations c8, c58, and c568 (variants of A containing B). The asymmetry can already be observed with the frequency of terms, which means that the reason lies in the compilation of terms. There are two terms for the first group, "completely contain" (12) and "contain" (14), and seven for the second (term numbers 4, 9, 13, 23, 24, 25, 26). After realizing this effect, the author tried to find the five missing terms for the first group, which was not as straightforward as the other way round. "A ist in B" ("A is in B"), e.g., came up quite naturally, while "A hat in sich B" ("A has inside B") sounds somewhat clumsy. But the more straightforward term "A enthält B" ("A contains B") is linguistically the inverse of another term: "A ist enthalten in B" ("A is contained by B"). It seems that there is a wider variety of terms for the first group. This could be due to a cognitive bias to focus on the contents of a container and not on the container itself. In everyday speech you rather say "there is coffee in the pot" instead of "the pot contains coffee" or "What is in the box?" instead of "What does the box contain?" although both alternatives are valid. The latter phrases sound more elaborate, at least in German.

To sum up, Table 4 compares the frequencies of GIS terms and most frequently selected terms for all GIS combinations. When deciding if an alternative is really a good candidate, significance and difference in frequency must be considered. In eight cases (combinations c6, c568, c2456, and c2345678 of ArcGIS, combinations c2, c568, c2345678 of GeoMedia, and combination c467 of both) greater frequency differences can be observed. The two atomic containment combinations (c7 and c8) were selected very rarely, which suggests dropping these combinations. Extracting from the table only combinations with significant terms leaves the following list:

c2	16: A is at B
c6	21: A coincides with B
c467	12: completely contain (A), entirely contain (G)
c568	13: are completely within (A), are entirely contained by (G), 25: A is completely inside B
c2345678	29: A is together with B

Incorporating these terms into the user interface, providing combination c1 if not available yet, laying less stress on combinations c7 and c8, and using synonyms are recommendations for user interface designers resulting from this survey.

Table 4. GIS terms and most frequently selected terms for GIS combinations

GIS combination	GIS term[10]	Abs. fr.[11]	Most frequently selected term[12]	Abs. fr.[13]
ArcGIS				
c3	1: are crossed by the outline of	19	**20: A intersects B**	**20**
c2456	5: touch the boundary of	6	**2: share a line segment with**	**13**
GeoMedia				
c2	7: meet	13	**16: A is at B**	**21***
c345678	10: overlap	15	6: intersect	15
			10: overlap	15
ArcGIS (A) and GeoMedia (G)				
c6	2: share a line segment with (A)	3	**21: A coincides with B**	**24***
	3: are identical to (A)	15		
	8: are spatially equal (G)	22*		
c7	12: completely contain (A), entirely contain (G)	3	12: completely contain (A), entirely contain (G)	3
c8	13: are completely within (A), are entirely contained by (G)	3	**25: A is completely inside B**	**7**
c467	14: contain (A, G)	18	**12: completely contain (A), entirely contain (G)**	**23***
c568	4: are contained by (A)	15	**13: are completely within**	**23***
	9: are contained by (G)	14	**25: A is completely inside B**	**23***
c2345678	6: intersect (A)	1	**29: A is together with B**	**25***
	11: touch (G)	8		

The study has some shortcomings. The selection of topological terms out of the list compiled from various sources should have been done by consensus of several people. This would have avoided having too many terms in the questionnaire.

The size effect with combination c1236 (see number 6) should be eliminated. In the studies of Knauff, Rauh, and Renz varying sizes were no problem for other relations like, for example, "disjoint" (c1) or "meet" (c2) [7, 14]. A possibility to control the size effect is to introduce various examples of each relation in different sizes.

In all likelihood there are more topological terms than those on our list. Maybe important terms are missing. Compiling the terms from several sources has minimized this risk, but nevertheless it is there. The problem is that the total of all topological terms is unknown. So one cannot know to which extent it is covered or prove completeness.

The sample is an ad hoc sample, which means it is not probabilistic. Although people with different education levels, professional fields, and ages were selected and the

---

[10] The numbers in front of terms are identical to those used in Table 2.

[11] Absolute frequency: number of subjects selecting this combination with the term.

[12] See footnote 10.

[13] See footnote 11.

author is not aware of a characteristic having a systematic influence on the topic of the study, this is possible. Random selection, however, facing the population of all native German speakers, is impractical.

5 Conclusions and Future Work

It could be shown that matches between topological terms and nine-intersection relations exist which are selected by the significant majority of people. The predominant number of them (nine out of ten) is not yet used in the two GIS products under consideration. This reveals potential for improving the wording at the user interfaces.

Beyond the naming issue it became obvious that the selection of combinations must be questioned as well. Both GIS provide combinations that did not show significant matches, which hints at a perceived need at the producer side. It should be clarified, if this need can be confirmed for the user side. This point can be put generally: the user needs ought to be taken into consideration more resolutely. The two systems together use 14 terms out of an unknown number of terms and ten combinations out of a total of 255. Who knows reliably if these are the appropriate ones for user tasks? The fact that the topological operator sets do not match raises doubts. This survey, even though limited due to its exploratory character, supports these doubts.

The present type of study aims at clarifying human concepts, but it does not look at how people use these concepts when accomplishing tasks. Approaches from both sides are needed to arrive at usable and useful operator sets.

This work presents an empirically compiled list of potentially useful topological operators. But it does not explain why these operators are useful. A theory would facilitate the identification of remaining cognitively adequate operators not contained in current GIS. Moreover, a theory would be beneficial for extending the finding for relations between two regions to lines and points.

Appropriate graphical representations are another field of work, and finally all aspects must be brought together. This will include a more natural test context, where subjects perform real tasks using real software.

Acknowledgements

Many thanks go to the 34 people who participated in the study. This work has benefited from discussions with Gunda Musekamp and Heiko Großmann as well as from the comments of three anonymous reviewers.

References

1. Albrecht, J.H.: Universal GIS Operations - A Task-Oriented Systematization of Data Structure-Independent GIS Functionality Leading Towards a Geographic Modelling Language. (1996)
2. Cohn, A.G., Hazarika, S.M.: Qualitative Spatial Representation and Reasoning: An Overview. Fundamenta Informaticae 1-2 (2001) 2-32

3. Egenhofer, M.J., Herring, J.R.: Categorizing Binary Topological Relations between Regions, Lines, and Points in Geographic Databases. Department of Surveying Engineering, University of Maine, Orono (1990)
4. Egenhofer, M.J., Herring, J.R.: A Mathematical Framework for the Definition of Topological Relationships. In: Brassel, K.E., Kishimoto, H. (eds.): 4th International Symposium on Spatial Data Handling. Department of Geography, University of Zurich, Zurich (1990) 803-813
5. Frank, A.U.: The Use of Geographical Information Systems: The User Interface Is the System. In: Medyckyj-Scott, D., Hearnshaw, H.M. (eds.): Human Factors in Geographical Information Systems. Belhaven Press, London (1993) 15-31
6. Furnas, G.W. et al.: The Vocabulary Problem in Human-System Communication. CACM 11 (1987) 964-971
7. Knauff, M., Rauh, R., Renz, J.: A Cognitive Assessment of Topological Spatial Relations: Results from an Empirical Investigation. In: Hirtle, S.C., Frank, A.U. (eds.): Spatial Information Theory: A Theoretical Basis for GIS, Laurel Highlands, Pennsylvania, USA. Lecture Notes in Computer Science (LNCS), Vol. 1329. Springer, Berlin (1997) 193-206
8. Landauer, T.K., Galotti, K.M., Hartwell, S.: Natural Command Names and Initial Learning: A Study of Text-Editing Terms. CACM 7 (1983) 495-503
9. Ledgard, H. et al.: The Natural Language of Interactive Systems. CACM 10 (1980) 556-563
10. Mark, D.M., Egenhofer, M.J.: Calibrating the Meanings of Spatial Predicates from Natural Language: Line-Region Relations. In: Waugh, T.C., Healey, R.G. (eds.): Advances in GIS Research, 6th International Symposium on Spatial Data Handling, Edinburgh, Scotland, UK. Department of Geography, University of Edinburgh, Edinburgh (1994) 538-553
11. Mark, D.M., Egenhofer, M.J.: Modeling Spatial Relations between Lines and Regions: Combining Formal Mathematical Models and Human Subjects Testing. Cartography and Geographical Information Systems 3 (1994) 195-212
12. Mark, D.M., Egenhofer, M.J.: Topology of Prototypical Spatial Relations between Lines and Regions in English and Spanish. In: Peuquet, D.J. (ed.) 12th International Conference on Automated Cartography, Charlotte, North Carolina, USA. 1995) 245-254
13. Randell, D.A., Cui, Z., Cohn, A.G.: A Spatial Logic Based on Regions and Connection. In: Principles of Knowledge Representation and Reasoning: Proceedings of the Third International Conference, Los Altos. Morgan Kaufmann, 1992) 165-176
14. Renz, J., Rauh, R., Knauff, M.: Towards Cognitive Adequacy of Topological Spatial Relations. In: Freksa, C. et al. (eds.): Spatial Cognition II - Integrating Abstract Theories, Empirical Studies, Formal Models, and Practical Applications. Lecture Notes in Computer Science (LNCS), Vol. 1849. Springer, Berlin (2000) 184-197
15. Riedemann, C.: Naming Topological Operators at GIS User Interfaces. In: Toppen, F., Painho, M. (eds.): 8th Conference on Geographic Information Science, Estoril, Portugal. Instituto Geográfico Português (IGP), 2005) 307-315

Spatial Relations Between Classes of Individuals

Maureen Donnelly[1] and Thomas Bittner[1,2]

[1] Department of Philosophy,
New York State Center of Excellence in Bioinformatics and Life Sciences,
University at Buffalo, Buffalo, NY 14260
{md63, bittner3}@buffalo.edu
[2] Department of Geography,University at Buffalo, Buffalo, NY 14261

Abstract. In the Spatial Data Transfer Standard and many other geographic standards and ontologies, we find statements such as (1) "waterfalls are parts of watercourses" and (2) "ecoregions of continental scale are parts of ecoregions of global scale", etc. In these examples, the terms "waterfall", "watercourse", "ecoregion of scale X", etc. refer to classes of individuals rather than to particular individuals. Since it is the purpose of these standards and ontologies to facilitate interoperability, it is important to give a clear semantics to statements like (1) and (2). For example, (1) should be understood to claim that every waterfall is part of some watercourse, but NOT that every watercourse has a waterfall as its part. In (2), by contrast, the term "part-of" has a stronger meaning: every ecoregion of continental scale is part of some ecoregion of global scale AND every ecoregion of global scale has some ecoregion of continental scale as a part. To overcome this kind of semantic heterogeneity, we propose a Mereotopology for Individuals and Classes (MIC) in which we define parthood, location, and connection relations among classes based on parthood, location, and connection relations between individuals. We then demonstrate the usefulness of this formal theory for making the logical structure of spatial information more precise. Although we focus here on the simplest and most pervasive of the spatial relations (parthood, location, and connection), the strategy employed in this paper can be used in analogous treatments of other kinds of relations among classes.

1 Introduction

Reasoning involving qualitative spatial relations such as parthood, location, and connection is central to many disciplines, including geography, biology, medicine, geology, ecology, and meteorology. Several different formal theories have been proposed in an effort to provide a rigorous foundation for this kind of reasoning (Asher and Vieu 1995), (Casati and Varzi 1999), (Cohn, Bennett et al. 1997), (Donnelly 2004). In this paper, we extend one such formal theory to encompass also the class-based spatial reasoning which is prevalent in biomedical ontologies such as the Foundational Model of Anatomy (FMA) and GALEN (Rosse and Mejino 2003), (Rector and Rogers 2002a), (Rector and Rogers 2002b), (Mejino, Agoncillo et al. 2003), (Rogers and Rector 2000). The relations presented in our theory can also be applied to geographic classes such as the entity types (*Beach*, *Coast*, *Shore*,

A.G. Cohn and D.M. Mark (Eds.): COSIT 2005, LNCS 3693, pp. 182–199, 2005.
© Springer-Verlag Berlin Heidelberg 2005

Woodland, etc) of the Spatial Data Transfer Standard (SDTS) (SDTS 1997) or ecological classes (*Ecoregion, Domain, Division*, etc) as used for example in (McNab and Avers 1994), (Cleland, Avers et al. 1997).

The theory developed in this paper, Mereotopology for Individuals and Classes (MIC), deals with two distinct sorts of entities—individuals and classes of individuals. By an *individual*, we mean here a concrete entity which occupies, at each moment of its existence, a unique spatial location[1]. Individuals can be either material (my liver, LaGuardia Airport, the state of New York) or immaterial (the cavity of my stomach, the interior of an airplane). Individuals are distinguished from *classes* (also called universals, kinds, or types) which may have, at each moment, multiple individual instances. Examples of classes are *Liver* (the class whose instances are individual livers), *Airport* (the class whose instances are individual airports), and *Federal State* (the class whose instances are individual federal states). (Throughout this paper, we use italics and initial capitals for class names.) Although classes may gain and lose instances over time (when, e.g., airports are constructed or dismantled), the class itself does not change its identity.

The formal theories proposed in (Asher and Vieu 1995), (Casati and Varzi 1999), (Cohn, Bennett et al. 1997), (Donnelly 2004) deal only with spatial relations among individuals. However, in many disciplines, it is common to also represent and reason about relations among classes. For example, in canonical anatomy, we find such assertions as "the stomach is continuous with the esophagus", "the right ventricle is part of the heart", or "the brain is contained in the cranial cavity". These assertions are claims about classes, not claims about specific individuals (for example, my stomach and my esophagus). They tell us that the specified anatomical classes (*Stomach, Esophagus*, and so on) stand in certain relations to one another. Similarly, in geography, we find assertions such as "a shore is a part of land which is in immediate contact with a body of water" or "an intersection is a junction of roads or tracks"[2], which describe relations among geographical classes (*Shore, Land Mass, Body of Water, Intersection, Road* etc) rather than relations among specific geographic individuals.

As is emphasized in (Smith and Rosse 2003), (Donnelly, Bittner et al. 2005), and (Smith, Ceusters et al. 2005), though class-level spatial relations depend on and should be defined in terms of individual-level spatial relations, it is crucial to distinguish class-level relations from the corresponding relations among individuals. This is because the logical properties of the class-level relations do not in general match those of the individual-level relations. Therefore, spatial theories are needed that deal both with spatial relations among individuals and with corresponding relations among classes. Also, as is shown in (Donnelly, Bittner et al. 2005), there are different types of class-level relations which can be defined in terms of a given binary relation among individuals. Since these different class-level relations have not been

---

[1] To simplify the discussion, we focus throughout the paper exclusively on *spatial* individuals. We will not deal here with *spatio-temporal* individuals such as my birth or the tenth eruption of Mount St. Helen. However, the theory developed in this paper can be applied also in domains consisting of four-dimensional individuals and classes of such individuals.

[2] These assertions are taken from the definitions of the entity types *Shore* and *Intersection* in the SDTS (1997).

clearly distinguished in scientific discourse, it is an important task of a formal theory of class-level spatial relations to clearly characterize distinct types of class relations.

Mereotopology for Individuals and Classes introduces distinct parthood, location, and connection relations for classes in terms of corresponding relations among individuals. MIC is based on a spatial theory for individuals which we call Mereotopology for Individuals (MI) and present in Section 2. In Section 3, we introduce classes and class-level relations. We examine the logical properties of the class-level relations of MIC in Sections 4 and 5.

MIC can be used to endow the class-level spatial relations which are already used in the FMA, GALEN, and other biomedical ontologies with a clear semantics and to support automated reasoning over these relations. MIC can also be used as a basis for introducing similar relations linking entity types in geographical terminologies such as the SDTS. Moreover, MIC can facilitate interoperability in environmental modeling by giving a clear semantics to terms used in ecosystem classifications along the lines proposed in (Sorokine, Bittner et al. 2004) and (Sorokine and Bittner 2005).

If desired, MIC can be extended to include stronger axioms or spatial relations which are particularly appropriate for geographic contexts, such earth-based directional relations (above, below) or qualitative distance relations (near to, far from).

2 Mereotopology for Individuals

MI is a weaker version of the theory called Layered Mereotopology which is developed in detail in (Donnelly 2004). MI is formulated in standard first-order predicate logic with identity. It is a time-independent theory which can be used to describe static spatial relations among individuals during a fixed time-frame. An important project for further work is to incorporate time and change into our theory. For some progress in this direction, see (Bittner, Donnelly et al. 2004).

We divide the presentation of MI into three parts, each of which focuses on one of the three primitive relations of MI.

2.1 Parthood Relations Among Individuals

MI uses one mereological primitive – the binary relation P where, on the intended interpretation, Pxy means

individual x is part of individual y.

The following relations among individuals are defined in terms of P:

(D1) $PPxy =: Pxy \,\&\, \sim Pyx$	(x is a proper part of y)
(D2) $Oxy =: \exists z\,(Pzx \,\&\, Pzy)$	(x and y overlap)
(D3) $DSxy =: \sim Oxy$	(x and y are discrete)

For example, Queens is a proper part of New York City, Canada overlaps North America's Atlantic coast, and my liver and my heart are discrete.

The mereological axioms for MI are as follows:

(P1)[3] Pxx   (every individual is part of itself)

(P2) Pxy & Pyx $\rightarrow$ x = y   (if x is part of y and y is part of x, then x and y
                                      are identical)

(P3) Pxy & Pyz $\rightarrow$ Pxz   (if x is part of y and y is part of z, then x is
                                      part of z)

(P4) ~Pxy $\rightarrow$ $\exists$z(Pzx & DSzy) (is x is not part of y, then x has a part
                                      that is discrete from y)

For example, since my left ventricle is part of my heart and my heart is part of my circulatory system, my left ventricle is part of my circulatory system (P3). Also, since New York City is not a part of Queens, New York City must have a part (e.g. Manhattan or the Bronx) which is discrete from Queens (P4).

From (P1)-(P4) it follows that PP is transitive and asymmetric, that O and DS are symmetric, that O is reflexive, and that DS is irreflexive. We can also prove the following theorems:

(PT1) Oxy & Pyz $\rightarrow$ Oxz   (if x overlaps y and y is part of z,
                                      then x overlaps z)

(PT2) Pxy & DSyz $\rightarrow$ DSxz   (if x is part of y and y is discrete from z,
                                      then x is discrete from z)

2.2 Location Relations

Location relations are more general than are mereological relations. Unlike mereological relations, location relations depend only on two individuals' locations. Thus, for example, a bolus of food may be located in my stomach cavity although it is never part of my stomach cavity. Similarly, a ship may be located in a harbor without being part of the harbor. Also, parts of the ecoregion called 'Humid Temperate Domain' in (McNab and Avers 1994), (Cleland, Avers et al. 1997) are located in the United States but are not parts of this socio-economic unit.

In MI, location relations are defined in terms of a function, r, which maps every individual to the unique spatial region it occupies throughout the given time-frame[4]. A spatial region is an immaterial individual which may be one, two, or three dimensional. For most applications, we assume that all spatial regions are fixed relative to the earth. Thus, for example, my stomach cavity, though immaterial, is not considered a spatial region. (But see (Donnelly 2005) for an alternative approach.) The following relations are defined in terms of the region function.

(D4) Loc-In(x, y) =: Pr(x)r(y)   (x is located in y)

(D5) PCoin(x, y) =: Or(x)r(y)   (x partially coincides with y)

(D6) NCoin(x, y) =: ~PCoin(x, y)   (x and y are non-coincident)

---

[3] Axioms specific to the mereological relations are labeled with a "P". Throughout this paper, initial universal quantifiers are omitted unless they are needed for clarity.

[4] We could describe movement only in a time-inclusive version of MI. Here, the region function would map a spatial individual and a time to a unique region.

For example, an airplane getting ready for take-off from LaGuardia Airport is located in the airport but is not part of (and does not share parts with) LaGuardia Airport. A storm system on the Gulf of Mexico may partially coincide with Florida, but does not share parts with Florida. These two examples show that non-coincidence is a stronger relation than discreteness – the airplane is always discrete from LaGuardia Airport, but the airplane and the airport are non-coincident only after the airplane leaves the airport.

Axioms for the region function and location relations are as follows.

(L1)[5] $r(r(x)) = r(x)$ (x's spatial region is its own spatial region)

(L2) $Pxy \rightarrow Loc\text{-}In(x, y)$ (if x is part of y, then x is located in y)

It follows from the axioms and definitions above that Loc-In is transitive, PCoin and NCoin are symmetric, Loc-In and PCoin are reflexive, and NCoin is irreflexive. We can also derive the following theorems.

(LT1) $Oxy \rightarrow PCoin(x, y)$ (if x and y overlap, then x and y partially coincide)
(LT2) $NCoin(x, y) \rightarrow DSxy$ (if x and y are non-coincident, then x and y
                            are discrete)
(LT3) $Loc\text{-}In(x, y)$ & $Pyz \rightarrow Loc\text{-}In(x, z)$ (if x is located in y and y is part of z,
                                then x is located in z)
(LT4) Pxy & $Loc\text{-}In(y, z) \rightarrow Loc\text{-}In(x, z)$ (if x is part of y and y is located in z,
                                then x is located in z)

For example, since i) my heart is located in my middle mediastinal space and ii) my middle mediastinal space is part of my thoracic cavity, it follows that: iii) my heart is located in my thoracic cavity (LT4). Similarly, if i) your car is located in Queens, then since ii) Queens is part of New York City, it follows that: iii) your car is located in New York City.

2.3 Connection Relations

A third primitive enables us to describe topological relations among individuals. On the intended interpretation, the connection relation C holds between individuals x and y if the distance between them is zero (where distance between extended individuals is here understood as the greatest lower bound of the distance between any point of the first individual and any point of the second individual). For example, Vermont and Canada are connected and non-coincident. Lake Erie and Canada are connected and partially coincident.

The following relations are defined using the connection relation.

(D7) $ECxy =: Cxy$ & $NCoin(x, y)$ (x and y are externally connected)
(D8) $SPxy =: {\sim}Cxy$ (x and y are separated)
(D9) $TPxy =: Pxy$ & $\exists z(Czx$ & $NCoin(z,y))$ (x is a tangential part of y)
(D10) $IPxy =: Pxy$ & ${\sim}TPxy$ (x is an interior part of y)

---

[5] Axioms specific to the region function are labeled with "L".

For example, Vermont and Canada are externally connected. Iowa and Canada are separated. Vermont is a tangential part of the United States. Iowa is an interior part of the United States.

Axioms for the connection relation are as follows.

(C1)[6] Cxx   (everything is connected to itself)

(C2) Cxy → Cyx   (if x is connected to y, then y is connected to x)

(C3) Loc-In(x, y) → ∀z(Czx → Czy) (if x is located in y, then everything
                       connected to x is also connected to y)

It follows from the axioms of MI, that EC and SP are irreflexive and symmetric, TP and IP are antisymmetric, and IP is transitive.

In addition, the following theorems can be derived.

(CT1) Cxy ↔ Cr(x)r(y)   (x and y are connected if and only if their
                                 regions are connected)

(CT2) Pxy → Cxy   (if x is part of y, then x and y are connected)

(CT3) SPxy → DSxy   (if x and y are separated, then they are discrete)

(CT4) SPxy → NCoin(x, y)   (if x and y are separated, then the are non-coincident)

(CT5) PCoin(x, y) → Cxy   (if x and y partially coincide, then x and y are
                                connected)

(CT6) PCoin(x, y) ∨ ECxy ∨ SPxy   (any two individuals either partially coincide,
                                are externally connected, or are separated)

(CT7) Cxy & Loc-In(y, z) → Cxz   (if x is connected to y and y is located in z then
                                x is connected to z)

(CT8) Loc-In(x, y) & SPyz → SPxz   (if x is located in y and y is separated from z,
                                then x is separated from z)

For example, since Interstate 70 is connected to Lambert Airport and Lambert Airport is located in St. Louis County, it follows that Interstate 70 is connected to St. Louis County (CT7).

3 Defining Relations on Classes

To define relations on classes, we expand MI to MIC, a theory whose domains are divided into two disjoint sorts of entities – individuals and classes. We retain the variables x, y, z for individuals and use capital letters from the beginning of the alphabet (A, B, C, D,...) as variables for classes. Note that classes are here treated as first-order members of the domain of quantification (albeit members of a distinguished sort), not as predicates ranging over individuals. We have chosen this approach because it fits most naturally with the treatment of classes and class relations in ontologies such as the FMA and GALEN where assertions are made about relations between anatomical classes but instances of these classes are not introduced.

---

[6] Axioms specific to the connection relation are labeled with "C".

All quantification in MIC is restricted to a single sort. Restrictions on quantification will be understood from conventions on variable usage.

3.1 Instantiation

All relations of MI (as well as the region function) remain restricted to individuals. To link individuals and classes, MIC includes the binary relation Inst which holds between an individual and a class where on the intended interpretation, Inst(x, A) means

individual x is an instance of class A.

For example, Inst(my heart, *Heart*), Inst(LaGuardia Airport, *Airport*), Inst(New York State, *Federal State*), and Inst(Humid Temperate Domain, *Ecoregion*).
MIC includes the following axioms for Inst:

(I1) $\exists$A Inst(x, A) (every individual is a member of some class)

(I2) $\exists$x Inst(x, A) (every class has some member)

We define the class subsumption relation, Is_a, in terms of Inst.

(D11) Is_a(A, B) =: $\forall$x(Inst(x, A) $\rightarrow$ Inst(x, B))   (class A subsumes class B)

For example, Is_a(*Heart, Organ*), Is_a(*Highway, Road*), Is_a(*Airport, Restricted Area*), Is_a(*Federal State, Socio Economic Unit*), and Is_a(*Domain, Ecoregion*). It follows from D11 that Is_a is reflexive and transitive.

3.2 Class-Based Spatial Relations

Spatial relations on classes are introduced in the following definition schemas where R is a meta-variable that can stand for any one of the spatial relations of MI (e.g. P, PP, Loc-In, etc.). (See also (Levesque and Brachman 1985), (Smith and Rosse 2003), and (Beck and Schulz 2003) for other discussions of these kinds of class relations.)

(DS1) R_{some}(A, B) =: $\exists$x$\exists$y(Inst(x, A) & Inst(y, B) & Rxy)

(DS2) R_{all-1}(A, B) =: $\forall$x (Inst(x, A) $\rightarrow$ $\exists$y(Inst(y, B) & Rxy))

(DS3) R_{all-2}(A, B) =: $\forall$y (Inst(y, B) $\rightarrow$ $\exists$x(Inst(x, A) & Rxy))

(DS4) R_{all-12}(A, B) =: R_{all-1}(A, B) & R_{all-2}(A, B)

(DS5) $R_{all-all}$(A, B) =: $\forall$x $\forall$y(Inst(x, A) & Inst(y, B) $\rightarrow$ Rxy)

R_{some} class relations are very weak. R_{some}(A, B) holds as long as at least one A stands in relation R to some B.

R_{all-1} class relations place restrictions on all instances of the *first* argument. R_{all-1}(A, B) tells us that something is true of all A's – each A stands in relation R to some B.

R_{all-2} class relations place restrictions on all instances of the *second* argument. R_{all-2}(A, B) tells us that something is true of all B's – for each B there is some A that stands in the R relation to it.

R_{all-12} class relations place restrictions on all instances of *both* arguments. R_{all-12}(A, B) tells us that something is true of all A's and something else is true of all B's—each A stands in the R relation to some B *and* for each B there is some A that stands in the R relation to it.

$R_{all-all}$ class relations are very strong. Like R_{all-12} relations, $R_{all-all}$ relations place restrictions on all instances of *both* arguments. But the restriction imposed by $R_{all-all}$ relations is much stronger than that imposed by the R_{all-12} relations. $R_{all-all}(A, B)$ tells us that every A stands in relation R to every instance of B.

As examples, we consider how such class-level relations are defined when R is the parthood relation (P).

P_{some} is the relation that holds between class A and class B if and only if some instance of A is part of some instance of B. For example, given that some forests are partially composed of oak trees $P_{some}(Oak, Forest)$ holds (even though not every oak tree is part of a forest and not every forest contains oaks).

P_{all-1} is the relation that holds between class A and class B if and only if every instance of A is part of some instance of B. For example, every instance of *Human Female Reproductive System* is part of some instance of *Human Being*. Thus, $P_{all-1}(Human Female Reproductive System, Human Being)$. Similarly, every waterfall is part of a watercourse. Thus, $P_{all-1}(Waterfall, Watercourse)$.

P_{all-2} is the relation that holds between class A and class B if and only if every instance of B has some instance of A as a part. For example, every instance of *Building* has an instance of *Wall* as a part. Thus, $P_{all-2}(Wall, Building)$. But notice that $P_{all-1}(Wall, Building)$ does NOT hold, since some walls (e.g. the Great Wall of China) are not part of any building. Also notice that $P_{all-2}(Human Female Reproductive System, Human Being)$ and $P_{all-1}(Waterfall, Watercourse)$ do NOT hold, since not all human beings have female reproductive systems and not all watercourses have waterfalls.

P_{all-12} is the relation that holds between class A and class B if and only if: i) every instance of A is part of some instance of B and ii) every instance of B has some instance of A as a part. For example, every instance of *Capital* is part of some instance of *Country* and every instance of *Country* has some instance of *Capital* as a part. Thus, $P_{all-12}(Capital, Country)$. Also, the P_{all-12} relation holds between the classes *Province*, *Division*, and *Domain* used in ecosystem classifications (McNab and Avers 1994), (Cleland, Avers et al. 1997). Every instance of the class *Division* is a part of some instance the class *Domain* and every instance of the class *Domain* has some instance of the class *Division* as a part, i.e., $P_{all-12}(Division, Domain)$. Similarly, $P_{all-12}(Province, Division)$ and $P_{all-12}(Province, Domain)$. By contrast, NONE of the following hold: $P_{all-12}(Human Female Reproductive System, Human Being)$, $P_{all-12}(Waterfall, Watercourse)$, $P_{all-12}(Wall, Building)$.

$P_{all-all}$ is the relation that holds between class A and class B if and only if every instance of A is part of every instance of B. $P_{all-all}$ is a very strong relation that is not useful in most contexts. $P_{all-all}(A, B)$ will generally be false except in unusual cases in which class B has very few instances. For example, if the SDTS entity type *Earth Surface* is understood as a class with exactly one instance (the earth's surface), then $P_{all-all}(Continent, Earth Surface)$ holds since every continent is part of this single instance of *Earth Surface*. Similarly, if we include the class *Ecoregion Of Planet Scale* in our ecosystem classification and the domain of this classification is restricted to the Earth, then $P_{all-all}(Domain, Ecoregion Of Planet Scale)$, $P_{all-all}(Division, Ecoregion Of Planet Scale)$, and so on hold.

3.3 Class-Level Detachment Relations

When R is any of the MI relations (P, PP, O, Loc-In, PCoin, C) which hold only between connected individuals, the $R_{\text{all-all}}$ relation is generally too strong to be of much use. The case is different when R is a detachment relation. By a *detachment relation*, we mean a relation which always holds between individuals x and y when x is a positive distance from y. The detachment relations in MI are DS (discreteness), NCoin (non-coincidence), and SP (separation). (Notice that SP, but not DS or NCoin, holds *only* between individuals which are a positive distance apart.) $R_{\text{all-all}}$ versions of the detachment relations are useful for geographical and anatomical reasoning.

$DS_{\text{all-all}}$ is the relation that holds between class A and class B if and only if every instance of A is discrete from every instance of B. For example, $DS_{\text{all-all}}(Sea, Island)$—no sea shares parts with any island (though seas and islands may be externally connected). Also, $DS_{\text{all-all}}(Airplane, Airport)$—no airplane shares parts with any airport (though airplanes may be located in airports).

$NCoin_{\text{all-all}}$ is the relation that holds between class A and class B if and only if every instance of A is non-coincident with every instance of B. For example, $NCoin_{\text{all-all}}(Sea, Island)$—no sea is coincident with any island. Also, $NCoin_{\text{all-all}}(Brain, Heart)$—no brain is coincident with any heart.

$SP_{\text{all-all}}$ is the relation that holds between class A and class B if and only if every instance of A is separated from every instance of B. For example, $SP_{\text{all-all}}(Ice Field, Desert)$—every ice field is a positive distance from every desert. Also, $SP_{\text{all-all}}(Brain, Heart)$—every brain is a positive distance from every heart.

On the other hand, R_{some}, $R_{\text{all-1}}$, $R_{\text{all-2}}$, and $R_{\text{all-12}}$ relations are not generally useful if the underlying relation R is a detachment relation. For example, the relation $NCoin_{\text{all-1}}$ holds between class A and class B if and only if every instance of A is non-coincident with some instance of B. Since most classes have instances in widely scattered locations, $NCoin_{\text{all-1}}(A, B)$ holds for nearly all classes A and B. Thus, e.g., $NCoin_{\text{all-1}}(Airplane, Airport)$ (every airplane is non-coincident with some airport) and $NCoin_{\text{all-1}}(Human Heart, Human Being)$ (every human heart is non-coincident with some human being).

4 Individual-Level Relations vs. Class-Level Relations

The purpose of this section is to consider the correlation between the logical properties of the class relations introduced in the previous section and the logical properties of the underlying MI relations.

4.1 Basic Properties of Binary Relations

The table below lists basic logical properties of binary relations. For each row, we assume that the MI relation R has the property listed under the heading. The remaining cells in the row tell us whether we can then prove in MIC that the class relations R_{some}, $R_{\text{all-1}}$, $R_{\text{all-2}}$, $R_{\text{all-12}}$, and $R_{\text{all-all}}$ also have the given property.

Table 1. Correlation between MI relations among individuals and MIC relations among classes

MI Relation	MIC	Class	Relations		
R is...	R_{some} is...?	R_{all-1} is...?	R_{all-2} is ...?	R_{all-12} is...?	$R_{all-all}$ is...?
reflexive	yes	yes	yes	yes	no
irreflexive	no	no	no	no	yes
symmetric	yes	no	no	yes	yes
asymmetric	no	no	no	no	yes
antisymmetric	no	no	no	no	yes
transitive	no	yes	yes	yes	yes

Table 1 shows that some, but not all, of the basic logical properties of the MI relations transfer to the defined class relations. For example, since theorems of MI require that PP is transitive, we can prove in MIC that PP_{all-1} is also transitive. But we cannot prove that PP_{all-1} is either irreflexive or asymmetric even though PP is irreflexive and asymmetric. Also, for a fixed MI relation R, the five class relations defined in terms of R will in general have different logical properties. For example, $Loc-In_{all-1}$ is transitive, but $Loc-In_{some}$ is not transitive. O_{all-12} is symmetric, but neither O_{all-1} nor O_{all-2} is symmetric. Thus, it is crucial that the different versions of class relations are always explicitly distinguished from one another as well as from the corresponding MI relations among individuals.

4.2 Simple Implications

Besides those listed in Table 1, some other important properties of MI relations carry over to the class relations. Given any MI relations R, S if

$$Rxy \rightarrow Sxy$$

is a theorem of MI, then

$$R_i(A, B) \rightarrow S_i(A, B), \quad i = \text{some, all-1, all-2, all-12, all-all}$$

are theorems of MIC. In particular, MIC has the theorems listed in Table 2. Notice that each row in the table corresponds to five class-level MIC theorems.

Table 2. Theorems of MIC, i = some, all-1, all-2, all-12, all-all

Label	MIC theorem	MI theorem
MICT1-5	$O_i(A, B) \rightarrow PCoin_i(A, B)$	LT1
MICT6-10	$P_i(A, B) \rightarrow Loc-In_i(A, B)$	L2
MICT11-15	$NCoin_i(A, B) \rightarrow DS_i(A, B)$	LT2
MICT16-20	$P_i(A, B) \rightarrow C_i(A, B)$	CT2
MICT20-25	$PCoin_i(A, B) \rightarrow C_i(A, B)$	CT5
MICT26-30	$SP_i(A, B) \rightarrow DS_i(A, B)$	CT3
MICT31-35	$SP_i(A, B) \rightarrow NCoin_i(A, B)$	CT4

For example, in MI the parthood relation among individuals entails the located in relation among individuals, i.e., $Pxy \rightarrow$ Loc-In(x, y). Thus, $P_i(A, B) \rightarrow$ Loc-In$_i(A, B)$ for i = some, all-1, all-2, all-12, all-all. In particular,

(MICT8) $P_{all-2}(A, B) \rightarrow$ Loc-In$_{all-2}(A, B)$.

For example, since every country has a capital as a part, every country has a capital located in it.

4.3 Composition of Relations

Not only simple implications, but also more complex implications which can be thought of as representing the *composition* of binary relations, carry over from MI to MIC. Let R, S, T be relations of MI. Suppose that

$$Rxy \ \& \ Syz \rightarrow Txz$$

is a theorem of MI. Then

$$R_i(A, B) \ \& \ S_i(B, C) \rightarrow T_i(A, C), \qquad i = \text{all-1, all-2, all-12, all-all}$$

are theorems of MIC. In particular, MIC has the theorems listed in Table 3. Notice that each row of Table 3 corresponds to four class-level MIC theorems.

Table 3. Theorems of MIC, i = all-1, all-2, all-12, all-all

Label	MIC theorem	MI theorem
MICT36-39	$P_i(A, B) \ \& \ P_i(B, C) \rightarrow P_i(A, C)$	P3
MICT40-43	$O_i(A,B) \ \& \ P_i(B,C) \rightarrow O_i(A,C)$	PT1
MICT44-47	$P_i(A,B) \ \& \ DS_i(B,C) \rightarrow DS_i(A,C)$	PT2
MICT48-51	Loc-In$_i(A, B) \ \& \ $Loc-In$_i(B, C) \rightarrow$ Loc-In$_i(A, C)$	transitivity of Loc-In
MICT52-55	Loc-In$_i(A, B) \ \& \ P_i(B, C) \rightarrow$ Loc-In$_i(A, C)$	LT3
MICT56-59	$P_i(A, B) \ \& \ $Loc-In$_i(B, C) \rightarrow$ Loc-In$_i(A, C)$	LT4
MICT60-63	$C_i(A, B) \ \& \ $Loc-In$_i(B, C) \rightarrow C_i(A, C)$	CT7
MICT64-67	Loc-In$_i(A, B) \ \& \ SP_i(B, C) \rightarrow SP_i(A, C)$	CT8

For example, Loc-In (x, y) & $Pyz \rightarrow$ Loc-In(x, z) is a theorem of MI. Thus,

(MICT54)   Loc-In$_{all-12}(A, B)$ & $P_{all-12}(B, C) \rightarrow$ Loc-In$_{all-12}(A, C)$

is a theorem of MIC. Using (MICT54), we can infer from

Loc-In$_{all-12}$(*Heart, Middle Mediastinal Space*)
(every heart is located in some middle mediastinal space and every middle mediastinal space has some heart located in it)

and

P_{all-12}(*Middle Mediastinal Space, Thoracic Cavity*)

(every middle mediastinal space is part of some thoracic cavity and every thoracic cavity has some middle mediastinal space as a part)

that

$$Loc\text{-}In_{12}(Heart, Thoracic\ Cavity)$$

(every heart is located in some thoracic cavity and every thoracic cavity has some heart located in it).

As another example, given:

$$Loc\text{-}In_{all\text{-}2}(Explosives, Ammunition\ Dump)$$
(every ammunition dump has explosives located in it)

and

$$P_{all\text{-}2}(Ammunition\ Dump, Military\ Base)$$
(every military base has an ammunition dump as part)

we can infer using (MICT50):

$$Loc\text{-}In_{all\text{-}2}(Explosives, Military\ Base)$$
(every military base has explosives located in it).

4.4 Properties of MI Relations That Do Not Transfer to the Class-Level Relations

Besides the basic properties listed in Table 1, many other properties of MI relations do not transfer to the class relations. In general, implications with a conjunction, negation, or existential quantification in the antecedent or with negation or existential quantification in the consequent do not transfer to the corresponding class relations[7]. In particular, most of the definitional implications of MI do not transfer to the class relations. For example, Cxy & $NCoin(x, y) \rightarrow ECxy$ follows immediately from the definition of EC in MI. But $C_i(A, B)$ & $NCoin_i(A, B) \rightarrow EC_i(A, B)$ (i = some, all-1, all-2, all-12) are not theorems of MIC. Thus, although

$$C_{all\text{-}1}(Island, Time\ Zone)$$
(every island is connected to some time zone)

and

$$NCoin_{all\text{-}1}(Island, Time\ Zone)$$
(every island is non-coincident with some time zone)

both hold, we cannot infer:

$$EC_{all\text{-}1}(Island, Time\ Zone)$$
(every island is externally connected to some time zone).

As another example, $SPxy \rightarrow \sim Cxy$ follows immediately from the definition of SP in MI, but $SP_i(A, B) \rightarrow \sim C_i(A, B)$ (i = some, all-1, all-2, all-12) are not theorems of MIC. Thus, although

$$SP_{all\text{-}1}(Continent, Sea)$$
(every continent is separated from some sea)

holds, we cannot infer:

---

[7] But as is shown in 4.3, implications of the form Rxy & $Syz \rightarrow Txz$ are exceptions.

$\sim C_{all-1}(Continent, Sea)$
(it is not the case that every continent is connected to some sea).

In addition, disjunctions do not in general transfer from MI to the class relations. R_{all-1}, R_{all-2}, R_{all-12}, and $R_{all-all}$ counterparts of MI theorem (CT6) are not theorems of MIC. For example, we cannot in MIC derive: $PCoin_{all-all}(Park, Road)$ $\lor$ $EC_{all-all}(Park, Road)$ $\lor$ $SP_{all-all}(Park, Road)$ (every park partially coincides with every road or every park is externally connected to every road or every park is separated from every road).

For some domains, it may be appropriate to strengthen MIC with additional axioms that require the class relations to retain more of the properties of the underlying relations among individuals. For example, we might add axioms requiring that P_{all-12} is antisymmetric and PP_{all-12} is asymmetric. Such axioms would allow us to infer, e.g., $\sim PP_{all-12}(Shore, Shoreline)$ from $PP_{all-12}(Shoreline, Shore)$.

But in most cases, it is appropriate that the logical properties of the MIC class relations do not exactly match those of the original MI relations. For example, although C is symmetric, we would not want the axioms of our formal theory to allow us to infer the false assertion:

$C_{all-1}(Road, Mobile\ Home\ Park)$
(every road is connected to some mobile home park)

from the true assertion:

$C_{all-1}(Mobile\ Home\ Park, Road)$
(every mobile home park is connected to some road).

5   Interaction Among Different Types of Class Relations

In this section, we present other important logical properties of MIC's class relations. We focus on interactions among the R_{some}, R_{all-1}, R_{all-2}, and R_{all-12} relations (subsection 5.1) and interactions between class-level spatial relations and the Is_a relation (subsection 5.2).

5.1   Interaction Among R_{some}, R_{all-1}, R_{all-2}, and R_{all-12} Relations

Important properties of the MIC class relations stem from the interaction among the R_{some}, R_{all-1}, R_{all-2}, and R_{all-12} relations.

For any MI relation R, the following implications hold in MIC:

(MICTS1) $R_{all-12}(A, B) \rightarrow R_{all-1}(A, B)\ \&\ R_{all-2}(A, B)$
(MICTS2) $R_{all-1}(A, B) \lor R_{all-2}(A, B) \rightarrow R_{some}(A, B)$

Thus, if R and S are MI relations such that the following implications hold in MIC

$R_i(A, B) \rightarrow S_i(A, B)$ i = some, all-1, all-2, all-12

then the following additional implications can be derived using theorem schemata (MICTS1) and (MICTS2):

$R_{all-12}(A, B) \rightarrow S_i(A, B)$	i = some, all-1, all-2

$R_{\text{all-1}}(A, B) \vee R_{\text{all-2}}(A, B) \rightarrow S_{\text{some}}(A, B)$	

For example using (MICT1-4), (MICTS1), and (MICTS2), we can prove the following MIC theorems:

MICT68-71	$O_{\text{all-12}}(A, B) \rightarrow PCoin_i(A, B)$	i = some, all-1, all-2
MICT72	$O_{\text{all-1}}(A, B) \vee O_{\text{all-2}}(A, B) \rightarrow PCoin_{\text{some}}(A, B)$	

Now assume that R, S, and T are MI relations such that the follow implications hold in MIC:

$$R_i(A, B) \,\&\, S_i(B, C) \rightarrow T_i(A, C), \quad i = \text{all-1, all-2}$$

Then using theorem schema (MICTS1), the following additional implications can be derived:

$R_{\text{all-12}}(A, B) \,\&\, S_i(B, C) \rightarrow T_i(A, C)$	i = all-1, all-2
$R_i(A, B) \,\&\, S_{\text{all-12}}(B, C) \rightarrow T_i(A, C)$	i = all-1, all-2

For example, the following theorems can be derived from (MICT53), (MICT54), and (MICTS1):

MICT73-74	$\text{Loc-In}_{\text{all-12}}(A, B) \,\&\, P_i(B, C) \rightarrow \text{Loc-In}_i(A, C)$	i = all-1, all-2
MICT75-76	$\text{Loc-In}_i(A, B) \,\&\, P_{\text{all-12}}(B, C) \rightarrow \text{Loc-In}_i(A, C)$	i = all-1, all-2

Using these and other theorems derived from (MICTS1), (MICT36-39), and (MICT48-59), we can construct the following composition table. Table 4 tells us which relation between class A and class C can be inferred from a given assertion about the relation between class A and class B (listed in row headings) in conjunction with an assertion about the relation between class B and class C (listed in the column headings). For example, given $P_{\text{all-2}}(A, B)$ (row 2) and $\text{Loc-In}_{\text{all-12}}(B, C)$ (column 6), it follows from the axioms of MIC that $\text{Loc-In}_{\text{all-2}}(A, C)$ must also hold.

A blank cell in the table tells us that, unless additional information is given, we cannot derive any assertion of the form $R_i(A, C)$ where R is one of the relations of MI. For example, from $\text{Loc-In}_{\text{all-1}}(A, B)$ (row 4) and $P_{\text{all-2}}(B, C)$ (column 2) we cannot in general make any inference about the relation of class A to class C. To see this, note that $\text{Loc-In}_{\text{all-1}}(\text{Prostate, Pelvic Cavity})$ (every prostate is located in a pelvic cavity) and $P_{\text{all-2}}(\text{Pelvic Cavity, Female Pelvis})$ (every female pelvis has a pelvic cavity as a proper part) both hold, but neither $\text{Loc-In}_i(\text{Prostate, Female Pelvis})$ nor $P_i(\text{Prostate, Female Pelvis})$ holds for any i = some, all-1, all-2, all-12, all-all.[8]

Similar composition tables can be constructed for other of the MIC relations.

---

[8] On the other hand, both $\text{Loc-In}_{\text{all-1}}(\text{Prostate, Male Pelvis})$ and $P_{\text{all-1}}(\text{Prostate, Male Pelvis})$ hold.

Table 4. Composition table for the relations P_i and Loc-In$_i$ for i = all-1, all-2, all-12

	$P_{all-1}BC$	$P_{all-2}BC$	$P_{all-12}BC$	Loc-In$_{all-1}BC$	Loc-In$_{all-2}BC$	Loc-In$_{all-12}BC$
$P_{all-1}AB$	$P_{all-1}AC$		$P_{all-1}AC$	Loc-In$_{all-1}AC$		Loc-In$_{all-1}AC$
$P_{all-2}AB$		$P_{all-2}AC$	$P_{all-2}AC$		Loc-In$_{all-2}AC$	Loc-In$_{all-2}AC$
$P_{all-12}AB$	$P_{all-1}AC$	$P_{all-2}AC$	$P_{all-12}AC$	Loc-In$_{all-1}AC$	Loc-In$_{all-2}AC$	Loc-In$_{all-12}AC$
Loc-In$_{all-1}$A B	Loc-In$_{all-1}AC$		Loc-In$_{all-1}AC$	Loc-In$_{all-1}AC$		Loc-In$_{all-1}AC$
Loc-In$_{all-2}$A B		Loc-In$_{all-2}AC$	Loc-In$_{all-2}AC$		Loc-In$_{all-2}AC$	Loc-In$_{all-2}AC$
Loc-In$_{all-12}AB$	Loc-In$_{all-1}AC$	Loc-In$_{all-2}AC$	Loc-In$_{all-12}AC$	Loc-In$_{all-1}AC$	Loc-In$_{all-2}AC$	Loc-In$_{all-12}AC$

5.2 Interaction Between Class-Level Spatial Relations and the Is_a Relation

Additional important properties of the MIC class relations stem from the interaction between the class spatial relations and the Is_a class subsumption relation. Table 5 lists theorem schemata of MIC describing the interaction between the R_{some}, R_{all-1}, R_{all-2}, R_{all-12}, and $R_{all-all}$ relations and the Is_a relation. Here, R can be replaced by any MI relation.

Table 5. Interaction between class-level spatial relations and the Is_A relation

$R_{all-1}(A, B)$ & Is_a(B, C) $\rightarrow R_{all-1}(A, C)$	$R_{all-12}(A, B)$ & Is_a(C, B) $\rightarrow R_{all-2}(A, C)$
$R_{all-1}(A, B)$ & Is_a(C, A) $\rightarrow R_{all-1}(C, B)$	$R_{all-all}(A, B)$ & Is_A(B, C) $\rightarrow R_{all-1}(A, C)$
$R_{all-2}(A, B)$ & Is_a(A, C) $\rightarrow R_{all-2}(C, B)$	$R_{all-all}(A, B)$ & Is_A(A, C) $\rightarrow R_{all-2}(C, B)$
$R_{all-2}(A, B)$ & Is_a(C, B) $\rightarrow R_{all-2}(A, C)$	$R_{all-all}(A, B)$ & Is_A(C, A) $\rightarrow R_{all-all}(C, B)$
$R_{all-12}(A, B)$ & Is_a(A, C) $\rightarrow R_{all-2}(C, B)$	$R_{all-all}(A, B)$ & Is_A(C, B) $\rightarrow R_{all-all}(A, C)$
$R_{all-12}(A, B)$ & Is_a(C, A) $\rightarrow R_{all-1}(C, B)$	$R_{some}(A, B)$ & Is_A(B, C) $\rightarrow R_{some}(A, C)$
$R_{all-12}(A, B)$ & Is_a(B, C) $\rightarrow R_{all-1}(A, C)$	$R_{some}(A, B)$ & Is_A(A, C) $\rightarrow R_{some}(C, B)$

MIC theorems for the class parthood relations corresponding to the schemata above are represented compactly in Table 6.

Table 6. Inferences from conjunctions of P_i and Is_a assertions

	Is_a(C, A)	Is_a(A, C)	Is_a(C, B)	Is_a(B, C)
$P_{some}(A, B)$		$P_{some}(C, B)$		$P_{some}(A, C)$
$P_{all-1}(A, B)$	$P_{all-1}(C, B)$			$P_{all-1}(A, C)$
$P_{all-2}(A, B)$		$P_{all-2}(C, B)$	$P_{all-2}(A, C)$	
$P_{all-12}(A, B)$	$P_{all-1}(C, B)$	$P_{all-2}(C, B)$	$P_{all-2}(A, C)$	$P_{all-1}(A, C)$
$P_{all-all}(A, B)$	$P_{all-all}(C, B)$	$P_{all-2}(C, B)$	$P_{all-all}(A, C)$	$P_{all-1}(A, C)$

The cells of Table 6 tell us i) which (if any) of the P_i relations must hold between A and C when a given P_i relation holds between A and B (listed in the row headings) and a given subsumption relation holds between B and C (listed in the column headings) or ii) which (if any) of the P_i relations must hold between C and B when a given P_i relation holds between A and B (row headings) and a given subsumption relations holds between A and C (column headings). For example, given $P_{all-2}(A, B)$ (row 3) and Is_a(C, B) (column 3), it follows that $P_{all-2}(A, C)$ must also hold. This corresponds to the MIC theorem:

$$P_{all-2}(A, B) \ \& \ Is\_a(C, B) \rightarrow P_{all-2}(A, C).$$

A blank cell indicates that, unless further information is given, no inference of the form $R_i(A, C)$, $R_i(C, A)$, $R_i(B, C)$, or $R_i(C, B)$ (with R a MI relation) can be made. For example, from $P_{all-2}(A, B)$ (row 3) and Is_a(C, A) (column 1), we cannot in general make any inference about the relation of A to C. To see this consider the example: $P_{all-2}(Building, Hospital\ Complex)$ (every hospital complex has a building as a part) and Is_a(*Cathedral, Building*) (every cathedral is a building), but no cathedral is part of any hospital complex.

Table 7 is analogous to Table 6 but represents inferences involving the Loc-In$_i$ relations rather than the P_i relations.

Table 7. Inferences from conjunctions of Loc-In$_i$ and Is_a assertions

	Is_a(C, A)	Is_a(A, C)	Is_a(C, B)	Is_a(B, C)
Loc-In $_{some}$(A, B)		Loc-In $_{some}$(C, B)		Loc-In $_{some}$(A, C)
Loc-In $_{all-1}$(A, B)	Loc-In $_{all-1}$(C, B)			Loc-In $_{all-1}$(A, C)
Loc-In $_{all-2}$(A, B)		Loc-In $_{all-2}$(C, B)	Loc-In $_{all-2}$(A, C)	
Loc-In $_{all-12}$(A, B)	Loc-In $_{all-1}$(C, B)	Loc-In $_{all-2}$(C, B)	Loc-In $_{all-2}$(A, C)	Loc-In $_{all-1}$(A, C)
Loc-In $_{all-all}$(A, B)	Loc-In $_{all-all}$(C, B)	Loc-In $_{all-2}$(C, B)	Loc-In $_{all-all}$(A, C)	Loc-In $_{all-1}$(A, C)

For example, from Loc-In$_{all-12}$(*Ovary, Cavity of Female Pelvis*) and Is_a(*Cavity of Female Pelvis, Cavity of Pelvis*), it follows (row 4/column 4) that Loc-In$_{all-1}$(*Ovary, Cavity of Pelvis*). On the other hand, no assertion of the form R_i(*Ovary, Cavity of Male Pelvis*) follows from Loc-In$_{all-1}$(*Ovary, Cavity of Pelvis*) and Is_a(*Cavity of Male Pelvis, Cavity of Pelvis*) (row 2/column 3).

6 Conclusions

A central goal in artificial intelligence is to create ontologies which encode the general background knowledge needed for organizing and using data in a specific domain such as medicine or geography. In recent years, much work has been invested in developing biomedical ontologies, such as the FMA and GALEN. These ontologies typically use class-level spatial relations similar to those of MIC for structuring information and (to a limited extent) automatically generating further assertions from assertions which have been manually inputted. However, the relations assumed in the biomedical ontologies have not been grounded in a formal theory. As a result, the logical properties and intended interpretations of these relations are not clear. It is shown in (Donnelly, Bittner et al. 2004) that some relations in the FMA and GALEN are used differently in different contexts and that the automated reasoning implemented in these ontologies does not always conform to valid inference forms. A reformulation of the biomedical ontologies' relations in terms of a formal theory such as MIC is necessary both for consistent automated reasoning within a given ontology and the alignment of different biomedical ontologies.

In addition, the class-level spatial relations of MIC (or an extension of MIC that also includes distance or orientation relations) can be used to link the entity types and included terms of the SDTS. This would transform the SDTS into a structured body of knowledge which can be used for automated reasoning. Many of the informal definitions of SDTS entity types already link geographic classes through spatial relations holding among their instances. For example, a shore is "that part of land in immediate contact with a body of water including the area in between high and low water lines", an oasis is "a small, isolated, fertile or green area in a desert region, usually having a spring or well", and ridge line is "line separating drainage basins" (SDTS 1997). By using these kinds of definitions as a basis for formal relations between entity types, we can make the meanings of the informal definitions more precise and make the information embodied in the informal definitions available for automated reasoning.

Acknowledgements

We gratefully acknowledge useful comments from Barry Smith, Alex Sorokine, David Mark, and three anonymous reviewers. Our work has been supported by the Wolfgang Paul Program of the Alexander von Humboldt Foundation, the EU Network of Excellence in Semantic Data Mining, and the Volkswagen Foundation Project "Forms of Life".

Bibliography

Asher, N. and L. Vieu (1995). Towards a geometry of commonsense: semantics and a complete axiomatization of mereotopology. Proceedings of IJCAI'95, Morgan Kaufmann: 846-852.

Beck, R. and S. Schulz (2003). Logic-based remodeling of the Digital Anatomist Foundational Model. AIMA Annual Symposium Proceedings: 71-75.

Bittner, T., M. Donnelly, et al. (2004). Individuals, universals, and collections: on the foundational relations of ontology. Proceedings of FOIS'04: 37-48.

Casati, R. and A. C. Varzi (1999). Parts and Places, Cambridge, MA: MIT Press.

Cleland, D. T., P. E. Avers, et al. (1997). National hierarchical framework of ecological units. Ecosystem Management Applications for Sustainable Forest and Wildlife Resources. M. S. a. H. Boyce, Alan W. New Haven, CT: 181-200.

Cohn, A. G., B. Bennett, et al. (1997). Qualitative spatial representation and reasoning with the Region Connection Calculus. Geoinformatica 1(3): 1-44.

Donnelly, M. (2004). A formal theory for reasoning about parthood, connection, and location. Journal of Artificial Intelligence 160: 145-172.

Donnelly, M. (2005). Relative places. Applied Ontology in press.

Donnelly, M., T. Bittner, et al. (in press). A formal theory for spatial representation and reasoning in biomedical ontologies. Artificial Intelligence in Medicine.

Levesque, H. and R. Brachman (1985). Fundamental Trade-off in Knowledge Representation and Reasoning. Readings in Knowledge Representation. R. Brachman and H. Levesque, Morgan Kaufmann, Los Altos: 41-70.

McNab, W. H. and P. E. Avers (1994). Ecological subregions of the United States, WO-WSA-5. http://www.fs.fed.us/land/pubs/ecoregions/index.html.

Mejino, J. L. V., A. V. Agoncillo, et al. (2003). Representing complexity in part-whole relationships within the Foundational Model of Anatomy. Proceedings of the American Medical Informatics Association Fall Symposium: 450-454.

Rector, A. L. and J. E. Rogers (2002a). Ontological issues in using a description logic to represent medical concepts: experience from GALEN: part 1 - principles. Methods of Information in Medicine.

Rector, A. L. and J. E. Rogers (2002b). Ontological issues in using a description logic to represent medical concepts: experience from GALEN: part II - the GALEN high level schemas. Methods of Information in Medicine.

Rogers, J. and A. Rector (2000). GALEN's model of parts and wholes: experience and comparisons. Proceedings of the AMIA Symp 2000: 714-718.

Rosse, C. and J. L. V. Mejino (2003). A reference ontology for bioinformatics: the Foundational Model of Anatomy. Journal of Biomedical Informatics 36:478-500.

SDTS (1997). American national standard for information systems - Spatial Data Transfer Standard (SDTS). http://mcmcweb.er.usgs.gov/sdts/standard.html

Smith, B., W. Ceusters, et al. (2005). Relations in biomedical ontologies. Genome Biology in **press**.

Smith, B. and C. Rosse (2003). The role of foundational relations in the alignment of biomedical ontologies. Proceedings of the 11th World Congress on Medical Informatics. M. Fieschi, E. Coiera and Y. J. Li: 444 - 448.

Sorokine, A. and T. Bittner (2005). Understanding taxonomies of ecosystems: a case study. Developments in Spatial Data Handling. P. Fisher: 559-572.

Sorokine, A., T. Bittner, et al. (2004). Ontological Investigation of Ecosystem Hierarchies and Formal Theory for Multiscale Ecosystem Classifications. Proceedings of GI-Science 2004.

CASL Specifications of Qualitative Calculi

Stefan Wölfl[1] and Till Mossakowski[2]

[1] Department of Computer Science, University of Freiburg,
Georges-Köhler-Allee, 79110 Freiburg, Germany
woelfl@informatik.uni-freiburg.de
[2] Department of Computer Science, University of Bremen, P.O. Box 330440,
28334 Bremen, Germany
till@tzi.de

Abstract. In AI a large number of calculi for efficient reasoning about spatial and temporal entities have been developed. The most prominent temporal calculi are the point algebra of linear time and Allen's interval calculus. Examples of spatial calculi include mereotopological calculi, Frank's cardinal direction calculus, Freksa's double cross calculus, Egenhofer and Franzosa's intersection calculi, and Randell, Cui, and Cohn's region connection calculi.

These calculi are designed for modeling specific aspects of space or time, respectively, to the effect that the class of intended models may vary widely with the calculus at hand. But from a formal point of view these calculi are often closely related to each other. For example, the spatial region connection calculus RCC5 may be considered a coarsening of Allen's (temporal) interval calculus. And vice versa, intervals can be used to represent spatial objects that feature an internal direction.

The central question of this paper is how these calculi as well as their mutual dependencies can be axiomatized by algebraic specifications. This question will be investigated within the framework of the *Common Algebraic Specification Language* (CASL), a specification language developed by the *Common Framework Initiative for algebraic specification and development* (COFI). We explain scope and expressiveness of CASL by discussing the specifications of some of the calculi mentioned before.

1 Introduction: Calculemus!

In the past 25 years qualitative spatial and temporal reasoning has evolved to a discipline in its own right within AI. Qualitative reasoning aims at describing the common-sense background knowledge on which our human perspective on the physical reality is based. The calculi, that is, formal languages and reasoning techniques, developed in this research area are of special interest for all application fields that rely on human-machine interaction in static or dynamically changing spatial environments. For example, some of these calculi may be implemented for handling spatial GIS queries efficiently and some may be used for navigating, and communicating with, a mobile robot.

One will hardly find an exact definition of the notion of qualitative reasoning, but the whole research area has been very much inspired by Hayes' naïve manifesto [15]. In fact, in different areas of mathematics and physics very expressive formalisms for reasoning about space and time have been developed. But from a computer scientist's

A.G. Cohn and D.M. Mark (Eds.): COSIT 2005, LNCS 3693, pp. 200–217, 2005.
© Springer-Verlag Berlin Heidelberg 2005

point of view most of these formalisms are too expressive. Since there is an inevitable trade-off between the expressiveness of a language and the computational costs for reasoning with its formulae, expressiveness can get a crucial point when these calculi are to be integrated in applications. Thus the fundamental idea of qualitative reasoning is to restrict the vocabulary of rich mathematical theories in such a way that diversified aspects of these theories are treated within distinguished decidable fragments with simple qualitative (i. e., non-metrical) languages.

From this starting point a large number of calculi for efficient reasoning about spatial and temporal entities have been proposed in the literature. The most prominent temporal calculi are the so-called point algebra, which deals with instants of a given linear flow of time, and Allen's interval algebra [1], which describes possible relations between intervals in linear flows of time (cf. Fig. 1). Analogous calculi have been proposed for more general classes of models such as branching flows of time [e. g., 5] or even structures, where flows of time are just required to satisfy the conditions of a partial order [e. g., 6]. Despite these one-sorted calculi, also many-sorted calculi have been proposed. For example, Vilain's point-interval calculus [24] deals with instants and intervals in linear flows of time and may be considered a combination of the point algebra and the interval algebra.

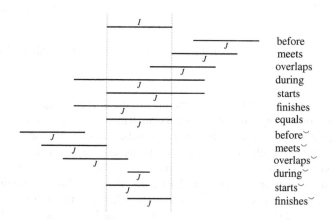

Fig. 1. Allen's interval relations

Examples of spatial calculi include mereotopological calculi (e. g., [2]), Frank's cardinal direction calculus [11], Freksa's double cross calculus [13], Egenhofer and Franzosa's 4- and 9-intersection calculi [9, 10], Ligozat's flip-flop calculus [16], and various region connection calculi proposed by Randell et al. [21], Cohn et al. [7], Düntsch et al. [8], and Gerevini and Renz [14]. It is interesting to see that even these few calculi employ concepts from a wide range of mathematical theories. Some of them are based on geometrical notions such as lines, half-planes, and angels, some describe relations between physical objects in terms of point set topology, and some include qualitative size information.

Interestingly, some spatial calculi are closely related to the temporal calculi mentioned previously. For example, the *2-point calculus* describes points of the plane and their relationships in terms of the point-to-point relations between their coordinates. This means, that one considers for each dimension one of the three possible point-to-point relations $<$, $=$, and $>$ between the point coordinates. The relation $(=,>)$, for instance, expresses the relation "north-to", $(>,>)$ corresponds to "north-east-to", etc. For this reason the 2-point calculus is often referred to as *cardinal direction calculus* in the literature. Analogously, the *rectangle calculus* describes possible relations between rectangles in the plane by comparing their coordinate projections in terms of the interval algebra.

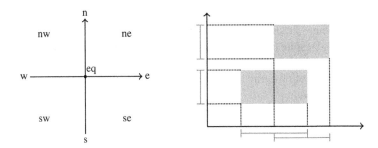

Fig. 2. Spatial calculi derivable from temporal algebras: The cardinal direction calculus and the rectangle calculus

To sum up this little discussion, researchers in the domain of qualitative reasoning face a vast, still increasing amount of calculi (much more than the rather incomplete list of calculi previously mentioned can indicate). On the other hand, many of these calculi are closely related to each other: some are simple extensions of others, some show similarities on the syntactic level, some have related classes of intended models, i. e., they are based on more or less the same background theory. Thus the guiding question of this paper is how to present qualitative calculi within a common framework in such a way that the mutual dependencies between them and their respective background theories becomes more transparent. In fact, transparency is an important issue for both avoiding redundancies and ensuring reusability of these calculi.

In our opinion, a naïve ontological classification system of qualitative or other calculi will not be able to fulfill these requirements in an adequate manner. A calculus classified as, say, "temporal" will always be a calculus about *temporal entities* such as instants or intervals, and can hardly be subsumed under the term "spatial calculus". Moreover, connections between temporal and spatial calculi like the ones previously mentioned seem inexpressible in any ontology of such calculi. For this reason we propose to present qualitative calculi by means of algebraic specification, as was already suggested by Frank [12]. More exactly, we explain how to develop such specifications within the *Common Algebraic Specification Language* (CASL).

The paper is organized as follows: In section 2 we briefly explain fundamental notions related to qualitative reasoning in more detail. Section 3 provides a short introduction into CASL and its extension HASCASL. Then in section 4 we discuss algebraic specifications of the concepts introduced in section 2 in an informal manner.

2   Qualitative Calculi

To exemplify the most fundamental ideas of qualitative reasoning, let us discuss the point algebra for linear time in more detail. From a model-theoretical perspective this point algebra aims at describing the class of *linear flows of time*, i. e., first order structures $\mathscr{T} = \langle T, < \rangle$ that are models of the following axioms:

Irreflexivity: $\forall x\, x \not< x$
Transitivity: $\forall xyz\, (x < y \wedge y < z \rightarrow x < z)$
Linearity: $\forall xy\, (x < y \vee x = y \vee y < x)$

From an ontological point of view, the point algebra takes instants of time as *primary objects*, and states that these entities are linearly ordered. If we shift consideration from these primary objects towards the relations between them, we can state the following observations: From $<$ being irreflexive and transitive, it follows that the relations $<$, $=$, and $>$ (where $>$ is just defined as the converse of relation $<$) are pairwise disjoint. Linearity guarantees that these relations are jointly exhaustive, i. e., for each pair of instants t and t', one of the relations $t < t'$, $t = t'$, or $t > t'$ holds. Speaking algebraically, the set $\{<, =, >\}$ forms a jointly exhaustive and pairwise disjoint (JEPD) system of relations.

Often temporal or spatial information is imprecise, for example, when we only have the information that instant t is not before instant t', or that instants t and t' are distinct. In this situation it becomes interesting to consider not only the *base relations* $<$, $=$, and $>$, but also arbitrary unions of them. Obviously, the system of unions of base relations defines an atomic Boolean algebra with the base relations as atoms. By the way, since the system of base relations is JEPD, unions of base relations may also be represented as *sets* of base relations.

Reasoning problems, then, are usually formulated as constraint satisfaction problems. A constraint network is a finite set of constraints where each constraint is a formula of the form $x R y$ with variables x and y (taking values in given domains D_x and D_y) and a relation R (a set of base relation) defined between the domains of x and y. Typical reasoning tasks are then to determine whether a constraint set is satisfiable, to check that some constraint is entailed by a constraint network, and to compute an equivalent minimal constraint set — it is not hard to see that all these reasoning tasks are equivalent under Turing reductions.

A crucial aspect for developing efficient algorithms for qualitative spatial and temporal reasoning is the fact that the underlying model classes usually contain infinite models. Hence, in order to test satisfiability of constraint networks, it is not feasible to enumerate all models until one finds a satisfying model. For this reason other techniques (such as path-consistency algorithms) must be applied for testing satisfiability. Many of these techniques, in turn, rely on semantically verified composition tables that list which relations are consistent when two base relations are composed. For

Table 1. The composition table of the point algebra for linear time

	<	>	=
<	<	<,=,>	<
>	<,=,>	>	>
=	<	>	=

example, Table 1 presents the composition table of the point algebra for linear time.[1] But a set of base relations together with a composition table satisfying some minimal conditions defines a *relation algebra* on the set of all unions of base relations. For this reason studying relation algebras has become a central aspect in the field of qualitative reasoning.

3 CASL and Friends

The Common Algebraic Specification Language (CASL) is a specification language, which was developed by the *Common Framework Initiative for Algebraic Specification and Development* (CoFI). CASL allows for writing algebraic specifications that can be expressed in a many-sorted first order language with partial function symbols. Basic CASL specifications consist of signature declarations and axioms characterizing the models to be described. These axioms, in turn, are first-order formulae or assertions regarding the definedness of partial function symbols. Going beyond first-order logic, CASL also provides constructs to state induction principles (called sort generation constraints) and datatype declarations. Furthermore, specifications may contain subsort declarations, whereby subsort inclusions are treated as embeddings. Finally, CASL also provides constructs for structured specifications, namely, translations, reductions, unions, and extensions of specifications.

In the sequel we will explain these concepts in more detail (for a full discussion see Bidoit and Mosses [4] and Mosses [19]).

3.1 Constructing Specifications

To start with, let us briefly explain the formal underpinnings of CASL specifications. As said before, CASL allows for specifying first order theories with partial function symbols. More precisely, CASL accepts languages with *(many-sorted) signatures* $\Sigma = \langle S, TF, PF, R \rangle$ such that:

---

[1] Note that there are (at least) two ways of reading such composition tables, the *extensional* and the *consistency-based* reading [cf. 3]. Following, we will only use the extensional reading, which means that the algebraic function of composing relations (as used, for example, in relation algebras) coincides with its set-theoretical characterization. In the case of the point algebra, for example, the consistency-based reading is correct for the class of linear flows of time, while the extensional reading is only correct for the class of dense linear flows of time without endpoints.

- S is a (finite) set of sorts.
- For each $(w, s) \in S^* \times S$, $TF_{w,s}$ and $PF_{w,s}$ are disjoint sets of total and partial function symbols, respectively (tuples $w \in S^*$ are referred to as *sort profiles*).
- For each $w \in S^*$, R_w is a set of relation symbols.

As usual, individual symbols can be introduced as 0-ary total function symbols. Accordingly, models of such signatures are many-sorted partial first-order structures: Given a signature Σ, a Σ-*model* is a structure consisting of non-empty carrier sets s^M (for each sort $s \in S$), partial and total functions $f^M : w^M \longrightarrow s^M$ (for each function symbol $f \in PF_{w,s}$ or $f \in TF_{w,s}$, respectively), and relations $r^M \subseteq w^M$ (for each relation symbol $r \in R_w$).

The most fundamental notion related to extensions, unions, and translations of specifications is that of a signature morphism. To explain this notion, let $\Sigma = \langle S, TF, PF, R \rangle$ and $\Sigma' = \langle S', TF', PF', R' \rangle$ be signatures. Then a *signature morphism* $\Sigma \to \Sigma'$ is a 4-tuple $\sigma = \langle \sigma^s, \sigma^t, \sigma^p, \sigma^r \rangle$ consisting of maps (families of maps, resp.):

- $\sigma^s : S \longrightarrow S'$,
- $\sigma^t_{w,s} : TF_{w,s} \longrightarrow TF'_{\sigma^s(w), \sigma^s(s)}$,
- $\sigma^p_{w,s} : PF_{w,s} \longrightarrow TF'_{\sigma^s(w), \sigma^s(s)} \cup PF'_{\sigma^s(w), \sigma^s(s)}$, and
- $\sigma^r_w : R_w \longrightarrow R'_{\sigma^s(w)}$.

That is, partial function symbols may be mapped to total function symbols, but not vice versa.

On the semantic level, signature morphisms inherit models from the target to the source signature. To see this, let $\sigma : \Sigma \longrightarrow \Sigma'$ be a signature morphism, and let M' be a Σ'-model. Then σ defines a Σ-model $M'|_\sigma$ (the σ-*reduct* of M') by

$$s^{M'}|_\sigma := \sigma^s(s)^{M'}, \quad f^{M'}|_\sigma := \sigma^{t/p}(f)^{M'}, \quad \text{and} \quad r^{M'}|_\sigma := \sigma^r(r)^{M'}.$$

We are now ready to explain some fundamental notions in more detail.

Basic Specification. The most simple kind of CASL specifications, called *basic specifications*, are asserted by the keyword **spec** and have the form

$$\textbf{spec } \textit{SpecName} = \text{Spec}$$

where *SpecName* is the name of the specification and Spec is a list of signature declarations and first order axioms.

Extension. *Extensions* have the form

$$\textit{Spec1} \textbf{ then } \textit{Spec2}$$

where *Spec1* is a specification or a specification name, and *Spec2* is a specification that extends the signature of *Spec1* and/or adds additional axioms. CASL allows to indicate the type of the extension: *Definitional extensions* are introduced by annotating the keyword **then** with **%def**. From a model-theoretical point of view, definitional extensions are justified when each model of *Spec1* can be uniquely extended to a model of the specification *Spec1-then-Spec2*. *Implied extensions* state some theorems of the original specification, i. e., *Spec1* and *Spec1-then-Spec2* have the same model class. This kind of extension is introduced by the annotated keyword **then** **%implies** and is considered well-formed only if *Spec1* and *Spec1-then-Spec2* have the same signature.

Union. It is possible to join two specifications, i. e., to build their *union*. The signature of a union of specifications *Spec1* and *Spec2* is just the union of the signatures of *Spec1* and *Spec2*. The models of the union are exactly those models of the union signature whose reducts are models of *Spec1* and *Spec2*, respectively. Unions of specifications obey the "same name, same thing" principle, which means that each symbol contained in both signatures has a single interpretation in each model of their union. Unions are declared by

$$Spec1 \textbf{ and } Spec2.$$

Translations. *Translations* are renamings of sort symbols and/or signature symbols and thus provide a signature morphism from the source specification into the specification resulting from the translation. The models of the translation are exactly those models of the result specification whose reducts along the morphism are models of the source specification. Translations are declared by

$$Spec \textbf{ with } \text{SymbolMappings}.$$

Reductions. CASL also provides constructs to restrict the signature of a given specification. It is possible to hide symbols or, alternatively, declare the symbols that are revealed. Reductions can be declared by

$$Spec \textbf{ hide } \text{Symbols} \quad \text{and} \quad Spec \textbf{ reveal } \text{SymbolMappings}.$$

Parameterization and Instantiations. CASL also allows parameterized specifications, which are written as:

$$\textbf{spec } Spec1[Spec2]\dots[SpecN] = \text{Spec}.$$

Sometimes it is necessary to instantiate a previously declared specification via a symbol mapping before it is used in a parameterized specification. This can be obtained by

$$Spec1[Spec2 \textbf{ fit } \text{SymbolMappings}].$$

Views. A *view* provides a signature morphism (defined by a symbol mapping) between two specifications. Actually, it is required to be a *theory morphism* (or *interpretation of theories*), which means that each model of the target specification induces (when reduced via the signature morphism) a model of the source specification. Views are declared by

$$\textbf{view } View : Spec1 \textbf{ to } Spec2 = \text{SymbolMapping}.$$

The possibility of specifying views is one of the distinguished features of CASL.

3.2 HASCASL

HASCASL (see [22]) is a higher-order extension of CASL based on the partial λ-calculus. The user declared sorts are used to generate *higher types* by closing the set of types under total and partial function spaces (written by $t_1 \rightarrow t_2$ and $t_1 \rightarrow ?t_2$, resp.). Predicates (written *Pred t*) are coded as partial functions into a singleton type (i. e., only the domain provides relevant information). In fact, we often will need HAS-CASL for specifying the model classes of qualitative calculi (e. g., for the real numbers, for metric and topological spaces). CASL's structuring constructs (union, translation, hiding, etc.) are independent of the underlying logical system and hence can be used for HASCASL as well.

3.3 Tools

HETS. The *Heterogeneous Tool Set (*HETS*)* [18], which is developed at the University of Bremen, Germany, is the main analysis tool for CASL and its extensions. HETS integrates a parser and a type-checker for heterogeneous specifications. A graphical interface allows for presenting the development graph (showing the specification structure) of CASL specifications as well as the logic graph presenting the underlying logic(s). It is possible to translate a CASL specification into XML-, LATEX-, and other formats. HETS also provides an interface to translate CASL specifications into Isabelle theory files. Of course, HETS also supports HASCASL specifications.

Isabelle. Isabelle is a generic proof assistant, which is developed by L. C. Paulson (University of Cambridge, UK) and T. Nipkow (Technical University of Munich, Germany). Isabelle provides a rich language for expressing mathematical formulae and contains tools for proving these formulae in a logical calculus. Its main application fields are the formalization of mathematical proofs and the formal verification of computer languages, protocols, computer hardware, and software specifications.

A main feature of Isabelle is that it is not restricted to a single formal calculus since it supports higher-order logic, axiomatic set theory, etc. We use Isabelle/HOL, the coding of higher-order logic in Isabelle. For a more comprehensive introduction we refer to Nipkow et al. [20].

4 Specifications of Qualitative Calculi

4.1 Relation Algebras

To start with, let us first discuss some specifications related to relation algebras. A *relation algebra* is a Boolean algebra with complement (its elements are referred to as *relations*) and with a distinguished element *id* (the *identity relation*), a unary total function ⌣ assigning to each relation its converse relation (algebraically, its *involution*), and a binary total function ∘ assigning to each pair of relations their composition. Note that here the term "relation" is used in an abstract manner, i. e., independently of its usual set-theoretical interpretation. Rather, these functions are characterized implicitly by the axioms listed in the following specification.

    spec RELATIONALGEBRA =
            BOOLEANALGEBRAWITHCOMPL **with sort** *Elem* ↦ *Rel*
    then
        ops *id* : *Rel*;
            __⌣ : *Rel* → *Rel*;
            __∘__ : *Rel* × *Rel* → *Rel*, assoc, unit *id*;

$\forall x, y, z : Rel$
- $(x^{\smile})^{\smile} = x$ %(inv_idempot)%
- $(x \sqcup y)^{\smile} = x^{\smile} \sqcup y^{\smile}$ %(inv_cup)%
- $(-x)^{\smile} = -x^{\smile}$ %(inv_compl)%
- $(x \circ y)^{\smile} = y^{\smile} \circ x^{\smile}$ %(inv_cmps)%
- $(x \circ y) \sqcap z^{\smile} = 0 \Rightarrow (y \circ z) \sqcap x^{\smile} = 0$ %(triangle)%

then %**implies**
$\forall x, y, z : Rel$
- $(x \sqcup y) \circ z = (x \circ z) \sqcup (y \circ z)$ %(cmps_cup_rdistrib)%
- $z \circ (x \sqcup y) = (z \circ x) \sqcup (z \circ y)$ %(cmps_cup_ldistrib)%
- $(x^{\smile} \circ -(x \circ y)) \sqcap y = 0$ %(RelAlg)%

end

We may define a partial order on a relation algebra in exactly the same way as we could introduce it for arbitrary Boolean algebras. This gives us some nice corollaries, which could easily be proven by Isabelle. For example, the composition of relations behaves monotonic with respect to the canonical partial order.

spec EXTRELATIONALGEBRABYPARTIALORDER[RELATIONALGEBRA] = %**def**
EXTBOOLEANALGEBRABYPARTIALORDER[BOOLEANALGEBRA]
with sort $Elem \mapsto Rel$
then %**implies**
$\forall x, y, x', y' : Rel$
- $x \leq x' \wedge y \leq y' \Rightarrow x \circ y \leq x' \circ y'$ %(cmps_monotonic)%
- $x \leq id \Rightarrow x^{\smile} = x$ %(inv_below_id)%

end

In the following, we will be mainly interested in *atomic relation algebras*. An *atom* of a Boolean algebra is a non-zero element x such that the zero element is the only element y with $y < x$. A relation algebra (or Boolean algebra) is said to be *atomic* if for each non-zero element x, there exists an atom a with $a \leq x$. Hence the unary predicate "Atom" defines a genuine subsort.

spec ATOMICRELATIONALGEBRA =
RELATIONALGEBRA
and ATOMICBOOLEANALGEBRA **with sort** $Elem \mapsto Rel$, $AtomElem \mapsto AtomRel$
end

As explained in section 2, often an (abstract) atomic relation algebra can be constructed from a set of base relations and a composition table. In CASL this procedure can be reconstructed as follows: First we define relations (i. e., sort *Rel*) as arbitrary sets of base relations such that base relations correspond to singleton sets of base relations.

spec SETREPRESENTATIONOFRELATIONS [**sort** $BaseRel$] = %**def**
local { SET [**sort** $BaseRel$ **fit** $Elem \mapsto BaseRel$]
        with $\_\_\cup\_\_ \mapsto \_\_\sqcup\_\_, \_\_\cap\_\_ \mapsto \_\_\sqcap\_\_, \_\_\subseteq\_\_ \mapsto \_\_\sqsubseteq\_\_$ }
within
        free type $Rel ::= $ **sort** $Set[BaseRel]$
        sort $BaseRel < Rel$
        ops $0, 1 : Rel;$

$- : Rel \rightarrow Rel$;
$\_\_\sqcup\_\_, \_\_\sqcap\_\_: Rel \times Rel \rightarrow Rel$

preds $\_\_\in\_\_: BaseRel \times Rel$;
$\_\_\sqsubseteq\_\_ : Rel \times Rel$

$\forall x : BaseRel; r : Rel$
- $x = \{x\}$
- $x \in 1 \wedge \neg x \in 0$
- $x \in -r \Leftrightarrow \neg x \in r$

then %**implies**
$\cdots$

end

view SETREPRESENTATION_AS_ATOMICBOOLEANALGEBRA [**sort** *BaseRel*] :
ATOMICBOOLEANALGEBRA
to SETREPRESENTATIONOFRELATIONS [**sort** *BaseRel*]
= *Elem* $\mapsto$ *Rel*, *AtomElem* $\mapsto$ *BaseRel*
end

In a second step we generate a relation algebra-like structure by extending the functions "composition" and "involution" (as defined for base relations) to total functions on all relations. In fact, not each composition table defines a relation algebra since many such constructed structures violate the associativity axiom of relation algebras [see, e. g., 17]. The view contained in the following small library states that we obtain a genuine relation algebra if the composition table satisfies certain conditions. For the sake of simplicity, our specification of composition tables includes both the the composition function and the involution function for base relations.

spec COMPOSITIONTABLE =
 sorts *BaseRel* < *Rel*
 ops *id* : *BaseRel*;
  $0, 1 : Rel$;
  $\_\_^{\smile} : BaseRel \rightarrow BaseRel$;
  $\_\_\circ\_\_ : BaseRel \times BaseRel \rightarrow Rel$;
  $-\_\_ : Rel \rightarrow Rel$;
  $\_\_\sqcup\_\_ : Rel \times Rel \rightarrow Rel$, assoc, idem, comm, unit 1
 $\forall x : BaseRel$
 - $x \circ id = x \wedge id \circ x = x$
 - $id^{\smile} = id$
 - $(x^{\smile})^{\smile} = x$
end

spec CONSTRUCTRELATIONALGEBRA [**sort** *BaseRel*] [COMPOSITIONTABLE] = %**def**
 SETREPRESENTATIONOFRELATIONS [**sort** *BaseRel*]
then %**def**
 ops *id* : *Rel*;
  $\_\_^{\smile} : Rel \rightarrow Rel$;
  $\_\_\circ\_\_ : Rel \times Rel \rightarrow Rel$;
 $\forall x, y : BaseRel; r, s : Rel$
 - $x \in r^{\smile} \Leftrightarrow x^{\smile} \in r$

- $x \in (r \circ s) \Leftrightarrow \exists y, z : BaseRel \bullet y \in r \wedge z \in s \wedge x \in (y \circ z)$

then %**implies**

    op $\_\_\circ\_\_ : Rel \times Rel \rightarrow Rel$, **unit** id;

end

view CONSTRUCTEDRELATIONALGEBRA_AS_ATOMICRELATIONALGEBRA
         [**sort** *BaseRel*] [GOODCOMPOSITIONTABLE]:
         ATOMICRELATIONALGEBRA
to   CONSTRUCTRELATIONALGEBRA [**sort** *BaseRel*] [GOODCOMPOSITIONTABLE]
= $Rel \mapsto Rel, AtomRel \mapsto BaseRel$
end

Let us now turn to the semantic level. First we define the concept *algebra of binary relations (BRA)*. Given a set X, an algebra of binary relations is a Boolean subalgebra of the set algebra of all binary relations on X that contains the identity relation and is closed with respect to involution and composition (in their usual set-theoretical meaning). Of course, each algebra of binary relations is a relation algebra. But contrary to Boolean algebras (cf. Stone's representation theorem), it is in general not the case that each relation algebra can be represented as an algebra of binary relations.

logic HASCASL

spec BINARYRELATIONS [**sort** *Elem*] = %**mono**
    SET
then %**mono**
    type $Relation ::= \mathrm{abs}(rep : Set(Elem \times Elem))$
end

spec SETALGEBRAOFBINARYRELATIONS =
    BINARYRELATIONS [**sort** *Elem*]
then
    type $Rel < Relation$
    ops $0, 1 : Rel$;
        $- : Rel \rightarrow Rel$;
        $\_\_\sqcup\_\_, \_\_\sqcap\_\_ : Rel \times Rel \rightarrow Rel$
    $\forall r, s : Rel$
    - $rep(0) = \emptyset$
    - $rep(1) = allSet$
    - $rep(r \sqcup s) = rep(r) \cup rep(s)$
    - $rep(r \sqcap s) = rep(r) \cap rep(s)$
    - $rep(-r) = rep(1) \setminus rep(r)$
end

spec ALGEBRAOFBINARYRELATIONS =
    SETALGEBRAOFBINARYRELATIONS
then
    ops $id : Rel$;
        $\_\_^{\smile} : Rel \rightarrow Rel$;
        $\_\_\circ\_\_ : Rel \times Rel \rightarrow Rel$;
    $\forall r, s : Rel; x, y : Elem$

- $(x,y) \in rep(r \circ s) \Leftrightarrow \exists z : Elem \bullet (x,z) \in rep(r) \wedge (z,y) \in rep(s)$
- $(x,y) \in rep(r^{\smile}) \Leftrightarrow (y,x) \in rep(r)$
- $(x,y) \in rep(id) \Leftrightarrow x = y$

then **%implies**

    ops $\_\_\circ\_\_ : Rel \times Rel \rightarrow Rel$, assoc, unit id

end

spec FULLALGEBRAOFBINARYRELATIONS [**sort** *Elem*] = **%def**
    { ALGEBRAOFBINARYRELATIONS **with type** *Rel* $\mapsto$ *Relation* }
end

view ALGEBRAOFBINARYRELATIONS_AS_RELATIONALGEBRA:
    RELATIONALGEBRA **to** ALGEBRAOFBINARYRELATIONS
end

In the following, we describe how a strong representation of an (abstract) atomic relation algebra can be constructed from a concrete interpretation of its atoms, i. e., its base relations. For this assume that we have a model for the base relations of the relation algebra, that is, a JEPD system of relations on a non-void set *Elem*. Of course, we want to extend this model in a canonical manner to a model of all relations. Relations are here exactly those binary relations on *Elem* that can be written as a (set-theoretical) union of base relations. In order to be a strong representation, the model we aim at must be an algebra of binary relations. But the fact that the system of relations is JEPD ensures only that the set of all unions of base relations forms a set algebra. This set algebra need not be an algebra of binary relation since, in general, it need not contain the identity relation, be closed with respect to involution, or closed with respect to composition. But if it does — a base relation model satisfying these conditions will be referred to as *closed for composition* — the canonical extension of the base relation model defines an atomic algebra of binary relations. In fact, the final view in the following list of specifications states that the concrete relation algebra defined in this way provides a strong representation of the abstract algebra.

spec JEPDBASERELMODEL =
    BINARYRELATIONS [**sort** *Elem*]
then
    type *BaseRel* < *Relation*
    $\forall x, y : Elem; \; r,s : BaseRel$
    - $\exists r : BaseRel \bullet (x,y) \in rep(r)$ %(JointlyExhaustive)%
    - $\neg r = s \Rightarrow rep(r) \cap rep(s) = \emptyset$ %(PairwiseDisjoint)%
end

spec RELATIONSFROMBASERELMODEL [JEPDBASERELMODEL] = **%def**
    type $Rel = \{x : Relation \bullet \exists X : Set(BaseRel) \bullet$
           $(\forall y,z : Elem \bullet (y,z) \in rep(x) \Leftrightarrow (\exists r : BaseRel \bullet r \in X \wedge (y,z) \in rep(r)))\}$
end

spec CONSTRUCTMODEL [COMPCLOSEDBASERELMODEL] = **%def**
    RELATIONSFROMBASERELMODEL [JEPDBASERELMODEL]

and ALGEBRAOFBINARYRELATIONS
then %def
    preds $\_\_ \in \_\_ : BaseRel * Rel$;
          $\_\_ \sqsubseteq \_\_ : Rel * Rel$
    $\forall x : BaseRel; r, r' : Rel$
    • $x \in r \Leftrightarrow rep(x) \subseteq rep(r)$
    • $r \sqsubseteq r' \Leftrightarrow rep(r) \subseteq rep(r')$
end

view RELATIONALGEBRA_FROM_BASERELMODEL [COMPCLOSEDBASERELMODEL]:
    RELATIONALGEBRA
to   CONSTRUCTMODEL [COMPCLOSEDBASERELMODEL]
end

4.2 RCC5, RCC8, and Allen's Interval Algebra

We are now ready to explain how spatial and temporal calculi (more exactly, their respective relation algebras) can be incorporated into the framework provided by the specifications presented in the previous section. Let us start with the region connection calculi RCC5 and RCC8. Both calculi describe relations between *regions*, which may be thought of as non-void, regular open (or alternatively, regular closed) subsets of a topological space. In RCC5 the following relations count as base relations: DR ("discrete"), PO ("partially overlap"), PP ("proper part"), PPi (the converse of PP), and EQ ("equal"). The set of RCC8 base relations is more fine-grained: DR splits into the relations DC ("disconnected") and EC ("externally connected") and PP into the relations TPP ("tangential proper part") and NTPP ("non-tangential proper part"). These relations can be defined in terms of the topological closure operation: the relation DC holds between regular open sets X and Y if their closures do not intersect, X NTPP Y holds if the closure of X is contained in Y, etc.

The following library shows how the (abstract) relation algebra of RCC5 can be defined via CASL specifications:

spec RCC5BASERELATIONS = %**mono**
    free type $BaseRel ::= dr \mid po \mid pp \mid ppi \mid eq$
end

spec RCC5COMPOSITIONTABLE =
    sort $BaseRel$
    ops dr, po, pp, ppi, eq: $BaseRel$
and   COMPOSITIONTABLE **with op** $id \mapsto eq$
then
    • $dr^{\smile} = dr$ %(sym_dr)%
    • $po^{\smile} = po$ %(sym_po)%
    • $pp^{\smile} = ppi$ %(inv_pp)%
    • $ppi^{\smile} = pp$ %(inv_ppi)%
    • $pp \circ pp = pp$ %(cmps_pppp)%

- $pp \circ ppi = 1$ %(cmps_ppppi)%
- $pp \circ po = pp \sqcup po \sqcup dr$ %(cmps_pppo)%
- $pp \circ dr = dr$ %(cmps_ppdr)%
- $ppi \circ pp = -dr$ %(cmps_ppipp)%
- $ppi \circ ppi = ppi$ %(cmps_ppippi)%

  $\ldots$

end

spec RCC5 =
    CONSTRUCTRELATIONALGEBRA [RCC5BASERELATIONS]
       [RCC5COMPOSITIONTABLE **fit op** *id:BaseRel* $\rightarrow$ *eq*]
end

view RCC5_AS_ATOMICRELATIONALGEBRA :
    ATOMICRELATIONALGEBRA **to** RCC5
=   *AtomRel* $\mapsto$ *BaseRel*
end

Obviously, the relation algebras of other qualitative calculi (such as RCC8 or Allen's interval algebra) can be specified in the very same manner. Moreover, we can declare natural views between these calculi as follows:

view RCC5_TO_RCC8 :
    RCC5
to   { RCC8 **then** %**def**
       ops *dr,pp,ppi* : *Rel*
       - $dr = dc \sqcup ec$
       - $pp = tpp \sqcup ntpp$
       - $ppi = tppi \sqcup ntppi$ }
=   **sort** *BaseRel* $\mapsto$ *Rel*
end

view RCC5_TO_ALLENIA :
    RCC5
to   { ALLENIA **then** %**def**
       ops *dr,po,pp,ppi,eq* : *Rel*
       - $dr = b \sqcup bi \sqcup m \sqcup mi$ %[Allen relations are denoted by the first]%
       - $po = o \sqcup oi$ %[letter of their names (cf. Fig. 1), i.e.,]%
       - $pp = d \sqcup s \sqcup f$ %[b: before, bi: before involuted, etc.   ]%
       - $ppi = di \sqcup si \sqcup fi$
       - $eq = e$ }
=   **sort** *BaseRel* $\mapsto$ *Rel*
end

It is worth mentioning that a corresponding view from RCC8 to ALLENIA is not valid: The function $dc \mapsto b \sqcup bi$, $ec \mapsto m \sqcup mi$, $po \mapsto o \sqcup oi$, $tpp \mapsto s \sqcup f$, $ntpp \mapsto d$, etc., only defies a view from RCC8 to Allen's interval algebra *as* Boolean algebras, but not as relation algebras. To put it another way, RCC8 does not form a subalgebra of Allen's interval algebra.

Let us now illustrate how the semantic level of such relation algebras can be presented via CASL specifications. The following library describes a specific class of RCC5 models, which is definable for the Euclidean plane (interpreted as a metric space). Its final view says that one obtains a model of RCC5 if we interpret the RCC5 base relations over open discs in the Euclidean plane. More precisely, it states that the canonical interpretation of RCC5 over open discs in the Euclidean plane provides a strong representation of RCC5.[2]

logic HASCASL

view EUCLIDEANPLANE_AS_METRICSPACE :
    METRICSPACE **to** EUCLIDEANPLANE
    . . .
end

spec RCC5OPENDISCBASERELMODEL[METRICSPACE] =
    SET
then op $openDisc(r : Real; x : Elem) : Set\ Elem = \lambda y : Elem \bullet dist(x,y) < r$
    type $OpenDisc = \{X : Set\ Elem \bullet \exists r : Real;\ x : Elem \bullet X = openDisc(r,x)\}$
then BINARYRELATIONS [**sort** $OpenDisc$]
then %**def**
    ops $drRel, poRel, ppRel, ppiRel, eqRel : Relation$
    type $BaseRel ::= ppRel \mid ppiRel \mid poRel \mid drRel \mid eqRel$
    $\forall x, y : OpenDisc$
    • $(x,y) \in rep(drRel) \Leftrightarrow x\,disjoint\,y$
    • $(x,y) \in rep(poRel) \Leftrightarrow \neg x \subseteq y \wedge \neg y \subseteq x \wedge \neg x\,disjoint\,y$
    • $(x,y) \in rep(ppRel) \Leftrightarrow x \subseteq y \wedge \neg x = y$
    • $(x,y) \in rep(ppiRel) \Leftrightarrow y \subseteq x \wedge \neg x = y$
    • $(x,y) \in rep(eqRel) \Leftrightarrow x = y$
end

spec RCC5OPENDISCMODEL[EUCLIDEANPLANE] = %**def**
    CONSTRUCTMODEL [RCC5OPENDISCBASERELMODEL [
        view EUCLIDEANPLANE_AS_METRICSPACE]
        fit sort $Elem \mapsto OpenDisc$]
end

view METRICSPACE_INDUCES_RCC5OPENDISCMODEL :
    RCC5
to   RCC5OPENDISCMODEL [EUCLIDEANPLANE]
=   **ops** $pp \mapsto ppRel, ppi \mapsto ppiRel, po \mapsto poRel, dr \mapsto drRel, eq \mapsto eqRel$
end

The development graph of these libraries, as output by HETS (a subgraph of it is depicted in Fig. 3), exhibits the mutual dependencies between the presented specifications. Dark arrows denote inclusions of theories; light arrows denote proof obligations (theory morphisms) generated by views.

---

[2] *Weak* representations of abstract relation algebras could be presented as CASL specifications as well, but not in terms of CASL views.

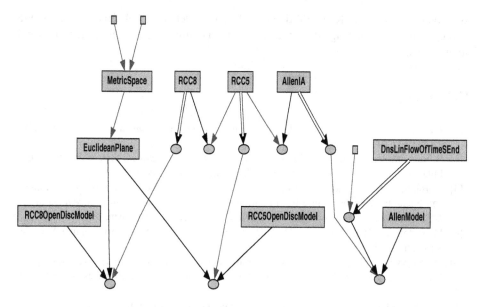

Fig. 3. Development graph of RCC5, RCC8, and Allen's interval algebra

5   Summary and Outlook

In this paper we discussed how qualitative calculi can be described via CASL specifications. We saw that CASL allows for an elegant representation of such calculi. Moreover, since the specifications presented here are built up in a modular way, we provide an easy interface for embedding other calculi into the CASL framework. Finally, CASL specifications ensure a high visibility of the mutual dependencies between qualitative calculi on both the syntactic and the semantic level, and hence may be considered superior to ontologies of such systems.

Because of lack of space, we could not show how these CASL specifications connect to theorem provers such as Isabelle. Furthermore, in this paper we could only explain a small fraction of algebraic theories related to spatial and temporal calculi. For example, the first-order theories of these calculi and Stell's Boolean connection algebras [23] should be integrated into CASL as well. Future work will also deal with the CSP languages of these qualitative calculi. In particular, we will investigate how these languages can be translated into MODALCASL, a sublanguage of CASL designed for specifying multi modal fragments of first order logic. Such translations would be particularly interesting for automated verifications of composition tables of the calculi mentioned in this paper.

Acknowledgments

This work was partially supported by the Deutsche Forschungsgemeinschaft (DFG) as part of the Transregional Collaborative Research Center SFB/TR 8 Spatial Cognition.

We would like to thank Klaus Lüttich and Bernhard Nebel for helpful discussions. We also gratefully acknowledge the reviewers' critical comments, as well as their hints and suggestions.

References

[1] J. F. Allen. Maintaining knowledge about temporal intervals. *Communications of the ACM*, 26(11):832–843, 1983.

[2] B. Bennett. *Logical Representations for Automated Reasoning about Spatial Relationships*. PhD thesis, School of Computer Studies, The University of Leeds, 1997.

[3] B. Bennett, A. Isli, and A. G. Cohn. When does a composition table provide a complete and tractable proof procedure for a relational constraint language? In *Proceedings of the IJCAI97 Workshop on Spatial and Temporal Reasoning*, Nagoya, Japan, 1997.

[4] M. Bidoit and P. D. Mosses. CASL User Manual. Lecture Notes in Computer Science. Springer, 2004.

[5] M. Broxvall. The point algebra for branching time revisited. In *Proceedings of the Joint German/Austrian Conference on Artificial Intelligence (KI-2001)*, pages 106–121, Sep 2001.

[6] M. Broxvall and P. Jonsson. Towards a complete classification of tractability in point algebras for nonlinear time. In *Proceedings of the 5th International Conference on Principles and Practice of Constraint Programming (CP-99)*, pages 129–143, Alexandria, VA, USA, 1999.

[7] A. G. Cohn, B. Bennett, J. M. Gooday, and N. Gotts. RCC: A calculus for region based qualitative spatial reasoning. *GeoInformatica*, 1:275–316, 1997.

[8] I. Düntsch, H. Wang, and S. McCloskey. Relation algebras in qualitative spatial reasoning. *Fundamenta Informaticae*, 39(3):229–249, 1999.

[9] M. J. Egenhofer. Reasoning about binary topological relations. In O. Günther and H.-J. Schek, editors, *Proceedings of the Second Symposium on Large Spatial Databases, SSD'91 (Zürich, Switzerland)*, Lecture Notes in Computer Science 525, pages 143–160. Springer, 1991.

[10] M. J. Egenhofer and R. D. Franzosa. Point set topological relations. *International Journal of Geographical Information Systems*, 5:161–174, 1991.

[11] A. U. Frank. Qualitative spatial reasoning with cardinal directions. In H. Kaindl, editor, *Proceedings of the Seventh Austrian Conference on Artificial Intelligence*, Informatik-Fachberichte 287, pages 157–167. Springer, 1991.

[12] A. U. Frank. One step up the abstraction ladder: Combining algebras - from functional pieces to a whole. In C. Freksa and D. M. Mark, editors, *Spatial Information Theory: Cognitive and Computational Foundations of Geographic Information Science, International Conference COSIT '99, Stade, Germany, August 25-29, 1999, Proceedings*, Lecture Notes in Computer Science 1661, pages 95–107. Springer, 1999.

[13] C. Freksa. Using orientation information for qualitative spatial reasoning. In A. U. Frank, I. Campari, and U. Formentini, editors, *Spatio-Temporal Reasoning*, Lecture Notes in Computer Science 639, pages 162–178. Springer, 1992.

[14] A. Gerevini and J. Renz. Combining topological and qualitative size constraints for spatial reasoning. In *Proceedings of the 4th International Conference on Principles and Practice of Constraint Programming*, pages 220–234. Springer, 1998.

[15] P. J. Hayes. The naive physics manifesto. In D. Michie, editor, *Expert Systems in the Micro-Electronic Age*. Edinburgh University Press, 1978.

[16] G. Ligozat. Qualitative triangulation for spatial reasoning. In A. U. Frank and I. Campari, editors, *Spatial Information Theory: A Theoretical Basis for GIS, International Conference COSIT '93, Marciana Marina, Elba Island, Italy, September 19-22, 1993, Proceedings*, Lecture Notes in Computer Science 716, pages 54–68. Springer, 1993.

[17] G. Ligozat and J. Renz. What is a qualitative calculus? A general framework. In C. Zhang, H. W. Guesgen, and W.-K. Yeap, editors, *PRICAI 2004: Trends in Artificial Intelligence, 8th Pacific Rim International Conference on Artificial Intelligence, Auckland, New Zealand, August 9-13, 2004, Proceedings*, Lecture Notes in Computer Science 3157, pages 53–64. Springer, 2004.

[18] T. Mossakowski. Heterogeneous specification and the heterogeneous tool set. Habilitation thesis, University of Bremen, 2005.

[19] P. D. Mosses, editor. CASL *Reference Manual*. Lecture Notes in Computer Science 2960. Springer, 2004.

[20] T. Nipkow, L. Paulson, and M. Wenzel. *Isabelle/HOL, A Proof Assistant for Higher-Order Logic*. Lecture Notes in Computer Science 2283. Springer, 2002.

[21] D. A. Randell, Z. Cui, and A. G. Cohn. A spatial logic based on regions and connection. In B. Nebel, W. Swartout, and C. Rich, editors, *Principles of Knowledge Representation and Reasoning: Proceedings of the 3rd International Conference (KR-92)*, pages 165–176. Morgan Kaufmann, 1992.

[22] L. Schröder and T. Mossakowski. HasCASL: Towards integrated specification and development of Haskell programs. In H. Kirchner and C. Reingeissen, editors, *Algebraic Methodology and Software Technology, 2002*, volume 2422 of *Lecture Notes in Computer Science*, pages 99–116. Springer-Verlag, 2002.

[23] J. G. Stell. Boolean connection algebras: A new approach to the Region-Connection Calculus. *Artificial Intelligence*, 122(1-2):111–136, 2000.

[24] M. B. Vilain. A system for reasoning about time. In D. L. Waltz, editor, *Proceedings of the National Conference on Artificial Intelligence. Pittsburgh, PA, August 18-20, 1982*, pages 197–201. AAAI Press, 1982.

A Spatial Form of Diversity

Christophe Claramunt

Naval Academy Research Institute, Lanvéoc-Poulmic,
BP 600, 29240 Brest Naval, France
claramunt@ecole-navale.fr

Abstract. Information theory has long provided a mathematical framework for statistically evaluating the diversity a given signal is able to produce. This paper applies information theory to multi-dimensional spaces, discusses the limits of conventional measures of diversity, and introduces a spatial component into the measure of diversity. This leads to a spatial form of diversity that provides a novel way to study spatial structures where categories and distances are amongst key factors in the analysis. The approach is exemplified with spatial configuration samples.

Keywords: Information theory, diversity, spatial diversity.

1 Introduction

The mathematical theory of communication (Shannon, 1948) has long been established as a fundamental support to study the diversity and capacity of information produced by a given discrete or continuous source. In Shannon's theory, information is considered objectively and independently of interpretation. This is closely associated with the diversity of message a given signal is able to produce. Surprisingly, and with the exception of the field of image processing, a few studies have explicitly evaluated to which extent Shannon's quantitative theory principles can be applied to space, in order to evaluate how much information and diversity a geographical system can produce. This is probably due to the fact that space is a specific form of multi-dimensional system where the different dimensions are intimately linked, while discrete systems studied by Shannon are made of messages decomposed into one-dimensional signals. This directly relates to a major difference between one-dimensional messages generated by a language of symbols (i.e. alphabet), and space described by either continuous or discrete representations.

One intriguing question is whether the notion of diversity as it is defined in the theory of information is, or not, influenced by some of the fundamental properties space generates and conveys. From a different perspective, it has been observed that information theory will be useful in clarifying some quantitative aspects of geographical information in the future (Tobler, 1997). The question we address in this paper is the evaluation of information diversity revealed by a spatial system, or in other words, the degree of uncertainty in the random selection of an entity in that system, all choices being equally likely, where an entity denotes a region or object of interest in that system. We do not consider the case of open systems where observation relates to experience and previous knowledge of comparable spatial systems, this

A.G. Cohn and D.M. Mark (Eds.): COSIT 2005, LNCS 3693, pp. 218–231, 2005.
© Springer-Verlag Berlin Heidelberg 2005

making the evaluation of diversity a far more complex issue as the one of strictly speaking assessing the diversity of a closed system. By analogy to Shannon's description of discrete sources of information, we restrict our study to discrete and not dense spatial systems.

Our work has been influenced by some of the ideas introduced by a recent attempt to combine Shannon's quantitative theory of information, cognitive perception with semantic categorisation (Haken and Portugali, 2003). These authors made a conceptual connection between the degree of information conveyed by a given spatial system, grouping and categorisation processes, and the degree of perception and memorisation by humans. They also intuit that the original Shannon's measure of diversity should be adapted to space, particularly by integrating a factor that evaluates the probability of finding two different entities at the same location. Our objective is to make an attempt to introduce an additional component to Haken and Portugali's systematic approach by specifically representing and measuring the influence of the spatial dimension on a measure of diversity, rather by evaluating it using probabilities. Although quantitative per nature, the objective of our work is to explore a computable representation that can favour the perception and interpretation of the diversity of spatial structures. As illustrated by Figure 1, our work can be positioned between quantitative theory of information and spatial systems.

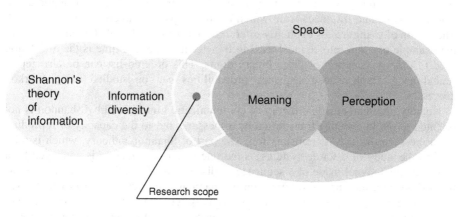

Fig. 1. From information to perception

The remainder of the paper is organized as follows. Section 2 describes the basic principles of information theory and the main components of Shannon's measure of diversity. Section 3 briefly surveys previous attempts in the extension of measures of diversity to spatial systems. Section 4 introduces a measure of spatial diversity, a comparative analysis with existing measures, and illustrates the potential of the spatial diversity with exemplified configurations. Finally, Section 5 concludes the paper and outlines further work.

2 Information Theory Basics

Information theory is considered as the first applied mathematical theory of information (Shannon, 1948), due to the earlier emergence of the communication and computation industry, and its early developments in code-breaking applications. Information theory is fundamentally the rigorous study of distinctions and their relations, inasmuch as they make a difference (MacKay, 1969). The notion of distinction is intimately related to cognitive perception, categories, and basic units of isolation (Spencer Brown, 1972). The more specific the information represented, the more information it requires to be classified, and the lower the level of abstraction is. Information theory is closely related to the notion of diversity (taken in the sense of the information content, not the meaning). Information is commonly understood as knowledge of facts acquired or derived from, e.g., study, instruction or observation. Information in this context is also considered as true and meaningful.

An important principle of information theory relies on the assumption that information can be represented as a string of binary digits via an isomorphic mapping where the properties of a given system are described as Boolean values. An ordered list of Boolean values supports classification and computing representations. Categorisations are equivalent to true/false propositions that support constructions of classifications and information theoretic functions such as truth functional logic. A categorisation is unambiguous although semantically restrictive. From a categorisation emerge classes and instances from which relationships and patterns can be analysed. The notion of string makes explicitly reference to an ordered list of Boolean values. In the example of a discrete channel used to transfer information, time is the dimension used to order succession of these binary digits. This ordered list can be also represented as a graph whose sequence probabilities can be studied using Markov processes.

Notably, the mathematical theory of communication introduced by Shannon is not oriented to the knowledge transmitted by a message, but to the capacity of reproducing a message through its transmission. The part of Shannon's theory, which is relevant to the study of discrete systems, is the one where information is conveyed by a message and a signal made of a sequence of discrete symbols. We retain two notions that introduce a measure of information in Shannon's theory: the *capacity* that gives an absolute measure of the information a given discrete channel is able to transmit per unit of time, and the *diversity*, also called entropy, which evaluates the variety of information produced by a given message. According to Shannon's theory, a discrete channel transmits information from a finite set of elementary symbols. The *capacity C* of such a discrete channel is given by

$$C = \lim_{T \to \infty} \frac{\log_2 S(T)}{T} \tag{1}$$

where $S(T)$ is the number of allowed signals of duration T. The *capacity C* is measured in bits as $\log_2$ is used in the expression.

Intuitively, the semantic information conveyed by a given instance or a true statement that materialises this instance is inversely related to probability. The *information content* of a statement i, denoted by *Cont(i)*, is defined as the complement of the *a priori* probability p_i that the statement i is true

$$Cont(i) = 1 - pi \tag{2}$$

where p_i is defined as the proportion of the total number Ni of entities of the class i over the total number N of entities, that is, $pi = \dfrac{Ni}{N}$.

Accordingly, the higher the probability p_i a statement i is true, the less informative it is. Simpson's *heterogeneity* S_{sim} is used to evaluate the probability that any entities selected at random would be different (Simpson, 1949). It is given as follows

$$S_{sim} = 1 - \sum_{i=1}^{n} pi^2 \tag{3}$$

where n is the number of classes

Another quantification of the information contained by a statement, although less intuitive but more expressive, is introduced by Shannon. The *information measure*, denoted by *Inf(i)*, of a given statement i corresponds to the observed fact that the output of most of information sources vary linearly in a logarithm scale, and to the intuition that adding one source input also doubles the information output (Hartley, 1928). The *information measure Inf(i)* is defined as follows

$$Inf(i) = log_2 \left(\frac{1}{1 - Cont(i)} \right) = -log_2 (pi) \tag{4}$$

The *information measure* is generalised to denote how much choice is involved in the selection of a statement, that is, the measure of information *diversity* or entropy H defined as (the term entropy comes from the fact that there is a clear relationship and isomorphism between Shannon's formulae and the definition of entropy in thermodynamics)

$$H = -K \sum_{i=1}^{n} pi \, log_2 (pi) \tag{5}$$

where K is a positive constant. The demonstrated maximum value of H for a given number of classes n is $H_{max} = K \, log_2 n$ when the distribution of classes is uniformly distributed (i.e. $pi = \dfrac{1}{n}$ for all i).

The minimum value of *diversity* is given by the case where only one class is present. For a given number of classes, *diversity* is maximum when each class is represented in equal proportions (evenness of the distribution); while for an evenness of distribution *diversity* increases with the number of classes (richness of the distribution). This logarithmic expression presents several additional and demonstrated advantages: the logarithm of the number of possible states, given an alphabet, increases linearly, H is a continuous function and additivity is respected (i.e. if a choice is broken down into several ones, *diversity* is equal to the weighted sum of the individual values of *H*). These properties were a prerequisite of the function to define, and Shannon demonstrated that the logarithmic form of this formula was the only way to conform them. The example below illustrates the notion of additivity using a discrete signal either (1) parameterised as the choice of a vowel V is broken into two possibilities (letters *A* or *E*), or (2) explicitly represented. The *diversity* values calculated exemplifies the additivity property with *K*=1.

(1)   *VBVVBV*   A vowel V being either *E* (75%) or *A* (25%) then
       $H = H(0.33, 0.66) + 0.66\,H(0.75, 0.25) = 1.44$

(2)   *EBEEBA*   then
       $H = H(0.5, 0.33, 0.16) = 1.44$

A notion related to diversity is the one of complexity. Complexity is a concept in between order and disorder, it is inversely correlated to organisation (Collier, 1990). Although a complex system should exhibit a high diversity, the reverse is not always true. A system can be diverse but also ordered, then revealing a low complexity and a relative degree of organisation. A challenging question relies on the measure of complexity. Amongst many contributions that followed information theory, algorithmic information theory provides important insights to the measure of complexity. In computing terms, complexity, also named *logical depth*, denotes the minimal computation steps required to compute an uncompressed string from its maximally compressed form. The concept of *logical depth* has been proposed by Bennett (1988) as a measure of organisation in a given sequence of information. A structure is deep if it is random but subtly redundant. Redundancy order is determined by the minimal number of elements in which a redundancy can be determined (order 1, pairwise, etc.). For example, a signal *EBEEBE* is of *logical depth* 3 as the primitive sequence reproduced is *EBE*, while the signal *EBEEBA* is of *logical depth* 6 as no primitive sequence shorter than the sequence can be identified. The first signal is more organised than the second as a form of redundancy does appear in it. Although algorithmic per nature, the *logical depth* provides a complementary quantitative measure to the one of diversity. However, the measure is still hampered by its difficulty to apply and compute in the case of large configurations, especially when these are multi-dimensional.

The large range of measures proposed by information theory and its extensions reflect the complexity of information. The ones which are algorithmic oriented, such as the measures of *capacity* and *logical depth*, encompass a computational evaluation of the degree of complexity of a given system. The measure of *diversity* has a much broader coverage of applications, due to its generality and unsophisticated nature. This explains largely the widespread of diversity in many application areas, some of them being spatially-related such as ecological and environmental studies. However, Shannon's measure of *diversity* was originally designed for a specific purpose, the one of transmission of mono-dimensional signals. Therefore, there is still an issue to investigate, that is, whether multi-dimensional systems should imply some adaptations of the measure of *diversity*, particularly in spatial systems where the multi-dimensionality component is very specific, and a factor of additional constraints to consider and apply.

3 Related Work

Although the measures of capacity and logical depth have not been yet applied to space, to the best of our knowledge, diversity has been widely used in spatial analysis. In ecological and environmental studies, diversity measures are employed to analyse the physiognomy of a landscape and the influence of spatial configuration on ecological functionality and biological diversity (Margalef, 1958; Brillouin, 1962; Menhinick, 1964; McIntosh 1967; Hurlbert, 1971; McGarigal and Marks, 1994; Smith and Wilson, 1996). Parametric or non parametric measures of diversity are derived from the number of classes and the distribution of these classes. The dominance index measures the extent to which one or a few category types dominate the landscape in terms of class distribution (O' Neill *et al*, 1988), while the abundance factor often used in ecological studies evaluates the number of elements of a given species (Tilman, 1996). Fragmentation and spatial heterogeneity indices evaluate the distribution of the number of patches per category, given a region of space (McGarigal and Marks, 1994). Nearest neighbour distances such as G, F and K functions evaluate the way events represented as points are clustered (Chakravorty, 1995). In cartography, quantitative measures have been also proposed to evaluate the diversity of symbols a cartographical representation produces (Bjørke, 1996; Li and Huang, 2002).

When applied to geographical systems, Shannon's measure of diversity has been extended using an integration of adjacency relationships of first order, as primitive relationships amongst the regions or local cells that compose a spatial system (regions for discrete representations of space, cells for continuous representations of space). These indices evaluate the relative degrees of interspersion, juxtaposition and contagion amongst several classes of region or local cells. They are based on an evaluation of the diversity of adjacency relationships. For discrete representations of space the measure of *adjacency aij* evaluates to which extent regions of a given class *i* are adjacent to regions of another class *j*.

$$aij = \frac{pi \times gij}{\sum_{j} gij} \tag{6}$$

where g_{ij} is the length of the edges between regions of classes *i* and *j*, p_i the proportion of regions of class *i*.

The measure of *contagion* gives the degree to which patches of the same attribute class are clumped into patches of the same attribute class (O'Neill *et al.*, 1988; Li and Reynolds, 1993), it is correlated with indices of attribute diversity and dominance (Riitters *et al.*, 1996). The *contagion* index is given as (Li and Reynolds, 1993)

$$C = 1 + \sum_{i}^{n} \sum_{j}^{n} aij \frac{log_2(aij)}{2 log_2(n)} \tag{7}$$

where *n* is the number of classes.

Contagion can be applied at either the local cell (interspersion and juxtaposition metrics) or patch, that is, region level (contagion metrics). The form given above is the most general form of the *contagion* (interspersion and juxtaposition measures are closely related). The *contagion* evaluates to which extent a given landscape is aggregated (i.e., higher values) or dispersed (i.e., lower values). *Contagion* is inversely correlated to *diversity*. For a given number of classes, the *contagion* is minimum when all classes are evenly distributed and equally adjacent to each others.

Despite its application to many environmental, ecological and cartographical studies, current measures of contagion suffer from several limitations, especially when applied to discrete systems. Firstly, the measures of contagion are independent of the relative dispersion and distances between the regions that compose the different classes, this leading to a lack of consideration of the underlying spatial structure and arrangements. Secondly, adjacency relationships introduce a quantitative bias as this measure does not take into account proximities of higher orders. Last but not least, adjacency is not always the most predominant spatial relationship of interest. In some cases, adjacency relationships might even not be defined when the system under study does not form a partition of space, or when partitioning space is not of interest from the domain of study considered. Most of these limitations come from the fact that these measures were initially applied to continuous representations of space (i.e. image-based views of space), where the objective was to analyse local variance of pixel distributions. Replacing the adjacency measure by an inverse function of distance is an alternative but a difference should be made between similar and different entities. Therefore, a key question to address is the exploration and specification of a measure of spatial diversity that can provide an alternative in situations where measures of contagion are not particularly adapted.

4   Information Theory and Space

4.1   Towards a Spatial Form of Diversity

Similarly to the definition given by information theory, a spatial measure of diversity should give the degree of uncertainty in selecting some entities of interest, taking into account the influence of space. A primordial question to address is how to measure the role played by space, if any. This implies to define a scope to qualify the extent and nature of interactions in discrete space. In a closed spatial system, interactions tend to happen between entities under the boundaries of the represented space, while there is no limit in the case of an open spatial system. While most of the studied spatial systems have some natural boundaries and are relatively closed per definition, part and result of their identity, the environment at large should not be ignored. This was a fundamental observation that led to the establishment of the First Law of Geography, a precept that introduces a primal order in space (Tobler, 1970):

> *Everything is related to everything else, but near things are more related than distant things.*

This law was introduced by Tobler to study the population growth of the Detroit Region, which depends not only of the initial population of this urban area, but also on the population of all other places. This explicitly models distance as a key factor that relates things in space, and favours the search for a conceptual bridge between Shannon's information theory and the way diversity should be evaluated in space. Taking the argument further, diversity should augment when the distance between different entities decreases, as diversity also should augment when the distance between similar entities increases. We convert these observations into two basic rules:

rule 1:   when different entities are closer, diversity increases
rule 2:   when similar entities are closer, diversity decreases

These rules give us a support for a tentative definition of a spatial measure of diversity. We introduce a first measure that evaluates the average distance between the entities of a same class. We call this measure the *intra-distance* d_j^{int} of a given class j. The second measure calculates the average distance between the entities of a given class and the entities of the other classes. We call this measure the *extra-distance* d_j^{ext} of a given class j. These measures are respectively defined as follows

$$d_j^{int} = \frac{1}{Nj \times (Nj-1)} \sum_{\substack{i=1 \\ i\in Cj}}^{Nj} \sum_{\substack{k=1 \\ k\neq i \\ k\in Cj}}^{Nj} d_{i,k} \text{ if } Nj > 1, \, d_j^{int} = \lambda \text{ otherwise} \qquad (8)$$

$$d_j^{ext} = \frac{1}{Nj \times (N-Nj)} \sum_{\substack{i=1 \\ i\in Cj}}^{Nj} \sum_{\substack{k=1 \\ k\notin Cj}}^{N-Nj} d_{i,k} \text{ if } Nj \neq N, \, d_j^{ext} = \lambda \text{ otherwise} \qquad (9)$$

where Cj denotes the set of entities of a given class j, Nj the number of entities of a given class j, N the total number of entities, $d_{i,j}$ the distance between two entities i and j, λ being a parameter taken relatively small.

This quantitative evaluation of the distance between similar and different entities supports the introduction of a new measure of diversity Hs, called *spatial diversity*. The usual coefficient K of Shannon's measure of diversity is replaced by a fraction that denotes the respective influence of the *intra-* and *extra-distances*:

$$Hs = -\sum_{i=1}^{n} \frac{d_i^{int}}{d_i^{ext}} pi \log_2 (pi) \qquad (10)$$

The *spatial diversity Hs* and the *diversity* are semi bounded by the real positive interval $[0,+\infty]$. For some given intra- and extra-distance values, Hs is maximum when the classes are evenly distributed. For a given distribution of classes, the *spatial diversity* increases when either the *intra-distance* augments, or the *extra-distance* decreases. Due to the fact that the coefficient K is not a constant any more, the additivity property of Shannon's diversity is not maintained as K is replaced by an expression which is not constant over the different classes.

4.2 Application to Basic Spatial Configurations

We introduce several elementary cases in order to support a comparative analysis of the *spatial diversity* with measures of *contagion* and *diversity*. In the spatial configurations of building layouts presented hereafter, and without loss of generality, a room represents an entity that belongs to a class of building units (i.e. two classes in the cases introduced below). A link between two rooms is equivalent to a one-step distance, λ set equal to the unit value. The distance $d_{i,j}$ is given by the shortest path in number of links between the rooms i and j. The entities of the two classes are distributed in constant proportions over the different cases, with the consequence that the *diversity C* remains constant.

In the first case presented in Figure 2, the *extra-distance* is constant over the two configurations, while the *intra-distance* is lower in the configuration exhibited by Figure 2a. The values given by the table in Figure 2 shows that when measures of adjacency do not differ in relative variation from measures of distance, *contagion* and *spatial diversity* outline convergent patterns. *Contagion C* and *spatial diversity Hs* confirm the higher diversity presented by the configuration presented in Figure 2a, as the rooms of the class B showed by Figure 2a are less distant than in the configuration exhibited by Figure 2b.

In the example presented by the second case (Figure 3), the *intra-distance* is constant over the two configurations, while the *extra-distance* is lower in the configuration exhibited by Figure 3b. Therefore, by application of rule 2, information diversity should be higher in the configuration of Figure 3b. This is verified by the *spatial entropy Hs*, but not by the *contagion C* which is almost constant over the two configurations. This is due to the fact that the *contagion* is derived from a logarithmic

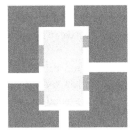

Fig. 2a.

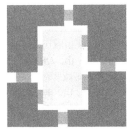

Fig. 2b.

	Fig. 2a	Fig. 2b
H	0.71	0.71
C	0.65	0.51
d_A^{int}	1	1
d_A^{ext}	1	1
d_B^{int}	2	1.33
d_B^{ext}	1	1
Hs	0.97	0.80

Room of class A

Room of class B

Link between two rooms

Fig. 2. Case 1

Fig. 3a.

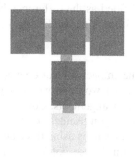

Fig. 3b.

	Fig. 3a	Fig. 3b
H	0.71	0.71
C	0.54	0.53
d_A^{int}	1	1
d_A^{ext}	2.5	1
d_B^{int}	1.66	1.66
d_B^{ext}	2.5	1
Hs	0.35	0.71

Room of class A

Room of class B

Link between two rooms

Fig. 3. Case 2

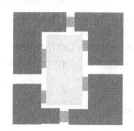

Fig. 4a.

Fig. 4b.

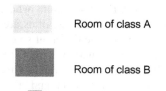

	Fig. 4a	Fig. 4b
H	0.71	0.71
C	0.54	0.54
d_A^{int}	1	1
d_A^{ext}	2.25	1.75
d_B^{int}	1.5	1.5
d_B^{ext}	2.25	1.75
Hs	0.37	0.44

Room of class A

Room of class B

Link between two rooms

Fig. 4. Case 3

based evaluation of the proportion of adjacency relationships between the regions of the two classes. In fact, the measure of *contagion* is not discriminant enough when the number of classes is low and the spatial configuration is small.

In the last case presented in Figure 4, similar entities are equally distant in the two configurations while different entities are less distant in Figure 4b than in Figure 4a. Therefore, information diversity should be higher in the configuration of Figure 4b, this being confirmed by the *spatial diversity* coefficient *Hs* only, while *diversity H* and *contagion C* are constant over the two configurations. Similarly to the previous case, the distance factor is not taken into account by the measure of *contagion C* in analysing the arrangement of the entities of class *B* with respect to the ones of class *A*.

The examples presented above are based on a semantic categorisation of room functions, and connectivity between them. These assumptions can be adapted to different sorts of configuration, preferably closed systems, where classification is application driven, and relationships between similar and different entities derived from either Euclidean or other forms of distance.

4.3 Spatial Diversity, Order and Additional Properties

The measure of *spatial diversity* is based on the assumption that classes and distances are important elements of its quantitative evaluation. One should expect from this measure the capability to study and qualify different arrangements of a given spatial structure. Two other notions related to Shannon's diversity are the measures of order and cohesion. The *spatial order* of a system can be defined as the difference between its *maximum spatial diversity* and its *spatial diversity* following (Landsberg, 1984)

$$Orders = Hs(max) - Hs(current) \qquad (11)$$

The concept of cohesion is more subtly defined as the impact a local change should have on the overall diversity of a given system, or in other words to which degree the diversity of a system is insensitive to fluctuations in the arrangements. Cohesion is correlated to order as the two notions evaluate the degree of compactness of a given system. Figure 5 compares *spatial diversity* with *spatial order* values using three configurations of some building units over a circular distribution. These configurations show increasing spatial diversity values from Figure 5.a to Figure 5.c, while *spatial order* values decrease, this reflecting the negative correlation between the two measures.

The examples of the case study show that there are different ways to analyse the diversity of a given spatial system. The measure of spatial diversity complements current measures of diversity in situations where distance is a key factor in the analysis of the way similar and different entities are interrelated. Multiple parameters are surely of interest as they outline different spatial patterns. Spatial diversity and order constitute a quantitative information content of the arrangement of a spatial system. The spatial diversity differs from Shannon's measure of diversity in several aspects. First, information quantitative theory and its extensions are mainly oriented to the analysis of mono- or multi-dimensional signals, not multi-dimensional spaces where structural properties have an influence on the notion of diversity. Secondly, current measures of contagions, as applied to geographical systems, integrate the spatial structure in a limited deepness as the adjacency relationship is not always the most appropriate parameter to take into account.

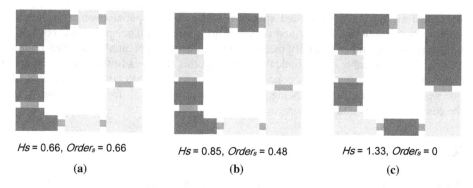

Hs = 0.66, $Order_s$ = 0.66

(a)

Hs = 0.85, $Order_s$ = 0.48

(b)

Hs = 1.33, $Order_s$ = 0

(c)

Fig. 5. Spatial diversity and spatial order

Although the interest of the measure of spatial diversity introduced has been illustrated by a large scale system and a built environment, the approach can be applied and adapted to different sorts of space. When appropriate, a distance threshold can be given to restrict the set of entities to consider when calculating intra- and extra-distances. Different measures of distance can be considered (e.g. contextual distance, cognitive distance) or even additional spatial parameters, either static or dynamic, depending on the phenomena studied. Additional semantic criteria can be also considered although this has not been originally integrated in the original measure of diversity (e.g. semantic importance of the different classes and entities represented).

From a cognitive point of view, diversity is intimately related to the notions of discernability and lisibility as defined in (Miller, 1956). In particular, the interpretation factor is an element of importance when studying the way human can perceive and categorise information and the degrees of diversity quantitatively evaluated. The relevance of the spatial diversity is influenced by the categorisation process and the type of distance considered for the problem to be solved. High numbers of categories make interpretation of a given spatial system more difficult to interpret. Hierarchical representations can help to organise the information perceived in organised structures, thus favouring the understanding of derived spatial diversities.

5   Conclusion

The aim of this preliminary study was to explore a representation of spatial information diversity that can act as a support for the analysis of spatial structures. The approach is based on Shannon's information theory that provides an elegant quantitative means for the evaluation of diversity in one-dimensional spaces. Although used in many geographically-related applications, we found that there is still a need to explore and quantify the constraints caused by an integration of the spatial dimension in a measure of diversity.

We introduce a measure of spatial diversity, extended from Shannon's measure of diversity on the one hand, and derived from the principles of the First law of Geography on the other hand. The properties and particularities of the spatial form of diversity over conventional measures of diversity and its extensions are examined and illustrated with some exemplified spatial configurations. Spatial diversity relationships with the notions of order and cohesion are discussed. The spatial diversity can be combined with other diversity measures to evaluate the potential of a given region for ecological, social and economical studies, but this deserves further validation and experimentation. Current work concerns extension of the theoretical foundations of our work and to empirically testing and implementing spatial diversity measures in the context of urban and environmental sciences.

References

Bennet, C. H., 1988, Logical depth and physical complexity. In *The Universal Turing Machine – A Half Century Survey*, R. Herken (ed.), Oxford University Press.

Bjørke, 1996, Framework for entropy-base map evaluation. *Cartography and Geographical Information Systems*, 23, 78-95.

Brillouin, L., 1962, *Science and Information Theory*, 2nd edition, Academic Press, Inc., New York.

Chakravorty, S., 1995, Identifying crime clusters: The spatial principles. *Middle States Geographer*, 28: 53-58.

Collier, J., 1990, Intrinsic Information. In *Information, Language and Cognition: Vancouver Studies in Cognitive Science,* Philip Hanson (ed.), University of Oxford Press, 390-409.

Haken, H. and Portugali, J., 2003, The face of the city and its information. *Journal of Environmental Psychology*, 23(4), 382-405.

Hartley, R. V. L., 1928, Transmission of information. *Bell System Technical Journal*, July, p.535.

Hurlbert, S. H., 1971, The non concept of species diversity: a critique and alternative parameters, *Ecology*, 52, 577-586.

Landsberg, P. T., 1984, Can entropy and order increase together?. *Physics Letters 102A*, 171-173.

Li, Z. and Huang, P., 2002, Quantitative measures for spatial information of maps. *International Journal of Geographical Information Science*, 16(7), 699-709.

Li, H. and Reynolds, J. F., 1993, A new contagion index to quantify spatial patterns of landscapes. *Landscape Ecology*, 8:155-162.

Margalef, R., 1958, Information theory in ecology. *General Systems*, 3, 36-71.

McGarigal, K. and Marks, B. J., 1994, *FRAGSTATS: spatial pattern analysis program for quantifying landscape structure*. Gen. Tech. Report PNW-GTR-351, USDA Forest Service, Pacific Northwest Research Station, Portland, OR.

McIntosh, R. P., 1967, An index of diversity and the relation of certain concepts to diversity. *Ecology*, 48, 392-404.

McKay, D. M., 1969, *Information, Mechanism and Meaning*, MIT Press, Cambridge, MA.

Menhinick, E. F., 1964, A comparison of some species individuals diversity indices applied to samples of field insects. *Ecology*, 45, 859-861.

Miller, G. A., 1956, The magic number seven, plus or minus two – some limits on our capacity for processing information. *The Psychological Review*, 63 (2), 81-97.

O'Neill, R.V., Krummel, J.R., Gardner, R.H., Sugihara, G., Jackson, B., De Angelis, D.L., Milne, B.T., Turner, M.G., Zygmunt, B., Christensen, S.W., Dale, V.H. and Graham, R.L., 1988, Indices of landscape pattern. *Landscape Ecology*, 1:153-162.

Riitters, K. H., O'Neill, R. V., Wickham, J. D. and Jones, B., 1996, A note on contagion indices for landscape analysis. *Landscape Ecology*, 11(4), 197-202.

Shannon, C. E., 1948, A Mathematical theory of communication. *The Bell System Technical Journal*, 27, 379-423 & 623-656.

Simpson, E. H., 1949, Measurement of diversity. *Nature*, 163:688.

Spencer Brown, G., 1972, *Laws of Forms*, New York, Bantam.

Tilman, D., 1996, Biodiversity: population versus ecosystem stability. *Ecology*, 77, 350-373.

Tobler, W.R., 1970, A computer model simulating urban growth in the Detroit Region. *Economic Geography*, 46, 234-240.

Tobler, W. R., 1997, Introductory comments on information theory & cartography. *Cartographic Perspectives*, 27 (Spring): 4-7.

Structure and Semantics of Arrow Diagrams

Yohei Kurata and Max J. Egenhofer

National Center for Geographic Information and Analysis,
Department of Spatial Information Science and Engineering,
Boardman Hall, University of Maine, Orono, ME 04469-5711, USA
{yohei, max}@spatial.maine.edu

Abstract. Arrows are major components of diagrams, where they are typically used to facilitate the communication of spatial and temporal knowledge. An automated interpretation of arrow diagrams would be highly desirable in pen-based interfaces. This paper develops a method for deducing possible interpretations of arrow diagrams, which is composed of a uni-directional arrow symbol and one or more components. Based on a study of the use of arrow diagrams, we classify their semantics into *properties*, *annotations*, *actions*, and *conjunctions*. Then, we discuss the structural requirements of arrow diagrams for illustrating each class of semantics, as well as the structural rules for adding optional components. Finally, we investigate all possible structures of simple arrow diagrams for each class of semantics and demonstrate that knowledge about the structure of an arrow diagram reduces the ambiguity of its interpretation.

1 Introduction

Diagrams are frequently used in people's daily communications. If computers could understand diagrams as well, then people could operate a computer system more intuitively, for instance, by sketching diagrams. Indeed, a number of pen-based systems that understand human-sketched diagrams have been developed in various fields, and their usefulness has been reported repeatedly (Oviatt 1996, Egenhofer 1997, Landay and Myers 2001, Davis 2002, Ferguson and Forbus 2002). Thus, computational diagram understanding is one of the highly prospective technologies for enriching human-computer interactions.

Arrows are major components of diagrams. Arrows appear in various types of diagrams, such as traffic signs, guideboards, route maps, flowcharts, and illustrations (Horn 1998, Wildbur and Burke 1998). One reason for such popularity is that arrows capture a large variety of semantics with their simple shape (Section 2). Another reason is that the existence of arrows encourages people to interpret causal and functional aspects in a diagram (Tversky *et al.* 2000). For instance, Fig. 1 contains only a few words and some arrow symbols over a background map, but people easily read the mechanism of a spatio-temporal process—the El Niño effect in the Southeastern Pacific Ocean indirectly influences the rise of tofu price in Japan. In this way, arrows are powerful tools that facilitate the communication of spatial and temporal knowledge in a static diagram.

A.G. Cohn and D.M. Mark (Eds.): COSIT 2005, LNCS 3693, pp. 232–250, 2005.
© Springer-Verlag Berlin Heidelberg 2005

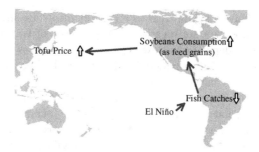

Fig. 1. A diagram with arrows, which illustrates a spatio-temporal process that the El Niño effect (i.e., sea temperature rise in the Southeastern Pacific Ocean) indirectly influence the rise of the tofu price in Japan

People can interpret such diagrams almost intuitively. Interpretation of arrow-containing diagrams is, however, a difficult task for computers due to the polysemy of arrows. For example, in Fig. 1, the arrow symbol departing from *El Niño* could be interpreted as a *spatial movement* (i.e., *El Niño is approaching South America*) or an *annotation* (i.e., *a fish species whose catches are declining is El Niño*), without any knowledge about *El Niño*. To avoid such misinterpretations, the current pen-based systems require their users to specify the meaning of every arrow (Forbus and Usher 2002) or restrict the meaning of arrows to a small set (Alvarado and Davis 2001, Landay and Myers 2001, Kurtoglu and Stahovich 2002). Consequently, people cannot illustrate their knowledge naturally in current pen-based systems. An automated interpretation of arrow-containing diagrams, therefore, remains a challenging problem for developing useful pen-based interfaces.

This paper develops a method for deducing possible interpretations of arrow-containing diagrams. Here, the *interpretation* of a diagram is referred to as (1) a process to determine the semantics of a diagram by reasoning, or (2) the determined semantics itself. The *semantics* of a diagram is referred to as a state or behavior of an entity or entities, which is represented by a diagram. A diagram contains symbols, each of which is assigned to an entity or its feature, which is called the *meaning* of the symbol. The semantics of a diagram is built from the meaning of each symbol in the diagram. We assume that the meaning of each symbol except arrows is already known.

For simplification, we do not consider the appearance of arrow symbols. Although the appearance of an arrow symbols is subject to a variety of visual variables (Bertin 1983), the arrow symbol alone rarely determine any specific meaning. Its semantic role is usually established when one or more arrow symbols depart from, traverse, or point to other elements in the diagram. Thus, we focus on the semantics associated with these related elements.

The combination of arrow symbols and the related elements is considered as a unit of syntax, and called *an arrow diagram*. Then, these arrow-related elements are called the *components* of the arrow diagram. An arrow diagram must have at least one arrow symbol and one component. As a first step, we consider the arrow diagram that contains only a uni-directional arrow symbol and its related elements. Bi-directional arrows are not considered, because they are regarded as a synthesis of two oppositely-directed arrow symbols. Also, independent arrow symbols, such as arrow-shaped traffic signs indicating curving roads and map symbols indicating north direction, are not discussed in this paper, because they have no components.

The remainder of this paper is structured as follows: Section 2 investigates various uses of arrow diagrams, and classifies their semantics into four different classes. Section 3 introduces a framework for the structure of arrow diagrams, based on the alignment of components and their semantic types. Section 4 discusses the structural requirements for illustrating each class of semantics, and Section 5 discusses the structural conditions of optional components. With these two bases, Section 6 investigates all structures of simple arrow diagrams that possibly illustrate each class of semantics, and demonstrates that the possible interpretations of arrow diagrams are determined simply by their structures. Finally, Section 7 presents conclusions and future work.

2 Semantics of Arrow Diagrams

What are arrows? Tversky (2001) defined an arrow as a special kind of line, with one end marked, inducing an asymmetry. Linearity and asymmetry are essential features of arrows. Asymmetry makes it possible to represent a direction or an order, whereas linearity makes it possible to represent a length, a path, or a connection. With these two features, arrow diagrams illustrate a large variety of semantics.

2.1 Use of Arrow Diagrams

One of the primary usages of arrow diagrams is to express a *direction*. Gombrich (1990) reported a very early example of an arrow diagram, where an arrow symbol was used to represent the direction of a water stream (Fig. 2a). Arrow diagrams may also refer to metaphorical directions. Upward directions are metaphorically associated with increase or improvement, whereas downwards directions are associated with decrease or debasement (Lakoff and Johnson 1980). Accordingly, upward and downward arrow symbols are used to illustrate those semantics.

An arrow symbol illustrating a direction has a diagrammatic freedom of length. This freedom allows the representation of a directed quantity, which is called a *vector*. A vector is a quantity that is specified by a direction and a magnitude.

Another traditional usage of arrow diagrams is to illustrate a *spatial movement*. Spatial movement is an event where an entity changes its spatial position continuously. Bertin (1983) showed a classic example that illustrates the paths of several expeditions in the Sahara Desert (Fig. 2b) and claimed that an arrow is the most efficient (and often the only) formula for representing a complex movement. The linearity and the asymmetry of an arrow symbol are appropriate features for illustrating the path and the, direction of the spatial movement respectively.

In geography, the flow of people, goods, or services between specific locations is called a *spatial interaction* (Bailey and Gatrell 1995). Since a spatial interaction is an aggregation of spatial movements, an arrow diagram can illustrate a spatial interaction just like a spatial movement, although the detailed route is often abbreviated due to the lack of the designer's concern (Fig. 2c). Monmonier (1990) reports that arrows are particularly useful for showing such spatial interactions as migration streams, spatial diffusion of ideas, migrations of tribes and refugees, and advance of armies. The scale of a spatial interaction is usually illustrated by the width of the arrow symbol, because people typically perceive the width of lines without a bias (Robinson *et al.* 1995). Spatial interactions can be generalized into *interactions*, which refer to the flow of a certain item from one entity to another entity, such that these two entities communicate indirectly. The communicating entities are not limited to locations; they can be other entities, such as people and physical objects (Fig. 2d).

Arrow diagrams are used not only in a spatial context, but also in a temporal context. For instance, timetables often contain arrow diagrams, each of which illustrates that something continues over a certain period (Fig. 2e). This semantics is called *temporal continuity*. People understand the concept of time with the aid of spatial metaphors (Lakoff and Johnson 1980). Temporal continuity is understood with the metaphor of travel, in the sense that something travels continuously over time instead of space. Therefore, arrow diagrams that originally describe travel in space (i.e., spatial movement) are naturally extended to describe a travel in time (i.e., temporal continuity).

Similarly, flowcharts often contain arrow diagrams, each of which captures a *temporal order* of two components (Fig. 2f). The order is distinctively expressed by the directionality of an arrow symbol. The connected two components in an arrow diagram refer to (1) two different entities or (2) two different states of an identical entity. In the former case, the arrow diagram may imply a *conditional relation* or a *causal relation* of the associated components, such that the proceeding component works as a precondition or a cause of the subsequent component. In the latter case, the arrow diagram may imply a *change*. A change is an event where an entity transforms its features, such as identity (Hornsby and Egenhofer 2000), appearance, name, and structure. For instance, a chemical equation, such as $2H_2+O_2 \rightarrow 2H_2O$, illustrates a set of materials transforming its molecular structures.

Arrow diagrams are also used in complicated illustrations, where each arrow symbol connects an element of the illustration with a short description (Fig. 2g). This use is called *labeling*. Arrow symbols or lines are often used for labeling when the illustration becomes messy if descriptions are placed directly on it. Although both an arrow symbol and a line can connect an element and its description, the directionality of the arrow symbol makes it clear that the description is assigned to the element, and accordingly people can distinguish the element and its description more easily.

Finally, the use of arrow diagrams to illustrate *relations* is a widespread convention in sketches (Forbus and Usher 2002). An arrow diagram visualizes the presence of a directed relation between two components. In mathematics, a set of directed binary relations is modeled as a directed graph, which is often visualized using arrow diagrams (Fig. 2h). Directed binary relations are a broad concept that includes temporal orders, conditional relations, causal relations, labeling, and so on.

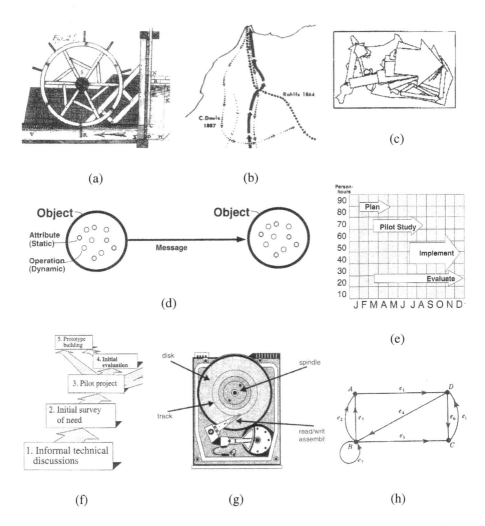

Fig. 2. Examples of arrow diagrams illustrating various semantics: (a) a direction (Gombrich 1990), (b) spatial movements (Bertin 1983), (c) spatial interactions (Tobler 1987), (d) an interaction (Worboys and Duckham 2004), (e) temporal continuities (Horn 1998), (f) temporal orders (Horn 1998), (g) labeling (Worboys and Duckham 2004), and (h) ordered relations (Lipschutz and Lipson 1997)

2.2 Classification of Arrow Diagram Semantics

For the following discussions, the semantics of arrow diagrams are classified into four different classes (Table 1).

First, the arrow-related semantics are divided into semantics that require only one component and semantics that require two or more components (Fig. 3). In the former

Table 1. Classification of the semantics of arrow diagrams

Example of semantics	Class	Definition
Direction, vector	Property	modification of a component by attaching an arrow symbol to the component
labeling	Annotation	modification of a component by connecting a description to the component
Spatial movement, temporal continuity, interaction	Action	a motion of one component that may be caused or cause an interaction with another component.
temporal order, conditional relation, causal relation, change, ordered relation	conjunction	association of components, where an arrow symbol does not express a motion

case, the arrow symbol is attached to a component and modifies it directly. Such semantics is called a *property*. Directions and vectors are examples of properties. Although an arrow diagram illustrating a property sometimes contains two components, only one of them is essential for illustrating that property (Section 5.2).

Among the arrow-related semantics that require two or more components, *labeling* is exceptional, because other semantics associate at least two independent components that do not modify each other, while labeling introduces one component only for modifying another component. Thus, labels are distinguished from the other semantics (Fig. 3) and referred to as an independent class of semantics called *annotation*. An annotation is defined as a modification of a component by connecting the component and its description. Annotations and properties are similar in the sense that both of them modify one component, although their representation styles are different.

The remaining arrow-related semantics are divided into *actions* and *conjunctions* (Fig. 3). Actions are the semantics where an arrow symbol illustrates a motion of one component that may be caused or cause an interaction with another component. Spatial movements, temporal continuities, and interactions are classified into actions, since these semantics refer to a spatial or temporal movement. On the other hand, a conjunction is an association of components, where an arrow symbol does not express a motion. Accordingly, the shape of an arrow symbol is often meaningful for an action, while it is not for a conjunction.

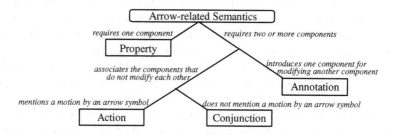

Fig. 3. Classification of arrow-related semantics

3   Structures of Arrow Diagrams

Arrow diagrams illustrate a large variety of semantics, which increases the difficulty of their interpretation. To tackle this problem, we first develop a method for making rough interpretation of an arrow diagram from its structural pattern alone. As the foundation of this method, this section summarizes the formal structures of arrow diagrams developed by Kurata and Egenhofer (2005).

The *components* of an arrow diagram are diagrammatic elements that an arrow symbol originates from, traverse, or points to. We classify the components of arrow diagrams into the following five types:

- An *object* takes an action, either independently (e.g., a person in Fig. 4a) or as a result of interaction (e.g., a bag in Fig. 4a).
- An *event* occurs in time, and is characterized by a set of changes. An event occurs over an interval (e.g., snow in Fig. 4b) or at an instant (e.g., a traffic accident in Fig. 4b).
- A *location* is a position in space. It may be a point (e.g., a place in Maine in Fig. 4c) or a homogeneous area (e.g., Maine).
- A *moment* is a position in time. It may be an instant (e.g., *8:20* in Fig. 5a) or a homogeneous interval (e.g., morning).
- A *note* is a short description that modifies an arrow symbol (e.g., *send* in Fig. 4a) or another component (e.g., *Mr. K* in Fig. 4a and *You are here* in Fig. 4c). A note and the modified component are placed adjacently to each other or connected with each other by an arrow symbol.

For simplification, an object, an event, a moment, a location, and a note are sometimes denoted as *O, E, M, L,* and *N*, respectively.

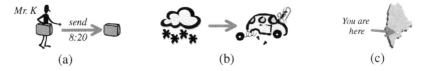

(a)   (b)   (c)

Fig. 4. Arrow diagrams that contain various types of components

Although a component may be mentioned by an icon, a text, or a specific position in a background drawing, this classification is not concerned with such a descriptive style of the component. We assume that automated interpreters of arrow semantics will distinguish these component types based on their knowledge base.

Components of an arrow diagram are located in front of the arrow's head, behind the arrow's tail, or along the arrow's body. We, therefore, consider that an arrow symbol identifies three different areas where the components of the arrow diagram are located (Fig. 5). These areas are called *component slots*, and they are further classified into a *tail slot*, a *head slot*, and a *body slot*. Each component in an arrow diagram is uniquely assigned to one of these three slots, thereby making the distinction of *tail components*, *body components*, and *head components*. The component slots need to be distinguished, because the same symbols, used in different slots, illustrate significantly different semantics (Kurata and Egenhofer 2005).

An arrow diagram has three slots, where five types of components may be located. The combination of the types of components in the three slots composes a certain pattern that is specific to every arrow diagram. These patterns are described as

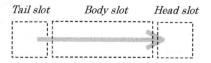

Tail slot   Body slot   Head slot

Fig. 5. Three component slots associated with an arrow symbol

($[M|E|L|O|N]^*$, $[M|E|L|O|N]^*$, $[M|E|L|O|N]^*$), where $[x]^*$ means empty or a sequence of any number of x, $x|y$ means x or y but not both, and the three elements in parentheses indicate the types of components in tail, body, and head slot, respectively. For example, the structures of arrow diagrams in Fig. 4a-c are described as (ON, MN, O), (E, $-$, E), and (N, $-$, L), respectively. These patterns capture fundamental structures of arrow diagrams, since they capture the alignment of components while abstracting the individual difference and the absolute location of the components.

4 Structural Requirements of Arrow Diagrams

This section discusses the structures of arrow diagrams that are required for illustrating certain semantics. The discussion follows the classification of the arrow-related semantics in Section 2.

4.1 Structural Requirements for Illustrating a Property

When an arrow diagram illustrates a property, its arrow symbol is tied to only one component. Consequently, all visual variables of the arrow symbol, such as length, width, shape, color, orientation, and pattern (Bertin 1983), can be controlled by its designer. Among these variables, length and orientation are predominant due to the linearity and asymmetry of arrow symbols. Arrow symbols are, therefore, appropriate for representing some feature that is related to a length, an orientation, or both. A length is, however, illustrated more easily by a line or a bar. Consequently, arrow symbols are exclusively used to illustrate (1) properties that are specified by an orientation (i.e., directions) or (2) properties that are specified by both an orientation and a length (i.e., vectors).

The component whose property is described by an arrow symbol is called a *subject*. The subject can be placed in any slot, since nothing conflicts with the subject (Fig. 6a-c). The different positions of the subject, however, yield slightly different semantics:

- if the subject is placed in the tail slot (Fig. 6a) the illustrated property may be related to an outgoing action,
- if the subject is placed in the body slot (Fig. 6b) the illustrated property may be related to a passing-through action, and
- if the subject is placed in the head slot (Fig. 6c) the illustrated property may be related to an incoming action.

Thus, a moving direction of a car, a wind direction, and a direction of an external force are distinctively illustrated by arrow diagrams with these different structures (Fig. 6a'-c').

The arrow diagrams in Fig. 6a'-c' indicate that an object, a location, and an event may represent a subject of a property (i.e., they can have a direction or a vector). A moment and a note cannot have these properties, since the moment is a one-dimensional concept and the note is a subsidiary component.

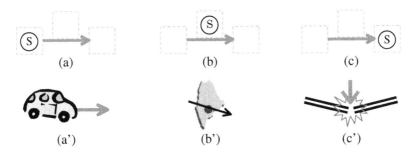

Fig. 6. (a-c) Basic structures of arrow diagrams for illustrating a property of a subject (S). (a'-c') Examples of the corresponding arrow diagrams, which illustrate: (a) a moving direction of a car, (b) a wind direction, and (c) a direction of an external force

4.2 Structural Requirements for Illustrating an Annotations

An arrow symbol may illustrate an annotation by connecting a component and its description (Fig. 7). Annotation and properties are similar in the sense that both modify a component. Thus, the annotated component is called a *subject* by analogy.

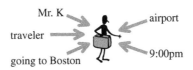

Fig. 7. Examples of arrow diagrams illustrating annotations. Annotations describe various features of a subject, such as name (*Mr. K*), category (*traveler*), status (*going to Boston*), spatial position (*airport*), and temporal position (*9:00am*).

When illustrating an annotation, an arrow diagram takes only one type of structure with regard to a description and a subject in order to specify that the description is assigned to the subject (Fig. 8).

Fig. 8. Basic structure of arrow diagrams for illustrating an annotation (D: description; S: subject)

The subject in annotations is represented by any type of component except for notes, since notes are subsidiary. The description is represented by a note, a location, or a moment. A location and a moment specify the spatial and temporal position of the subject, respectively (e.g., *airport* and *9:00am* in Fig. 7), whereas a note describes other feature of the subject, such as name, category, and status (*Mr. K, traveler*, and *going to Boston* in Fig. 7).

4.3 Structural Requirements for Illustrating an Action

In this section, we show how different types of actions can be modeled with arrow diagrams eliciting different types of structures.

In a primitive sense, an arrow is a flying weapon that moves in space. Naturally, an arrow diagram is used to illustrate a *spatial movement* of an entity. In addition, an arrow diagram is analogically used to illustrate a *temporal continuity* of an entity, because an entity travels in time, just like it travels in space.

The movement of an entity may accompany another action. If there is another entity in the way of the moving entity, it implies that the moving entity gets into contact with this entity. This type of semantics is called an *encounter*. An encounter is an action where an entity (*mover*) physically or conceptually moves, and eventually has a contact with another entity (*receiver*). Encounters are described by such verbs as *approach, enter, join, conflict, receive, consume, mount*, and *import*.

An encounter refers to two entities getting together. Conversely, a *division* occurs when two entities become separated. It is an action where an entity (*mover*) moves away from another entity (*sender*). Divisions are described by such verbs as *exit, withdraw from, branch off, leave, produce, uninstall*, and *export*.

The third possibility is the combination of a division and an encounter, that is, an action where an entity (*mover*) moves away from another entity (*sender*), and then gets into contact with the third entity (*receiver*). This semantics is called a *ditransitive action*, since it is analogous to a ditransitive verb in linguistics, such as *send, give, show, tell, explain, sell*, and *buy*. Ditransitive verbs refer to two types of objects: (1) *direct objects*, which receive the subject's action directly, and (2) *indirect objects*, which receive the subject's action indirectly. The direct objects intermediate the subject and the indirect objects. If this intermediation accompanies physical or conceptual movement, the semantics can be represented diagrammatically with an arrow diagram as a ditransitive action.

Any of the above five types of actions requires a mover. In addition, an encounter requires a receiver, a division requires a sender, and a ditransitive action requires both a sender and a receiver. The sender and the receiver must be placed in the tail and the head slot, respectively, since arrow's tail and head imply the initial and final positions of a motion. Meanwhile, the mover should be placed in any slot that is not occupied by the sender or the receiver, so that the mover can be visually distinguished from them. Thus, this classification of actions yields eight different structures of arrow diagrams with regard to a mover, a sender, and a receiver (Table 2).

A mover is represented by an object or an event, because objects and events are the only components that may move in space or time. A sender is represented by an object, an event, or a location. For example, an industrial plant (object), deforestation (event), and a volcano (location) can send out air pollutants (mover). Similarly, a receiver is represented by an object, an event, or a location. For example, a famous statue (object), a festival (event), and a historic site (location) can receive a tourist (mover). Arrow diagrams can be applied to illustrate all of these scenarios.

Table 2. Basic structures of arrow diagrams for illustrating actions with regard to a mover (Mv), a sender (Sd), and a receiver (Rc)

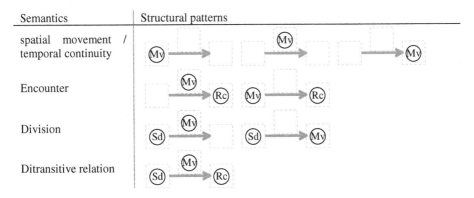

Semantics	Structural patterns
spatial movement / temporal continuity	
Encounter	
Division	
Ditransitive relation	

Spatial movement also needs at least one component in an arrow diagram that specifies its origin, route, or destination. Similarly, temporal continuity needs at least one component that specifies its start-time or end-time. The origin, route, and destination are represented by locations in the tail, the body, and the head slot, respectively (Fig. 9a), while the start-time and end-time are represented by moments in the tail and the head slot (Fig. 9 b). These components are referred to as *adverbial components* (Section 5.2).

(a)   (b)

Fig. 9. Arrow diagrams illustrating an action contains adverbial components, which specify (a) the origin, the route, and the destination, and (b) the start-time and the end-time of the action

4.4 Structural Requirements for Illustrating a Conjunction

An arrow diagram may illustrate a conjunction, where multiple components are associated without referring to a motion. For example, *Lobster→Maine* is a conjunction where lobster is associated with Maine. Every conjunction has a certain theme that justifies the association (Table 3). In *Lobster→Maine*, lobster and Maine are associated, for instance, under the theme of local specialty. The theme may be specified in a caption, a legend, or an *adverbial component* that modifies the arrow symbol (Section 5.2), or may be infered from the context.

Since an arrow symbol is oriented, the associated components are naturally ordered. Therefore, if a certain rationale is available, the associated components should be aligned in a meaningful way. The underlying theme often provides such a rationale as temporal order (e.g., from old to new), spatial order (e.g., from high to low, from front to back, and from part to whole), logical order (e.g., from presumption to conclusion), or thinking order (e.g., first comes into head, first placed) (Table 3).

Table 3. In conjunctions, components are associated under a certain theme and aligned under a certain rational

Associated Components	Theme	Rationale of Alignment	Representation
Lobster, Maine	local specialty	Thinking order	Lobster→Maine
Plan, Do, See	work process	temporal order	Plan→Do→See
Niagra Falls, Lake Ontario	water flow	spatial order (high to low)	Niagra Falls→Lake Ontario
Maine, New England	geographic attribution	spatial order (part to whole)	Maine→New England
Snow, Delay	causal relation	logical order	Snow→Delay

The associated components are placed in the tail and the head slot of an arrow diagram, such that these components look equally associated, as well as the order among these components are highlighted (Fig. 10). Any type of component, except notes, can be the associated components, as long as people identify the theme that associates these components.

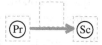

Fig. 10. Basic structure of arrow diagrams for illustrating conjunctions (Pr, Sc: the associated component that proceeds and succeeds in an order, respectively)

5   Optional Components of Arrow Diagrams

The previous section discussed the structural requirements for illustrating each class of semantics. In addition to the components requested by these requirements, arrow diagrams may have the following two types of optional components.

5.1   Adjective Components

An *adjective component* of an arrow diagram modifies an individual component of the arrow diagram. It is analogous to adjectives in linguistics, which modify an individual noun. The adjective component describes a feature of the component, such as its category, name, scale, spatial position, and temporal position—just as descriptions in annotations do.

The component and its adjective component are placed adjacently to each other and, therefore, they are always placed in the same slot (Fig. 11). An adjective component is represented by a location, a moment, or a note, each of which illustrates the place, time, and other features of the modified component.

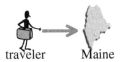

traveler   Maine

Fig. 11. An arrow diagram that contains adjective components (*traveler* and *Maine*)

5.2   Adverbial Components

An *adverbial component* of an arrow diagram modifies the main semantics that the arrow diagram illustrates. It is analogous to adverbs and adverbial phrases in linguistics, which modify a verb. Properties, actions, and conjunctions may have adverbial components in arrow diagrams, whose functions are as follows:

- An adverbial component of a property describes a feature of the property, such as category, name, or scale (Fig. 12a-c).
- An adverbial component of an action describes a feature of the action, such as (1) origin, route, or destination (Fig. 9a), (2) start-time or end-time (Fig. 9b), or (3) category, measure, overall spatial position, overall temporal position, or scale (Fig. 12d-f).
- An adverbial component of a conjunction describes a feature of the conjunction, such as associating theme, sequential condition, sequential probability, overall spatial position, or overall temporal position (Fig. 12g-i).

Adverbial components are usually optional, but spatial movement requires at least one adverbial component that specifies the origin, the route, or the destination, and temporal continuity requires at least one adverbial component that specifies the start-time or the end-time (Section 4.3).

An adverbial component is represented by a location, a moment, or a note, depending on the feature that it describes:

- a location represents origin, route, destination, or overall spatial position,
- a moment represents start-time, end-time, or overall temporal position, and
- a note represents other features.

An adverbial component is usually placed in the body slot, especially around the centerof an arrow symbol, so that it appears visually that the adverbial component is devoted to not a part but the whole of the arrow symbol (Fig. 12). Exceptionally,

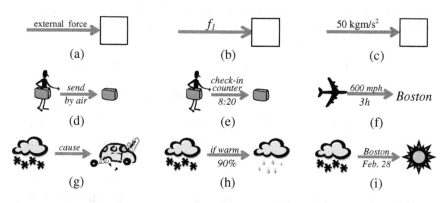

Fig. 12. Arrow diagrams, illustrating (a-c) a property, (d-f) an action, and (g-i) a conjunction. Each arrow diagram contains adverbial components which describe: (a) property's category, (b) property's name, (c) property's scale, (d) action's category and measure, (e) overall spatial and temporal position of the action, (f) action's scales, (g) underlying theme of the conjunction, (h) sequential condition and sequential probability of the conjunction, and (i) overall spatial and temporal position of the conjunction.

locations representing an origin and a destination are placed in the tail and the head slot, respectively, since arrow's tail and head imply the initial and final positions of a motion. Similarly, moments representing a start-time and an end-time are placed in the tail and the head slot, respectively. In any case, an adverbial component must be placed in an empty tail slot, an empty head slot, or an empty part of the body slot; otherwise, the adverbial component is misinterpreted as an adjective component.

6   Deducing Possible Interpretations from Structures

semantics and the structural conditions for adding optional components, we can consider the structures of arrow diagrams that possibly illustrate each class of semantics. One problem is that arrow diagrams take countless structures, since each slot can contain an arbitrary slot Now that we have both the structural requirements for illustrating each class of number of components. For simplification, here we consider only *simple arrow diagrams*, which have at most one component per each. Since each of the three slots has six possible patterns (an object, an event, a location, a moment, a note, and an empty component) but an arrow diagram must have at least one component, simple arrow diagrams may take only $2^3 - 1 = 215$ different patterns of structures. Thus, we can exhaustively investigate all structures of arrow diagrams that possibly illustrate each class of semantics. For example, to illustrate a *property*, one of the three slots must contain a subject, which is represented by an object, an event, or a location. In addition, an optional component (i.e., an adverbial component), which is represented by a note, may be place in the body slot if it is empty. Accordingly, there are 15 structures of simple arrow diagrams that possibly illustrate a property (Table 4). Similarly, we can enumerate 12 structures for annotations, 98 structures for actions, and 64 structures for conjunctions.

Table 4. All structural patterns of arrow diagrams illustrating a property

| | | Adverbial component | |
		–	Note (*N*)
Subject	Object (*O*)	$(O, -, -) (-, O, -) (-, -, O)$	$(O, N, -) (-, N, O)$
	Event (*E*)	$(E, -, -) (-, E, -) (-, -, E)$	$(E, N, -) (-, N, E)$
	Location (*L*)	$(L, -, -) (-, L, -) (-, -, L)$	$(L, N, -) (-, N, L)$

Based on this result, we can judge whether a given arrow diagram has a possibility of illustrating each class of semantics. Among the 215 structures of simple arrow diagrams, 82 structures have no corresponding class, 81 structures correspond to exactly one class, and 52 structures correspond to multiple classes (Fig. 13).

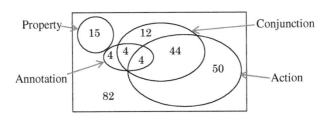

Fig. 13. The correspondence of 215 structures of simple arrow diagrams to the four classes of semantics

An arrow diagram with one of the 82 structures that has no corresponding class is automatically judged as a meaningless diagram (Fig. 14a), and an arrow diagram with one of the 81 structures that correspond to exactly one class leads to a unique class of interpretation (Fig. 14b). Fig. 13 highlights that an arrow diagram illustrating a property is always interpreted uniquely, while an arrow diagram illustrating other semantics (especially an annotation or a conjunction) is often ambiguous.

(a)   (b)

Fig. 14. (a) An arrow diagram, whose structure (N, O, L) has no corresponding class of semantics, is judged as meaningless, and (b) an arrow diagram, whose structure (L, O, L) corresponds to exactly one class of semantics (action), leads to a unique class of interpretation

Table 5 shows the structures of arrow diagrams that correspond to multiple classes of semantics. The structures S_1 and S_2 indicate that (1) an arrow diagram that annotates a component by a location or a moment (Fig. 15a$_1$) and (2) an arrow diagram that illustrates a conjunction whose proceeding component is a location or a moment (Fig. 15a$_2$) cannot be distinguished by their structures alone. An arrow diagram is, however, uniquely interpreted as a conjunction if the head component cannot be located at the spatial or temporal position that is specified by the tail component (for instance, Fig. 15a$_2$ cannot be interpreted as an annotation because *Lake Ontario* is not located in *Niagara Falls*). Otherwise, we need to judge whether the diagram has a certain theme for associating these components.

Table 5. Structures of arrow diagrams corresponding to multiple classes of semantics

Structures		Number of structures	Semantics
S_1	$(L|M, -, L|M)$	4	annotation and conjunction
S_2	$(L|M, -, O|E)$	4	annotation, action, and conjunction
S_3	$(O|E, -|L|M|N, O|E|L|M)$, $(L|M, L|M|N, O|E)$	44	action and conjunction

San Jose
 ↖Costa Rica

(a$_1$)

(b$_1$)

Paris→

(c$_1$)

Niagra ⇒ Lake
Falls   Ontario

(a$_2$)

(b$_2$)

Paris➡

(c$_2$)

Fig. 15. Three pair of arrow diagram whose semantic classes are different but whose structures are same. The structures of arrow diagrams illustrating (a$_1$) an annotation and (a$_2$) a conjunction are both $(L, -, L)$, those illustrating (b$_1$) an action and (b$_2$) a conjunction are both $(O, -, O)$, and those illustrating (c$_1$) an annotation and (c$_2$) an action are both $(L, -, O)$.

Similarly, arrow diagrams illustrating an action or a conjunction, whose structures are in S_2 or S_3, cannot be distinguished by their structures alone (Fig. 15b$_1$ and 15b$_2$). The large number of these ambiguous structures shows the importance of their distinction. An arrow diagram is uniquely interpreted as a conjunction if neither its tail component nor its head component is supposed to move (for instance, Fig. 15b$_2$ cannot be interpreted as an action because both the house and the burning house are immovable). Otherwise, we need to judge whether the diagram has a certain theme for associating these components and whether the action possibly illustrated in the arrow diagram is a typical scenario of actions.

We further need the distinction of arrow diagrams illustrating annotations and actions, whose structures are S_2 (Fig. 15c$_1$ and 15c$_2$). An arrow diagram is uniquely interpreted as an annotation if the tail component is not supposed to move (for instance, Fig. 15c$_1$ does not illustrate an action because the Eiffel Tower is immovable). Otherwise, their distinction may depend on the context.

7   Conclusions and Future Work

This paper presented the classification of arrow-related semantics, the structural requirements for illustrating each class by arrow diagrams, and the structural conditions for adding optional components. Based on these settings, we demonstrate that 82 structures of simple arrow diagrams are automatically judged as meaningless, 81 structures leads to a unique class of semantics, and 52 structures lead to two or three possible classes of semantics. This indicates that knowledge about the structure of an arrow diagram reduces the ambiguity of its possible interpretations.

The knowledge about the structure is still useful for deducing more detailed interpretation of the arrow diagram. For example, Section 4.3 indicated that five types of actions, each of which has different structural requirements. Thus, the structural differences of arrow diagrams are probably useful for distinguishing these different types of actions.

In addition to the structures, various clues lead people to a unique and more detailed interpretation of an arrow diagram. For example, intrinsic mobility, movable space, and movable direction of each component are critical clues for judging whether the component is supposed to move (Fig. 16). This knowledge is, then, useful for judging whether an arrow diagram may illustrate an action, and if so, which type of action it may illustrate (i.e., *encounter* or *division*). Similarly, the caption or the adverbial component of an arrow diagram may specify or imply a theme for association, which is useful for judging whether an arrow diagram may illustrate a conjunction, and if so, which type of conjunction the arrow diagram may illustrate.

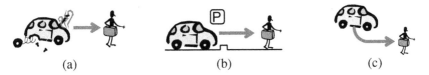

(a)   (b)   (c)

Fig. 16. Thanks to the knowledge about (a) intrinsic mobility, (b) movable space, and (c) movable direction of the car, arrow diagrams that are typically interpreted as *a person leaves a car*, not as *a car approaches a person*

Making use of such available clues comprehensively, we are now challenging to develop more sophisticated method for interpreting arrow diagrams. This method is expected to enhance the usability of pen-based systems, such that its user can communicate with the systems more naturally by drawing an arrow-containing diagram. In addition, since an arrow diagram is popular and often the simplest tool for the representation of spatial and temporal knowledge, to reveal the mechanism of arrow diagrams should lead to further understanding of people's communication about space and time.

Acknowledgments

This work was partially supported by the National Geospatial-Intelligence Agency under grant number NMA201-01-1-2003. Yohei Kurata is further supported by a University of Maine International Tuition Scholarship. Max Egenhofer's work is further supported by the National Science Foundation under grant numbers EPS-9983432 and IIS-9970123; the National Geospatial-Intelligence Agency under grant numbers NMA201-00-1-2009, and NMA401-02-1-2009.

References

Alvarado, C. and Davis, R. (2001) Resolving Ambiguities to Create a Natural Sketch Based Interface. *17th International Joint Conference on Artificial Intelligence (IJCAI-01)*, Seattle, WA. pp.1365-1374.

Bailey, T. and Gatrell, A. (1995) *Interactive Spatial Data Analysis*. Essex, UK: Longman.

Bertin, J. (1983) *Semiology of Graphics: Diagrams, Networks, Maps*. Madison, WI: University of Wisconsin Press.

Davis, R. (2002) Sketch Understanding in Design: Overview of Work at the MIT AI Lab. *AAAI Spring Symposium on Sketch Understanding*, Menlo Park, CA, AAAI Press. pp.24-31.

Egenhofer, M. (1997) Query Processing in Spatial-Query-by-Sketch. *Journal of Visual Languages and Computing*. 8(4): 403-424.

Ferguson, R. and Forbus, K. (2002) A Cognitive Approach to Sketch Understanding. *AAAI Spring Symposium on Sketch Understanding*, Menlo Park, CA, AAAI Press. pp.67-72.

Forbus, K. and Usher, J. (2002) Sketching for Knowledge Capture: A Progress Report. *7th International Conference on Intelligent User Interfaces*, San Francisco, CA, ACM Press. pp.71-77.

Gombrich, E. (1990). Pictorial Instructions. In *Images and Understanding: Thoughts About Images-Ideas About Understanding*. Barlow, H., Blakemore, C. and Weston-Smith, M. (eds.), Cambridge, UK: Cambridge University Press, pp.26-45.

Horn, R. (1998) *Visual Language: Global Communication for the 21st Century*. Bainbridge Island, WA: MacroVu, Inc.

Hornsby, K. and Egenhofer, M. (2000) Identity-Based Change: A Foundation for Spatio-Temporal Knowledge Representation. *International Journal of Geographical Information Science*. 14(3): 207-224.

Kurata, Y. and Egenhofer, M. (2005) Semantics of Simple Arrow Diagrams. *AAAI Spring Symposium on Reasoning with Mental External Diagram: Computational Modeling and Spatial Assistance*, Menlo Park, CA, AAAI Press.

Kurtoglu, T. and Stahovich, T. (2002) Interpreting Schematic Sketches Using Physical Reasoning. *AAAI Spring Symposium on Sketch Understanding*, Melon Park, CA, AAAI Press. pp.78-85.

Lakoff, G. and Johnson, M. (1980) *Metaphors We Live By*. Chicago, IL: University of Chicago Press.

Landay, J. and Myers, B. (2001) Sketching Interfaces: Toward More Human Interface Design. *IEEE Computer*. 34(3): 56-64.

Lipschutz, S. and Lipson, M. L. (1997) *Schaum's Outline of Theory and Problems of Discrete Mathematics*. New York: McGraw-Hill.

Monmonier, M. (1990) Strategies for the Visualization of Geographic Time-Series Data. *Cartographica*. 27(1): 30-45.

Oviatt, S. (1996) Multimodal Interfaces for Dynamic Interactive Maps. *Conference on Human Factors in Computing Systems (CHI '96)*, New York, ACM Press. pp.95-102.

Robinson, A., Morrison, J., Muehrcke, P., Kimerling, A. and Guptill, S. (1995) *Elements of Cartography*. New York: John Wiley & Sons Inc.

Tobler, W. (1987) Experiments in Migration Mapping by Computer. *The American Cartographer*. 14(2): 155-163.

Tversky, B. (2001). Spatial Schemas in Depictions. In *Spatial Schemas and Abstract Thought*. Gattis, M. (ed.), Cambridge, MA: MIT Press, pp.79-111.

Tversky, B., Zacks, J., Lee, P. and Heiser, J. (2000) Lines, Blobs, Crosses and Arrows: Diagrammatic Communication with Schematic Figures. *Theory and Application of Diagrams (Diagram 2000)*, Edinburgh, UK, Anderson, M., Cheng, P. and Haarslev, V. (eds.), Lecture Note in Artificial Intelligence. Berlin: Springer. pp.221-230.

Wildbur, P. and Burke, M. (1998) *Information Graphics: Innovative Solutions in Contemporary Design*. New York: Thames & Hudson.

Worboys, M. and Duckham, M. (2004) *GIS: A Computing Perspectinve*. Boca Raton, FL: CRC Press.

Cognitive Maps Are over 60

Juval Portugali

Environmental Simulation Laboratory (ESLab), Porter School of Environmental Studies,
Department of Geography and the Human Environment, Tel Aviv University, Tel Aviv
Tel:+972(0)3-640-8661, Fax: +972(0)3-640-6243
juval@post.tau.ac.il
http://www.eslab.tau.ac.il

Abstract. The notion cognitive map (CM) is regarded as highly valuable but also somewhat obscured and as a consequence highly contested. This paper examines several forms of obscurity and suggests that in order to clarify the notion CM it would be useful to treat it not in terms of a single meaning entity, but in terms of kinds of cognitive maps. This is exemplified by introducing several new forms of CM.

1 Introduction

The notion cognitive map (CM) is due to Tolman (1948) who in a set of experiments conducted during the 1930s and 1940s has shown that animals and humans have the capability to construct in their minds a representation of the external extended environment they experience. The notion CM is over 60 years old now. During these years the concept has attracted a large number of scholars from a wide spectrum of research domains and disciplines, it stood at the center of cognitive geography, environmental perception and spatial behavior, it entailed concepts such as systematic distortions, to name but few that come to mind. On the other hand, the notion CM has considerably influences the hermeneutic traditions of human and humanistic geographies – mainly through Lowenthal's (1961) seminal paper. More recently it was used by Jameson in his postmodernist writings and through him the concept re-entered social geography (Gregory, 1994).

Despite its attraction, the term was never fully and clearly defined with the consequence that different scholars mean different things when referring to CMs. Thus, Okeef and Nadel (1978) refer to a certain part of the brain – the hippocampus, place-cells and the like. For cognitive geographers and psychologists the term meant a name or a metaphor for internally represented information, etc. Concluding a paper about systematic distortion, Tversky (1992) notes that "cognitive maps are impossible figures", while Kitchin and Blades (2002) write that the concept is highly contested.

The suggestion in this paper is that the obscurity of the notion CM is, among other things, a consequence of (1) the development of ideas in cognitive science with respect to the very nature of cognition and cognitive processes; (2) the related development of research about kinds of memory and (3) the fuzzy boundary between the notion CM and the cognitive process of categorization.

A.G. Cohn and D.M. Mark (Eds.): COSIT 2005, LNCS 3693, pp. 251–264, 2005.
© Springer-Verlag Berlin Heidelberg 2005

2 Obscurities

2.1 Obscurity Due to Different Ontologies

Our view of cognition has changes considerably: Tolman has developed his CM in the context of behaviorism. At a later stage and in retrospect Tolman's CM was treated as one of the first cracks in the dominancy of behaviorism (Gardner, 1987). In the 1950s and 1960s the study of cognitive maps was dominated by classical cognitivism and its information processing approach, while more recently paradigms that come under headings such as *embodied* or *situated* cognition seem to challenge the classical view.

As will be emphasized below, these new ideas have already started to effect studies of cognitive maps. What is not yet clear is that the approaches that challenge the classical view imply, in fact, different ontologies as to what is cognition and by implication a CM. The result is that when using the term CM, students approaching it from the above different perspectives assume, in fact, different things: some see it as an internal representation in the classical way, others in terms of affordances while still others in terms of experiential realism and so on.

2.2 Obscurity Due to Different Kinds of Memory

The notion CM is usually defined as a specific form of memory, namely, memory for extended space. The entity 'memory', as is well known, is very general and rich and can thus mean several things. Roediger III, Marsh and Lee (2001) have noted in this connection, that "memories come in multifarious forms, and within each form there are multiple dimensions." In their study entitled "kinds of memory" they suggest a classification of memory types that includes declarative vs. procedural memory, explicit vs. implicit, conscious vs. unconscious, voluntary vs. involuntary, retrospective vs. prospective, long-term (LTM) vs. working memory (WM), autobiographic, semantic, implicit forms of memory and several other kinds.

CM as a form of memory is not independent of other forms of memory. This fact has not escaped the attention of students of cognitive mapping and we do find some (though, not enough) discussions about the relations between CM and other forms of memory: LTM vs. WM, episodic vs. procedural and so on (see below). What typify these studies is that they put CM as a form of memory alongside other forms of memory. But the relationships between CM and the various forms of memory goes beyond that: In some cases the relationships are such that they affect the very nature of the cognitive maps themselves with the implication that when using the notion CM some students in fact mean LT CM, while other WM CM, still others autobiographic CMs and so on.

2.3 Obscurity Due to Categorization

One of the central premises of CM studies is that people construct in their mind an image of their environment and that this image, i.e. CM, participates in determining their behavior in the environment. For example, people's "image of the city" within which they live significantly influences their movement in the city and the decisions they take in it: where to live, work, shop and so on. In a recent paper it has been shown that people's "image of a city", that is to say, their image of the category and concept "city", plays a similar role (Portugali, 2004). It has further been shown, that many cognitive maps have category-like properties (Portugali, 2005) with the implication that when talking about a CM of a city, for instance, it is not always clear whether one speaks about a cognitive map of a city, that is, about the category city and a cognitive map of a specific city.

2.4 Toward a Clarification

In light of the above it is no wonder that the concept CM is highly contested: different scholars mean different things when they employ the concept. The obscured nature of the concept CM has recently led to two opposing reactions. On the one hand, Kitchin and Freundschuh (2000) and Kintchin and Blades (2002) suggest further research and elaboration. On the other hand, Roberts (2001, 16) argues "that the term cognitive map may have lived its usefulness." It has played its part, helped get rid of S-R behaviorism, provided impetus for brain research and studies on spatial behavior, but now it is time for it to go. In place of cognitive map theory Roberts (ibid) suggests putting a number of spatial representation mechanisms that operate hierarchically.

This paper is in line with Kitchin and Freundschuh and Kitchin and Blades view and in opposition to Robers'. It suggests – following Oscar Wilde's famous phrase – that "the rumors about the death of CMs are rather pre-mature". That we have not as yet started to explore the richness enfolded in the notion CM. That it's many possible meanings and forms suggest, in fact, a research agenda that might keep us busy for the next 60 years. The aim of this papers is thus to trigger a discussion that will clarify this rich research agenda. This will be done by reference to the three forms of obscurity discussed above. Accordingly, I'll first discuss different ontologies regarding cognition and will suggest a distinction between classical, embodied and SIRN CMs. Next, I'll discuss some forms of memory and suggest a distinction between autobiographic, prospective, long-term and working CMs. Finally I'll discuss categorization and a distinction between categorical vs. specific CMs.

3 Classical, Embodied and SIRN CM

Since the emergence of cognitive science in the mid 1950s (Gardner, 1987) our view of cognition has changes considerably. This section discusses some of these changes and the implications thereof to the notion CM.

3.1 Classical CM

In the 1950s the study of cognitive maps was dominated by classical cognitivism and its information processing approach. According to this view, there is a clear-cut separation between the cognitive system in the brain, the body within which the brain is located and the world outside. Cognition, according to this view is essentially the manipulation of symbols on the hardware of the brain. Accordingly, CMs were perceived as essentially static symbols, internal representation of the external extended environment. In response to a certain need or task, the person consults this internal representation and on the basis of this consultation takes decision and sends instructions to the body how to act. This view that has dominated CM studies in the 1960s and 1970s is still dominant today.

3.2 Embodied CM

Classical cognitivism and its information processing approach are not as dominant as they used to be in the past. New paradigms, such as PDP (Parallel Distributed Processing) and neo-connectionism (Rumelhart et al, 1986), pragmatist environmental cognition (Freeman 1999), experiential realism (Johnson, 1987; Lakoff, 1987), situated cognition (Calencey, 1997) and embodied cognition (Varela et al, 1994), seem to seriously challenge the classical view. The notion of embodied cognition (Varela et al, 1994) commonly serves also as an umbrella term to these challenging views and so it will be used here.

According to these views, cognition is *embodied* in the sense that mind and body are not independent from each other but form a single integrated cognitive system, and in the sense that many cognitive capabilities are derived from the bodily experiences in the environment. Accordingly, perception and action are not two independent faculties, but two facets of a single *action-perception* integrated system, and cognition is *situated* (Calensey, 1997), that is, intimately related to the environment within which it takes place. A CM according to this view is an ad-hoc entity – an event created in the brain in relation to a certain bodily action situated in a specific environment.

3.3 SIRN CM

The notion of SIRN (Synergetic Inter-representation Networks) that was suggested as an approach to cognition and cognitive mapping (Portugali, 1996; Haken and Portugali, 1996; Portugali, 2002) is in line with the embodied views. It adds to them a special focus on artifacts – the products of the action-perception system and the objects that form the environment within which cognition is situated. More specifically, according to SIRN action produces two types of externally represented artifacts: *bodily artifacts*, such as speech or dance, and *stand-alone artifacts* such as texts, tools, artworks and the like. Such externally represented artifacts form an integral element of the cognitive system, so that instead of the dual action-perception system one should put a triple *action-perception-production* system.

SIRN views CM in a way similar to the embodied view, that is, an ad-hoc entity – an event created in the brain in relation to a certain bodily action situated in a specific environment. However, it emphasizes that this event evolves as a play between internal representations that are 'sub-events' constructed by the mind, and external representations that are events constructed in the world. Such a play gives rise to an inter-representation network that in a process of circular causality constructs the world outside and inside.

3.4 Models of Cognitive Mapping and Kinds of CM

The above new ideas have already started to effect cognitive maps studies. Thus, in a collection on *The Construction of Cognitive Maps* (Portugali, 1996), Gopal as well as Ghiselli-Crippa et al suggest connectionist and neural network approaches to CM,

Heft suggests a Gibsonian interpretation, Couclelis an experiential realism approach , while Portugali and Haken and Portugali introduce SIRN. More recently Kitchin and Blades (2002) have surveyed the various approaches as they appear in the context of geographical and psychological studies of CM.

In their book Kitchin and Blades (ibid) refer to the above approaches as "Models of cognitive mapping", which indeed is what they are. But beyond being different models of cognitive mapping, the above approaches suggest also different views regarding the ontological status of cognitive maps with the implication that when using the term CM, students approaching it from the above different perspectives assume, in fact, different things: some see it as an internal representation in the classical way, others in terms of affordances while still others in terms of experiential realism etc.

My aim in this section is not to convince one to adopt a certain position with respect to the notion CM, but to be aware that the meaning of the term changes when employed by proponents of the various approaches. In other words, that it may be useful to distinguish between classical CM, embodied CM and SIRN CM.

4   Forms of Memory and Forms of CM

"There is now general agreement among memory researchers that it is appropriate to think of memory as a collection of distinctly different types of memories rather than one undifferentiated memory" (Thompson, et al 1996). In this section I discuss three types of memory (autobiographic, prospective and long- vs. working memories) and the way they are related to CM. The choice is personal – these are types of memory that I currently study in relation to CM. The aim, therefore, is not to produce a complete correspondence between memory types and the notion CM, but to examine a few examples. These examples illustrate, so I argue, that it might be useful 'to think of the notion CM as a collection of distinctly different types of CM rather than one undifferentiated memory type'.

4.1   Autobiographic Memory

The notion of autobiographic (AB) memory refers to the way memory for everyday personal experiences is related to the passage of time (Schacter, 1996, 73). What differentiates autobiographic memory from *retrospective memory* (Neisser, 1982) that refers to past events in general is that autobiographic memory is associated with a person's subjective, specifically significant, memories.

Studies on memory indicate that "the general rule [is] that memories become gradually less accessible with the passage of time" Schacter (1996, 76) and that the shape of this time decay curve is 'extraordinarily well described by a power function' (Crovitz and Schiffman 1974). This applies to autobiographic memory too.

What happens to the forgotten memory that decays with time? Most psychologists would hold that it is lost forever. Others would claim that it is stored somewhere, just waiting for the proper cue to invoke it and bring it to the fore. Neurobiological studies tend to support the dominant view in psychology. They indicate that as time passes, there may be a diminution in the strength of connections among the neurons that represent particular experiences. At a biological level some engrams or cells might literally fade away over time (Schacter , ibid., 78).

Memorizing events from the distant past is related also to the nature of the cues. Empirical studies indicate that for about a year after the occurrence of an episode it "can be accessed readily – with virtually any cue" … with time the range of cues … progressively narrows (Schacter , ibid., 79). The neurobiological explanation for this is that "soon after an event, the engram is a rich source of information … Relatively little retrieval information is needed to elicit the appropriate engram .." (Schater, ibid).

The above should be taken in conjunction with an important property of AB memory, namely, that it is in a way a story we tell ourselves and others over and over again. This explains why some favorite past personal episodes that we tend to recount frequently, are very easy to access regardless of which particular cue elicit it (ibid., 79-80). But when we deal with an episode from the past that we have not recounted many times, the engram of the event is more impoverished source of information, and considerable cueing may be needed.

In the latter case, because the engram is so impoverished, recollective experience may be determined more heavily by salient properties of the cue, which itself has strong associations and meaning in memory. This leads to another important property of AB memory, namely, that it is a dynamic entity that evolves in line with the SIRN process and as a consequence might change over time. AB memory, as noted, is a story we tell over and over again. Story telling is a typical example for the SIRN process of a play between internal and external representations: as we tell our story, the story itself and with it the memory of what really happened, might gradually change too. This was shown by Bartlett's (1932/1961) serial reproduction scenarios in his classic *Remembering* that provided the paradigmatic case-study to the notion of SIRN (Portugali, 2002).

4.1.1 ABC Maps

My suggestion is to define a class of cognitive maps that will be termed *autobiographic cognitive maps,* or in short, *ABC maps*. Such maps are part of each person's autobiographic memory and as such refer to the way memory for everyday personal experiences in space is influenced by the passage of time. In light of what has been said above about autobiographic memory in general one can propose, first, that ABC Maps are dynamic entities. It depends when and where in one's biography we ask to draw the maps. That is, which cues are functioning at this moment and place so that a specific map of the past is produced.

As an example for the research potential of ABC maps consider Figures 1 and 2. They are taken from a study (Fenster 2000; Portugali, in preparation) in which Ethiopian Jews who immigrated to Israel during the 1990s were asked to produce two drawings: One (Fig. 1) describing the neighborhood and home left in Ethiopia and another (Fig. 2) describing the current neighborhood and home in Israel. The differences are dramatic: the drawing of the home left in Ethiopia is full of details, affection, and nostalgia; it is what humanistic geographers (e.g. Relph, 1976) would call *place*. The drawing of the current home is rather alienated with very few details and no signs of affection; it provides a good example to what Relph (ibid) has defined as *placelessness*.

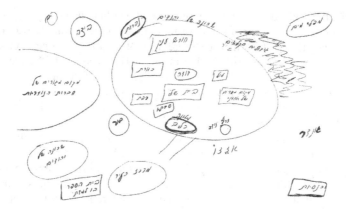

Fig. 1. A drawing of past neighborhood

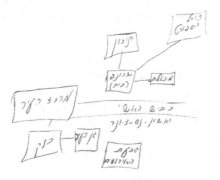

Fig. 2. A drawing of present neighborhood

Looking at the above (and similar other) pairs of drawings from the perspective of SIRN, it can be said that these are not simply past "maps" retrieved at will out of a person's autobiographic atlas, stored in that person's memory. Rather, they are ad-hoc entities that emerge out of the dynamics of the experiments we have conducted and the present daily situation of the respondents. The cue that triggered their response was the externally represented sentence 'please make two drawings: one of your past environment and another from your present living environment'. This very requirement seems to invite a comparison and indeed the results we got where in fact statements about the present in comparison with the past. Neither drawing in each pair can be looked at in isolation, independent of its twin drawing. Both are, from the outset, relational – the past as perceived from the present situation.[1]

4.2 Pprospective Memory (PM)

The notion of *prospective memory* (PM) was developed in recent years in order to study a specific kind of memory that involves the realization of delayed intentions.

---

[1] See in this connection Lowenthal's (1985) *The Past is a Foreign Country.*

For example, the realization of an intention to call a colleague at 3:00 pm, or to buy a newspaper when encountering a kiosk, or to shop in a grocery on the way home back from work. PM thus refers to the future use of memory – one "remembers to remember" and then to perform (Sellen et al, 1997). Cases of PM are commonly divided into *time-dependent* and *cue-dependent* PM (Brandimonte et al, 1996; Marsh & Hicks, 1998). The above example to call a colleague is typical of time-dependent PM, while the example about the intention to by a newspaper is typical of cue-dependent PM.

An important feature of PM is that during the time lag between the moment of intention and the moment of performance of the PM task, the person temporarily "forgets" the task and can thus engage his/her memory in other tasks (Smith, 2003). Then, at the proper time, place, or event, one suddenly remembers. The common interpretation is that in this sudden event the remembered PM task is retrieved from long-term store to working memory.

Similarly to other forms of memory, the PM task might be remembered and performed as intended, but also forgotten or remembered and performed too late/early (Graf and Uttl, 2001). The common interpretation is that this is a consequence of the complex processes associated with the transfer of a delayed intention/plan from long-term to working memory.

To the best of my knowledge the concept 'space' is absent in PM studies. One often finds time dependent PM, cue/event dependent PM but not as yet *space-dependent PM*. An exception is a recent paper "A synergetic approach to cue-dependent prospective memory" (Haken and Portugali, 2005) in which the authors indicate possible links between the notions of PM and CM.[1] In what follows I further elaborate on this possibility suggesting that space is related to PM in two basic circumstances: in cases of *cognitive map dependent PM* and in cases of *prospective CM*.

4.2.1   Cognitive Map Dependent PM

The suggestion here is that cognitive maps play an important role in many PM tasks and that this is so due to the association between space and time and space and event/cue.

First, in relation to time: one dimensional space, that is to say, distance, is intimately related to time. Time dependent PM assumes that humans have some kind of "internal alarm clock": you need to get up at 5 am, you set the alarm clock to 4:55 and in 4:50 you wake up spontaneously before the alarm rings. In a similar way imagine a scenario by which you are starting a 2000 km trip and that your car's petrol meter is out of order (which is the case in my old 1960 Volvo for several years now). You fill up the tank and hit the road, knowing that a full tank can take you, say, 700 km. After about 650 km you suddenly remember: you enter the first patrol station, fill the tank and forget about the whole thing for the next several hundred km. What happens here is that you internal alarm clock has been transformed into an "internal distance alarm."

But there is another interpretation to the above scenario: You know that your tank can take you some 700 km but you are also familiar with the route: your cognitive

---

[2] In a study currently under way Mintz, et al test whether the navigation in 3D virtual environments featuring different visual load suppresses or reinforces time-dependent PM task.

map of it is very detailed and accurate. As a consequence, when you pass a certain landmark on the road – a settlement or a prominent mountain – you suddenly remember: this landmark is about 600 km from my point of origin. On the face of it one could say that this landmark acted as an ordinary cue so that what we have here is an ordinary cue-dependent PM. But note the difference: the precondition for this landmark to function as a cue is your cognitive map of the rout. What we have here, therefore, is cognitive map dependent PM.

4.2.2 Prospective CM

Imagine in your mind a long and complex route, say, from one side of a metropolitan area to the other. You'll realize that this is a difficult task as from time to time you loose the sequence and the image gets rather obscured. Now compare this to a situation in which you practically stand at the beginning of the route or at any point along it, trying to imagine the next points that are not visible from were you stand. You'll realize that the points next to your current location come much clearer in your mind. It is as if the point you are standing acts as a cue that reminds you of the next few points. In terms of SIRN the explanation would be that what we have here is a play between internal and external representations. Each externally represented cue confirms our location and retrieves further information from long-term to working memory so that the road unfolds in its details before our "mind's eye". The CM constructed in working memory is "prospective" in the sense that it indicates your prospective route and action.

Prospective CM can be seen as a special case of CM-dependent PM. For example, imagine a typical PM task such as to shop in a supermarket on your way home. At a certain junction on your way home you suddenly remember the task; not because you practically see a supermarket, but because in this junction one needs to get off the main road in order to get to a supermarket. Such a junction acts as a 'secondary, indirect PM cue' due to what has been described above, namely, it triggers a prospective CM within which a supermarket is located.

4.3 Long-Term and Working CM

Several authors have already dealt with the link between long-term versus working memory (LTM vs. WM) and cognitive maps (Kitchin and Blades 2002, and detailed bibliography there). The general view is that cognitive maps are essentially part of LTM that in order "to be put into practice ... [must] ... be reconstructed in working memory" (ibid, 80). Some suggest that such a recall is achieved through a system of 'frames', while others use the notion of task specific "spatial mental models" that are constructed in working memory (ibid). Regarding the various terms (frame, mental models, etc.) that refer to the constructed frame or spatial mental model in WM, I think it would be useful to unify them under the term *working cognitive map*, or in short, *W-cognitive map* and thus make a distinction between *W-* and *LT- cognitive maps.*[2] There are several reasons for this suggestion.

First, in many experiments it is not always clear to what the notion CM refers: to the information internally represented in LTM, or to externally represented information that was constructed (as frame or spatial mental model) in WM. For example,

---

[3] Note that the ABC maps and prospective CM discussed above are cases of W-cognitive maps.

experiments on systematic distortions in cognitive maps typically start by asking subjects a question about spatial relations. The question triggers a process of information retrieval from LTM to WM that gives rise to an answer that we term "cognitive map". But what exactly is this cognitive map? Spatial information as it is stored in LTM or as constructed in WM? The distinction between LT- and W-cognitive maps will provide researchers with an impetus to be explicit about this issue.

Second, one of the milestones in the study of WM is Miller's (1956) "The magic number seven, plus or minus two: Some limits on our capacity for processing information". Employing Shannon theory of information (Shannon and Weaver, 1949), Miller showed that our capacity for processing one-dimensional information in WM is limited to about 2.5 bits (Plus/minus seven items). In a recent study on "The face of the city is its information", Haken and Portugali (2003) discuss the implications of Miller's work to cognitive maps. They show that our brain has several tricks by which it can overcome the Miller's limitation: one is by grouping and categorization – a possibility already noted by Miller. As shown by Haken and Portugali (ibid), some of the systematic distortions effects in CMs are directly related to this process. For example, Stevens and Coupe's (1978) "due to hierarchy" (Tversky, 1992).

A second 'trick' is by means of the SIRN processes and the play between internal and external representations. A case in point might be the task of learning a route in order to later navigate along it without using a map. Let say there are 100 decision points along the route. Bringing the whole route into WM (that is, creating a working CM for the entire route) is beyond the navigator's capability. According to SIRN, as the navigator moves along the route every cue encountered triggers a prospective CM that "lights" its subsequent (+-7) decision points and so on.

5   C- Versus s-Cognitive Maps

Consider the following scenario: an agent comes to a new city on which s/he has no prior knowledge, in order to invest in building a building there. What the agent has in mind is an image of "a city" (that is, of the concept and category "city") that is based on the agent's previous experience. This image of the category city serves our agent as the basis for the first set of decisions taken in the new place and as means to learn the structure of the new city. Our agent learns the new city by interpreting the incoming information from the new city by means of his/her previous knowledge of cities. As our agent moves around the new city s/he creates a cognitive map of this specific city.

The above scenario is taken from CogCity – an agent base urban simulation model (Portugali, 2004). This study suggests calling the image of the category "city", with which our agent comes to the city, *categorical-* or *c-cognitive map* and the image of the new city *specific-* or *s-cognitive map*. As can be seen, the two notions are similar to each other in that both are an internal representation of a city. They differ from each other in their very definition, namely, in that the s-cognitive maps refer to specific cities, while c-cognitive maps to the category or concept of a city.

The similarity between c- and s-CM is not surprising. It reflects, on the one hand, the fact that one's category and image of a city is derived from one's experience and knowledge of many specific cities. On the other, that a CM has many category-like properties (Portugali, 2004). This is due to the fact that CMs usually refer to very large areas. For example, the notion "Mediterranean" might be seen and employed as a cognitive map but also as a category so that one often talk about Mediterranean food, behavior, landscape and so forth (Portugali, 2004a).

The point to emphasis is that in the absence of specific information about *the* city one can still behave successfully in the city by using one's information about *a city*, namely about the category city. Or, put differently, in the absence of s-cognitive map of a certain city one can still function successfully in the city by using one's c-cognitive map of it.

6 Concluding Notes

This paper suggests several reasons – 'forms of obscurity' – that are responsible for the obscured and contested nature of the notion CM. It further suggests that in order to overcome these obscurities it will be useful to treat the notion CM in terms of kinds

Table 1. Forms of obscurity of kinds of cognitive maps

Forms of obscurity	Description	Kinds of CMs	Description
Obscurity due to different ontologies	When using the term cognitive map (CM), students approaching it from different ontological perspectives [classical cognitivism, embodied or situated cognition, synergetic inter-representation networks (SIRN) etc.] assume, in fact, different things: some see it as an internal representation in the classical way, others in terms of affordances or experiential realism, while still others in terms of SIRN and so on.	Classical CM	An essentially static symbol – an internal representation of the external extended environment.
		Embodied CM	An ad-hoc entity – an event created in the brain in relation to a certain bodily action situated in a specific environment.
		SIRN CM	Accepting the embodied view, it adds that CMs evolve as a play between internal and external representations that constructs the world outside and inside.

Table 1. *(continued)*

Obscurity due to different kinds of memory	When using the notion CM some students in fact mean long-term (LT) CM, while others working memory (WM) CM, still others autobiographic CMs and so on.	ABC map	A dynamic entity – the spatial part of a person's autobiographic memory, referring to the way memory for everyday personal experiences in space is influenced by the passage of time.
		CM depn't PM	As in time dependent PM, CM dependent PM assumes that humans have an "internal clock" and internal "distance alarm."
		Prospective CM	Due to the association between space and time and space and events/cues, cognitive maps participate in many time and cue dependent PM tasks.
		LT-CM	Cognitive maps constructed in long-term memory
		W-CM	Cognitive maps constructed in working memory
Obscurity due to categorization	When talking about a CM of a city, for instance, it is not always clear whether one speaks about a CM of the category city and a cognitive map of a specific city.	C-CM	Cognitive map of a certain spatial or environmental category (e.g. of a city)
		S-CM	Cognitive map of a specific spatial or environmental entity (e.g. of a certain city)

of CMs rather than as a single-meaning entity. Finally, the paper exemplifies this suggestion by introducing in a preliminary manner several kinds of CM. In concluding the paper it would be constructive to illustrate how the various forms of obscurity and kinds of CMs are interrelated. This is done in Table 1 that explicates what until now was implicit in the discussion, namely, the way the various forms of obscurity give rise to the various kinds of CMs.

It must be emphasized that the aim of this short and preliminary discussion is not to produce an inclusive description of all possible forms of obscurity and kinds of CMs. Rather the aim is to indicate the potential of thinking about the notion cognitive map as a multiplicity of forms and to initiate a discussion about such an approach. Further work is needed, firstly, to study in details the kinds of CMs discussed above, and secondly, to introduce and discuss new ones.

References

Bartlett, F.C. 1932/1961. *Remembering*. Cambridge: Cambridge university press.

Brandimonte, M., Einstein, G. O. & McDaniel, M.A. (Eds.). 1996. *Prospective Memory: Theory and Applications*, Mahwah NJ, Lawrence Erlbaum.

Edelman, G. 1992. *Bright Air Brilliant Fire: on the Matter of the* Gardner, H. 1987. *The Mind's New Science*. New York: Basic books.

Fenster T. 2000. Ashakenazi man, Ethiopic woman: between centralized to social-gender planning". *Pnim* 13, 54-59 (Hebrew).

Freeman, W.J. 1999. *How Brains Make Up Their Minds*. London: Weidenfeld & Nicolson.

Gardner, H. 1987. *The Mind's New Science*. New York: Basic books.

Gibson, J.J. 1979. *The Ecological Approach to Visual Perception*. Boston: Houghton Mifflin.

Graf, P. & Uttl, B. 2001. Prospective memory: a new focus for research. *Consciiousness and Cognition* 10, 437-450.

Gregory, D. 1994. *Geographical Imaginations*, (Blackwell , Cambridge, MA)

Haken, H., and Portugali, J. 1996. Synergetics, Inter-representation networks and cognitive maps. Pp. 45-67 in *The construction of cognitive maps*, edited by J Portugali. Dordrecht: Kluwer academic publishers.

Haken, H. and Portugali, J. 2003. "The face of the city is its information", *Journal of Environmental Psychology* 23, 385-408.

Haken, H. and Portugali, J. 2005. A Synergetic Interpretation of Cue-Dependent Prospective Memory. Cognitive Processing (In print).

Johnson M. 1987. *The Body in the Mind: the bodily basis of meaning, imagination, and reason*. (The university of Chicago press, Chicago)

Kitchin, R. and Blades, M. 2002. *The Cognition of Geographic Space*. (I.B.Tauris, London)

Kitchin, R. and Freundschuh, S. (Eds.). 2000. *Cognitive Mapping*. Routledge, London

Kohonen T. 1989. *Self-organization and Associative Memory*. (Springer, Berlin)

Lakoff G. 1987. *Women, Fire and Dangerous Things: what can categories reveal about the mind*. (The University of Chicago press, Chicago)

Lowenthal, D. 1961. "Geography, experience, and imagination: towards a geographical epistemology", *Annals of the association of american geographers* **51**:241-260.

Lowenthal, D. 1985. *The Past is a Foreign Country*. (Cambridge University press, Cambridge)

Lynch, K. 1960. *The Image of the City*. (MIT press, Cambridge Mass.)

Marsh, R. L., Hicks, J. L., & Watson, V. 2002. The dynamics of intention retrieval and coordination of action in event-based prospective memory. *Journal of Experimental Psychology: Learning, Memory, and Cognition*, 28, 4, 652-659.

Miller, G.A. 1956. "The magic number seven, plus or minus two: some limits on our capacity for processing information", *The Psychological Review* **63**(2): 81-97.

Minsky, M. 1977. Frame system theory. Pp 355-376 in *Thinking: Readings in Cognitive Science*, edited by Johson-Laird, P.N. and Wason, P.C. Cambridge: Cambridge University Press.

O'Keefe, J. and Nadel, L. 1978. *The Hippocampus as a Cognitive Map*, Oxford: Clarendon

Portugali, in preparation. ABC maps.

Portugali, J. 2004a. The Mediterranean as a cognitive map. *Mediterranean Historical Review* 19,2, 17-25

Portugali, J. 2004. Toward a cognitive approach to urban modelling. *Environment and Planning B: Planning and Design*, 31, 589-613

Portugali, J. 2002. The seven basic propositions of SIRN (Synergetic Inter-Representation Networks). *Nonlinear Phenomena in Complex Systems* 5(4) 428-444.

Portugali, J. 1999. *Self-Organization and the City*. Heidelberg: Springer.

Portugali, J. 1996. Inter-representation networks and cognitive maps. Pp. 11-43 in *The construction of cognitive maps*, edited by J. Portugali. Dordrecht: Kluwer academic publishers.

Portugali, J. (Ed.). 1996. *The construction of cognitive maps*. Dordrecht: Kluwer academic publishers.

Roberts, W.A. 2001. Spatial representation and the use of spatial codes in animals. Pp 15-44 in *Spatial Schemas and Abstract Thought*, edited by M. Gattis. Cambridge: MIT press.

Roediger III, H.L., Marsh, E.J. and Lee, S.C. 2001. *Kinds of Memory*. http//media.weley.com

Relph, E.C. 1976. *Place and Placelessness*. (Pion, London)

Sellen, A. J., Louie, G., Harris, J. E., & Wilkins, A. J. 1997. What brings intentions to mind? A study of prospective memory. *Memory*, 5, 4, 483-507.

Shannon CE, and Weaver W. 1949. *The Mathematical Theory of Communication*. (University of Illinois Press, Urbana).

Smith, R.E. 2003. The cost of remembering to remember in event-based prospective memory: investigating the capacity demands of delayed intention performance. *Journal of Experimental Psychology: Learning, Memory, and Cognition*, 29, 3, 347-361.

Stevens, A., & Coupe, P. 1978. Distortions in Judged Spatial Relation. *Cognitive Psychology* 10, 422-437

Tolman, E. 1948. Cognitive maps in rats and men. *Psychological review* 56:144-155.

Thompson, C.P., Skowronski. J.J., Larsen, S.F. and Betz GTE, A. 1996. *Autobiographical Memory: Remembering What and Remembering When*. Lawrence Erlbaum Associates, Inc

Tversky, B. 1996. Distortions in cognitive maps. Pp. 131-138 in *Geography, Environment and Cognition*, edited by Portugali, J. *Geoforum* 23(2),.

Varela, F.J, Thompson, E. and Rosch, E. 1994. *The Embodied Mind*. (MIT press, Cambridge)

Categorical Methods in Qualitative Reasoning: The Case for Weak Representations

Gérard Ligozat

LIMSI-CNRS, Université Paris-Sud,
F-91403 Orsay Cedex, France
ligozat@limsi.fr

Abstract. This paper argues for considering qualitative spatial and temporal reasoning in algebraic and category-theoretic terms. A central notion in this context is that of weak representation (WR) of the algebra governing the calculus. WRs are ubiquitous in qualitative reasoning, appearing both as domains of interpretation and as constraints. Defining the category of WRs allows us to express the basic notion of satisfiability (or consistency) in a simple way, and brings clarity to the study of various variants of consistency. The WRs of many popular calculi are of interest in themselves. Moreover, the classification of WRs leads to non-trivial model-theoretic results. The paper provides a not-too-technical introduction to these topics and illustrates it with simple examples.

1 Introduction

Looking at qualitative spatial and temporal reasoning from the point of view of algebra and category theory is not a new enterprise: a pioneering work in this respect was Ladkin and Maddux's study of binary constraint networks [12], which dealt mainly with Allen's calculus. Subsequently, Hirsch [11] studied general problems of constraint satisfaction with respect to general representations of relation algebras.

The common — orthodox view of qualitative reasoning in this particular perspective can be expressed quite briefly: (a) a qualitative reasoning uses some abstract *relation algebra* A; (b) reasoning is concerned with a domain of interpretation D which is a *representation* of A.

Our main goal in this paper is to argue for a shift of this basic perspective and to explain and illustrate what can be gained from this change. As first explained in [16], we claim that: (a) a qualitative reasoning involves a *non-associative algebra* [21] A which, in some cases, is a relation algebra; (b) reasoning is concerned with a domain of interpretation D which is a *weak representation* of A.

In other words, we propose two changes. More about each of them:

1. *From relation algebras to non-associative algebras.* This point has to do with a fact of life. In practice, most qualitative reasoning systems are based on a finite partition of D × D (called a set of JEPD relations in the local jargon). Now it is a fact (first proved in [16]) that this data naturally defines a

A.G. Cohn and D.M. Mark (Eds.): COSIT 2005, LNCS 3693, pp. 265–282, 2005.
© Springer-Verlag Berlin Heidelberg 2005

non-associative algebra, through the notion of weak composition, a natural approximation in this context to actual composition. Non-associative algebras constitute a generalization of the classical notion of relation algebra, and the point here amounts to acknowledging that the more general notion is the adequate one. Only in particular cases is this non-associative algebra a relation algebra.

2. *From representations to weak representations.* In our view, this second point is fundamental. We think that weak representations, which are a more general notion than representations, should occupy the center stage in qualitative spatial and temporal reasoning. A first reason is that, in a qualitative reasoning system based on a set of JEPD relations on a domain D, the domain D is in all cases a weak representation of A, as shown in [16]. Only in particular cases is it a representation. But this is only part of the picture.

We contend that there are indeed excellent reasons for a broad use of the notion of weak representation. Those reasons can be summarized by four keywords: ubiquity, clarity, usefulness, and fruitfulness.

Ubiquity: The notion of weak representation appears implicitly in many qualitative spatial or temporal calculi, since most of those calculi are based on a set of JEPD binary relations. Key concepts in a calculus, such as the domain of interpretation, or the notion of scenario, are particular avatars of the notion. This means that in a sense dealing with such a calculus means implicitly dealing with weak representations.

Clarity: As has been remarked before [20], there has been some confusion in the literature about various notions of consistency, partly because qualitative reasoning has developed on the image of CSP paradigm, and because of the fact that for classical calculi such as Allen's calculus this identification was without serious drawbacks. The algebraic point of view, and the consideration of weak representations makes easier the separation between what is of a purely algebraic nature, such as path-consistency or algebraic closure, and the notions which pertain to specific interpretations of the calculi — actually, specific weak representations — such as various notions of k-consistency or global consistency. As a bonus, it clarifies the role of those WRs which are in fact representations.

Usefulness: Once WRs have been recognized, what are they useful for (besides the intellectual gratification this provides)? Firstly, considering WRs as worthy of interest in themselves leads to considering what sorts of objects they are for some of the best known calculi: the fact that the corresponding structures have been considered independently, and that some classical paradigms (an example is the theory of dimension of partial orderings) can be considered as translations in an alternative language of questions about WRs, is a proof that the notion of WR is not a futile will-o'- the-wisp, but a notion with actual content. Secondly, and more importantly, defining and studying the properties of the category of WRs of the algebra of a given qualitative calculus is the key to treating consistency as what it is in essence: the mere existence of a morphism.

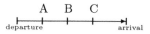

Fig. 1. Reasoning about a boat race

Fruitfulness: The fruitfulness of the approach has already been apparent in the past: in [13], the proof of the $\aleph_0$-categoricity of a whole class of calculi was based on a classification of all WRs of the corresponding algebras (which include Allen's algebra). This result was extended to a larger class of calculi based on linear orderings in [19]. In the recent past, the possibility of relating different calculi based on the same abstract algebra, and of comparing WRs of algebras related between them by some morphism has been used [6,24]. Hirsch [11,7] has also stressed the interest of adopting the wider point of view of algebra for the study of satisfiability in qualitative reasoning. We believe that further progress is to be expected in this direction, which will also be relevant for combining different formalisms.

2 What Is a Qualitative Spatial or Temporal Calculus?

Imagine you are reporting on a boat race which takes place on a river (Fig. 1). There is a point of departure, and a point of arrival for the race. At first approximation, the river can be assimilated to a line between the two points, and each boat as a point which moves along the line.

You know that there is a set of competitors, say Abe, Beth, Cess, Dee, Elf, on boats A, B, C, D, E. The race has been on for some time. You have a partial view of the situation, because you missed some boats when they passed the stretch of river you can observe. Now you have to report to your employer.

You: – C is ahead of B, and A is *behind* B.

               – I also know that D is *behind* B; and E is on the same level as D.

Employer: – OK!

You: – I have just been told that C is *behind* E.

Employer: – This can't be true!

You have been experiencing a very simple case of qualitative spatial reasoning related to the so-called Point Algebra (PA). In your particular case, this goes as follows:

- You are dealing with a domain D of interpretation, an oriented segment of line which represents the possible locations of the boats along the river.
- You consider three qualitative relations between elements of the domain, and you can use a finite language of relation symbols, say $\prec$ (*behind*), $\succ$ (*ahead of*), *eq* (*on the same level*) to refer to them. The three relations that you have chosen are such that, for any pair of locations, exactly one of the three relations holds.
- You use the relation symbols to express the knowledge you have about a (finite) set of elements in D, also called a (finite) *configuration* in D. For instance, the contents of what you first reported can be represented by:

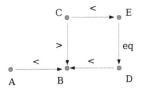

Fig. 2. A network representing the (inconsistent) report on the boat race

$C \succ B$, $A \prec B$, $D \prec B$, E *eq* D. Then you added: $C \prec E$.

This type of knowledge is often represented by using *constraint networks*:[1] in your case, the network would have five nodes corresponding to the five boats, and each oriented edge would be labeled by the information you have about the corresponding pair. For instance, the arc from A to B would be labeled $\prec$, and so on (Fig. 2).

- The calculus is used to make inferences. These inferences are based on simple facts about the relations you are considering: firstly, $X \succ Y$ is equivalent to $Y \prec X$; in other terms, the two relations denoted by $\prec$ and $\succ$ are mutually *converse*;[2] then, both $\prec$ and $\succ$ are transitive: if $X \prec Y$ and $Y \prec Z$, then $X \prec Z$, and similarly for $\succ$. On the other hand, if $X \prec Y$ and $Y \succ Z$, all cases $X \prec Z$, $X = Z$, and $X \succ Z$ are possible. This is expressed in terms of *composition*:[3] for instance, composing $\prec$ with itself implies $\prec$, composing $\prec$ with $\succ$ leaves all three possibilities.
- A fundamental problem about reporting is the question of *consistency*:[4] is the information I gave to my boss consistent? Can there really be five boats on the river such that what I claim to be true is indeed true? this is often expressed in terms of instantiations: can each of the names (or variables, a term used for the nodes of the network) be instantiated as an actual location on the river in such a way that all my contentions are true?
- The question of consistency is related to the notion of *scenario*,[5] meaning here, informally, the choice of a definite relation between each pair of boats such that the properties of composition and converse hold. Clearly, an instantiation results in a scenario.

Unfortunately for you, here, consistency does not hold. Here is your employer's proof: you claim that $D \prec B$ and that $C \succ B$. By converse, the latter means that $B \prec C$. By composition, I conclude that $D \prec C$. But, on the other hand, D *eq* E and $C \prec E$. Composing *eq* with $\prec$ implies $\prec$, hence $C \prec D$, which is in contradiction with $D \prec C$.

---

[1] Refer to definition 3 for a formal definition of a constraint network.
[2] Let R be a (binary) relation on D, that is, a subset of $D \times D$. Then the converse $R^{\smile}$ of R is defined by: $R^{\smile} = \{(x, y) \in D \times D \mid (y, x) \in R\}$.
[3] The composition $R \circ S$ of two relations R and S is defined by: $R \circ S = \{(x, y) \in D \times D \mid (\exists z \in D)\ (x, z) \in R \land (z, y) \in S\}$.
[4] Refer to definition 4 for a formal definition.
[5] Refer to definition 5.

Let us summarize at this point. What we have been using is the Point Algebra PA.[6] It has a set $B = \{\prec, \succ, eq\}$ of three basic relation symbols, which denote what is called a set of JEPD (jointly exhaustive, pairwise disjoint) binary relations in some domain D. In mathematical parlance, the relations denoted by the symbols in B constitute a *partition*[7] of $D \times D$. One of the symbols, here eq, denotes equality. For each symbol b in B, the converse of the relation denoted by b is denoted by some — possibly the same — symbol in B. Finally, composing two basic relations implies that some (possibly non trivial) disjunction of basic relations holds.

Partial information about configurations in D can be expressed by using networks whose nodes stand for elements of D and whose oriented edges are labeled by subsets of B. Let $A = 2^B$ denote the set of all subsets of B.

A basic problem, given a network, is to determine whether it is consistent, that is, whether the nodes can be instantiated as elements of A so that the constraints expressed by the labels hold.

A particular instantiation can be described by a scenario.

3 Yet Another Qualitative Reasoning Formalism: A Do-It-Yourself Kit

Now you can feel emboldened enough by your first experience with PA to try and build your own qualitative reasoning system. This is how you proceed:

- You choose some domain D. The elements of the domain are your spatial or temporal entities. Mathematically, a binary relation on D is just a subset of $D \times D$. You choose a finite collection $(R_i)_{i \in I}$ of non-empty relations which constitute a partition of $D \times D$. You only have to take care of two things:
   1. One of the R_is, say R_0, is the identity relation.
   2. For each R_i, its converse $R_i^\smile$ belongs to the collection, that is, there is exactly one $j \in I$ such that $R_i^\smile$ is R_j.
- You choose a symbol r_i for each relation R_i. Let B be the corresponding set of symbols.[8] They are called basic relations. The set A of all subsets of B is just called the set of relations.
- You now have a set of labels you can use to label finite networks and represent in this way any finite configuration of elements in D.
- What about inference? The *composition* of two of your relations R_i and R_j gives a new relation $R_i \circ Rj$, which is some subset of $D \times D$, hence you (may) get new information. Of course, you cannot expect that the resulting set is exactly the union of some of your R_is. If you still do, think of your previous experience in the boat race example: composing *ahead of* with *behind* yields

---

[6] Actually, PA is usually interpreted in terms of time points: the domain is the time line, and $\prec$, $\succ$, and eq stand for *before*, *after*, and *equals* respectively.

[7] A partition of a set U is a collection of non-empty subsets of U which are pairwise disjoint and whose union is U.

[8] We usually write eq, instead of r_0, for the identity relation.

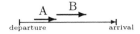

Fig. 3. Intervals on a line: Allen and INDU calculus

pairs in all three relations *ahead of*, *behind*, and *equals*. But consider the starting point d of the race. It is in relation *equals* with itself. But there is no y on the path $[d, a]$ from departure to arrival such that $d \succ y$ and $y \prec d$, that is, (d, d) does not belong to the composition of *ahead of* with *behind*, although it belongs to the identity relation. Hence the composition of $\succ$ with $\prec$ does not contain the whole of the identity relation. If you want to keep track of the inferences using your basic relations, you have to use what is called the *weak composition*[9] of two relations R_i and R_j. Intuitively, it corresponds to the strictest constraint you can express in terms of the relations you started with.

- Now you are done! You have now a full-blown qualitative (spatial and/or temporal) reasoning formalism. Congratulations!

3.1   Enjoy! A Quick Tour of Some Formalisms

Now you know what all this is about, enjoy your freedom and rediscover some of the calculi around.

Allen's Calculus. [1] Now you are interested in the boats as extended entities: a boat is an extended object which extends from prow to stern (Fig. 3). Since this is a race, each boat goes in the same direction, but you would like to be able to express finer distinctions, such as e.g. a boat overtaking another: her prow comes to the same level as the stern of the other, then it overlaps the other, and so on. What you have here is a (spatial) interpretation of Allen's Interval algebra (IA). Your domain is the set of all pairs of locations on the line representing the river. A boat defines such a pair, also called an interval. Here you have a set of 13 basic relations. Composition was first described by Allen's table.

The INDU Calculus. [22] Even though in your race all boats are equal as competitors, some boats are longer than others. You would like to keep track of the relative lengths of two boats. So you will for instance write that boat A meets boat B and that A is shorter than B (Fig. 3), which you write $A \, m^< \, B$. This is the INDU calculus, where basic relations are pairs of an Allen relation and a comparison relation with three values. Of course, some combinations are excluded: if X starts Y, for instance, X has to be shorter than Y. All in all, this gives a set of 25 basic relations.

Directed Intervals. [24] Come back to Allen's calculus, and forget about the race. Now your boats can go upstream or downstream (Fig. 4). You now have

---

[9] The weak composition, denoted by $R_i \diamond R_j$, of two relations R_i and R_j is defined by:
$R_i \diamond R_j = \bigcup_{k \in J} R_k$ where $k \in J$ if and only if $(R_i \circ R_j) \cap R_k \neq \emptyset$.

two versions of overlapping, one for a boat overtaking another, and one for two boats which are in the process of crossing one another, going in opposite directions. This yields 26 basic relations. The corresponding calculus has been defined and studied by Renz.

The Cardinal Direction Calculus. [17] Here your boats are at sea, on a 2D area. You zoom out and only consider them as points on the surface of the sea (Fig. 5). The basic relations are the cardinal directions North, South, East, West, and intermediate directions North West, and so on. You need of course equality, hence a total of nine basic relations. This is the cardinal direction calculus, your first example of a spatial calculus in dimension greater than one!

The Calculus of Spatial Congruence. [6] Zoom in again, and look at your boats as 2D shapes on the surface of the sea. Cristani has defined what he calls the congruence calculus between such shapes. Here you have an algebra with four basic relations, namely *eq* (two shapes can be exactly superposed), $\prec$ (one can be embedded in the other), $\succ$ (the converse relation), and $\|$ (neither can be embedded in the other). For instance, in Fig. 6, we have $A \prec B$, $A \prec C$, and $B \| C$. As an abstract algebra, the resulting algebra M_4 is the algebra of partial orderings also studied by Anger, Mitra and Rodriguez [2] and Broxvall and Jonsson [4].

4   Revisiting the Calculus: Weak Representations

4.1   The Algebras of Qualitative Reasoning

We now come back to the general discussion and consider a qualitative calculus as defined in Section 3.

In abstract terms, an algebra A is a set together with a set of operations on it such that some axioms are satisfied. For instance a Boolean algebra is a set together with a top element, a bottom element, a join and a meet for every pair of elements, and a complement satisfying a set of axioms (refer e.g. to [11] for the algebraic notions). In particular, the set of subsets of a given set is a concrete Boolean algebra. This applies to B, the set of basic relation symbols, so that the set A of subsets of B is a Boolean algebra. Now this set has also operations of converse and of (weak) composition, as well as an element which is a neutral element for composition. In order to avoid any confusion, we will from now on denote by ; the operation induced by weak composition on the algebra A of subsets of B.

The following result is proved in [16]:

Proposition 1. *The algebra* A *is a non-associative algebra (NAA) in the sense of Maddux [21,11].*

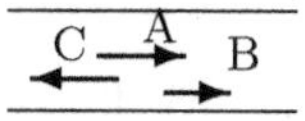

Fig. 4. Upstream and downstream: directed intervals

Fig. 5. Using cardinal directions   **Fig. 6.** Congruence: $A \prec B$ and $A \prec C$

Definition 1. *A non-associative algebra* A *is a 8-tuple* $A = (A, +, -, \mathbf{0}, \mathbf{1}, ;, \smile, \mathbf{1}')$ *satisfying the following conditions:*

1. $(A, +, -, \mathbf{0}, \mathbf{1})$ *is a Boolean algebra.*
2. $\mathbf{1}'$ *is a constant,* $\smile$ *a unary operation, and* ; *a binary operation such that, for any* $a, b, c \in A$:

   (a) $(a^\smile)^\smile = a$ (b) $\mathbf{1}' ; a = a ; \mathbf{1}' = a$ (c) $a ; (b + c) = a ; b + a ; c$
   (d) $(a + b)^\smile = a^\smile + b^\smile$ (e) $(a - b)^\smile = a^\smile - b^\smile$ (f) $(a ; b)^\smile = b^\smile ; a^\smile$
   (g) $(a ; b).c^\smile = 0$ *if and only if* $(b ; c).a^\smile = 0$

A non-associative algebra is a relation algebra *[25] if the operation* ; *is associative.*

Associativity. In the more classical cases, such as the Point Algebra or Allen's algebra, the resulting algebra is a relation algebra. But this is not the general case at all. For instance, associativity does not hold for the INDU algebra [15].

In the context of qualitative reasoning, the first (to our knowledge) non-associative algebra considered in the literature was Egenhofer and Rodríguez' "integrated container-surface algebra" [9].

Maddux [21] introduced intermediate classes of non-associative algebras between relation algebras (**RA**) and general non-associative algebras (**NA**), namely weakly associative (**WA**) and semi-associative (**SA**) algebras. These classes form a hierarchy: **NA** $\supseteq$ **WA** $\supseteq$ **SA** $\supseteq$ **RA**.

In particular, semi-associative algebras are those non-associative algebras which satisfy the following condition: for all a, $(a ; \mathbf{1}) ; \mathbf{1} = a ; \mathbf{1}$.

Egenhofer and Rodríguez' container-surface algebra is semi-associative. So is the INDU algebra. More generally, it is proved in [16] that the algebra resulting from a set of JEPD relations is semi-associative provided that the basic relations are serial.[10]

Concrete and Abstract Algebras. It is important to clearly make the distinction between the abstract algebra A whose elements are subsets of the basic relations, that is, subsets of abstract symbols, and the interpretations of these symbols in terms of actual binary relations on the domain D. In particular, while most operations at the abstract level correspond to concrete operations in terms of sets, the operation of composition (denoted here by ;) at the abstract level only corresponds to weak composition (denoted by $\diamond$) at the concrete level.

---

[10] A relation R is *serial* if the following condition holds: $(\forall x \in D)(\exists y \in D)$ such that $(x, y) \in R$.

4.2   Domains as Weak Representations

Let us take another look at the situation, starting now from the algebras. We have:

- A domain D, and a map of interpretation $\varphi : B \to 2^{D \times D}$ which interprets each symbol r_i in terms of a binary relation R_i.
- A map φ which extends naturally to A, yielding a homomorphism of Boolean algebras.
- $\varphi(eq)$ is the identity relation on D.
- For each $r_i \in B$, $\varphi(r_i^\smile) = \varphi(r_i)^\smile$
- For all pairs $(r_i, r_j) \in B \times B$, $\varphi(r_i \,;\, r_j) \supseteq \varphi(r_i) \circ \varphi(r_j)$

Definition 2. *A pair* (D, φ) *satisfying the conditions above is called a* weak representation *of* A.

The preceding discussion, then, shows that a qualitative calculus is defined by two data: a non-associative algebra A, and a weak representation (D, φ) of A.

A Note About the Terminology. In the context of abstract algebra, the term of representation expresses the idea that an abstract object is represented in concrete terms. Here concreteness means that symbols for binary relations appear as actual binary relations. But the representation is weak, since in particular the abstract composition only corresponds to a weak notion of composition of binary relations. This is precisely what the last condition means: $\varphi(r_i \,;\, r_j)$ is the weak composition of the two basic relations R_i and R_j, while $\varphi(r_i) \circ \varphi(r_j)$ is the actual composition of R_i and R_j as binary relations. Hence we know that the former contains the latter, but there is no equality in general.

4.3   Pictures of the World: Scenarios as Weak Representations

What is a qualitative reasoning formalism used for? Answer: it is used for representing and reasoning about some configurations in a "real" world. What kind of pictures does it allow to take? As already discussed above, finite configurations of objects are described by constraint networks. If complete qualitative knowledge is assumed, the labels of the network are well-defined basic relation symbols. Hence they are *atomic* networks.

Definition 3. *A constraint network on* A *is a pair* (N, ν), *where* N *is a finite set of nodes, and* $\nu : N \times N \to A$ *associates a constraint* $\nu(i, j) \in A$ *to each pair of nodes* (i, j). *The network is* atomic *if* $\nu(i, j)$ *contains exactly one basic relation for each pair* (i, j). *It is* normalized *if* $\nu(i, i) = eq$ *for all* $i \in N$, *and* $\nu(j, i) = \nu(i, j)^\smile$ *for all* (i, j). *It is* algebraically closed *(or a-closed) if, for every triple of nodes* (i, j, k) *we have* $\nu(i, j) \,;\, \nu(j, k) \supseteq \nu(i, k)$.

Recall here the definition of consistency of a network with respect to a WR:

Definition 4. *A network* (N, ν) *on* A *is consistent with respect to a WR* (D, φ) *of* A *if there is a map* $h : N \to D$ *such that for each pair* $(i, j) \in N \times N$ *there is a basic element* b *in* $\nu(i, j)$ *such that* $(h(i), h(j)) \in \varphi(b)$.

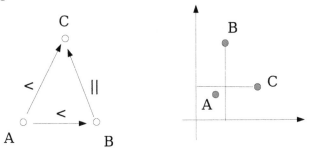

Fig. 7. A scenario for M_4 (left) and an interpretation of it in $\mathbf{Q} \times \mathbf{Q}$ (right)

In plain English, this means that the actual relation between $h(i)$ and $h(j)$, the interpretations in D of the variables i and j is one among those specified by the label $\nu(i,j)$.

Since it derives from an actual set of objects in the domain, and because the relations in the domain are (weakly) described by the algebras, it is fairly clear that the network describing a configuration is normalized, atomic, and algebraically closed. This is precisely what is called a scenario:

Definition 5. *A* scenario *is a normalized, a-closed atomic network.*

As an example, the left part of Fig. 7 is a scenario for the algebra M_4 of partial orderings (or of the congruence calculus). The right part is an interpretation of this scenario in the partial ordering $\mathbf{Q} \times \mathbf{Q}$ (the ordering used here is the product ordering[11]).

Consider a scenario (N, ν). The labeling function $\nu : N \times N \to \mathsf{B}$ associates a basic relation to each pair of nodes. Consider the inverse map $\nu^{-1} : \mathsf{B} \to 2^{N \times N}$. In plain English, this map lists, for each basic relation b, the set of pairs of nodes which are labeled by b.

We claim that the pair (N, ν^{-1}) is a weak representation of A. Indeed, ν^{-1} is a Boolean algebra homomorphism which commutes with the converse operation and moreover: $\nu^{-1}(r_i \,;\, r_j) \supseteq \nu^{-1}(r_i) \circ \nu^{-1}(r_j)$. Indeed, this is just another way of expressing the a-closure property.

We know that a picture of a configuration is a scenario. Here we use the term "picture" in an informal way: what we mean is that, given the language of representation provided by a calculus in terms of networks and basic relation symbols, taking the picture of a configuration yields a scenario as a result. We come later to the basic problem of satisfiability: when, conversely, is a scenario a picture of some configuration in the domain?

4.4   Beyond Weakness: Representations

We have met two instances of weak representations: first as domains of qualitative calculi, then as scenarios. Both indeed represent the algebra, in the sense

---

[11] This means that for $x = (x_1, x_2)$ and $y = (y_1, y_2)$ in $\mathbf{Q} \times \mathbf{Q}$, we have $x \leq y$ if and only if $x_1 \leq y_1$ and $x_2 \leq y_2$.

that the symbols of binary relations are interpreted as actual binary relations. However they do so only in a weak sense, because firstly, some basic relations may be represented by the empty relation, and secondly, because only weak composition, not actual composition, is represented

If a weak representation has to represent an algebra as concrete relations, two extra properties have to hold:

1. Each basic relation has to be represented as non-empty relation: this implies that φ is injective.[12]
2. The abstract composition has to coincide with *bona fide* composition, that is: $\varphi(r_i \,;\, r_j) = \varphi(r_i) \circ \varphi(r_j)$.

The two properties characterize the notion of representation:

Definition 6. *A* representation *of* A *is a weak representation of* A *satisfying the two properties (1) and (2).*

As mentioned above, representations rather than weak representations used to be the norm in many calculi. The usual interpretation of Allen's algebra, in terms of intervals on $\mathbf{Q}$ or $\mathbf{R}$ happen to be representations. But this is not the general case: determining which of the weak representations of $RCC8$, for instance, are representations, is an important and difficult task. On the other hand, interpreting Allen's algebra in terms of intervals on the integers, for instance, makes perfect sense. But the resulting WR is no longer a representation: for instance, although $\mathsf{p}\,;\,\mathsf{p} = \mathsf{p}$, hence $\varphi(\mathsf{p}\,;\,\mathsf{p}) = \varphi(\mathsf{p}) \supseteq \varphi(\mathsf{p}) \circ \varphi(\mathsf{p})$, we do not have equality, as the following example shows: the pair of intervals $((1,2),(3,4))$ belongs to $\varphi(\mathsf{p})$, but not to $\varphi(\mathsf{p}) \circ \varphi(\mathsf{p})$, since we cannot find an interval (i,j) of integers such that $(1,2)$ precedes (i,j) and (i,j) precedes $(2,3)$.

We think that weak representations, rather than representations, should be the main focus of attention in qualitative reasoning. The preceding discussion showed that they appear quite naturally in at least two different guises: as domains of interpretation, and as scenarios. In the following sections, we first examine some instructive examples, then we proceed to studying WRs in their own right.

4.5   Weak Representations of Some Well-Studied Formalisms

WRs of the Point Algebra. As a matter of fact, WRs of PA are well-known objects, namely, linear orderings. Indeed, consider a weak representation (D, φ). Let $R = \varphi(\prec)$. Then the axioms of a WR imply that R, its converse $R^\smile$ and $\Delta = \{x, y) \in \mathsf{D} \times \mathsf{D} \mid x = y\}$ are three disjoint relations whose union is $\mathsf{D} \times \mathsf{D}$, and that $\varphi(\prec\,;\,\prec) = R$ contains $\varphi(\prec) \circ \varphi(\prec) = R \circ R$. This means that R is transitive, that it does not intersect its converse, and that for any pair (x, y) either $x \prec y$, or $x = y$, or $x \succ y$. In other words, R defines a strict linear

---

[12] Let b_1, b_2 be two distinct basic relation symbols. Because $b_1 \cdot b_2 = 0$, $\varphi(b_1)$ and $\varphi(b_2)$ have an empty intersection. If neither is empty, they are distinct binary relations. Hence φ is injective on B, hence also on A.

ordering on D. Conversely, any strict linear ordering can be viewed as a weak representation of PA.

Our original example (the stretch of river, cf. Fig. 1) was such a linear ordering. Notice that there is no condition of finiteness for the domain.

WRs of the Algebra of Partial Orders. The algebra of partial orders M_4 has four basic relations. A weak representation of M_4 is in essence the same concept as a partial order. Specifically, consider such a weak representation (U, φ). Let R be the binary relation $\varphi(\prec)$ on U. Then it is easily checked that (U, R) is a partially ordered set.

Conversely, *any* partial order R on a set U provides a weak representation (U, φ) of M_4: define $\varphi(eq)$ as the diagonal relation Δ on U, $\varphi(\prec)$ as R, $\varphi(\succ)$ as $R^\smile$, and $\varphi(||)$ as $U \times U \setminus (\Delta \cup R \cup R^\smile)$.

The two constructions sketched above are in fact reverse to each other.

What about representations? A direct proof is given in [15] of the fact that when $U = \mathbf{Q} \times \mathbf{Q}$ with the product order, then (U, φ) is a *representation* of M_4, which implies that M_4 is in fact a relation algebra, a fact which can also be directly checked.

WRs of Allen's Algebra. As already observed, take any linear ordering $(T, \prec)$ and consider the set of intervals $I(T)$ on T: $I(T) = \{\underline{x} = (x_1, x_2) \in T \times T \mid x_1 < x_2\}$

Then the pair $(I(T), \varphi_{Allen})$ is a weak representation of IA, where φ_{Allen} is defined in the usual way: for instance, $\varphi_{Allen}(\mathsf{m}) = \{(\underline{x}, \underline{y}) \in I(T) \times I(T) \mid x_2 = y_1\}$ (x ends where y begins), and so on.

This applies in particular to $\mathbf{Q}$, the set of rational numbers, to $\mathbf{R}$, the real numbers, to $\mathbf{Z}$, the set of integers. As is well known, the resulting weak representation is a representation if and only if T is dense and has no first or last element, which applies to $\mathbf{Q}$ and $\mathbf{R}$.

The preceding construct, however, does not yield all possible weak representations. This is apparent from a counting argument: if T is finite with m elements, then $I(T)$ has $m(m-1)/2$ elements. But it is easy to describe WRs with a number of elements which cannot be expressed in this way: a simple example is the two-element WR defined by $\mathsf{D} = \{a, b\}$, and $\varphi(o) = \{(a, b)\}$, $\varphi(o^\smile) = \{(b, a)\}$, and $\varphi(r) = \emptyset$ for all other basic relations r except eq.

But, as shown in [13], any WR of IA can be embedded in a WR associated to a linear ordering, and there is a smallest such WR, called the closure of the given WR (see section 6).

WRs of the INDU Calculus. The standard interpretation of the INDU calculus in terms of intervals on $\mathbf{R}$ is a weak representation of the INDU algebra. As observed in [3], some weak representations associated to scenarios are not consistent in the sense that there is no instantiation of their variables in terms of intervals in $\mathbf{R}$. The same phenomenon also arises for other calculi, such as the cyclic interval calculus.

WR of the Cardinal Direction Calculus. The set of points in the Euclidean plane with the natural interpretation of cardinal directions[13] provides a weak

---

[13] We mean the interpretation defined as follows: given two points (x, y) and (x', y') in the plane, (x', y') is north of (x, y) if $x' = x$ and $y' > y$, south of (x, y) if $x' = x$ and

representation of the Cardinal direction algebra which is in fact a representation. More generally, for any linear ordering T, the set $T \times T$ of pairs provides a WR, which can be called the product representation on T. Thus the standard representation is the product WR on R.

Here again, many finite representations are not of the product type. For instance, consider the three-element scenario which describes the configuration in Fig. 5, where $N = \{A, B, C\}$, and $\varphi(nw) = \{(A, B)\}, \varphi(n) = \{(A, C)\}, \varphi(ne) = \{(B, C)\}$. This is not a product WR, but there is a 6-element WR which is the smallest product WR containing it [19].

5   A Categorical Perspective: WRs as Objects

Let us take a look back. We started with a general recipe for building qualitative calculi, and we discovered that the inferential part of the calculus is governed by a particular kind of algebra, which is in the general case a non-associative algebra, in the sense of Maddux, and in several classical cases even a relation algebra in the sense of Tarski. When expressed in terms of networks, configurations are described by scenarios, which are a-closed, atomic networks. Then we discovered that both the original domains of the calculus and scenarios are the same kind of object, called WRs of the algebra associated to the calculus.

Because WRs seem to be such ubiquitous objects, it is worth studying them in their own right, especially since WRs of some standard calculi are well-known objects.

A way of giving WR a full-citizenship is to consider them as objects of a category, and studying the resulting category. We do not need to know much here about category theory, except for the moment that a category has objects and morphisms, that morphisms can be composed and that some general properties hold for composition.

5.1   The Category of WRs of an Algebra

Fix some non-associative algebra A, and consider all WRs of this particular algebra. These will be the objects of the category.

We still have to define morphisms between two objects (U, φ) and (V, ψ) which are WRs of A. The intuition behind a morphism is that it is some transform which respects the structure of the objects. In our case, the structure is given by the map from the algebra to the domain. We define a morphism as a map $i : U \to V$, with qualifications. Such a map sends pairs in U to pairs in V, so that a binary relation on U is mapped to a binary relation on V. Recall that, in a WR, each pair of the domain has a well-defined label which is a basic relation symbol of the algebra. Now the condition for structure-preservation becomes obvious: if a pair (u_1, u_2) in U is labeled by some basic relation b in A, then its image $(i(u_1), i(u_2))$ in V has to be labeled by the same symbol. This can be written as: $(\forall b \in B)\ (x, y) \in \varphi(b) \to (i(x), i(y)) \in \psi(b)$.

$y' < y$. It is west if $y' = y$ and $x' < x$, east if $y' = y$ and $x' > x$, north-west if $x' < x$ and $y' > y$, and so on.

Use the very illuminating (at least to this author) view of a WR as a partition of $U \times U$ labeled by basic relations of A. The condition implies that each cell of the partition of $U \times U$ is mapped into the cell in $V \times V$ which has the same label. In particular, the map i has to be injective, because a pair in U which is not labeled by the identity element cannot be mapped to a pair labeled by the identity element: using i, U can be considered as a subset of V.

Examples. A simple example is provided by the WRs of PA, that is, linear orderings. A morphism is a simply a strictly increasing map.

We also implicitly used the notion of morphism while discussing the WRs of Allen's algebra, or of the Cardinal direction algebra.

Connoisseurs of category theory will notice that if i is a morphism from $\mathcal{U}$ to $\mathcal{V}$, then i makes the first object a sub-object of the second, in a technical sense.

5.2 Arrows and Consistency

Why bother with arrows in the category of WRs? The answer is that it gives us the key for understanding the notion of consistency in a very general sense.

Consider again the problem of determining whether a given network is consistent. We limit ourselves to considering scenarios, because ultimately all known methods for determining consistency try to reduce the problem to the consideration of sub-networks of the given networks which are scenarios.

What does it mean for a scenario (N, ν) to be consistent with respect to a given calculus? We already know that a calculus is a pair $(\mathsf{A}, \mathcal{W})$, where $\mathcal{W}$ is a WR of A. Now $\mathcal{N}$, as a scenario, is also a WR of A. A short time of reflection will convince you that:

Proposition 2. *A scenario $\mathcal{N}$ is consistent with respect to $\mathcal{W}$ if and only if there is a morphism from $\mathcal{N}$ to $\mathcal{W}$.*

The beauty of this definition is its general character. In categorical terms, it says that a scenario is consistent with respect to a WR if it can be mapped to a sub-object of the WR. Moreover, it shows that in fact consistency is a notion which applies to any pair of WRs, but is none other than the notion of sub-object.

The Case of the Point Algebra. In this particular case a network is just a finite linear ordering, and it is consistent with respect to another if and only if the other has at least as many elements. Not a very fascinating result!

More generally, a linear ordering is consistent with respect to another if it can be embedded in it. This opens up the fascinating domain of the theory of linear orderings and their structure. The following example provides a glimpse into what can happen: think for instance of the natural numbers N with their usual ordering, and of the same with the reverse ordering: both are weak representations of the Point algebra, but neither of them is embeddable in the other, because the second has a greatest element, and the first has none.

The Case of the Algebra of Partial Orders. The cases of the algebra M_4 of partial orders is more complex: in this particular case, consistency is related to the dimension theory of partial orderings.

We need the following notions: if (P, R) and (Q, S) are two partial orders, a map $h : P \to Q$ is an embedding of (P, R) into (Q, S) if h is injective, and $(u, v) \in R$ implies $(h(u), h(v)) \in S$. It is a full embedding if for all $u, v \in P$ $(u, v) \in R$ if and only if $(h(u), h(v)) \in S$.

Then it is fairly obvious to check that $h : P \to Q$ induces a morphism from (P, R) to (Q, S), as WRs of M_4, or in other terms that (P, R) is consistent with respect to (Q, S), if and only if h is a full embedding.

The dimension of a partial order (P, R) is defined as the least integer n such that (P, R) can be fully embedded in $\mathbf{Q}^n$. In other words, a M_4 scenario (that is, a finite partial order) has a dimension, which is the smallest n for which it is consistent with respect to $\mathbf{Q}^n$.

The Case of Allen's Algebra. Let us apply our new knowledge to that case, and consider first the WR whose domain is $I(\mathbf{Q})$. This WR is in fact a representation. Then it is known that any scenario is consistent with respect to this representation. But we can also think of more restricted calculi based on $I(T)$, for a linear ordering T. Then a scenario with n nodes is consistent if and only if T is "big enough", which will be true if T has more than $2n$ elements (the exact boundary depends on the precise structure of the scenario, of course).

5.3   WRs as a Tool for Clarification

A first lesson taught by the preceding discussion is that it is good policy to separate two notions of algebra when discussing qualitative reasoning: the first notion is about an abstract algebra, the abstract algebra governing a specific calculus, which is a NAA and in many cases a RA; the second notion is the "concrete" algebra of relations generated by the interpretations of the symbols of the abstract algebra in a specific WR.

The second lesson has to do with the existence, in general, of many WRs of the same algebra. Hence a given consistency problem only makes sense for a given algebra em and a well-defined WR of that algebra. Usually in the literature, the term of algebra refers to both the abstract algebra and the WR. This in our sense is a cause for confusion.

A case in point is the RCC5 algebra, aka the containment algebra [23,12]. When the first term is used, it is in relation to some topological (actually set-theoretic, cf. [18]) interpretation. The second term is used when the intended interpretation is in terms of intervals. But both terms refer to the same abstract algebra. As Düntsch has rightly pointed out [8], many RCC calculi are based on interpretations of the RCC5 composition table which are not representations of the algebra. That is, they are only WRs. In its avatar as the containment algebra, the algebra is a sub-algebra of Allen's algebra. So it can be interpreted in terms of intervals on the time line. Ladkin and Maddux [12] showed that there are scenarios which are not consistent with this interpretation. But they are consistent with many other WRs of the very same algebra, such as regions in the plane, subsets, or intervals on a circle.

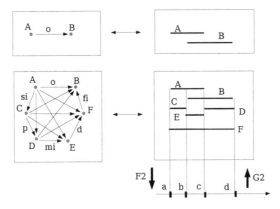

Fig. 8. A two element WS (top) and its closure (bottom)

6   Classifying WRs

In this section, we relate the preceding discussion to the use of the classification of the WRs of specific algebras for studying the properties of a calculus. A first example was the proof of the $\aleph_0$-categoricity of the generalized interval calculi [13], which generalizes the result for Allen's algebra. This result was extended to many formalisms based on linear orderings in [19].

In what follows, we try to give an intuitive idea of the method. To make things simple, the discussion is illustrated by the particular case of Allen's calculus, which is about 2-intervals in this context.

A n-interval is by definition an increasing sequence of length n in a linear ordering T. Similarly to the particular case of Allen's calculus (which corresponds to $n = 2$), the qualitative calculus about n-intervals is governed by a relation algebra $\mathbf{A_n}$.

Obviously, the case $n = 1$ corresponds to the point algebra. Now it is shown in [14] that there are two functors (mainly, the equivalent of maps for categories) between the category $\mathcal{C}_n$ of WRs of $\mathbf{A_n}$ and the category $\mathcal{C}_1$ of WRs of $\mathbf{A_1}$, that is, the category of linear orderings.

The first functor G_n takes an object in $\mathcal{C}_1$, that is a linear ordering, and associates to it all n-intervals (all strictly increasing sequences of length n) in it (the part of the construction concerned with arrows is easy and we leave it out).

The second functor F_n takes an object in $\mathcal{C}_n$, that is a WR of $\mathbf{A_n}$, and associates a linear ordering to it (intuitively, this is the set of all implicit points in the n-intervals involved in the WR).

In order to get the intuitive meaning of these results, consider Fig. 8. On the upper left-hand side is a WR of Allen's algebra with two elements A and B related by $A \circ B$. On the upper right-hand side is a concrete instantiation of the WR (or scenario) in terms of two intervals.

The functor F_2 associates to this WR a weak representation of the Point Algebra, i.e. a linear ordering, namely here, the set $\{a, b, c, d\}$ of endpoints, with $a \prec b \prec c \prec d$.

The functor G_2 takes a linear ordering, for instance the 4-element set above, and outputs a WR of Allen's algebra, namely, that of all intervals built on the 4-element set. Here we get 6 intervals A, B, C, D, E, F corresponding to $[a, b]$, $[a, c]$, and so on. The corresponding network with 6 nodes is shown on the lower left-hand side (part of the labels are left out for clarity). It contains the original one, and is in fact the smallest one which makes use of all end-points.

Going back to the the general case, it is also shown in [13] that in F_n is left adjoint to G_n, which implies in particular two consequences: (a) any WR $\mathcal{U}$ of $\mathbf{A_n}$ has a smallest WR containing it, its closure, which is computable as $G_n(F_n(\mathcal{U}))$ and which is closed; this is precisely the phenomenon we described for the example; (b) the two functors F_n and G_n define an equivalence of categories between the closed WRs of $\mathbf{A_n}$ and linear orderings; in Allen's case, this amounts to the fact that, if you consider networks with all intervals implicitly present made explicit, then you can talk entirely in terms of endpoints.

As a consequence of those results, it is shown that all countable representations of $\mathbf{A_n}$ are isomorphic. A consequence of this fact is that the corresponding theory is decidable.

7   Conclusions

The main character in this paper is the notion of a weak representation. We showed how it emerges naturally when constructing any qualitative calculus based on a finite set of JEPD relations, in at least two guises: on the one hand, the domain of interpretation is a WR of the algebra governing the calculus, and on the other hand scenarios, the objects which describe finite configurations in the domain, are themselves WRs. This realization motivated the introduction of the category of WRs of a given algebra, with at least two advantages: firstly, consistency is interpretable in terms of morphisms in the category; then, in many cases at least, a classification of all weak representations can be made, which leads to non-trivial results about the calculus, a strategy which was first used in [13] and pursued in [19].

This paper opens perspectives for further study. In particular, a lot of interest is currently being devoted to the study of hybrid formalisms which integrate several aspects of qualitative reasoning, such as combining topological reasoning in the RCC style with temporal reasoning à la Allen [10], or temporalizing various formalisms [5]. Our conviction is that developing such formalisms can profit a lot from the use of the general and clarifying methods provided by algebra and category theory.

References

1. J. F. Allen. Maintaining knowledge about temporal intervals. *Comm. of the ACM*, 26(11):832–843, 1983.
2. F.D. Anger, D. Mitra, and R.V. Rodriguez. Temporal Constraint Networks in Nonlinear Time. In *Proc. of the ECAI-98 Workshop on Spatial and Temporal Reasoning (W22)*, pages 33–39, Brighton, UK, 1998.

3. P. Balbiani, J.-F. Condotta, and G. Ligozat. On the Consistency Problem for the INDU Calculus. In *Proceedings of TIME-ICTL-2003*, Cairns, Australia, 2003.

4. M. Broxvall and P. Jonsson. Towards a complete classification of tractability in point algebras for nonlinear time. In *Principles and Practice of Constraint Programming*, pages 129–143, 1999.

5. J.-F. Condotta, S. Tripakis, and G. Ligozat. Ultimately periodic qualitative constraint networks. 2005.

6. M. Cristani. The complexity of reasoning about spatial congruence. *J. Artif. Intell. Res. (JAIR)*, 11:361–390, 1999.

7. M. Cristani and R. Hirsch. The complexity of constraint satisfaction problems for small relation algebras. *Artif. Intell.*, 156(2):177–196, 2004.

8. I. Düntsch. Relation algebras and their application in qualitative spatial reasoning. Technical Report CS-03-07, Brock University, St. Catarines, Ontario, 2003.

9. M. Egenhofer and A. Rodríguez. Relation algebras over containers and surfaces: An ontological study of a room space. *Spatial Cognition and Computation*, 1:155–180, 1999.

10. A. Gerevini and B. Nebel. Qualitative spatio-temporal reasoning with RCC-8 and Allen's interval calculus: Computational complexity. In *Proc. of ECAI 2002*, pages 312–316, 2002.

11. R. Hirsch and I. Hodkinson. *Relation Algebras by Games*. Number 147 in Studies in Logic and the Foundations of Mathematics. North Holland, 2002.

12. P. B. Ladkin and R. D. Maddux. On Binary Constraint Problems. *Journal of the ACM*, 41(3):435–469, May 1994.

13. G. Ligozat. Weak Representations of Interval Algebras. In *Proc. of AAAI-90*, pages 715–720, 1990.

14. G. Ligozat. On generalized interval calculi. In *Proc. of AAAI-91*, pages 234–240, 1991.

15. G. Ligozat, D. Mitra, and J.-F. Condotta. Spatial and temporal reasoning: beyond Allen's calculus. *AI Communications*, 17(4):223–233, 2004.

16. G. Ligozat and J. Renz. What is a qualitative calculus? a general framework. In *Proc. of PRICAI'04, LNCS 3157*, pages 53–64, Auckland, New Zealand, 2004.

17. G. Ligozat. Reasoning about cardinal directions. *Journal of Visual Languages and Computing*, 1(9):23–44, 1998.

18. G. Ligozat. Simple models for simple calculi. *Proc. of COSIT 199, Lecture Notes in Computer Science 1661*, pages 173–188, 1999.

19. G. Ligozat. When Tables Tell It All: Qualitative Spatial and Temporal reasoning Based on Linear Orderings. *Proc. of COSIT 2001, Lecture Notes in Computer Science 2205*, pages 60–75, 2001.

20. G. Ligozat and J. Renz. Problems with local consistency for Qualitative Calculi. In *Proceedings of ECAI'04*, pages 1047-1048, Valencia, Spain, 2004.

21. R. Maddux. Some varieties containing relation algebras. *Trans. Amer. Math. Soc.*, 272(2):501–526, 1982.

22. A. K. Pujari, G. Vijaya Kumari, and A. Sattar. INDU: An Interval and Duration Network. In *Australian Joint Conf. on Artificial Intelligence*, pages 291–303, 1999.

23. D. Randell, Z. Cui, and T. Cohn. A spatial logic based on regions and connection. In B. Neumann, editor, *Proc. of KR-92*, pages 165–176, San Mateo, CA, 1992.

24. J. Renz. A spatial odyssey of the interval algebra: 1. Directed intervals. In *IJCAI*, pages 51–56, 2001.

25. A. Tarski. On the calculus of relations. *Journal of Symbolic Logic*, 6:73–89, 1941.

On Internal Cardinal Direction Relations

Yu Liu, Xiaoming Wang, Xin Jin, and Lun Wu

Institute of Remote Sensing and Geographic Information Systems,
Peking University, Beijing 100871, P.R. China
liuyu@urban.pku.edu.cn, wangxiaoming@pku.edu.cn,
terrible2@sina.com, lwu@urban.pku.edu.edu.cn
http://www.geosoft.pku.edu.cn

Abstract. Internal Cardinal Direction (ICD) relations can be considered as the refinement to the contains/within topological relation. It is widely used to describe the position of an object in a region. In this paper, three ICD models with varying degrees of details are presented -- ICD-5, ICD-9 and ICD-13. In each of these, the notion of a "middle part" is defined using Minimum Bounding Rectangles (MBR). Then focusing on ICD-9, three representation methods are discussed. They are major portion, point set of intersections and proportions of intersections respectively. The ICD-9 model is validated by a cognitive experiment, which helped to determine the size of the middle part in ICD-9 and validate the MBR-based partition method. Based on a psychological theory about vague predicates, conceptual neighborhood and intersection of ICD relations are also discussed briefly.

1 Introduction

Qualitative Spatial Reasoning (QSR) abstracts from metrical details of the physical world and enables computers to make predictions about spatial relations, even when precise quantitative information is not available [Cohn 2001]. Several kinds of useful spatial relations have been studied in the literatures e.g., topological relations [Bennett 1997] [Cui 1993] [Egenhofer 1991] [Randell 1992] [Renz 1999], cardinal direction relations [Goyal 1997] [Ligozat 1998] [Goyal 2000] [Skiadopoulos 2001] and qualitative distance relations [Hernández 1995] [Hernández 1997] [Gahegan 1995]. Compared with qualitative distance relations, topological relations and cardinal direction relations are more often used in spatial cognition and spatial reasoning. For example, the combination of topological relation and cardinal direction relation is usually used to describe the information in small-scale spaces, such as room spaces [Hernández 1995]. Qualitative positional information in large-scale space is commonly represented by the combination of cardinal direction relation and qualitative distance relation between a target object and a reference object [Clementini 1997] [Frank 1992] [Zimmermann 1993]. This method can be regarded as a refinement to disjoint topological relations [Mark 1999].

In practical applications, there exists another kind of direction relations named "Internal Cardinal Direction (ICD)" used to qualitatively represent positional information in commonsense geographic world. These are contrasted with the conventional cardinal direction relations which we call "External Cardinal Direction (ECD)" relations.

A.G. Cohn and D.M. Mark (Eds.): COSIT 2005, LNCS 3693, pp. 283–299, 2005.
© Springer-Verlag Berlin Heidelberg 2005

In contrast to ECD relations, ICD relations mainly deal with the situation when one spatial object is within another areal object. In many verbal expressions, ICD and ECD are differentiated clearly. Consider the following two phrases:

> He lives in the west of England.
> Korea is to the west of Japan.

Clearly the propositions "in the west of" and "to the west of" have different meanings. The former implies "within" (or "contains") relation while the latter indicates a "disjoint" relation.

Existing work related to ICD mainly focused on hierarchical spatial reasoning (HSR) where a bigger object is conceived as the container of lots of smaller objects. Using the transitive relation of containment, a hierarchy of containers can then be formed. For example, if A is the container of B, B is the container of C, A must be the container of C. HSR is based on the economic principle to use the least detailed representations sufficient to answer a question. Ever since HSR is validated by experiments [Stevens 1978] [Hirtle 1985], HSR has been used in areas such as data structures [Yashino 1991] [Hernández 1994] [Timpf 1997] [Leung 1999], wayfinding [Timpf 1992] [Car 1994] and direction relations reasoning [Papadias 1997]. However, few researches have studied how HSR can be refined by ICD which provides more details on contains/within topological relation.

The objective of this paper is to develop and explore a cognition-accordant formal model of ICD relations. As argued in [Duckham 2001], there are two perspectives related to QSR, i.e. computational perspective and cognitive perspective. In order to develop more cognitively plausible models, these two perspectives should be integrated. Following this direction, cognitive experiments were used to validate the ICD models presented in this paper. After the description of ICD given in section 2, section 3 presents three ICD models which are ICD-5, ICD-9 and ICD-13 respectively. The cognitive experiment to evaluate ICD-9 is presented in section 4. The paper ends with a conclusion and agenda for future research.

2 Description of ICD Models

In this section, a multi-level framework of ICD is developed. Firstly, we analyze the ICD relations in daily life to develop three ICD models with varying degree of details. The granularity selection factors are then presented.

2.1 Multi-level Descriptions of ICD

Propositions similar to "a is inside b", such as Peking University is located in city of Beijing, can qualitatively determine the position of a based on b in everyday life. The container a is the reference object (RO) in which the target object (TO) b is located. Meanwhile, b plays a role of containee. Actually, the containers, together with their name, form a common geo-referencing system, although it is somewhat coarse and not metric [Longley 2001]. Since such a referencing system is not precise enough, we usually use more detailed ICD relations to refine the position in above propositions. For example, we might say "Los Angeles is in the West of U.S.A.", or "City Xi'an is in the middle of China" to represent their positional information.

By investigating the common-sense geographic world, thirteen items below could be identified to describe the ICD relations: east (*I_E*), west (*I_W*), south (*I_S*), north (*I_N*), northeast (*I_NE*), northwest (*I_NW*), southwest (*I_SW*), southeast (*I_SE*), middle (*I_M*), middle east (*I_ME*), middle west (*I_MW*), middle south (*I_MS*), and middle north (*I_MN*). Distinguished from ECD, the symbols are prefixed with "I". Based on these relations, container is partitioned into corresponding parts and forms a geo-referencing system with higher resolution. In theory, if all these relations are applied, the positioning precision can be increased thirteen times.

[Mennis 2000] argued that spatial knowledge includes two types of cognitive categorization: taxonomy (superordinate-subordinate relationships) and partonomy (part-whole relations). ICD relation plays an important role to represent partonomy knowledge. Since qualitative representations should only make as many distinctions as necessary in a given context, we may identify three levels of ICD-based partonomy knowledge with varying granularities. According to the numbers of chosen atomic relation items, ICD relations can be modeled by three levels of resolutions. They are marked with ICD-5, ICD-9 and ICD-13 respectively.

1) ICD-5 is the most coarse and basic model. It includes five atomic direction relations (Fig. 1-a), which are *I_M*, *I_N*, *I_E*, *I_S* and *I_W*. Because of its simplicity, ICD-5 is cognized and employed by human beings in very early days. For example, in ancient Chinese philosophy, these five basic directions are already identified. Moreover, they are matched to the five basic elements in the world (soil, water, wood, fire and metal) respectively.

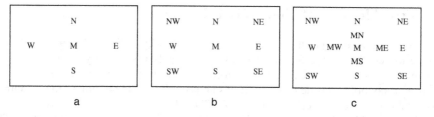

Fig. 1. Models of Internal direction relations (**a.** ICD-5 **b.** ICD-9 **c.** ICD-13)

2) Compared with ICD-5, ICD-9 adds four combined directions, i.e. *I_NE*, *I_NW*, *I_SE* and *I_SW* (Fig. 1-b). These four directions are applied to depict the positions between north and east, north and west, south and east or south and west.

3) When ICD-13 is applied, the other four directions, *I_MN*, *I_ME*, *I_MS* and *I_MW*, are included (Fig. 1-c). They increase precision of the positioning system inside the container.

It can be seen that ICD-9 is a refinement of ICD-5 while ICD-13 is a refinement of ICD-9. Depending on the ICD model used, the same containee will be related to the container differently.

2.2   Factors Affecting the Choice of ICD Models

In order to describe the position inside a container, any of the three ICD models discussed in Section 2.1 can be applied. The choice depends on the following two aspects:

1) Scale of the container object

When the container is large in size, e.g. it is a metropolitan or a country (geographic space), ICD-9 or ICD-13 can be applied to partition the whole region. When the container's size gets smaller (environmental space), ICD-5 or ICD-9 might be sufficient. The definitions and differences of environmental space and geographic space are introduced in [Montello 1993].

2) Spatial distribution characteristics of targets objects

The spatial distribution characteristics of TOs also affect the choice of ICD models. If the distribution is clustered, then models with lower detail level may be suitable; otherwise, a more detailed model should be used. Fig. 2 illustrates situations where ICD-5, ICD-9 and ICD-13 are appropriate to the situation (a), (b), and (c) respectively.

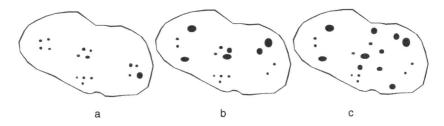

a   b   c

Fig. 2. Spatial distribution of TOs affecting the choice of ICD models

3   Formal Representations of ICD Relations

ICD relations form a partition of the container. Usually, ICD-5, ICD-9 and ICD-13 divide the container into 5, 9 and 13 one-piece parts respectively. However, if the container is concave, an ICD relation part may consist of two or more separate polygons. For any object inside one part, it has an according ICD relation to the container. To establish a formal representation of ICD relations, the following two aspects should be considered: 1) the range of the middle part, and 2) the partition method. The second question partially relies on the first one, i.e. the shape of middle part determines the acceptable partition method. In the following part of this section, the range of the middle part is discussed at first, followed by the partition solutions to ICD-5, ICD-9 and ICD-13. Lastly, the formal ICD-9 model is defined in detail.

3.1   Definition of Middle Part

Let a be a connected region object. "Connected" means that for every pair of points in a there exists a line joining these two points such that all the points on the line are also in it. Generally, the middle part of a should satisfy the following three conditions:

1) the shape of the middle part is related to the shape of the container a;
2) usually, the part is inside a, especially when a is a convex polygon;
3) the intuitive center of a should be inside the middle part.

To find such an appropriate region in the container is somewhat difficult. The central point o of a can be determined through different approaches, for example, cen

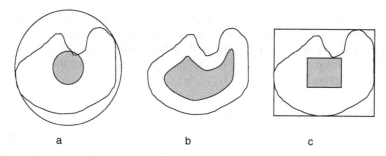

Fig. 3. Middle part of a region (**a**. circumcircle based middle part **b**. inter buffer zone based middle part **c**. MBR based middle part)

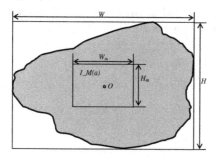

Fig. 4. MBR based middle part of a region

troid of a, center of a's MBR (minimum bounding rectangle) or center of the circumcircle of a are all reasonable. Furthermore, based on different definitions of central point, the shape of middle region might also vary. Fig. 3 demonstrates the three optional cases, where circumcircle, inner buffer zone and MBR are applied respectively.

These three cases all have their advantages and disadvantages. The case of (b) insures the middle part inside the container, but it is difficult to calculate and complicated to determine the other ICD parts. In case (a), the shape of the container is neglected. Compared with (a) and (b), the MBR-based method (Fig.3-c) considers the shape of the container along two major axes, and it is easy to calculate the middle region. Additionally, a potential advantage of this solution is that it enables ICD relations to be integrated with MBR-based ECD relations more easily. Thus, it is chosen to define the ICD relations in this paper.

As shown in Fig. 4, the shape of an MBR-based middle part of a is a rectangle similar to a's MBR. It is denoted by $I\_M(a)$. The MBR of a and $I\_M(a)$ have an identical center o. Assume the width and height of the MBR are W and H, and those of $I\_M(a)$ is W_m and H_m, where 'm' stands for 'middle'. Because of the similarity characteristic, we have the equation:

$$\frac{W_m}{W} = \frac{H_m}{H}$$

Let this ratio be ρ, it determines the size of $I\_M(a)$ relative to a. Varying ρ will yield different concrete ICD-models. The most appropriate value of ρ is likely to be domain or context specific and will need to be determined experimentally. In the fourth part of this paper, we will try to find a proper value of ρ for ICD-9 model based on a cognitive experiment.

3.2 Multiple Partition Levels of ICD

Now that the middle part of the container has been determined, the other parts can be defined using conventional cardinal direction relations. For example, if a place b is to the north of $I\_M(a)$, then we could infer that b is in the north of a, i.e. b has the ICD relation I_N to a. Thus, the container can be partitioned based on appropriate ECD models, which include cone-based model, project-based model [Frank 1991] double-cross model [Freksa 1992] and MBR-based model [Goyal 2000] [Goyal 2001] [Skiadopoulos 2001] [Skiadopoulos 2004], etc. (Fig. 5).

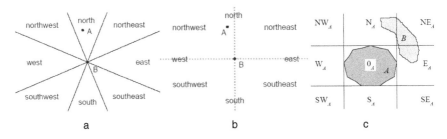

Fig. 5. Three cardinal direction relation models (**a.** cone-based model **b.** project-based model **c.** MBR-based model)

In practice, the partition includes two spatial operators. The MBR of the container is first divided into a series of sectors. Then these sectors are intersected with the container to generate ICD parts. Since the concrete shape of the container might be intricate, some parts might be disconnected and some parts might even be empty.

For the ICD-5 model, the cone-based model is suitable because only four directions should be distinguished. As shown in Fig. 6-a, the MBR of *a first* separates out the middle part and a peripheral band. Then, the peripheral band is further divided

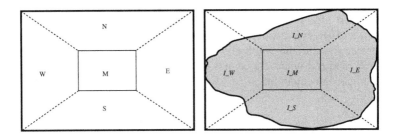

Fig. 6. Partition of ICD-5 model

into four parts using a cone-based ECD model. To be simple, a diagonal line connecting the corners of MBR and $I\_M(a)$ is drawn to separate two neighboring sectors. It is not necessary for the angle to be $\pi/4$, $3\pi/4$, $5\pi/4$ or $7\pi/4$. (Fig. 6-b).

Compared with ICD-5, ICD-9 is more straightforward. By elongating four edges of $I\_M(a)$ and intersecting the edges of container's MBR, the MBR is divided into 9 cells (Fig. 7). Obviously, the MBR-based ECD model is adopted here.

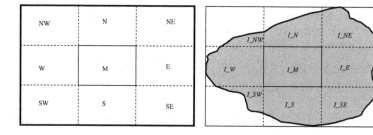

Fig. 7. Partition of ICD-9 model

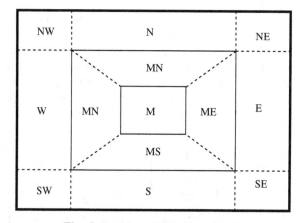

Fig . 8. Partition of ICD-13 model

In order to determine a place within the container more precisely, the middle part in ICD-5 or ICD-9 should be further partitioned to form a more detailed positioning system, which is three-layered. There are four available approaches by combining ICD-5 and ICD-9. The first one is dividing the middle part of ICD-9 model using ICD-9. It is called ICD-9-9 and includes 17 parts. Analogously, the other three solutions are ICD-5-5, ICD-9-5, and ICD-5-9. They include 9, 13 and 13 parts respectively. As mentioned in section 2.1, the last one is more acceptable, because it makes the areas of all parts comparable. For brevity, we name it ICD-13.

In the ICD-13 model, the middle part includes five sub-parts based on the ICD-5 model. They are $I\_M(a)$, $I\_MN(a)$, $I\_ME(a)$, $I\_MS(a)$ and $I\_MW(a)$. Then applying ICD-9 model, the peripheral band is partitioned into the other 8 sectors, i.e. $I\_N(a)$, $I\_NE(a)$, $I\_E(a)$, $I\_SE(a)$, $I\_S(a)$, $I\_SW(a)$, $I\_W(a)$ and $I\_NW(a)$ (Fig. 8).

3.3 Representation of ICD-9 Relations

Compared with ICD-5 and ICD-13, the ICD-9 model is more common and generic in commonsense geographic world. It is thus discussed in detail in the following part.

Let region a be the container. According to the ICD-9 model, a is divided into 9 parts denoted by $I\_N(a)$, $I\_NE(a)$, $I\_E(a)$, $I\_SE(a)$, $I\_S(a)$, $I\_SW(a)$, $I\_W(a)$, $I\_NW(a)$ and $I\_M(a)$. Let b be the containee inside a. If b occupies one of the above parts, then b accordingly has an ICD relation to a. For example, in Fig. 9-a, b is in the east of a, i.e. b has the I_E relation to a. In this case, the ICD relation is atomic.

But when b is a region (or line) object and occupies more than one cell of the partition, the ICD relation is complex. Assuming b is areal and covers n ($2 \leq n \leq 9$) parts of a, then the intersection part of b and one part of a is $b \cap R_i(a)$, where

$$R_i \in \{I\_N, I\_NE, I\_E, I\_SE, I\_S, I\_SW, I\_W, I\_NW, I\_M\}$$

and $1 \leq i \leq n$. Clearly, $\bigcup_{i=1}^{n}(b \cap R_i(a)) = b$. According to these intersection parts, three different approaches are available to represent the ICD relation between b and a: major portion, point set of the intersections, and proportion of the intersections.

1. Major portion approach

Using this method, the areas of intersection regions should be calculated and sorted. The portion related to maximum area determines the ICD relation between b and a. For example, in Fig. 9-b, b covers four parts, and the intersection portion between b and $I\_W(a)$ has the maximum area. Hence, we could tell that b is in the west of a. This approach is simple and usually consistent with spatial cognition; however, it has two disadvantages: 1) it is less precise, and 2) it is not appropriate for the case when there isn't a distinctly maximum intersection region.

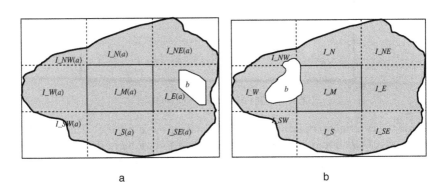

a   b

Fig. 9. ICD-9 based relation definition

2. Point set based approach

This approach is widely used to model external cardinal directions [Goyal 2000] [Skiadopoulos 2001]. It includes two concrete methods. Using a formula-based method (as opposed to the matrix-based approach), an ICD relation is defined as: b

R_1:...: R_n a, where $2 \leq n \leq 9$, R_i belongs to ICD-9. The following two conditions should be satisfied:

1) $\forall 2 \leq i \leq n, b \cap R_i(a) \neq \varnothing$

2) $\forall i \neq j, R_i \neq R_j$

For instance, in Fig. 9-b, we have b $I\_NW:I\_N:I\_W:I\_M$ a.

Another way is the ICD relation matrix illustrated in Fig. 10-a, the size of which is 3*3. Each element in it is $\varnothing$ or $\neg\varnothing$ according to related point set intersection. Fig. 10-b describes the relation in Fig. 9-b. Obviously, the formula-based method can also be applied for ICD-5 and ICD-13, but the matrix method is only suitable for ICD-9.

$$\begin{pmatrix} I\_NW(a)\cap b & I\_N(a)\cap b & I\_NE(a)\cap b \\ I\_W(a)\cap b & I\_M(a)\cap b & I\_E(a)\cap b \\ I\_SW(a)\cap b & I\_S(a)\cap b & I\_SE(a)\cap b \end{pmatrix} \begin{pmatrix} \neg\varnothing & \neg\varnothing & \varnothing \\ \neg\varnothing & \neg\varnothing & \varnothing \\ \varnothing & \varnothing & \varnothing \end{pmatrix}$$

$$\quad\quad\quad\quad a \quad\quad\quad\quad\quad\quad\quad\quad\quad\quad\quad\quad b$$

Fig. 10. ICD-9 relation matrix

3. Area weight based approach

This approach is an extension of the second approach and has been introduced in [Goyal 2001]. It is a more detailed description of ICD relations based on the proportion of how much of the target object falls into each part. The formula-based method for this approach is defined as: b R_1(overlapped proportion):...:R_n(overlapped proportion) a, where the overlapped proportion of R_k is the ratio of area($R_k(a) \cap b$) and area(b). For example, the detailed formula representation of ICD-9 relation in Fig. 9-b is b $I\_NW(0.05):I\_NE(0.03):I\_M(0.15):I\_W(0.77)$ a. The details of the ICD-9 matrix method is presented in Fig. 11-a. Fig. 11-b gives actual matrix values for the relation in Fig. 9-b. Clearly, all non-zero elements in the matrix sum up to 1.

$$\begin{pmatrix} \dfrac{area(I\_NW(a)\cap b)}{area(b)} & \dfrac{area(I\_N(a)\cap b)}{area(b)} & \dfrac{area(I\_NE(a)\cap b)}{area(b)} \\ \dfrac{area(I\_W(a)\cap b)}{area(b)} & \dfrac{area(I\_M(a)\cap b)}{area(b)} & \dfrac{area(I\_E(a)\cap b)}{area(b)} \\ \dfrac{area(I\_SW(a)\cap b)}{area(b)} & \dfrac{area(I\_S(a)\cap b)}{area(b)} & \dfrac{area(I\_SE(a)\cap b)}{area(b)} \end{pmatrix} \begin{pmatrix} 0.05 & 0.03 & 0 \\ 0.77 & 0.15 & 0 \\ 0 & 0 & 0 \end{pmatrix}$$

$$\quad\quad\quad\quad\quad\quad a \quad\quad\quad\quad\quad\quad\quad\quad\quad\quad\quad\quad\quad\quad b$$

Fig. 11. Detailed ICD-9 relation matrix

Note that if b is a line object, the length of b and $R_k(a) \cap b$ could be adopted instead of area in the first and the third representations.

Besides these representation approaches, there still is a problem left. Assuming that a point (or line) object happens to locate on the boundaries between the sectors, what is the proper ICD relation? In [Renz 2004], the similar cases are defined using additional relations. In this paper, because complex ICD relations are permitted, we specify this case as a complex relation, i.e. the containee belongs to both sectors.

4 Cognitive Experiment Based Validation for ICD-9 Model

In this part, a cognitive experiment focusing on ICD-9 is designed and implemented. The purpose of the validation experiment includes: 1) validating the rationality of MBR-based partition method and 2) determining the value of ρ.

4.1 Experiment and Data Preprocessing

A computer program is developed to facilitate the cognitive experiment. When the program is running, a group of points with stochastic position inside a given polygon is generated and displayed together with the polygon in turn. When a point appears on the screen, each subject is asked the following two questions:

1) Which part of the polygon does the point belong to?
2) How well does your answer fit for this situation?

There are 9 choices for the first question: north, east, south, west, middle, north-east, south-east, south-west and north-west. They correspond to the 9 atomic ICD-9 relations. Meanwhile, three distinctions are provided for the second question: very well, well and fairly. These three answers are quantified into three grades, i.e. 3, 2 and 1. After the subject has chosen the answer and pressed the "OK" button, the next point is displayed. The program continues until all points have been presented.

The subjects of this experiment are students from department of geography of Peking University. They all know the common sense of "north at the top" well, so the north arrow, which might disturb the subjects' decision, is not presented. Forty eight students participated in the experiment, each of which was test for 20 points. All together, 960 points were tested and 925 points were considered as valid. 35 points were removed due to the error associated with them. The error was caused by carelessness of the subjects. They submitted the same answer for two points appearing in succession although they were significantly distant. This usually means that he (or she) pressed "OK" without changing the answer to a new point.

The preprocessed points are distributed in the polygon evenly. This makes the collected data statistically credible. Fig. 12 visualizes the experiment result, where the small arrows and circles represent the directions answered by the subjects. The length of an arrow depends on the degree of agreement on how well the point fits for the direction. From the visualization, we could find intuitively that MBR-based middle part is rather acceptable.

4.2 Determination of ρ

Assuming the MBR-based method is acceptable, an ideal value of ρ should make the following two propositions correct for ICD-9 model:

1) If a point is inside the middle part, the subjects should answer "middle" to the first question.
2) If the subjects choose "middle" for a point, it should be inside the middle part.

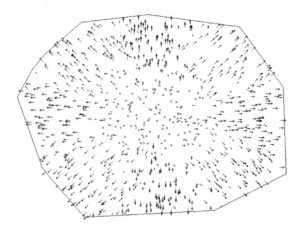

Fig. 12. Visualization of the experiment result data

Obviously, these two statements are conflicting. A smaller middle part might make the first proposition prone to be satisfied, while a large middle part would support the second one. Therefore what we need is a moderate value of ρ.

Referring to the concepts of recall and precision in information retrieval (IR) realm [Schamber 1990], two indices are identified to quantify the propositions. Assuming ρ is specified, two sets of points can be defined: S_{mp} and S_{mc}. Where S_{mp} is the set of points in the middle part and each element in S_{mc} is the point that the subjects answer "middle" to. Let N_{mp}, N_{mc} and N_{mm} be cardinalities of S_{mp}, S_{mc} and $S_{mc} \cap S_{mp}$. Then we define two indices as:

$$P_1 = N_{mm} / N_{mp} \text{ (precision)}$$

$$P_2 = N_{mm} / N_{mc} \text{ (recall)}$$

Based on the second answer in the experiment, we could calculate a weighted sum of these points. Considering the weighted sum, the indices evolve into P_{w1} and P_{w2}. Because weight indicates the degree of subject's agreement to a direction, P_{w1} and P_{w2} are more reasonable to validate the above propositions than P_1 and P_2.

Generally, P_{w1} is a decreasing function of ρ, while P_{w2} is an increasing function, i.e. P_{w1} and P_{w2} are inversely related. Hence, a proper ρ should make $(P_{w1}+P_{w2})/2$ achieve the maximum value. Table 1 lists the values of P_{w1}, P_{w2} and $(P_{w1} + P_{w2})/2$ when ρ varies from 1/12 to 8/12 based on the experiment result. Fig. 12 displays the functions between P_{w1}, P_{w2} and ρ. These two curves intersect when $\rho=1/3$. That makes $(P_{w1} + P_{w2})/2$ maximum. As contrasts, the cases without considering weight are also presented. As a rule, the values with weight are larger than the values without weight. This is because the points with lower weight usually locate near the boundary of middle part.

Table 1. The values of P_{w1}, P_{w2} and $(P_{w1} + P_{w2})/2$ changes according to ρ (unit: percent)

ρ	1/12	2/12	3/12	4/12	5/12	6/12	7/12	8/12
P_{w1}	100	97.05	95.38	85.43	70.07	51.83	37.58	27.85
P_{w2}		11.40	33.22	62.41	86.57	96.64	100.00	100.00	100.00
$(P_{w1}+P_{w2})/2$	55.70	65.14	78.89	86.00	83.35	75.91	68.79	63.92
P_1	100	94.87	95.18	81.94	65.72	49.32	36.77	28.46
P_2	8.22	25.34	54.11	80.82	95.89	100.00	100.00	100.00
$(P_1+P_2)/2$	54.11	60.11	74.65	81.38	80.80	74.66	68.35	64.23

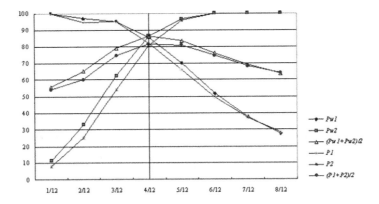

Fig. 12. Curves of the functions between P_{w1}, P_{w2}, $(P_{w1} + P_{w2})/2$, P_1, P_2, $(P_1 + P_2)/2$ and ρ

As shown in Table 1 and Fig. 12, a proper value of ρ is close to 1/3 for the ICD-9 model. It is a sound conclusion because $\rho = 1/3$ means that the MBR is divided equally and the ICD parts are comparable. For the same reason, the middle part in ICD-5 model should be larger than that in ICD-9. However, ICD-13 has a smaller middle part.

4.3 Evaluation of ICD-9 Model

With a specific value of ρ, the other eight parts in the ICD-9 model can be determined. The values of above indices for them are also calculated (table 2).

As shown in table 2, the values of these indices are rather high (average of the third column is 84.65%), so the ICD-9 model could be constructed based on the condition $\rho=1/3$.

If a point is inside one part (e.g. north), but the subject's choice for it is not the corresponding relation (e.g. east, middle, west, or even south), it is called a "mistaken" point. In Fig. 14, the mistaken points are drawn and we outline their distribution

Table 2. Indices of the other eight parts in ICD-9 when $\rho=1/3$

	P_{w1}	P_{w2}	$(P_{w1}+P_{w2})/2$	P_1	P_2	$(P_1+P_2)/2$
I_E	91.49%	79.93%	85.71%	89.52%	74.60%	82.06%
I_W	89.70%	83.94%	86.82%	85.71%	77.78%	81.75%
I_S	93.55%	66.16%	79.86%	93.90%	62.10%	78.00%
I_N	96.83%	71.81%	84.32%	95.91%	68.61%	82.26%
I_SE	83.55%	92.61%	88.08%	78.02%	92.21%	85.12%
I_SW	66.30%	95.77%	81.04%	59.17%	94.67%	76.92%
I_NE	82.16%	95.60%	88.88%	75.95%	92.31%	84.13%
I_NW	67.32%	97.71%	82.52%	62.26%	97.06%	79.66%

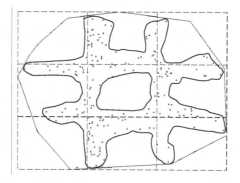

Fig. 14. Spatial distribution of mistaken points in ICD-9

region. This region surrounds the boundary of ICD-9, indicating that the MBR-based ICD-9 model is intuitively correct.

4.4 Conceptual Neighborhood and Intersection of ICD Relations

Mistaken points imply that ICD relations are vague and imprecise as one class of qualitative spatial relation. In general, when a group of points are near the border between two neighboring ICD sectors, their ICD-relations to the container are gradually changed. The boundary is fuzzy and there exists a zone around it. Referring to [Smith 2000], it belongs to fiat boundaries. Mistaken points usually locate inside this border zone. There exist two possible psychological views to interpret this phenomenon [Bonini 1999]:

1) Truth-gap theory

According to truth gap theory, the relation between points in the border zone and container does not belong to any atomic ICD-9 relation. A more detailed ICD model such as ICD-13 is needed.

2) Truth-glut theory

Applying this theory, the mistaken points might have two or more ICD relations to the container. For example, a point has I_N and I_NE relation at the same time.

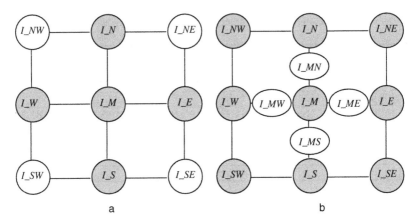

Fig. 15. Conceptual neighborhood graph of ICD-9 and ICD-13 (**a.** ICD-9 **b.** ICD-13)

Worboys [2001] has pointed out that nearness relations are accordant to the truth-gap theory validated by an experiment. An experiment is also needed to determine which view is more appropriate for ICD relations. Based on truth-glut theory, the intersection of two neighboring relations should not be empty. Meanwhile, truth-gap theory implies that there are some "gaps" in a specific ICD model. However, these gaps might be represented by an ICD model with higher resolution. In some sense, boundary zone, gap, and intersection have the same meanings. We thus believe that a combined relation in ICD-9, e.g. I_NE, is a conceptual intersection of its components in ICD-5, i.e. I_N and I_E. The same relation applies to I_ME in ICD-13 and I_M, I_E in ICD-9. Similar to the method presented in [Goyal 2001], Fig. 15 depicts the conceptual neighborhood graph of ICD-9 and ICD-13, where the gray circles indicate that the relations are also included in lower detailed ICD model, and the white circles denote that the relations are conceptual intersections of relations in lower-leveled model.

5   Conclusions and Future Research

This paper introduced and discussed internal cardinal direction (ICD) relations which can be applied to refine the spatial survey knowledge when two objects have contains/within topological relations. The main contributions of this paper are as follows:

1) The cardinal direction relations were introduced as well as their characteristics;
2) Three ICD models with varying levels of details were discussed, which are ICD-5, ICD-9, and ICD-13;
3) A formal model of ICD-9 was developed;
4) A cognitive experiment was designed to validate the ICD-9 model.

Although ICD relations are very useful to establish a qualitative referencing system, they also have inherent shortcomings. The most notable one is that they are not closed under inverse. For example, if a and b have a specific ICD relation, the relation between b and a will not be ICD any more. This property makes ICD relations significantly different from the other spatial relations. ICD relations themselves could not form an integral algebraic system. They should be integrated with the other relations (e.g. topological relations) in QSR.

Based on the above consideration and the findings reported in this paper, future research will focus on the following directions: 1) ICD relation involved qualitative reasoning; 2) further discussion on the middle part in ICD-5 and ICD-13; 3) computability of ICD relations; 4) research for the psychological nature of ICD relation; 5) combination of ICD and ECD to represent positional information and 6) detailed effect of ICD relations to express spatial knowledge.

Acknowledgements

The authors are grateful for early discussions with Dr. Yong Gao, Center for Information Science, Peking University, on formalization methods, and to Dr. Zhenji Gao, our colleague in the lab, for his help with the experiment design. We also would also like to thank the three anonymous reviewers for their comments, and Professor Tony Cohn (University of Leeds) and Dr. Xiaohang Liu (San Francisco State University), for their comments and help to polish the English.

References

[Bennett 1997] Bennett, B.: Logical Representations for Automated Reasoning about Spatial Relations. PhD thesis, School of Computer Studies, University of Leeds (1997).

[Bonini 1999] Bonini, N., Osherson, D., Viale, R. and Williamson, T.: On the Psychology of Vague Predicates. Mind and Language, 14(1999) 373-393.

[Car 1994] Car, A., Frank, A.: Modeling of the Hierarchy of Space Applied to Large Road Networks. In: Nievergelt, J., et al. (eds.): IGIS'94:Geographic Information Systems. Lecture Notes in Computer Science, Vol. 884. Springer-Verlag, Berlin (1994) 15-24.

[Clementini 1997] Clementini, E., Felice, P., Hernández, D.: Qualitative Representation of Positional Information. Artifical Intelligence, 95(1997) 317-356.

[Cohn 2001]Cohn, A.G, Hazarika, S. M.: Qualitative Spatial Representation and Reasoning: An Overview. Fundamenta Informaticae, 46(2001) 1-29.

[Cui 1993] Cui, Z., Cohn, A.G., Randell, D.A.: Qualitative and Topological Relationships in Spatial Databases. In: Abel, D. J., Ooi, B. C. (eds.): Proceedings of the Third International Symposium on Advances in Spatial Databases. Lecture Notes in Computer Science, Vol 692. Springer-Verlag, London (1993) 296-315.

[Duckham 2001] Duckham, M., Worboys, M.: Computational Structure in Three-valued Nearness Relations. In: Montello, D. R. (eds.). Spatial Information Theory: Foundations of Geograpic Information Science. Lecture Notes in Computer Science, Vol, 2005. Springer-Verlag, Berlin (2001) 76-91.

[Egenhofer 1991] Egenhofer, M. J.: Reasoning about Binary Topological Relationships. In: Switzerland, Z., Gunther, O., Schek, H. J. (eds.). Second Symposium on Large Spatial Databases. Lecture Notes in Computer Science, Vol. 525. Springer-Verlag, London (1991) 143-160.

[Faltings 1995] Faltings, B.: Qualitative Spatial Reasoning Using Algebraic Topology. In: Frank, A. U., Kuhn, W. (eds.): Spatial Information Theory: A Theoretical Basis for GIS. Lecture Notes in Computer Science, Vol. 988. Springer-Verlag, Berlin (1995) 17-30.

[Frank 1992] Frank, A. U.: Qualitative Spatial Reasoning about Distances and Directions in Geographic Space. Journal of Visual Languages and Computing. 3(1992) 343-371.

[Gahegan 1995] Gahegan, M.: Proximity Operators for Qualitative Spatial Reasoning. In: Frank, A. U., Kuhn, W. (eds.): Spatial Information Theory: A Theoretical Basis for GIS. Lecture Notes in Computer Science, Vol. 988. Springer-Verlag, Berlin (1995) 31-44.

[Goyal 1997] Goyal, R., Egenhofer, M. J.: The Direction-Relation Matrix: A Representation for Directions Relations between Extended Spatial Objects. In: the annual assembly and the summer retreat of University Consortium for Geographic Information Science, June 1997.

[Goyal 2000] Goyal R., Egenhofer M. J.: Cardinal Directions between Extended Spatial Objects. IEEE Transactions on Data and Knowledge Engineering, (in press), 2000. Available at http://www.spatial.maine.edu/~max/RJ36.html.

[Goyal 2001] Goyal, R., Egenhofer, M. J.: Similarity of Cardinal Directions. In: Jensen, C.S., Schneider, M., Seeger, B., Tsotras. V. J. (eds.): Advances in Spatial and Temporal Databases. Lecture Notes in Computer Science, Vol. 2121. Springer-Verlag, Berlin (2001) 36-55.

[Hernández 1995] Hernández, D., Clementini, E., Felice, P. D.: Qualitative Distance. In: Frank, A. U., Kuhn, W. (eds.): Spatial Information Theory: A Theoretical Basis for GIS. Lecture Notes in Computer Science, Vol. 988. Springer-Verlag, Berlin (1995) 45-57.

[Hernández 1997] Hernández, D.: Qualitative vs. Fuzzy Representation of Spatial Distance.In: Freksa, C., Jantzen, M., Valk, R. (eds.): Foundations of Computer Science: Potential Theory – Cognition. Lecture Notes in Computer Science, Vol. 1337. Springer-Verlag, Berlin (1997) 389-398.

[Hernández 1994] Hernández, D. (ed): Qualitative Representation of Spatial Knowledge. Lecture notes in Artificial Intelligence, Vol. 804. Springer-Verlag, New York (1994).

[Hirtle 1985] Hirtle, S. C., Jonides, J.: Evidence of hierarchies in Cognitive Maps. Memory & Cognition. 13 (1985) 208-217.

[Leung 1999] Leung, Y., Keung, K S., He, J. Z.: A Generic Concept-based Object-oriented Geographical Information System. International Journal of Geographical Information Science. 13 (1999) 475-498.

[Ligozat 1998] Ligozat, G.: Reasoning About Cardinal Directions. Journal of Visual Languages and Computing. 9 (1998) 23-44.

[Longley 2001] Longley, P. A., Goodchild, M. F., Maguire, D., Rhind, D. W.: Geographic Information Systems and Science. John Wiley & Sons, Ltd. (2001) 79-96.

[Mark 1995] Mark, D., Comos, D., Egenhofer, M., Freundschuh, S., Gould, M., Nunes, J.: Evaluating and Refining Computational Models of Spatial Relations through Cross-linguistic Human-subjects Testing. In: Frank, A. U., Kuhn, W. (eds.): Spatial Information Theory: A Theoretical Basis for GIS. Lecture Notes in Computer Science, Vol. 988. Springer-Verlag, Berlin (1995) 553-568.

[Mark 1999] Mark, D. M.: Spatial Representation: A Cognitive View. In: Maguire, D. J., Goodchild, M. F., Rhind, D. W., and Longley, P. (eds.): Geographical Information Systems: Principles and Applications, Second edition, Vol. 1. John Wiley & Sons, New York (1999) 81-89.

[Mennis 2000] Mennis, L. M., Peuquet, D. J., Qian, L.: A Conceptual Framework for Incorporating Cognitive Principles into Geographical Database Representation. International Journal of Geographical Information Science. 14 (2000) 501-520.

[Montello 1993] Montello, D.: Scale and Multiple Psychologies of Space. In: Frank, A.U., Campari, I. (eds.): Spatial Information Theory: A Theoretical Basis for GIS. Lecture Notes in Computer Science, Vol. 716. Springer-Verlag, Berlin (1993) 312-321.

[Papadias 1995] Papadias, D., Theodoridis, Y., Sellis, T., Egenhofer, M.: Topological Relations in the World of Minimum Bounding Rectangles: A Study with R-trees. In: Carey, M., Schneider, D. (eds.): Proceedings of ACM SIGMOD. San Jose, CA (1995) 92-103.

[Papadias 1997] Papadias, D., Egenhofer, M.: Algorithms for Hierarchical Spatial Reasoning. Geoinformatica. 1 (1997) 251-273.

[Randell 1992] Randell, D., Cui, Z., Cohn, A.: A Spatial Logic Based on Regions and Connection. In: Principles of Knowledge Representation and Reasoning: Proceedings of the Third International Conference (KR'92). Morgan Kaufmann (1992) 165-176.

[Renz 1999] Renz, J., Nebel, B.: On the Complexity of Qualitative Spatial Reasoning: A Maximal Tractable Fragment of the Region Connection Calculus. Artificial Intelligence. 12 (1999) 95-149.

[Renz 2004] Renz, J., Mitra, D.: Qualitative Direction Calculi with Arbitrary Granularity, In: Zhang C., Guesgen H.W., Yeap W.K. (eds.): 8th Pacific Rim International Conference on Artificial Intelligence (PRICAI'04), Auckland, New Zealand, Lecture Notes in Artificial Intelligence, Vol. 3157. Springer-Verlag, Berlin (2004) 65-74.

[Schamber 1990] Schamber, L., Eisenberg, M. B., Nilan, M. S.: A re-examination of relevance: toward a dynamic, situational definition. Information Processing & Management. 26 (1990) 755-776.

[Sistla 1994] Sistla, A. P., Yu, C., Haddad, R.: Reasoning About Spatial Relationships in Picture Retrieval Systems. In: Jorge, B., Jarke, M., Zaniolo, C. (eds.): Proceedings of VLDB, Morgan Kaufmann (1994) 570-581.

[Skiadopoulos 2001] Skiadopoulos, S., Koubarakis, M.: Composing Cardinal Direction Relations. In: Jensen, C.S., et al. (eds.): SSTD-2001. Lecture Notes in Computer Science, Vol. 2121. Springer-Verlag, Berlin (2001) 299-317.

[Smith 2000] Smith, B., Varzi, A. C.: Fiat and Bona Fide Boundaries. Philosophy and Phenomenological Research. 60 (2000) 401-420.

[Stevens 1978] Stevens, A., Coupe, P.: Distortions in Judged Spatial Relations. Cognitive Psychology. 10 (1978) 422-437.

[Timpf 1992] Timpf, S, Volta, G.S., Pollock, D.W., Egenhofer, M.J.: A Conceptual Model of Wayfinding Using Multiple Levels of Abstraction. In: Frank, A.U., Campari, I., Formentini, U. (eds): Theories and methods of spatio-temporal reasoning in geographic space. Lecture Notes in Computer Science, Vol. 639. Springer-Verlag, Berlin (1992) 348-367.

[Timpf 1997] Timpf, S., Frank, A.U.: Using Hierarchical Spatial Data Structures for Hierarchical Spatial Reasoning. In: Hirtle, S.C, Frank, A.U. (eds): Spatial Information Theory: A Theoretical Basis for GIS. Lecture Notes in Computer Science, Vol. 1329. Springer-Verlag, Berlin (1997) 69-83.

[Worboys 2001] Worboys, M.F.: Nearness Relations in Environmental Space. International Journal of Geographical Information Science. 15 (2001) 633-651.

[Yashino 1991] Yashino, R.: A Note on Cognitive Map: An Optimal Spatial Knowledge Representation. Journal of Mathematical Psychology. 35 (1991) 371-393.

[Zimmermann 1993] Zimmermann, K.: Enhancing Qualitative Spatial Reasoning- Combining Orientation and Distance. In: Frank, A.U., Campari, I. (eds.): Spatial Information Theory: A Theoretical Basis for GIS. Lecture Notes in Computer Science, Vol. 716. Springer-Verlag, Berlin (1993) 69-76.

Dynamic Collectives and Their Collective Dynamics

Antony Galton

SECSM, University of Exeter, Exeter EX4 4QF, UK
A.P.Galton@ex.ac.uk

Abstract. Dynamic collectives are discrete dual-aspect phenomena: they may present themselves from different viewpoints as either objects or events, and they arise from the collective action of groups of individual elements. In this paper we outline a formal theory of dynamic collectives and discuss their relevance to a range of GIScience concerns, including the identities and lifestyles of geographical entities, the architecture of information systems, and the relation between information systems and modelling systems.

1 Preamble

A recurrent theme in recent deliberations concerning Geographic Information Science is the perceived gulf between, on the one hand, the low-level observational data that constitutes the 'raw material' of our science, and on the other hand, the high-level conceptual schemes through which we as humans interpret, understand, and use that data. A priority for GIScience is to find appropriate ways and means of bridging that gulf. A recent crisp expression of this situation is the *pyramid framework* of [1], which comprises a low-level *Data Component* encompassing the three perspectives of location, time and theme, and a high-level *Knowledge Component*, which handles objects and the classification of those objects into categories organised in terms of superordinate-subordinate (taxonomic) and part-whole (partonomic) relationships. This framework supersedes the earlier Triad framework of [2], which recognised the three poles of What, Where and When, without reference to higher or lower levels of organisation. The earlier 'What' component has been divided into a low-level part—the 'theme' element of the Data Component—and a high-level part, encompassing all of the Knowledge Component. These two parts can themselves be aligned with the well-known distinction between field-based and object-based representations: at the lower level, a theme 'behaves as a spatiotemporal field of measurement' (e.g., temperature), whereas at the higher level, an object is 'a geographic conceptual entity that has a unique and cohesive identity and is related to a specific combination of observational data stored in the location, time, and theme perspectives' [1].

The pyramid framework thus posits a clean separation between raw observations and conceptualised knowledge derived from them. As such, it has performed the valuable service of drawing our attention to the fact that there is indeed a gross difference in character between, say, a mass of observations concerning temperature, wind-speed and barometric pressure across a certain region over a certain interval of time, and a description of a storm system with its particular history, motion, causes and effects; and

A.G. Cohn and D.M. Mark (Eds.): COSIT 2005, LNCS 3693, pp. 300–315, 2005.
© Springer-Verlag Berlin Heidelberg 2005

that a geographical information system worthy of the name should somehow encompass representations of both kinds as well as handling the relationships between them.

None the less, the cleanness of this separation can be called into question. The contrast is between 'raw' observations in terms of location, time, and theme, and a conceptual understanding in terms of objects, their properties, and their relations. But how raw is 'raw'? Consider (1) a spatio-temporal field of numerical values collected by some automatic sensor (e.g., on a satellite or weather station), (2) a record of individual speeding events collected by a speed-camera which is triggered by individual instances of a vehicle exceeding the speed limit at a particular place and time, (3) data collected by a human observer of (a) the frequency of vehicles passing through a particular road intersection or (b) the distribution of different species of tree growing in some wooded area. Any of these could count as 'raw' data to be entered into an information system. But proceeding down the list it becomes progressively more interpreted, or conceptualised, already at the point of data entry. When a person observes that there is, say, an oak tree growing at a particular spot, there has *already* been a transition from low-level observation (e.g., of a particular distribution of colours and textures) to the interpretation of those observations as an oak tree—but all this happens within the observer before the information is entered into the system. Unless we are prepared to extend the scope of the pyramid framework to encompass the cognitive processes involved in the collection of observations by humans, or, alternatively, to restrict it to data collected automatically by mechanisms with no capacity to interpret what they 'see', we have to admit that the 'theme' aspect of the data component cannot be restricted to low-level field-type observations but must sometimes include observations already structured and interpreted in terms of some possibly quite elaborate conceptual scheme involving fully-fledged objects and not just attribute values at various spatio-temporal locations.

The distinction between high-level and low-level thus becomes a relative one: *higher*-level and *lower*-level. The observation of an oak-tree is higher-level relative to the observations of particular leaf-shape, bark-texture and subtler nuances of form and posture which give rise (perhaps unconsciously) to the identification of the tree as an oak; but a set of individual tree identifications are lower-level in the context of the subsequent identification of a stand of woodland as being of a particular ecological type, say, or as having a certain economic value.

Two topics from the preceding discussion provide the context for the present paper. One is the element of *time*: of the three elements recognised in [1] as constituting the Data Component, this is the one which is generally regarded as especially problematic for GIScience, perhaps because of the relatively belated recognition of its importance. The second topic is the relationship between higher and lower (or more and less conceptualised) levels of representation: I shall in particular be concerned with aggregate phenomena which in some sense exist at a higher level of conceptualisation than their smaller-scale constituents—but which are none the less real for that.

In the next section I recall the notion of *dual-aspect phenomena* introduced in [3], and single out a particular class of such phenomena which I call 'dynamic collectives'. In §3 I discuss informally some of the properties of dynamic collectives, then in §4 I develop a formal vocabulary for describing them; in §5 I use the formal apparatus to analyse one of the phenomena in detail. In §6 I look beyond this rather straighforward analysis to consider more delicate issues of definition, identity, life, and motion of dynamic collectives. Finally, in §7 I make some general observations concerning the kinds of information systems within which it may be appropriate to model the phenomena considered here.

2   Dual-Aspect Phenomena

Elsewhere [3,4] I have drawn attention to a class of phenomena whose spatio-temporal complexity is such that they naturally present themselves in very different guises—as objects, events, or processes—when observed from different points of view. A hurricane, for example, is from one point of view *something that happens*—an event or occurrent—but it can also be viewed as something which comes into existence, moves along a trajectory, perhaps changing in character as it does so, and then ceases to exist, all of which are hallmarks not of occurrents but of continuants—in everyday terms, objects. In [3], I called such phenomena *dual-aspect*, thereby drawing attention to their object-event duality.[1] They include such examples as floods, wildfires, storms, epidemics, processions, traffic hold-ups, queues, flocks, swarms, and plagues. Not all of these are at a truly geographical scale, but all come within the sphere of interest of geography.

It is worthy of note that, in focussing on the dual (or multiple) aspects of these phenomena, I may seem to be calling into question the validity of the philosophical distinction between *continuants*, which endure through time, and *occurrents*, which are extended in time. This distinction is enshrined in, for example, the SNAP/SPAN ontological framework of [5]. Given that objects are traditionally classed as continuants and events as occurrents, and the categories of continuant and occurrent are mutually exclusive, it should not be possible for *one and the same phenomenon* to be viewed now as an object, now as an event.

To handle this apparent anomaly, we may observe that from a fully four-dimensional perspective, as advocated by, for example, [6], the world, as the totality of space-time, is populated neither by continuants nor by occurrents, but simply 'four-dimensional chunks of matter'—which I have elsewhere [4] called *hyperobjects*. A hyperobject is simply the material content of some region of space-time. To view a hyperobject as a continuant, we must focus on its instantaneous slices perpendicular to the time axis—'snapshots'. We must say that each of these snapshots of the hyperobject *is* the continuant at a particular stage of its history. The continuant can undergo change—meaning that successive snaphots of the corresponding hyperobject can have different properties; but in the continuant point of view it is one and the same continuant which has these different properties at different times (if they were different entities, this would not be an example of change). By parts of the continuant are meant, not these different snapshots (which each present the whole continuant as it is at one time) but rather the spatially separated parts of any one snapshot. Since different snapshots may have different sets

---

[1] Subsequently [4], recognising an additional process view, I renamed them *multi-aspect phenomena*. However, in this paper I focus only on the object/event duality and hence retain the term 'dual-aspect'. Since writing [4] I have come to regard the process aspect as problematic; this will be explored in further work.

of parts, it is possible for a continuant to gain or lose parts. Contrasting with this, a hyperobject may equally be viewed as an occurrent—essentially the life or history of the continuant already described—by focussing instead on its temporal extent, perhaps picking out salient temporal parts of this as significant episodes in the lifetime of the continuant.

In the present paper I focus on an observation made in [4], that many dual-aspect phenomena involve large numbers of similar units acting in a more or less coordinated way. The phenomenon does not consist merely of the aggregation of individuals; the dynamic aspect, their collective behaviour, is an essential component. For this reason I call these discrete dual-aspect phenomena *dynamic collectives*. To model them in such a way that we can retrieve both their object-like and their event-like aspects, it is necessary to espouse, at least to a certain extent, the four-dimensionalist view, while at the same time being hospitable to the language of continuants and occurrents whose proper home is a more traditional '3-plus-1'-dimensional model. The technical details of how this is handled are presented in §4. The nature of the constituent units of a dynamic collective varies from case to case, and it may not be equally helpful to analyse all dual-aspect phenomena in this way. Flocks, swarms, crowds, traffic events, etc, are naturally conceptualised as dynamic collectives; but floods, storms, and the like, while ultimately these too arise from the action of large numbers of molecules constituting the water or air, are for many purposes more naturally thought of as continuous fluid masses.

Dynamic collectives are relevant to the earlier discussion of higher and lower level perspectives because they simultaneously embody two levels of analysis: the level at which the collective is viewed as a whole, with its own dynamics and its own descriptive vocabulary; and the level at which it resolves into the individual behaviours of the components, to which a different set of conceptual tools may be appropriate. Observations of the behaviour of the individuals constitute the 'raw data' from which knowledge of the collective as a whole emerges through conceptualisation.

If we grant that these dual-aspect phenomena are of importance to GIScience, then we need a set of conceptual tools which will enable us to do justice to their distinctive characters. Two kinds of approach are possible here. On the one hand, we could focus on how higher-level phenomena emerge from their lower-level constituents—this approach naturally lends itself to implementation through visualisation of e.g., flocking phenomena, using computer animation (cf. the 'boids' of [7]). On the other hand, we could take a more abstract, conceptual approach, and attempt to clarify in precise logico-mathematical terms the essential ingredients and relationships exhibited by such phenomena; such an analysis may not directly lend itself to computer implementation, but will provide an abstract specification to which any practical implementations should be answerable. It is this latter approach which I shall take here.

3   Dynamic Collectives and Their Participants

The phenomena I am concerned with include such examples as

- Protest marches and processions. I discussed these in some detail in [3,4].
- Guided tours. Consider two separate parties being taken on a tour of an art gallery. Each party goes from room to room, at times forming compact groups, at other

times more spread out and mingling with other visitors to the gallery. The parties might encounter one another in a particular room, their members mingling, interacting, etc, perhaps transferring from one party to the other. Then the two parties continue on their separate ways.

– Traffic flows at a roundabout. If a roundabout has three exits A, B, and C, there are nine potential flows of traffic, A to B, A to C, B to A, B to C, C to A, C to B, A to A, B to B, and C to C. Members of all these flows mingle on the roundabout, and they cannot be distinguished, short of interrogating the drivers.

For the purpose of modelling dynamic collectives, a key notion is the *participation* in the large-scale phenomenon by its constituent small-scale elements. To analyse this feature with maximum clarity, let us consider first a much simpler example, which is not really a dual-aspect phenomenon at all but in which the notion of participation still plays a key role, namely a *string quartet*. At any one time, a quartet comprises four musicians: a cellist, a viola-player, and two violinists. However, if one of the members leaves, the quartet can continue in existence so long as a replacement can be found; and typically in such cases the quartet retains its name and its identity. Like Theseus' ship, one can readily envisage a situation in which the quartet ends up with none of its original members. Suppose that initially our quartet has members $\{A, B, C, D\}$ but that at successive later times its membership evolves as $\{A, B, C, D'\}$, $\{A, B', C, D'\}$, and then $\{A, B'', C, D'\}$. Viewed historically, the quartet cannot be equated with any set of individuals. The full set of participants in the quartet over its lifetime is $\{A, B, B', B'', C, D, D'\}$, but the history of the quartet is not the sum of the histories of this set of people, and at no time in the life of the quartet does this set of people coincide with its membership. Thus while it can be said that in a certain sense a quartet is ontologically dependent [8] on its members, the specific nature of this dependence is complicated by the fact that its membership varies with time. If the quartet is denoted Q, then, viewing it as a continuant, we may write things like $members(Q, t) = \{A, B', C, D'\}$, where $members$ is thus a time-varying function, or *fluent*. The history of the quartet involves the histories of its membership, but there is no one set of members such that it is the sum of *their* histories. On the other hand, the life of the quartet may be regarded as the aggregation of a set of *episodes* from the lives of the various people who are at different times members of it. The quartet is not an individual continuant in the same sense as the people who belong to it are, but it is what we might call a *notional* continuant.

Broadening the discussion to dynamic collectives more generally, a number of observations should be made here. First, the life of the collective consists of an aggregation of individual histories, each one recording the participation by a single individual over an interval of time. Second, an element only participates in a dynamic collective for part of its lifetime; so the participation by the element in the collective is an *episode* in its lifetime. Third, an element need not participate in the collective throughout the duration of the collective's existence—indeed it may participate intermittently more than once, joining, leaving, then rejoining. These observations point the way towards the formal analysis in §4. I assume a *base level* at which the primitive elements are the individuals capable of participating in a dynamic collective—e.g., people or vehicles. Although in this context these primitive elements constitute 'raw data', they may, from a different perspective, already be highly structured objects, another instance of the relativity of 'high-level' versus 'low-level' alluded to earlier.

In the proposed analysis, a 'raw observation' or 'brute fact' consists of the presence of a particular atomic individual in a particular location at a particular time. I notate this $pos(c, t) = s$, where spatial location s is the position of individual c at time t. I use a function pos rather than a relation because an individual cannot be in more than one place at a given time. I use 'c' for the individual as a reminder that an individual is a continuant, that is, an object which endures, existing as a whole at each time of its life.

4   Formal Analysis of Dynamic Collectives

I assume a primitive ontology of spatial points $s \in S$, times $t \in T$, and individual continuants $c \in C$. These elements correspond to the location, time, and theme aspects of the data component of the pyramid framework [1], although as discussed above, an ontology of individuals presupposes a higher-level of conceptualisation than is envisaged for the thematic aspect. Compare also the *three-domain model* of [9], which 'represents geographic information according to the three primary components of semantics, space, and time'. It represents these components in 'three independently managed data domains: the semantic domain, spatial domain, and temporal domain' — rather than, for example, treating semantic and temporal information as attributes of spatial objects.[2]

The three domains are related by means of the mapping $pos : C \times T \to 2^S$ which records the position of each continuant at each time. A position is represented as a *set* of spatial points, doing duty for a possibly extended region; this also allows us to write $pos(c, t) = \emptyset$ if c does not exist at time t. Of course, not every set of points will be a possible position for a continuant, and it is normal to restrict the available positions to sets of points with certain topologically 'nice' properties. In practice, also, our knowledge of the function pos will be restricted to some finite sample of instances; some form of interpolation procedure—e.g., linear interpolation—will have to be used to 'fill in the gaps'.

In terms of pos I define some functions for describing the life of an individual:

- $lifeline(c) = \{\langle s, t \rangle \in S \times T \mid s \in pos(c, t)\}$.
  The set of spatio-temporal positions occupied by a continuant in the course of its existence.
- $lifetime(c) = \{t \in T \mid pos(c, t) \neq \emptyset\}$.
  The time interval during which the continuant exists (i.e., has a position).
- $epi(c, t_1, t_2) = \{\langle s, t \rangle \in lifeline(c) \mid t_1 \leq t \leq t_2\}$.
  An *episode* in the lifeline of a continuant, i.e, any portion of that lifeline delineated by a start time and end time (it is assumed that $(t_1, t_2) \subseteq lifetime(c)$).

An episode, viewed as an occurrent, could be regarded as an *event* in the life of an individual, but it is more normal to reserve the latter term for those episodes which have some especial significance for the individual concerned, or which can be picked out as instances of some salient generic description.

---

[2] The choice of the term 'semantic' here is unfortunate—after all, spatial and temporal terms have meaning (semantics) as well. 'Thematic' would be more appropriate.

I now define a dynamic collective in terms of its lifeline, which is specified by means of a collection $\mathcal{C}$ of episodes from two or more individual lifelines, with the condition that any episode which is a proper subset of an episode in $\mathcal{C}$ is also an episode in $\mathcal{C}$. To *specify* such a collection, of course, we need only give its maximal contributory episodes, i.e., only include both $epi(c, t_1, t_2)$ and $epi(c, t'_1, t'_2)$ if either $t_2 < t'_1$ or $t'_2 < t_1$. I call this collection a *collective dynamic*; it represents the event-like aspect of the dual-aspect phenomenon. The object-like aspect, i.e., the dynamic collective itself, is that notional (or higher-level) continuant whose lifeline is the aggregation of the episodes in the collective dynamic $\mathcal{C}$. This could be denoted $lifeline^{-1}(\bigcup \mathcal{C})$, but for convenience I shall use the notation $\mathcal{C}^*$, which is stipulated to obey the rule $lifeline(\mathcal{C}^*) = \bigcup \mathcal{C}$. Note that whereas $\bigcup \mathcal{C}$ is a hyperobject, $\mathcal{C}^*$ is a continuant, and it is therefore possible to apply to the latter, but not the former, functions and predicates whose domain of definition is restricted to continuants—these include, in particular, time-varying properties such as *pos*.

As defined here, a collective dynamic could consist of an arbitrary collection of episodes selected from the lifetimes of an arbitrary set of individuals. In practice, we will mainly be interested in collections whose members we have some good reason to consider as belonging together. In §6, I shall consider what form such reasons might take.

Properties of dynamic collectives and collective dynamics, and how they are related to their constituent individuals, may be described using the following functions:

- $participants(\mathcal{C}) = \{c \in C \mid \exists t_1, t_2(epi(c, t_1, t_2) \in \mathcal{C})\}$.
  The set of individual continuants from whose lifelines a collective dynamic contains at least one episode.
- $members(\mathcal{C}^*, t) = \{c \in C \mid \exists t_1, t_2(t_1 \leq t \leq t_2 \land epi(c, t_1, t_2) \in \mathcal{C})\}$.
  The members of a dynamic collective at a given time.
- $lifetime(\mathcal{C}^*) = \{t \in T \mid participants(\mathcal{C}, t) \neq \emptyset\}$.
  The set of all times at which the collective has at least one participant. The relationship between *participants* and *members* can now be given by the rule

$$participants(\mathcal{C}) = \bigcup \{members(\mathcal{C}^*, t) \mid t \in lifetime(\mathcal{C}^*)\}.$$

- $participation(c, \mathcal{C}^*) = \bigcup \{e \in \mathcal{C} \mid \exists t_1, t_2(e = epi(c, t_1, t_2))\}$.
  The aggregation of those episodes which c contributes to $\mathcal{C}$.

Note that there is nothing in the definition of a dynamic collective to imply that its lifetime must be a convex interval; and indeed it may well be thought desirable to leave open the possibility that the history of a dynamic collective can have temporal gaps. This would enable us to model the phenomenon of 'reincarnation' by which a collective is reconstituted after having gone temporarily out of existence. If it is felt that reincarnation in this sense should be ruled out of court, then an additional condition could be added to restrict the possible collections of episodes that can count as collective dynamics.

The function *lifetime* gives the temporal extent of the life of a dynamic collective, but its spatial extent is more problematic. Figure 1 (left) shows a set of points; to this set are added, in the centre, the perimeter of its convex hull, and at the right, one of the infinitely many non-convex hulls that can be defined over the set. To represent the spatial region occupied by the points, we could simply take the set of points themselves; but we may want something less detailed, some simply-specified region sufficient to indicate the area over which the points are distributed, or their broad-brush configuration. The convex hull, though easy to compute and uniquely defined, does not always give a satisfactory answer: in Figure 1 the points occupy a roughly 'C'-shaped area, but this information is not apparent from the convex hull. The non-convex hull does retain the 'C'-shape discerned in the original set of points; but non-convex hulls have the disadvantage that there is no unique way to define them. They form an interesting study in their own right, but that lies outside the scope of this paper.

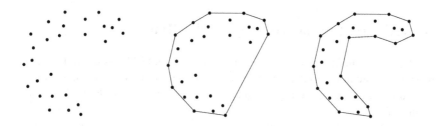

Fig. 1. A flock of points, with their convex hull and a non-convex hull

Here I shall make do with a rather underspecified notion of the position of a dynamic collective at a time, using a generic notion of 'hull': that is, if X is a set of points then $hull(X)$ is *some* suitably defined hull of X—maybe the convex hull, maybe one of the possible non-convex hulls, or maybe simply the set X itself. Hence we may put

– $pos(\mathcal{C}^*, t) = hull\left(\bigcup\{pos(c, t) \mid c \in participants(\mathcal{C}^*, t)\}\right)$
   The position of a dynamic collecive at a time.

This in turn allows us to say when a collective occupies a given location. If we want to know when a procession passes a particular viewpoint, for example, then instead of the collection of all the individual times at which the participants in the procession pass that point, we may prefer to be given an interval beginning when the first participant reaches that point and ending when the last one does. Or we may prefer a non-convex interval if the convex interval contains 'empty' periods during which it is inappropriate to say that the procession is currently passing the viewpoint. All this is taken care of by the possible variation in $pos(\mathcal{C}^*, t)$ brought about by the under-determination of the *hull* function used in its definition. Thus we have:

– $incidence(\mathcal{C}^*, s) = \{t \in T \mid s \in pos(\mathcal{C}^*, t)\}$.
   The times at which a dynamic collective is at a given location.

The use of *hull* allows us to model the *spatial interpenetration* of two (or more) dynamic collectives. An interpenetration episode in the lives of collectives $\mathcal{C}_1^*$ and $\mathcal{C}_2^*$ can be defined as

$$\{\langle s, t\rangle \mid s \in pos(\mathcal{C}_1^*, t) \cap pos(\mathcal{C}_2^*, t) \wedge t \in i\},$$

where i is a maximal convex interval over which the condition $pos(\mathcal{C}_1^*, t) \cap pos(\mathcal{C}_2^*, t) \neq \emptyset$ holds. Which episodes in the lives of the collectives are thereby to count as interpenetration episodes will depend on the *hull* function used in defining the position of a collective at a time. Different *hull* functions represent different granularities, and what counts as interpenetration at one level of granularity may not do so at another.

Functions such as *pos* and *lifetime*, which apply to individual continuants as well as to dynamic collectives, allow us to describe the higher-level phenomena in similar terms to how we describe their low-level constituents. A collective $\mathcal{C}^*$ is here being treated as a kind of individual, with its own life and motion. This in turn opens up the possibility of describing and reasoning about collectives even when we lack detailed information about their constituents.

5   Case Study: Interpenetrating Guided Tours

Figure 2 shows a simple instance of the art gallery example. The space of the gallery is represented by the horizontal axis: there are just three rooms, 1, 2, and 3, and distances are measured in metres from the left-hand extremity of room 1 at 0m to the right-hand extremity of room 3 at 30m. The space is portrayed as one-dimensional; the reader's imagination can supply the missing dimensions! The passage of time is represented by the vertical axis, spanning a period of 20 minutes. There are two groups of visitors, A and B. From the diagram it can be seen that group A starts in Room 1 and then visits Rooms 2 and 3 in succession, whereas group B starts in Room 3 and then visits rooms 2 and 1. The two groups visit Room 2 at the same time. There are eight individuals, labelled $a, \ldots, h$. Group A initially consists of individuals a, b, c and d, while B comprises f, g, and h. However, d switches allegiance from A to B while both groups are in Room 2, and b separates from A, remaining in Room 2 while the rest of the group proceed to Room 3. Another individual, e, is initially in Room 2, not attached to any group, but joins group A when they come into the room.

For present purposes, the universe is restricted to the space $S = [0, 30]$ and time $T = [0, 20]$. The fact that in reality the individuals have lifetimes extending beyond these limits is ignored. I assume that the individuals occupy point locations; hence, $pos(x, t)$ will be a singleton set (e.g., $pos(a, 0) = \{2\}$). The lifeline of an individual can be represented as, e.g.,

$$lifeline(d) = \{\langle 8, 0 \rangle, \langle 8, 2 \rangle, \langle 12, 5 \rangle, \langle 12, 7 \rangle, \langle 13, 11 \rangle, \langle 13, 14 \rangle, \langle 9, 16 \rangle, \langle 9, 18 \rangle, \langle 7, 20 \rangle\}$$

where positions at times intermediate to those represented here are interpolated linearly (e.g., $\langle 11, 15 \rangle$). The lifetime of every individual is the full temporal extent, e.g., $lifetime(d) = [0, 20]$. A lifeline can be split into episodes in many different ways; for example, the lifeline of individual d includes the episode

$$epi(d, 3.5, 15.5) = \{\langle 10, 3.5 \rangle, \langle 12, 5 \rangle, \langle 12, 7 \rangle, \langle 13, 11 \rangle, \langle 13, 14 \rangle, \langle 10, 15.5 \rangle\}$$

representing the time that that individual spent in Room 2.

Each of the groups can be represented as a dynamic collective. As explained in §4, the route to defining a dynamic collective is first to define its collective dynamic as a

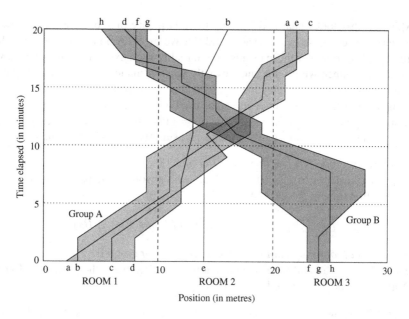

Fig. 2. The Art Gallery Example

collection of episodes, and then to introduce the collective itself as that notional continuant whose lifetime is the the union of those episodes. Thus the collective dynamics
corresponding to groups A and B are specified by the episode-sets

$$\mathcal{A} = \{epi(a, 0, 20), epi(b, 0, 12), epi(c, 0, 20), epi(d, 0, 11), epi(e, 8.5, 20)\}$$
$$\mathcal{B} = \{epi(d, 14, 20), epi(f, 0, 20), epi(g, 0, 20), epi(h, 0, 20)\}$$

Groups A and B themselves are then modelled as the corresponding continuants $\mathcal{A}^*$
and $\mathcal{B}^*$.

From these definitions of the groups as dynamic collectives we can easily read off
their participants:

$$participants(\mathcal{A}) = \{a, b, c, d, e\}, \quad participants(\mathcal{B}) = \{d, f, g, h\}.$$

and also the members of a group at a given time, e.g.,

$$members(\mathcal{A}^*, 5) = \{a, b, c, d\}, \quad members(\mathcal{A}^*, 16) = \{a, c, e\}.$$

The participation of d in group B is $participation(d, \mathcal{B}^*) = \{epi(d, 14, 20)\}$, and the
location of group B at time 16 is $pos(\mathcal{B}^*, 16) = hull(\{9, 11, 12, 15\})$. If we define
$hull(X) \equiv X$, then $pos(\mathcal{C}, 16)$ is just the set of four points $\{9, 11, 12, 15\}$. But if
$hull(X)$ is the convex hull of X, then $pos(\mathcal{B}^*, 16) = [9, 15]$—as indicated by the
shaded regions in the diagram. The time during which group A passed from room 1 to
room 2 is given by $incidence(\mathcal{A}^*, 10) = \{3.5, 4.8, 5.33, 9.5\}$, these being the times at
which d, c, a, b respectively crossed the boundary between the rooms. Again, this set of
times could be represented by their convex hull, the interval $[3.5, 9.5]$.

Although this looks neat and precise, there is plenty of scope for vagueness and uncertainty. For example, when exactly did d leave group A and join group B? I have specified precise times for these events, but in reality such exact times may be unavailable to us (e.g., since we do not know exactly when d made the decision to leave one group and join the other), or they may be by nature indeterminate.

Consider now how these concepts can be used for answering more complex queries. I introduce three new terms $R_1 = (0, 10)$, $R_2 = (10, 20)$, $R_3 = (20, 30)$, representing rooms 1, 2, and 3. These are spatial regions, i.e., subsets of S, not continuants.

- *Who was in group* A *when they were in room 3?*
  This query asks us to find all individuals who at some time participated in A and were at that time in Room 3. We require

$$\{x \mid \exists t(x \in members(\mathcal{A}^*, t) \wedge pos(x, t) \subset R_3)\}.$$

- *When was all of group* A *in the same room as all of group* B*?*
  This asks us to find all times t such that for some room R_i, the positions of groups A and B are both contained in R_i. We require

$$\{t \in T \mid (pos(\mathcal{A}^*, t) \cup pos(\mathcal{B}^*, t) \subseteq R_1) \vee (pos(\mathcal{A}^*, t) \cup pos(\mathcal{B}^*, t) \subseteq R_2) \vee (pos(\mathcal{A}^*, t) \cup pos(\mathcal{B}^*, t) \subseteq R_3)\}$$

- *At what times did members of group* A *intermingle with members of group* B*?*
  This is asking for the times at which the two collectives interpenetrate, given by

$$\{t \mid pos(\mathcal{A}^*, t) \cap pos(\mathcal{B}^*, t) \neq \emptyset\}.$$

As suggested earlier, the times returned by this query will depend on the *hull* function used for defining the position of a collective at a time.

As dual-aspect phenomena, the guided tours can also present themselves as events. A curator sitting at the doorway between rooms 1 and 2 would experience group A as a passage of individuals from room 1 into room 2; this event is followed about five minutes later by a passage of individuals moving from room 2 into room 1, the curator's experience of group B. Each of these events comprises four subevents, each being the passage of a single individual through the doorway. At any time between 3.5 and 9.5, the curator can say the group A is moving from room 1 into room 2. At 4.8 he can say that individual c moves from room 1 to room 2. Since individuals are treated as point-like, one individual's crossing the threshold between the rooms is a point event, whereas a whole group's crossing may be a durative event.

6   Identity, Life and Motion of Dynamic Collectives

As so far defined, a collective dynamic could consist of an arbitrary collection of episodes selected from the lifetimes of an arbitrary set of individuals; the corresponding dynamic collective might then be an extremely odd construct. This is not what we had in mind in setting up the concept! When is it reasonable to consider a set of episodes as constituting the lifeline of a dynamic collective? Some collectives, we may say, are more 'natural' or better-motivated than others. The well-known distinction between *bona fide* and *fiat* objects [10] is relevant here. It seems that we might distinguish between bona fide and fiat dynamic collectives. A bona fide collective exists by virtue of some relevant 'genuine' connection amongst its members, whereas a fiat collective is one that is *deemed* to exist without there being any reasons that would be sufficient to make it a collective without such deeming.

What counts as a 'genuine' connection sufficient to make a collection of episodes a bona fide collective dynamic? More exactly: *In virtue of what is x a member of collective X at time t?* Many answers are possible here; only some of them provide 'genuine' connection:

- Causal interaction: x belongs to X because its action affects and is affected by the actions of other individuals in X.
- Common causal influence: x belongs to X because its actions are influenced by some external cause which also affects other individuals in X.
- Position and motion: x belongs to X because it moves along with the other members of X, in proximity to them.
- Position: x belongs to X by virtue of its proximity to other members of X.
- Motion: x belongs to X because its motion is coordinated with that of other members of X, without necessarily being close to them.
- Intention on the part of x: x deems itself to belong to X.
- Intention on the part of members of X: x belongs to X because the other members of X deem it to be so.
- Intention on the part of some third party: x belongs to X because someone who is not a member of X (the 'fiat-giver') deems it to be so.

These reasons are not independent: e.g., the fact that x's motion is coordinated with that of the rest of X may be the result of causal interactions amongst the members of X, or of some common external cause, or of x's intentions. But they do seem to form a rough spectrum from the most bona fide reasons for considering a set of episodes to form a collective dynamic to the most fiat reasons.

The distinction between bona fide and fiat collectives is problematic if the participants are humans, in which case third-party fiats can themselves influence behaviour. Consider schoolchildren mixing together in the playground; then the teacher orders them to assemble in their classes. Each class is a collective, and it is so for 'fiat' reasons, i.e., the children in question have been deemed by the school to belong in the same class. But the act of deeming, and the resultant knowledge on the part of each child as to which class it has been assigned to, constitute sufficient causal force to make the collective bona fide. A truly fiat collective would be one where the existence of the fiat does not affect the participants in any way, either because they are not rational beings and hence cannot be aware of the fiat, or because they have not been informed of the fiat.

As something of an aside here, it might be mentioned that the formal characterisation of dynamic collectives allows them to be used to model phenomena very different from the dual-aspect phenomena which motivated the theory. Consider, for example, the entity to which we refer by the epithet 'President of the United States'. At different times this epithet refers to different people; but at any one time it only refers to one person. So the President of the United States is not a collective in the normal sense of the term. Nonetheless, we could model it as that continuant whose life history consists of episodes such as $epi(AbrahamLincoln, 4/3/1861, 15/4/1865)$ and $epi(RonaldReagan, 20/1/1981, 20/1/1989)$—in other words, a dynamic collective in the technical sense. If we let $\mathcal{P}$ denote this collection of episodes (considering the presidency from its inception to the present day—its future being, of course, unknown), then we can put, for example,

$$participants(\mathcal{P}) = \{GeorgeWashington, JohnAdams, \ldots, GeorgeWBush\}$$

and

$$members(\mathcal{P}^*, 1/1/2001) = \{BillClinton\}.$$

A distinctive feature of this kind of 'collective' is that the function $members$ always returns either a singleton set or $\emptyset$.

The discussion of reasons for considering a collective to exist leads naturally to a consideration of the creation, modification, and destruction of collectives: in short, their *lifestyles* [11]. There have been a number of studies of identity-based change, beginning with [12]. Of particular relevance here is [13], in which 'composite objects' are discussed. Two kinds of composite object are distinguished: those, such as artefacts, which are assemblies of components put together in a pre-ordained structure, and those which are mere collections. Our dynamic collectives are more akin to the latter kind.

A collective may come into existence in various ways. Pre-existing individuals may come together—either by chance or in response to some attractive force—in a situation where they begin to act in a coordinated way. The birth of the collective is the start of the coordinated action. This may be spread out in time, so collectives do not necessarily have sharp starting points. Another possibility is creation *ex nihilo*: individuals are created in a context where their actions are coordinated from the start. A fiat collective is created when it is deemed that henceforth (or until time t, or until condition X holds, or 'for the foreseeable future') these individuals will constitute a collective. The start point is not marked by any intrinsic discontinuity. Destruction of a collective can similarly happen because its constituent individuals are destroyed, because they disperse or cease to interact causally, or because the collective is deemed no longer to exist.

In the formal account, it is simple enough to say that the dynamic collective $\mathcal{C}^*$ starts life at the earliest start point of any of the episodes in its collective dynamic $\mathcal{C}$, i.e., at

$$t_0 = \inf\{t \mid \exists c \in C, t' \in T(epi(c, t, t') \in C)\}.$$

This leaves out of consideration, however, the idea of a 'spread out' coming-to-be of the collective. To model this, we should need to define what it means for the collective to be 'fully' in existence. This will vary from case to case: e.g., for some types of collective there may be a minimum membership condition for full existence. If this condition first obtains at a time t_1 later than the t_0 defined above, then the 'coming into existence' event is durative, lasting through the whole interval $[t_0, t_1]$.

There are other lifestyle events to consider. *Merging* occurs where two or more collectives come together—through causal interaction, or in the fiat case they are deemed to count as one from henceforth. This is not the same as the creation of a 'metacollective' whose constituents are individual collectives. *Splitting* is the reverse of this, the creation of new collectives by subdividing an existing collective. *Spawning* occurs when a subsidiary collective splits off from a main collective; the individuals in the new collective may be newly created from those of the parent collective, or already exist as individuals in the parent collective; this depends on whether the type of individual allows for spawning of new individuals. *Growth* may occur through recruitment of individuals from outside the collective, or through procreation within the collective; *reduction* or *diminution* is the reverse of this, with two corresponding flavours. *Steady-state* existence of a collective occurs when any growth is balanced by commensurate diminution. All of these changes can be described in terms of the mathematical formalism established earlier. This leads to the possibility of developing procedures for detecting the occurrence of the various lifestyle changes that dynamic collectives can undergo.

7   Epilogue: Information Systems *vs* Modelling Systems

A background assumption underlying all of the above discussion is that ultimately we want to be able to develop methods for handling dual-aspect phenomena in the context of a geographical (or more generally spatial) information system. This raises some important questions about what we should expect of such an information system.

Consider the case of traffic systems. A traffic system involves a complex interaction of different kinds of causes. Some of these are continuously acting (e.g., the necessity for individuals to travel to and from their places of work each week-day), while others are discrete 'one-off' events (e.g., an accident). All the operative causes interact to produce effects which can, in principle, be inferred from a knowledge of the causes. The inference involved here will usually be statistical, certainly quantitative, and requires detailed understanding of the dynamics of the system under consideration and ability to solve the relevant equations, numerically if not analytically. In principle all this could be incorporated in a computer program which models the dynamics of a traffic system and simulates its behaviour—let us call this a *modelling system*. It could be used to *predict* the future, to *explain* the past, or to *explore* 'what if' scenarios (e.g., to determine the likely effects of installing a new set of traffic lights, or converting a crossroads into a roundabout).

In general, the adequate handling of dynamic collectives will require access to approriate modelling systems. Such systems would not be examples of what is normally understood by an 'information system'—in a nutshell, they do too much computation! An information system is more like a *repository* of information; the computational capabilities it provides should be sufficient to allow us to retrieve the information in various combinations and to provide different ways of representing it. It is natural that such a system should be capable of a certain amount of inference, but how much?

Suppose we have an information system (IS) linked to a modelling system (MS). The IS will hold information about events, processes, objects, fields, etc, across both space and time, and will include information about what happens as a result of what. But this latter information will actually be computed by the MS and fed back into the IS, which will have no ability to perform the necessary calculations itself. The MS is not handling *information* so much as *data*—many low-level facts about the values of various physical quantities in different locations (mainly field data rather than object data). The high-level descriptions of this in qualitative terms that are meaningful to the human user will be the preserve of the IS. Here there will be a greater emphasis on objects than on fields, though not to the exclusion of the latter. An important problem is how to derive high-level, qualitative, largely object-based information from low-level, quantitative, largely field-based data: we are back at the vision of the pyramid framework of [1] discussed at the beginning of the paper. It is a problem which crops up in different guises in many areas of computing (e.g., pattern recognition, image understanding).

We may distinguish three different kinds of problem:

1. MS problems: how to model systems and simulate their behaviour, using low-level numerical data and scientific or mathematical techniques.
2. IS problems: how to represent the various facets of those systems in ways that are meaningful from a human point of view and can be related directly to the concepts we use in our everyday interaction with them.
3. Linkage problems: how to get from the data handled and generated by the MS to the information handled by the IS.

How far can work on IS problems be pursued in isolation from MS problems? There must surely be some interaction between the two, but despite this I believe that they are partly separable. Some IS issues originate solely at the IS level without reference to anything that would be meaningful to the MS: for example, the definition of regions in terms of administrative and cadastral boundaries does not emerge from low-level data by means of a mathematically definable process, and similar remarks would apply to dynamic collectives for which the fiat element plays a prominent part of their definition. There are also humanly meaningful concepts (for example, 'neighbourhood') which are the product of phenomena at various levels but not primarily ones which could be handled using an established body of mathematical theory.

Modelling systems research and information systems research require different kinds of vision and expertise. An unfortunate consequence is that practitioners of the two areas do not talk to each other often enough, and when they do they tend not to understand one another. In this paper I have largely focussed on an IS perspective on dual-aspect phenomena; but at many points questions are opened up which can only be answered through taking on board an MS perspective. Similar remarks apply, I believe, to a substantial proportion of work that is presented at conferences such as COSIT.

References

1. Mennis, J.L., Peuquet, D.J., Qian, L.: A conceptual framework for incorporating cognitive principles into geographical database recognition. International Journal of Geographical Information Science **14** (2000) 501–520
2. Peuquet, D.J.: It's about time: A conceptual framework for the representation of temporal dynamics in geographic information systems. Annals of the Association of American Geographers **84** (1994) 441–461
3. Galton, A.P.: Desiderata for a spatio-temporal geo-ontology. In Kuhn, W., Worboys, M.F., Timpf, S., eds.: Spatial Information Theory: Foundations of Geographical Information Science, Berlin, Springer (2003) 1–12 Vol. 2825, Lecture Notes in Computer Science.

4. Galton, A.P.: Fields and objects in space, time, and space-time. Spatial Cognition and Computation **4** (2004) 39–67
5. Grenon, P., Smith, B.: SNAP and SPAN: Towards dynamic spatial ontology. Spatial Cognition and Computation **4** (2004) 69–104
6. Heller, M.: The Ontology of Physical Objects: Four-dimensional Hunks of Matter. Cambridge University Press, Cambridge (1990)
7. Reynolds, C.W.: Flocks, herds, and schools: A distributed behavioural model. Computer Graphics **21** (1987) 25–34
8. Simons, P.: Parts: a Study in Ontology. Clarendon Press, Oxford (1987)
9. Yuan, M.: Representing geographic information to support queries about life and motion of socio-economic units. In Frank, A., Raper, J., Cheylan, J.P., eds.: Life and Motion of Socio-economic Units. Taylor and Francis (2001) 217–234
10. Smith, B.: Fiat objects. Topoi **20** (2001) 131–148
11. Medak, D.: Lifestyles. In Frank, A., Raper, J., Cheylan, J.P., eds.: Life and Motion of Socio-economic Units. Taylor and Francis (2001) 139–153
12. Al-Taha, K., Barrera, R.: Identities through time. In Ehlers, M., Steiner, D.R., Johnston, J.B., eds.: Proceedings of International Workshop on Requirements for Integrated GIS. Environmental Research Institute of Michigan (1994) 1–12
13. Hornsby, K., Egenhofer, M.J.: Identity-based change operations for composite objects. In Poiker, T., Chrisman, N., eds.: Proceedings of the 8th International Symposium on Spatial Data Handling, International Geographical Union (1998) 202–213

A Linguistics-Based Framework for Modeling Spatio-temporal Occurrences and Purposive Change

Jeff T. Howarth and Helen Couclelis

Department of Geography,
University of California,
Santa Barbara, California, USA
{howarth, cook}@geog.ucsb.edu

Abstract. We present a linguistics-based approach for modeling spatio-temporal change and in particular, purposive change, as in the change in land uses. We extend Talmy's theory of force dynamics in language by means of the Aristotelian distinction between constitutive, agentive, and telic dimensions in things, to derive a framework for describing different kinds of occurrences and changes, both purposive and not. The framework, which can span any number of spatio-temporal scales and granularities, may be seen as a tentative ontology of change that highlights the role of goal-directed action. We illustrate our argument by means of excerpts from a historic text detailing a cattle-ranching operation.

1 Introduction

Transcending the mapping roots of GIS has been a major objective of GIScience for well over a decade. Related efforts are characterized by a growing interest in the representation of change and time as these pertain to geographic phenomena. The reasons for the emphasis on these abstract concepts are practical as well as conceptual. Practically, there is an urgent need to be able to trace, anticipate and represent the multitude of significant changes taking place in the geographic world, for purposes of scientific study, policy making or management. Even if the mappable pattern remains the signature product of today's sophisticated GIS, we want to know how that pattern got to be the way it is, and what it may become next. Conceptually, it is clear by now that there is more to the representation of change than the addition of a time axis: indeed, the formulation of novel change-oriented models and ontologies has been an active research area in GIScience for some time.

GIS-based models of land use exemplify how cartographic data models constrain representations of change. By representing land use as an attribute of space at a particular time, cartographic data models can only represent it as a physical phenomenon, that is, as change in the attributes and geometry of space at different times. That representation is quite different from how people actually change space as they use it, through actions aiming to achieve particular goals, and does not help GIS users to understand how this use, these actions and goals, contribute to change over time. By analyzing how natural language represents actions and change, this paper seeks to bring queries on human actions and goals and their spatio-temporal

A.G. Cohn and D.M. Mark (Eds.): COSIT 2005, LNCS 3693, pp. 316–329, 2005.
© Springer-Verlag Berlin Heidelberg 2005

consequences into a GIS. This paper builds on a paper presented at GIScience 2004 (Howarth & Couclelis, 2004) and develops a framework to represent change that is based on a small set of concepts that underlie lexical representations of motion, change, and goal-oriented action. The framework contributes towards an informal ontology of occurrences, and provides a means to examine how goal-oriented actions in time and space distribute human attention among different aspects of reality.

Our approach draws on Talmy's (2000a) theoretical work on force dynamics in language, and uses a historical text on ranching activities on Santa Cruz Island, California. Cognitive linguistics allows us to frame this descriptive text as a window into two different worlds: a 'projected-world' consisting of entities that are expressible as linguistic terms (signs) and can partake in linguistic relations (Jackendoff, 1985), and a 'real-world' consisting of entities such as the stone (or the stuff that we might agree to call a stone) that Dr. Johnson, a lexicographer, kicked to demonstrate the reality of (Quine, 1960:3). With textual descriptions as sources of geographic information, we rely on a symbolic system of signs and their relations to understand aspects of the 'real-world' that time, or space, make inaccessible to us. Furthermore, we use this symbolic account to derive the spatial and temporal dynamics of the 'real-world', and by doing so to understand how these dynamics reflect both the goals of agents and the constraints and possibilities afforded by their environment. At the same time, we are aware that the text that we rely on represents the 'projected-world' of its author, who catalogued things and happenings for a purpose, who focused attention on particular aspects of his or her 'real-world', and whose conceptions of the world may not mirror our own.

We proceed as follows: Section 2 reviews the background of previous research on natural language in GIScience, particularly in relation to the representation of change and the human use of space. In addition, we discuss certain limitations of the commonly used analogy between occurrences, or phenomena that span time, and spatial objects, and of taxonomic-based ontologies of geographic phenomena. Section 3 develops our conceptual framework, based on cognitive linguistics. We define states of relations between lexical terms, identify sixteen 'projected-world' primitive situations and changes in situations that underlie lexical representations of action and motion, and connect these primitive situations with relations expressing goal-oriented action. This provides a means to examine the intentional context of lexical descriptions of entities, situations, and change. Section 4 discusses the informal ontology of this linguistics-based approach, pointing towards a novel way of modeling land use that connects fine-grained actions to broader occurrences and change in time and space.

2   Background

2.1   Language and GIScience

There is a strong tradition in GIS data modeling, and in applied science more generally, to only concern oneself with what is observable and measurable (Smith, 1997). However, as both scientists and philosophers have pointed out, what is observable and measurable in a given context is a function of the conceptual and technical apparatus available for measurement and representation (Carnap, 1969; Eddington, 1939). Language in particular, whether formal or natural, both enables and constrains what can be expressed within a given domain (Wittgenstein, 1953). Clearly, many important things in the geographic world can be observed and measured that GIS has no means of expressing. For instance, when people observe a particular setting, such as a working ranch, they see mappable structures such as fences, roads, gates, and water troughs, but they also see day-to-day actions involving those structures, such as people driving on roads, checking fences, opening and shutting gates, and moving animals from place to place so that they may come to water or eat fresh grass. These actions provide the reasons why the structures and other adaptations of the land exist in the first place, and why they appear in the landscape, change over time and perhaps eventually disappear. Yet, unlike the structures themselves, the associated actions are only beginning to be represented in GIS data models (Worboys & Hornsby, 2004). These actions are, however, immediately understandable to an observer and are easily described verbally, in written documents or in speech. Even more problematic than the representation of actions is the representation of the reasons for these actions, in particular the fact that actions are means that people use to achieve particular ends. Again, this is something that natural language - but not GIS - can easily express. An ability to translate natural-language descriptions of (spatio-temporal) actions and their purposes into a language directly understandable by GIS can complement ordinary cartographic representations and contribute towards the development of more effective spatio-temporal ontologies for dealing with human-induced landscape changes. Observable actions in time and space are indeed at the basis of more abstract notions such as land use, of the functions associated with particular forms of spatial organization, and ultimately, of human purpose – all of them notions that underlie so many important kinds of observable, measurable change.

An interest in natural language studies has long been connected with research in GIScience. The work of several cognitive linguists (Herskovits, 1986; Lakoff, 1987; Lakoff & Johnson, 1980; Talmy, 1983) was very influential in parts of the GIScience community especially in the early 1990s (Mark & Frank, 1991), and has left a research legacy that continues to this day. Much of that interest was motivated by the realization that natural language is a gateway to understanding how people conceptualize space and spatial relations (Mark, 1988). From this perspective, natural language texts are important empirical data sources for the study of common-sense models of spatial phenomena (Egenhofer & Mark, 1995; Mark, Egenhofer, & Hornsby, 1997).

Several researchers have used natural language texts as a source of geographic information. McGranaghan (1991) matched written descriptions of botanical records with locations on a topographic map. Others (Kuhn, 2001; Soon & Kuhn, 2004) have used natural language texts to develop task-based ontologies of objects based on the notion of affordances (Gibson, 1979). That work connects a user's intention with the function of synchronic entities (what an object or 'surface' affords a user or organism) but lacks an explicit representation of time. In addition, the approach defines relations between actions based largely on lexical entailment (Fellbaum, 1999), rather than the syntactic and temporal context of their use. Perhaps closest to the work presented here, Cheylan & Lardon (1993) derived activities from a shepherd's diary, mapped these events onto spatial representations, and developed a conceptual framework for spatio-temporal reasoning in GIS. Our research overlaps with that work, but is distinguished by our focus on a small set of primitive concepts that underlie lexical expressions, as well as on notions of purpose and agency.

2.2 Time, Objects, Events, Purpose

In GIScience, research on the representation of change has been closely associated with research on time: indeed, change happens in time, and time brings change. Early research efforts on temporal databases often focused on formal approaches, including temporal logics. Allen's (1983) interval-based calculus, based on thirteen primitive relations, remains a robust framework for temporal reasoning. Many early efforts in temporal GIS, however, used an instant-based temporal addressing system, adding a time dimension orthogonal to space to locate a feature at a point in time. This extends the traditional cartographic model that represents the physical configuration of space at a single moment of time to a sequence of configurations or 'time slices'. Change, or 'transactions', such as the motion, deformation, growth, shrinkage, appearance or disappearance of objects, result from geometric and attribute differences between time slices (Claramunt & Theriault, 1995, 1996; Hornsby & Egenhofer, 2000; Langran, 1992). This approach derives diachronic entities, such as processes that span time and underlie change, from finite differences in synchronic entities, thus conflating change and representational error, and filtering processes by the granularity of the time step used (Chrisman, 1998).

More recent theoretical efforts in spatio-temporal ontology (Galton, 2003; Grenon & Smith, 2004) and data modeling (Worboys & Hornsby, 2004) have moved the ontological status of spatial change and diachronic phenomena from derivatives to complements of synchronic entities. Explicit in these approaches is an analogy between objects and diachronic phenomena, such as events or processes. From a practical standpoint, treating events as objects is desirable because it facilitates implementation with object-oriented programming languages, which are available in many current GIS. The development of object classes and relations are central to these implementation languages. Zacks, Tversky, & Iyer (2001) argue that people perceive both partonomic and taxonomic structure in events, in analogy to the hierarchical organization of object categories (Tversky & Hemenway, 1984). However, these results are based on experiments that involve fairly low-level, physical actions: assembling a saxophone, making a bed, etc. Zacks & Tversky (2001:7) suggest that fine-grained events are generally distinguished by physical acts, while coarse-grained objects reflect the goals, plans, or intentions of the events' participants. However, it is questionable whether these more abstract levels of human actions will be perceived by an observer as objective events: they will more likely be inferred based on the observer's knowledge of causality or the intentions of agents, and the resulting event hierarchies will necessarily be subjective.

Subjectivity is inherent in taxonomic-based ontologies because taxonomies are based on the class membership criteria chosen by the taxonomist. For instance, ecology has many different criteria for basing decisions on what relationships are important for observation, including organism, population, community, landscape, and biome (Allen & Hoekstra, 1992). Indeed, Fonseca, Martin, & Rodriguez (2002) noted that classifications of 'eco-ontologies' are inherently subjective because the components of a self-organized system are both means and ends for other components and for the whole. This removes the 'neutral ground' on which the semantics of realist geo-ontologies must be built. What is promising about affordance-based ontologies is also what limits them: they match the function of an object to the intention of a user, but not of many different kinds of users. Thus an affordance-based ontology of roads will be useful for road-drivers, but maybe not road-builders, and certainly not for wildlife managers who view roads as barriers, rather than conduits, to movement. While taxonomic-based approaches to ontologies can be very useful in some contexts, they do have limitations that restrict their validity in other cases.

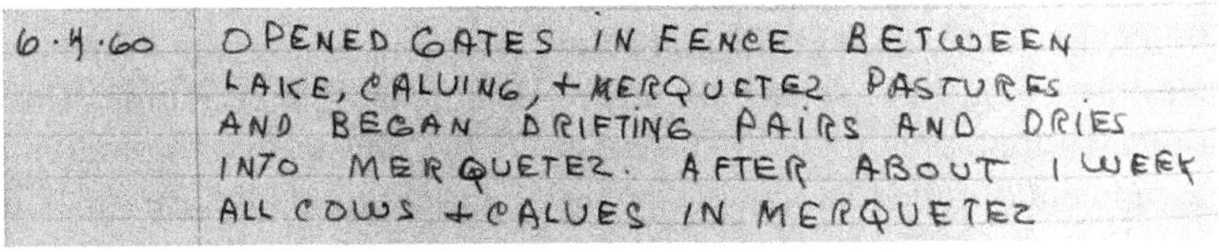
6·4·60 OPENED GATES IN FENCE BETWEEN
LAKE, CALVING, + MERQUETEZ PASTURES.
AND BEGAN DRIFTING PAIRS AND DRIES
INTO MERQUETEZ. AFTER ABOUT 1 WEEK
ALL COWS + CALVES IN MERQUETEZ

Fig. 1. Excerpt from the Santa Cruz Island Company Cattle Books

3 A Linguistic Perspective on Action, Agency, Purpose, and Spatio-temporal Change

3.1 Linguistic Representation of Action

This section demonstrates how the representation of spatial objects and relations may be expanded so as to include notions of *action*, *agency* and *purpose*. For this we draw on research in natural language and on occasion, philosophy. An entry from the Santa Cruz Island Company's Cattle Book[1] will provide an example throughout this discussion (Fig. 1).

We will consider four kinds of information expressed (or implied) in natural language:

1. Situations, constituted of things in relation to one another
2. Means, in particular actions and activities, for changing situations
3. Ends, or the reasons why means are applied to change situations
4. Things, their groupings and distinguishing characteristics

This sequence corresponds to the sequence of *constitutive*, *agentive*, *telic*, and *distinguishing* dimensions of things as discussed by Moravcsik (1975) following Aristotle, and as applied by Bibby & Shepherd (2000) to the modeling of land use in a GIS context. The constitutive dimension concerns the parts (material and immaterial) that something is made of; the agentive dimension concerns the animate or inanimate agent responsible for something happening; and the telic dimension addresses "the

[1] The text was written by Dr. Carey Stanton, president of the Santa Cruz Island Company (1954-1987), and was made available for this research courtesy of Lyndal Laughrin, Reserve Manager, Santa Cruz Island Reserve.

notion of a purpose that an agent might have in performing an act, and that of a built-in aim or function in terms of which we specify certain activities" (Moravcsik, 1975:627). Note that, as this quote suggests, agentive and telic factors (i.e., means and ends) may be nested within each other.

3.2 Constitutive Dimension

We first parse the text into syntactic constituents and then map these constituents to their semantic roles (Figure 2). These semantic roles are based on the notion that there are a limited number of abstract semantic relationships between a verb and its arguments (Allen, 1995:242). We then define *Situations* that consist of an *Activity* (a verb representing some 'happening') and the associated roles of *Figure, Ground, Agent,* and *Time.* In sentence 1 (S1), verbs identify two different happenings. In the first, the noun 'gates' complements the verb and the prepositional phrase 'in fence between Lake, Calving, Merquetez Pastures' gives a spatial reference. We can map these syntactic elements to their semantic roles: Activity (opened), Figure (gates), and two nested Grounds (in fence, between pastures). Similarly, the second verb 'drifting' (used in the text as a transitive verb) maps to an Activity, the noun phrase 'pairs and dries' maps to Figure, and the prepositional phrase 'towards Merquetez' maps to Ground. The Agent (the thing that caused the gate to open and caused the cattle to drift) is omitted. Sentence 2 (S2) omits both an Agent and an Activity. It consists of a prepositional phrase denoting Time ('after about a week'), a noun phrase denoting a Figure ('all cows and calves'), and a second prepositional phrase as Ground ('in Merquetez').

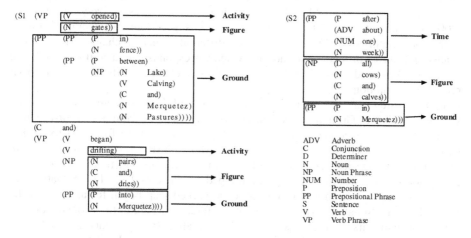

Fig. 2. Syntactic structure of text mapped to semantic roles

Linguists have linked prepositional phrases that ground the location of Figures to geometric idealizations (Herskovits, 1986; Talmy, 2000b) or image-schemata (Johnson, 1987; Lakoff, 1987). Here, we draw on these idealizations and introduce the notion of *forces* that underlie these locative expressions, as developed by Talmy (2000a), in particular, the *relative force* and the *intrinsic tendency* of projected-world entities. For instance, the phrase 'gates in fence' evokes the *container* image-schema, where the relative strength of the container (fence) is greater than that of the contained (gates) since the latter can 'leak' if opened. Further, animals in an enclosure, as in "cows and cattle in Merquetez", also brings up the *container* schema, and here too the relative strength of the contained (cattle) is less than that of the container (Merquetez). In addition to relative force, objects have internal tendencies towards either motion or rest. For instance, cattle have an internal capacity to be active, while gates and pastures are intrinsically passive.

3.3 Agentive Dimension

The interplay between a Figure's internal tendency and its force relative to the Ground's defines a set of sixteen potential Situations (Fig. 3). We represent the Figure with a circle and the Ground with a line. The first four (a-d) define elemental Situations, while the others (e-p) represent transitions. In the first four Situations (a-d), the Figure's internal tendency and its force relative to Ground both remain consistent: they are static in the sense that they do not change, though they may represent either a state-of-rest or state-of-motion. For instance, Situation (a) represents a Figure with an internal tendency towards motion that is greater than the force of Ground, and this represents a state-of-motion (e.g., 'the cattle walking on the road'). Situation (b), in contrast, represents a state-of-rest, where a Figure's internal tendency towards rest is greater than the force of Ground (e.g. 'the gate is hanging on the fence'). Situations (c) and (d) represent situations where a Ground's force is greater than the internal tendency of the Figure, and this causes the Figure to be in either a state of rest or motion that is contrary to its internal tendency (e.g. 'the cattle are in the enclosure' and 'the jeep rolled down the hill', respectively). In Talmy's (2000a) treatment, these static situations can change. Here, we modify Talmy's framework slightly in order to represent two different aspects of change. The first is a change in external propensity to motion, which corresponds with a shift in the relative force of figure and ground, and the second is a change in the figure's internal propensity to motion.

In four Situations (e-h), the internal tendency of Figure is invariant while its strength relative to Ground shifts. These represent changes from a state-of-rest to a state-of-motion or vice-versa. Situation (e) represents a change from rest to motion ('cattle were released from the corral'), while Situation (f) represents a change from motion to rest ('the jeep crashed into the ravine'). Situation (g) represents a change from motion to rest ('the sheep were trapped in the corral'), while Situation (h) represents a change from rest to motion ('the parked jeep began to lurch down the hill'). In four Situations (i-l), the relative force of Figure to Ground remains consistent, while its internal tendency shifts. Situation (i) represents a shift from a tendency towards motion to rest (e.g. 'the bull was tranquilized'). Situation (k) represents a shift in the internal tendency that results in a change from a state-of-rest to a state-of-motion ('the dead sheep tumbled down the hill'). Situations (j & l) could refer to inanimate objects that have an internal capacity to move, e.g., vehicles ('the broken-down jeep was repaired' and 'we pop-started the jeep on the hill'). Each of these shifts may be related to an Agent that affects change by altering the force dynamics between Figure and Ground.

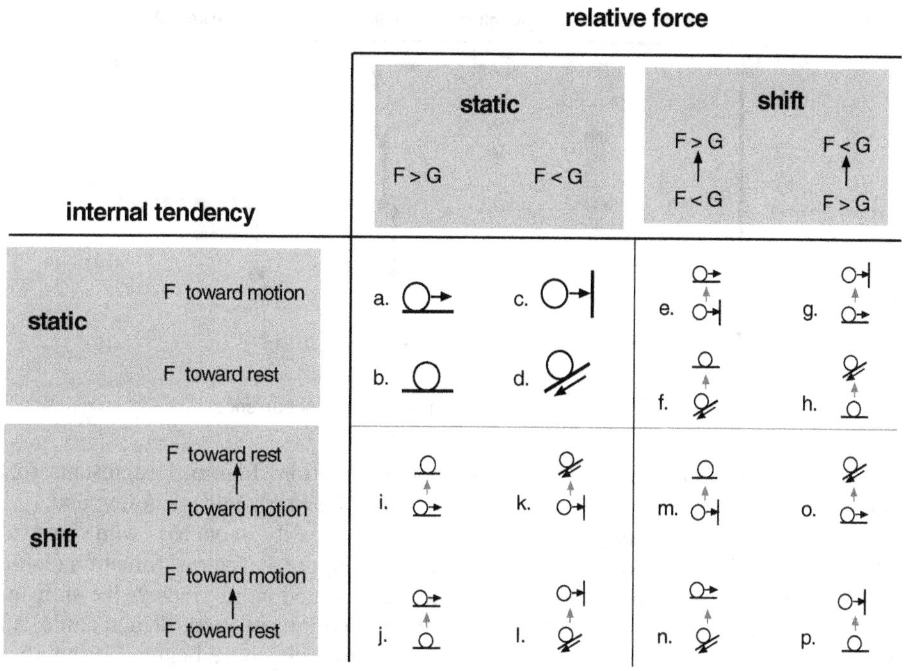

relative force

Fig. 3. Force dynamics, elemental concepts, and change

The four remaining situations (Fig. 2m-p) represent simultaneous shifts in both the Figure's internal tendency and its strength relative to Ground. While these can be logically derived from our framework, whether they will be commonly expressed in language remains unclear.

3.4 Telic Dimension

Lastly, we model telic relations, or the relations between means and ends. The telic relation may be implicit. In the text excerpt (Fig. 1), there is no explicit statement of the goal of drifting the cattle from one pasture to another. However, there is an implicit relation between the two verbs 'open' and ''drift' (Fig. 4). The first alters the force relations between cattle and the Gestalt of gate-fence-enclosure, 'releasing' the active figure. This allows cattle to 'drift', a situation where the internal tendency of the Figure is unencumbered by the Ground. The shift from the origin situation is not explicitly mentioned in the description, but is implicit in Situation (e) (Fig. 3). This is an example where language enables reasoning with incomplete information. Situation (e) represents a change from Situations (c) to (a). Thus opening gates is a means to a state-of-motion, where the cattle are drifting. Our example concludes with another state-of-rest, where cattle are contained by the Merquetez Gestalt, and this implies another force schema, that which 'restrains' (g) cattle movement, when the gates are again closed along the fence of the Merquetez Pasture. This implied Situation (g) is a means to this end. Thus each Situation can be represented as both a means to its

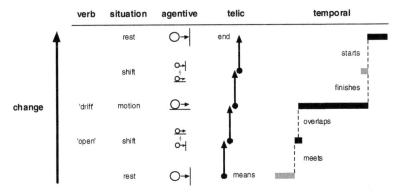

Fig. 4. Dimensions of the June 4, 1960 occurrence

subsequent Situation, and goal of its precedent Situation. Figure 4 represents this graphically, with closed circles denoting means and closed triangles denoting end.

The temporal relations of this telic chain are easily modeled with interval relationships following Allen (1983) and depicted in Figure 4. The rest Situation (cattle in Lake), omitted by the description (and hence represented in grey) *meets* the shift-in-balance 'opened gates'. This shift *overlaps* with the motion situation 'drifted cattle', in the sense that more than one gate was opened and cattle may have begun drifting after the first, but before the last, gate was opened. The shift-in-balance 'closing gates' *finishes* the motion situation and *starts* the final (in this occurrence) rest situation.

Following Allen (1984), Situations can also be distinguished into parallel classes depending on whether or not they are brought about by purposive Agents: when caused by such Agents, the situation is an Action, otherwise the Situation is an Event. Our example, however, does not explicitly identify Agents, and this reflects the author's criteria at a high level: it tracks the activities related to the management of livestock, and not the employees charged with acting out those tasks. But to develop this discussion, let us assume that a Vaquero was employed to open the gates. The subsequent Situation 'drifting cattle' appears at first glance somewhat ambiguous as to whether or not cattle move on their own accord, because it is not clear whether 'drift' is used here as a transitive or intransitive verb. Indeed, vaqueros do not cause cattle to move: the animals move themselves, as the 'drifting' reflects a situation where the internal tendency of the Figure is no longer constrained by the relative force of Ground. Yet the cattle do not just drift somewhere, but rather, the vaqueros direct their internal tendency towards a goal: Merquetez. In other words, telic relations to preceding and subsequent situations define drifting as part of a higher-level action.

The pairing of activities with outcomes develops the nested granularity of the telic dimension (Fig. 4). At a physical level, the verb 'drifting' means the physical activity of cattle moving themselves (Situation a of Figure 3), but the entire sequence of

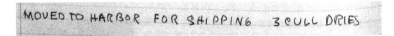

Fig. 5. Example of action associated with higher-level activity from text

activities (open, drift, close) can be generalized at a higher-level, where the verb 'drifting' encapsulates all of these physical activities. At this level, 'drifting' can describe a situation where cattle are passive and move due to the stronger relative force of Ground (Situation d of Figure 3). Indeed, many descriptions in the Cattle Books characterize activities, not at the level of physical actions, but at a more abstract, goal-oriented level. In the example shown in Figure 5, we can model the first situation as the Activity 'moved', the Figure 'cull dries', and the Ground 'to harbor'. As indicated by the preposition 'for', 'moved' has a telic link to the Activity 'shipping'. At this level, the verbs 'moved' and 'shipping' both encapsulate a series of physical actions, such as opening and closing gates, driving cattle on a pier, loading them on a schooner, etc. These actions, encapsulated by the activity 'move', are a means for the activity 'shipping'.

3.5 Distinguishing Dimension

The categories of things mentioned in the text are a reflection of the criteria by which the author selects the aspects of the world that are relevant to his or her interests at that time. This is where the duality between the 'projected-world' and 'real-world' originates. In our example (Fig. 1), 'gates', 'fence', and 'Merquetez' are projected-world things that can be mapped to measurements of real-world objects (gates and fence), their constituent wholes (enclosures), and their named identity (Merquetez). The terms 'pairs', 'dries', 'cows' and 'calves' are also projected-world things that stand for 'real-world' things: the living and breathing creatures that existed on Santa Cruz Island on June 4, 1960. These terms relate to the category 'cattle', and each term attends to different aspects of this class, such as sex, kinship, fecundity, and age. In our example, the terms distinguish aspects of fecundity and kinship ('pairs' and 'dries') that are at a finer granularity than the sex and age of the animals ('cows' and 'calves'). This suggests some implicit information for the June 4 activity; for instance, that calves have not yet been weaned.

There are fundamental aspects of the distinguishing dimension that this paper does not address. These include: (1) the appearance and disappearance of types and tokens, such as the birth and death of animals or the introduction of a new technology, (2) a change in the type of tokens, such as from calves to cows, (3) a change in the quantity of tokens, such as in the number of animals moved from pasture to pasture. These are examples of generation/corruption, qualitative, and quantitative change that a complete spatio-temporal ontology must include (Couclelis, 1998), whether in the real-world or the projected-world or both. Here, we limited the discussion to the aspects of objects that can be distinguished through the windows of attention framed by the author of a specific historical text.

4 Discussion

This research proposes a tentative, linguistics-based ontology of action and change, and points towards a novel way of modeling land use change in GIS. Here we summarize the ontology underlying our model, leaving its formalization for a separate paper. A fundamental aspect of this ontology is the adoption of Jackendoff's (1985) distinction between the 'projected-world' of language and the 'real-world' that the language is about, and the use of the former to help model the latter. The ontology also distinguishes events and actions, rejoining previous work in AI on actions and change (Allen, 1984) that itself draws largely on the philosophy of language (Mourelatos, 1978).

Our approach distinguishes between different aspects of causation, particularly with respect to motion. In the elemental force relations (Fig. 2), a Figure either realizes its internal tendency towards rest or motion, or the greater relative force of Ground causes the Figure to rest or move. This distinguishes two basic types of causation, where the rest or motion of the Figure occurs (1) 'despite' or (2) 'because (of)' the relative force of Ground (Talmy, 2000a). We have extended this framework to include a particular kind of qualitative change, where an Agent alters either the Figure's internal tendency toward movement or rest, or the force of the Figure relative to Ground (Fig. 2i-l).

In this framework, telic relations between occurrences derive from the logical structure of the underlying schematic, or causative, elements (Fig. 2). While our approach is compatible with the lexical entailment used in affordance-based ontologies (Kuhn, 2001), our focus on the deeper structure of language can provide a logical framework to distinguish different domains of occurrences, such as the physical domain where cattle move themselves and the projected-world of land use where cattle passively drift. The logical framework may also improve temporal reasoning with incomplete knowledge, as illustrated in the representation of two unstated situations in our example. Additionally, this approach may help avoid the 'historical fallacy' of interpreting historical text with the modern meanings of terms (Golledge, personal communication).

Our focus on the granularity of occurrences, the nesting of activities as the means and ends of actions, also contributes towards the development of partonomies and taxonomies of actions. Each description in the text represents one level of granularity that is constitutive of activities and actions at a finer level, while itself a part of higher-level occurrences, such as 'shipping' relative to 'moved' (Fig. 5). While that granularity reflects the text author's subjective attention to particular aspects of the real-world, it also reflects the real-world's objective conditions at that time. For instance, activities such as breeding, moving animals between pastures, and shipping reflect the constraints and opportunities presented by animal physiology, precipitation patterns, topography, market conditions, etc. The temporal and spatial patterns of the author's attention to occurrences can thus lead to the reconstruction of unstated real-world occurrences at the time and place where the text was written.

By now, we have come full circle. Even though our linguistic framework builds on projected-world entities and situations, the properties of their real-world counterparts are never out of sight. For example, early on, we made an *a priori* assumption that cattle have an internal tendency towards motion, while gates, fences, and enclosures have internal tendencies toward rest. This realist distinction between organic and inorganic things is so intuitive that it may have passed unnoticed. This is the way things should be. Our objective here was not the radical translation of language (Quine, 1960), but rather, to find how we might translate descriptions of action and change that are encoded in natural language into cartographic- and GIS-based models. The goal is to improve what we can ask and understand about action and change in the real-world by analyzing what language has to say about the projected-world. Our conceptual framework rests on this duality. Also, we have suggested how, where purposive human action is concerned, seemingly insignificant occurrences can add up to spatio-temporally significant outcomes. Opening gates and letting cattle through may seem a long way from the kind of activity that changes the human use of the land over the broad scales of space and time commonly represented in GIS. Still, once we are able to connect occurrences with specific relations of means to ends, we can move up the chain of activities and actions that eventually, over longer time periods, lead to visible macro-traces on the land.

5 Summary

In this paper we outlined a linguistics-based framework for the representation of purposive action in space and time. Key elements of the framework are Jackendoff's distinction between a real-world and a projected-world of language, Talmy's theory of force dynamics, and the Aristotelian notion of agentive and telic dimensions in things. The framework suggests a tentative ontology of occurrences, activities, and actions that may contribute to more efficient modeling of change in GIS, in particular change intended to meet a human purpose, as in the case of land use. Drawing on text describing a historical cattle-ranching operation, we distinguish sixteen elementary linguistic force relations between projected-world 'figure' and 'ground', and associate these with agentive and telic perspectives. We also show how very fine-grained purposive actions, such as opening cattle gates, may be used as building blocks to help infer purposes and spatio-temporal changes at much higher granularity. The framework, meant to derive as much information as possible directly from source textual material, presents an alternative to certain recently developed spatio-temporal ontologies that either model occurrences as objects, or propose conceptual taxonomies based on class membership criteria imposed by the taxonomist.

Acknowledgements

This material is based upon work supported under a National Science Foundation Graduate Research Fellowship and a Mildred E. Mathias Graduate Student Research Grant from the University of California Natural Reserve System. A student travel scholarship from Intergraph Mapping and Geospatial Solutions supported travel to GIScience2004 for the presentation of an earlier version of this work. The authors are also grateful to the three anonymous reviewers for their constructive comments.

References

Allen, J. (1995). *Natural language understanding* (2nd ed.). Redwood City, Calif.: Benjamin/Cummings Pub. Co.

Allen, J. F. (1983). Maintaining Knowledge About Temporal Intervals. *Communications of the ACM, 26*(11), 832-843.

Allen, J. F. (1984). Towards a General-Theory of Action and Time. *Artificial Intelligence, 23*(2), 123-154.

Allen, T. F. H., & Hoekstra, T. W. (1992). *Toward a unified ecology*. New York: Columbia University Press.

Bibby, P., & Shepherd, J. (2000). GIS, land use, and representation. *Environment and Planning B-Planning & Design, 27*(4), 583-598.

Carnap, R. (1969). *The logical structure of the world*. Berkeley: University of California Press.

Cheylan, J.-P., & Lardon, S. (1993). Towards a conceptual data model for the analysis of spatio-temporal processes: the example of the search for optimal grazing strategies. In I. Campari (Ed.), *Spatial information theory : a theoretical basis for GIS : European conference, COSIT'93, Marciana Marina, Elba Island, Italy, September 19-22, 1993 : proceedings* (Vol. 716, pp. 158-176). Berlin ; New York: Springer-Verlag.

Chrisman, N. R. (1998). Beyond the Snapshot: Changing the Approach to Change, Error, and Process. In R. G. Golledge (Ed.), *Spatial and temporal reasoning in geographic information systems* (pp. 85-93). New York: Oxford University Press.

Claramunt, C., & Theriault, M. (1995). Managing Time in GIS: An Event-Oriented Approach. In A. Tuzhilin (Ed.), *Recent Advances in Temporal Databases: Proceedings of the International Workshop on Recent Advances in Temporal Databases, Zurich, 17-18 September 1995* (pp. 23-42). Berlin: Springer-Verlag.

Claramunt, C., & Theriault, M. (1996). Towards Semantics for Modelling Spatio-Temporal Processes with GIS. In M. Molenaar (Ed.), *Advances in GIS Research II: Proceedings 7th International Symposium on Spatial Data Handling* (pp. 27-43): Taylor and Francis.

Couclelis, H. (1998). Aristotelian Spatial Dynamics in the Age of Geographic Information Systems. In R. G. Golledge (Ed.), *Spatial and temporal reasoning in geographic information systems* (pp. 109-118). New York: Oxford University Press.

Eddington, A. S. S. (1939). *The philosophy of physical science*. New York: The Macmillan Co. Cambridge Eng. The University Press.

Egenhofer, M. J., & Mark, D. M. (1995). Naive geography, *Spatial Information Theory* (Vol. 988, pp. 1-15).

Fellbaum, C. (1999). *WordNet: an electronic lexical database* [electronic resource]. Cambridge Mass: MIT Press.

Fonseca, F., Martin, J., & Rodriguez, M. A. (2002). From Geo- to Eco-ontologies. In D. M. Mark (Ed.), *Geographic Information Science: Second International Conference, GIScience 2002, Boulder, CO, USA, September 25-28, 2002, Proceedings* (Vol. 2478, pp. 93-107). Berlin: Springer.

Galton, A. (2003). Desiderata for a Spatio-temporal Geo-ontology. In S. Timpf (Ed.), *Spatial information theory : foundations of geographic information science : international conference, COSIT 2003, Ittingen, Switzerland, September 2003 : proceedings* (pp. 1-12). Berlin; New York: Springer.

Gibson, J. J. (1979). *The ecological approach to visual perception*. Boston: Houghton Mifflin.

Grenon, P., & Smith, B. (2004). SNAP and SPAN: Towards Dynamic Spatial Ontology. *Journal of Spatial Cognition and Computation, 4*(1), 69-103.

Herskovits, A. (1986). *Language and spatial cognition : an interdisciplinary study of the prepositions in English*. Cambridge: New York.

Hornsby, K., & Egenhofer, M. J. (2000). Identity-based change: a foundation for spatio-temporal knowledge representation. *International Journal of Geographical Information Science, 14*(3), 207-224.

Howarth, J. T., & Couclelis, H. (2004, October 20-23, 2004). *Towards a Linguistic Model for a Historical GIS of Land Use Change*. Paper presented at the GIScience 2004 Extended Abstracts and Poster Summaries, Aldephi, Maryland, U.S.A.

Jackendoff, R. S. (1985). *Semantics and cognition* (1st paperback ed.). Cambridge, Mass.: MIT Press.

Johnson, M. (1987). *The body in the mind : the bodily basis of meaning, imagination, and reason*. Chicago: University of Chicago Press.

Kuhn, W. (2001). Ontologies in support of activities in geographical space. *International Journal of Geographical Information Science, 15*(7), 613-631.

Lakoff, G. (1987). *Women, fire, and dangerous things : what categories reveal about the mind.* Chicago: University of Chicago Press.

Lakoff, G., & Johnson, M. (1980). *Metaphors we live by.* Chicago: University of Chicago Press.

Langran, G. (1992). *Time in geographic information systems.* London ; New York: Taylor and Francis.

Mark, D. M. (1988). *Cognitive and linguistic aspects of geographic space : report of a workshop.* Santa Barbara, CA: National Center for Geographic Information and Analysis.

Mark, D. M., Egenhofer, M. J., & Hornsby, K. (1997). *Formal Models of Commonsense Geographic Worlds: Report on the Specialist Meeting of Research Initiative 21* (97-2). Santa Barbara, California: National Center for Geographic Information and Analysis.

Mark, D. M., & Frank, A. U. (1991). *Cognitive and linguistic aspects of geographic space.* Dordrecht: Boston.

McGranaghan, M. (1991). Matching Representations of Geographic Locations. In A. U. Frank (Ed.), *Cognitive and linguistic aspects of geographic space* (pp. 387-402). Dordrecht: Boston.

Moravcsik, J. M. (1975). AITIA as generative factor in Aristotle's Philosophy. *Dialogue, 16*(4), 633-638.

Mourelatos, A. P. D. (1978). Events, processes, and states. *Linguistics and Philosophy, 2,* 415-434.

Quine, W. V. (1960). *Word and object.* Cambridge Technology Press of the Massachusetts Institute of Technology.

Smith, B. (1997). Barry Smith's Perspective. In K. Hornsby (Ed.), *Formal Models of Commonsense Geographic Worlds: Report on the Specialist Meeting of Research Initiative 21* (pp. 48). Santa Barbara, California: National Center for Geographic Information and Analysis.

Soon, K., & Kuhn, W. (2004). Formalizing user actions for ontologies, *Geographic Information Science, Proceedings* (Vol. 3234, pp. 299-312).

Talmy, L. (1983). How Language Structures Space. In L. P. Acredolo (Ed.), *Spatial orientation* (pp. 225-282). New York: Plenum Press.

Talmy, L. (2000a). Force Dynamics in Language and Cognition, *Toward a cognitive semantics. Volume I, Concept structuring systems* (pp. 409-470). Cambridge, Mass.: MIT Press.

Talmy, L. (2000b). How Language Structures Space, *Toward a cognitive semantics. Volume I, Concept structuring systems* (pp. 177-254). Cambridge, Mass.: MIT Press.

Tversky, B., & Hemenway, K. (1984). Objects, Parts, and Categories. *Journal of Experimental Psychology-General, 113*(2), 169-193.

Wittgenstein, L. (1953). *Philosophical investigations.* New York: Macmillan.

Worboys, M., & Hornsby, K. (2004). From Objects to Events: GEM, the Geospatial Event Model. In H. J. Miller (Ed.), *GIScience 2004* (Vol. 3234, pp. 327-343). Berlin Heidelberg: Springer-Verlag.

Zacks, J. M., & Tversky, B. (2001). Event structure in perception and conception. *Psychological Bulletin, 127*(1), 3-21.

Zacks, J. M., Tversky, B., & Iyer, G. (2001). Perceiving, remembering, and communicating structure in events. *Journal of Experimental Psychology: General, 130*(1), 29-58.

Ordering Events for Dynamic Geospatial Domains

Suzannah Hall[1] and Kathleen Hornsby[2]

[1] Maine Department of Transportation,
16 State House Station, Augusta, ME 04333, USA
[2] Department of Spatial Information Science and Engineering,
National Center for Geographic Information and Analysis,
University of Maine, Orono, ME 04469-5711, USA
suzannah.hall@maine.gov
khornsby@spatial.maine.edu

Abstract. Complex scenarios of events occurring in dynamic geospatial domains can be simplified or summarized in the form of linear orders. Sets of possible linear orders of events can be automatically generated based on the temporal relations that hold between events and refined further according to the particular semantics associated with the temporal relations. This paper discusses how the spatial aspects of events contribute to generating plausible linear orders of events. Locations of events, for example, are useful for determining the spatial relevance of events in an order. In addition, we explore how the presence of certain patterns of event locations can be used for further filtering, this time pruning improbable orders from the set of possible orders. The result is a methodology for automatically generating linear orders of events from partial orders that exploits the spatial and temporal relationships associated with events occurring in geospatial domains.

1 Introduction

Important aspects of geospatial domains include the dynamic actions and activities that occur over time and space. There has been considerable interest in modeling these actions and activities as *events* for information system design, for example, extracting events for text processing [1-2], as a basis for developing reasoning and planning systems [3-4], and for modeling dynamic geospatial domains [5-10]. Complementary to the research on events in geospatial domains, is research by psychologists on the perception of events, in particular how people segment activities into events [11-13]. In this work, we assume Galton's [14] definition of event that refers to a discrete element or individual of change that can be distinguished from all other elements of change. This definition of event is differentiated from the concept of a *state*, which takes place over a period of time but does not result in any specific change. Computational tools are needed to represent, manage, and analyze events. This paper focuses on an approach for the abstraction of events, in particular the development of automated methods for generating simplified linear orders of events from partial orders. These methods exploit the spatio-temporal relationships and semantics associated with events that occur in a geospatial domain.

A.G. Cohn and D.M. Mark (Eds.): COSIT 2005, LNCS 3693, pp. 330–346, 2005.
© Springer-Verlag Berlin Heidelberg 2005

Research on modeling the dynamic aspects of geospatial domains has highlighted how a solely *object*-oriented perspective misses the contribution of dynamic aspects by modeling happenings or events implicitly, often in a static snapshot style [10]. Recently, Grenon and Smith [8] have presented a formal ontology for describing geospatial objects and geospatial processes or events. Their work on SNAP and SPAN ontologies treat continuants and occurrents respectively, presenting a framework that handles both kinds of elements and allows for a representation of the relations that can exist between these ontologies. The Geospatial Event Model (GEM) [9], presents a formal model for reasoning about objects, events, and settings (extents in space and time) in geospatial domains.

Information systems need to manage not only large numbers of events but also the relations that exist between events. Events that occur in geospatial domains, such as weather or traffic events, often assume complex patterns. Events may be simultaneous, may occur directly after each other, may overlap or be related according to any of the possible thirteen temporal interval relations [15]. Research by cognitive scientists has shown the cognitive adequacy of Allen's temporal logic based on interval relations (see, for example, [16]). For many reasoning tasks, users require a simple, *linear order* of events, where for every pair of events, A and B, either A *is before* B or B *is before* A, or both. The case where A *is before* B and B *is before* A describes the case where A occurs *at the same time as* B. This linear ordering helps us to understand and communicate in a simpler fashion how events occur over time. The automatic generation of linear orders of events and the implications of the ordering procedure, involve a number of open research questions such as:

- How are subsequent event intervals placed in an order?
- Do the linear orders generated retain the semantics of the original event scenario?

When the relations in an event database are collapsed to a set containing only *before* and *equals* relations, the result is an abstraction of the events, presented as a linear order. The results of the ordering process bring their own questions, for example:

- Are all orders equally plausible? Which orders should be presented to a user?
- How can information about event locations be used for ordering events?

This paper focuses in particular on these last two questions and explores the use of spatial information relating to events for these kinds of abstractions. The remainder of the paper is structured as follows: Section 2 presents an approach for generating linear orders of events based on a set of events and the temporal relations that hold between them. Section 3 introduces how event locations can be used to further improve the generated linear orders. Event locations must be at the same granularity or translated to the same granularity in order to determine whether any event locations can be considered as spatial outliers, and therefore candidates for removal from an order. Hierarchical clustering is the technique applied in section 4 to identify whether any spatially irrelevant events exist in a set of events, while Section 5 describes the role of patterns of locations of events and shows how the identification of such patterns can be used to further refine the generated orders of events. Conclusions and future work are presented in Section 6.

2 Ordering Events Based on Temporal Information

In this paper, we use as a prototypical example, a set of nine events relating to a morning commute to work. The nine events include: Susan leaves home, Susan joins a carpool, a vehicle passes a truck, Susan's vehicle slows for construction, Susan's vehicle stops to wait for a school bus, Susan's vehicle follows a school bus, Susan's vehicle navigates the traffic circle, the bus turns off the road, and Susan arrives at work. We assign each of the events a shorter name for more compact referencing as follows: *LeaveHome, JoinCarpool, PassTruck, SlowsForConstruction, WaitForSchoolBus, FollowBus, NavigateCircle, BusTurnsOff,* and *ArriveAtWork*. In this work, we assume that events occurring in a dynamic geospatial domain are modeled in the form of *event-relation combinations*, that is, two events are related to each other by one of Allen's [15] thirteen temporal interval relations. An example of an event-relation combination is *LeaveHome before JoinCarpool*. A scenario of dynamic occurrences, therefore, can be modeled as a set of event-relation combinations. The example scenario of events is described by the following set of event-relation combinations:

LeaveHome *before* JoinCarpool
PassTruck *overlaps* SlowsForConstruction
WaitForSchoolBus *during* SlowsForConstruction
WaitForSchoolBus *meets* FollowBus
FollowBus *equals* NavigateCircle
BusTurnsOff *ends* FollowBus
NavigateCircle *before* ArriveAtWork

A set of event-relation combinations captures the variety of temporal relationships that exist between the events in a geospatial domain. Reasoning about events, however, commonly requires assembling the events into a sequence or *order of events* such that a temporal pattern of events becomes distinguishable [17] and therefore more understandable. A linear order of events describes the case where for every pair of events, *A* and *B*, either *A is before B* or *B is before A*, or both. Linear orders are common, for example, in the case of real-time sensor data where a single stream of data is produced [18]. In this work, linear sequencing is applied as an abstraction tool for understanding and communicating in a simpler fashion how events occur over time. In order to summarize the events in a dynamic domain that may involve any of the thirteen temporal interval relations (*equals, before, meets, starts, ends, overlaps, during,* and their converses), therefore, the set of event-relation combinations is translated to a set involving only *before* and *equals*.

2.1 Mapping Event Relations to *before* and *equals*

To translate from the original set of event-relation combinations where any relation may hold, to a set where only *before* and *equals* hold, a set of mapping rules is applied. These rules are based on the relationship between the start points and end points of two events related by any of the thirteen relations. To distinguish the translated relations from the relations in the original set of event-relation combinations, the relations in the linear description are denoted as *l_before* and *l_equals*. Given A *R* B, where A and B are events related by the temporal relation *R*, if the start point of A precedes the start point of B, then the relation maps to A *l_before* B. If the start point of B precedes the start point of A, then the relation maps to B *l_before* A. In cases where the start points of A and B are simultaneous, the end points of the two events are considered instead, and if the end point of A precedes the end point of B, then the relation maps to A *l_before* B, while if the end point of B precedes the end point of A, the relation maps to B *l_before* A. Where the start points and end points of A and B are both simultaneous, the relation between A and B is *equals*. The *equals* relationship is preserved in this translation of event relations, and in this case, A *R* B maps to A *l_equals* B.

Given the relations that hold for events in the set of event-relation combinations, the application of these mapping rules to the prototype example yields a revised set of event-relation combinations:

LeaveHome *l_before* JoinCarpool
PassTruck *l_before* SlowsForConstruction
SlowsForConstruction *l_before* WaitForSchoolBus
WaitForSchoolBus *l_before* FollowBus
FollowBus *l_equals* NavigateCircle
FollowBus *l_before* BusTurnsOff
NavigateCircle *l_before* ArriveAtWork

It is relevant to underscore the point that our goal with these mapping rules is to generate a linear order or sets of orders that have as many of the original events as possible occupying some unique position in an order. It might be possible to consider alternative bases for mapping, for example, the conceptual neighborhood graph for temporal interval relations [19]. On the basis of conceptual neighbors, however, *during* relations would map to their conceptual neighbor, *equals*, and in scenarios where there are multiple *during* relations (e.g., A *during* B, C *during* B, and E *during* B), the resulting orders would show all these events as being simultaneous with each other (i.e., A=B=C=E) as opposed to holding (possibly) different and unique positions in an order.

2.2 Generating Linear Orders by Applying a Topological Sort

The revised set of event-relation combinations provides a foundation for expressing the events in a linear order. However, it is seldom the case that the application of the linear order mapping rules will yield a single linear order. This is due to the fact there are almost always uncertainties or unknowns in the ordering (e.g., it is not known for certain whether event JoinCarpool is *before* PassTruck or *after* PassTruck). For these cases where partial orders exist, it is possible to generate *a set of possible orders* of the given events, each of which reflects a possible linear ordering between events. The set of possible orders is derived automatically using a topological sorting algorithm. In general, given a set of ordered pairs of elements where the pair *(A,B)* implies that element *A* precedes element *B*, a topological sorting algorithm returns the set of all possible orders of elements such that the relations between elements in the set of ordered pairs are not violated [20]. The translated set of event-relation combinations is used as the basis for the topological sorting algorithm.

The topological sort takes as input a valid topological sort, i.e., an initial order of events. This initial order is generated by evaluating the set of event-relation combinations and identifying any event (or set of events) *A* for which there are no relations of the form *B before A*. That is, there is no event or set of events *B* that explicitly precedes the event or set of events *A*. *A* is placed at the head of a list *S*. The event-relation combinations that do not involve *A* are evaluated in the same way, and the event(s) that are not explicitly preceded by any event or set of events are appended onto *S*. This process is repeated until all events represented by the set of event-relation combinations are in the list *S*. For the example set of event-relation combinations, the list *S* is *LeaveHome, PassTruck, JoinCarpool, SlowsForConstruction, WaitForSchoolBus, (FollowBus=NavigateCircle), BusTurnsOff, ArriveAtWork*. Note that in this list (and all orders generated from it) the equality relation between FollowBus and NavigateCircle is reflected by these two events being enclosed in parentheses within the list. Equality relations between events are preserved in the linear ordering process, and any events that are *equal* to each other occupy the same position in a linear order.

The result of running the topological sorting algorithm on a translated set of event-relation combinations and the initial order of events is a set *O*, containing all possible linear orders of the events in the set, given the temporal information available. For our example, the set *O* contains 56 possible orders of the nine events. A subset of these are shown below:

O=

{*1   LeaveHome   PassTruck   JoinCarpool   SlowsForConstruction   WaitForSchoolBus (FollowBus=NavigateCircle) BusTurnsOff ArriveAtWork*

2   LeaveHome   PassTruck   JoinCarpool   SlowsForConstruction   WaitForSchoolBus (FollowBus=NavigateCircle) ArriveAtWork BusTurnsOff

3   LeaveHome   PassTruck   SlowsForConstruction   JoinCarpool   WaitForSchoolBus (FollowBus=NavigateCircle) BusTurnsOff ArriveAtWork

4   LeaveHome   PassTruck   SlowsForConstruction   JoinCarpool   WaitForSchoolBus (FollowBus=NavigateCircle) ArriveAtWork BusTurnsOff

5   LeaveHome   PassTruck   SlowsForConstruction   WaitForSchoolBus   JoinCarpool (FollowBus=NavigateCircle) BusTurnsOff ArriveAtWork

6   LeaveHome   PassTruck   SlowsForConstruction   WaitForSchoolBus   JoinCarpool (FollowBus=NavigateCircle) ArriveAtWork BusTurnsOff

7 LeaveHome PassTruck SlowsForConstruction WaitForSchoolBus (FollowBus=NavigateCircle) JoinCarpool BusTurnsOff ArriveAtWork

...

54 PassTruck SlowsForConstruction WaitForSchoolBus (FollowBus=NavigateCircle) BusTurnsOff LeaveHome ArriveAtWork JoinCarpool

55 PassTruck SlowsForConstruction WaitForSchoolBus (FollowBus=NavigateCircle) BusTurnsOff ArriveAtWork LeaveHome JoinCarpool

56 PassTruck SlowsForConstruction WaitForSchoolBus (FollowBus=NavigateCircle) ArriveAtWork BusTurnsOff LeaveHome JoinCarpool }

2.3 Plausibility of Event Orders

Although the results of applying a topological sort is a set of orders where the explicit temporal relationships that hold for the event-relation combinations are preserved, some orders are more cognitively plausible than others. For example, in the original set of event-relation combinations, WaitForSchoolBus *meets* FollowBus. Based on the relation *meets*, we would expect FollowBus to be the immediate successor of

WaitForSchoolBus in a resulting linear order. Not all orders, however, meet this expectation. For example, consider the following two orders:

{ *PassTruck   SlowsForConstruction   WaitForSchoolBus   LeaveHome   (FollowBus=NavigateCircle)*
ArriveAtWork JoinCarpool BusTurnsOff
PassTruck   SlowsForConstruction   WaitForSchoolBus   LeaveHome   (FollowBus=NavigateCircle)
BusTurnsOff JoinCarpool ArriveAtWork }

In both of these orders, WaitForSchoolBus and FollowBus are separated by the event LeaveHome. It is also known from the original set of event-relation combinations that WaitForSchoolBus takes place *during* SlowForConstruction. We would thus expect SlowForConstruction to be the immediate successor of WaitForSchoolBus. Other orders, however, show the event LeaveHome as occurring between SlowForConstruction and WaitForSchoolBus. For example,

{ *PassTruck   SlowsForConstruction   LeaveHome   WaitForSchoolBus   (FollowBus=NavigateCircle)*
ArriveAtWork JoinCarpool BusTurnsOff
PassTruck   SlowsForConstruction   LeaveHome   WaitForSchoolBus   (FollowBus=NavigateCircle)
BusTurnsOff JoinCarpool ArriveAtWork }

Given the variability in orders in the set *O*, therefore, and the potentially large number of possible orders generated by the topological sort, it is important to have methods available for the meaningful refinement of the set, where a smaller set of plausible orders is returned to a user [2]. The plausibility of a linear order depends largely on the degree to which it maintains the semantics of the original temporal relations present in the set of event-relation combinations. A method of filtering orders in which the semantics of the original temporal interval relations are not well preserved, has the benefit of returning to a user only those orders that are plausible according to certain rules.

Semantics of the temporal relations are used in the development of a set of constraints that, when applied to the set of possible orders, eliminate orders that contradict the semantics of the original temporal relations [2]. For example, a constraint is applied such that for any two events A and B, where A *during* B, no intermediate events may come between A and B in a linear order (i.e., B must be the immediate successor to A in the order). Any orders in *O* in which there are intermediate events between two events related by *during* are eliminated from *O*. The result is a new set of plausible orders, O_1, which is a subset of the initial set *O*. For the full set of possible temporal semantic constraints as well as a more detailed discussion of the derivation of these constraints, we refer readers to Hornsby and Hall [2].

The set O_1 that results when semantic constraints are applied to the set *O* for the morning commute example contains only 10 orders:

$O_1 =$
{ *1   LeaveHome   PassTruck   JoinCarpool   SlowsForConstruction   WaitForSchoolBus,*
(FollowBus=NavigateCircle) BusTurnsOff ArriveAtWork
2   LeaveHome   PassTruck   SlowsForConstruction   WaitForSchoolBus   (FollowBus=NavigateCircle)
BusTurnsOff JoinCarpool ArriveAtWork
3   LeaveHome   PassTruck   SlowsForConstruction   WaitForSchoolBus   (FollowBus=NavigateCircle)
BusTurnsOff ArriveAtWork JoinCarpool
4   LeaveHome   JoinCarpool   PassTruck   SlowsForConstruction   WaitForSchoolBus
(FollowBus=NavigateCircle) BusTurnsOff ArriveAtWork

5   *PassTruck   LeaveHome   JoinCarpool   SlowsForConstruction   WaitForSchoolBus* (*FollowBus=NavigateCircle*) *BusTurnsOff ArriveAtWork*
6   *PassTruck   LeaveHome   SlowsForConstruction   WaitForSchoolBus   (FollowBus=NavigateCircle)* *BusTurnsOff JoinCarpool ArriveAtWork*
7   *PassTruck   LeaveHome   SlowsForConstruction   WaitForSchoolBus   (FollowBus=NavigateCircle)* *BusTurnsOff ArriveAtWork JoinCarpool*
8   *PassTruck   SlowsForConstruction   WaitForSchoolBus   (FollowBus=NavigateCircle)   BusTurnsOff* *LeaveHome JoinCarpool ArriveAtWork*
9   *PassTruck   SlowsForConstruction   WaitForSchoolBus   (FollowBus=NavigateCircle)   BusTurnsOff* *LeaveHome ArriveAtWork JoinCarpool*
10   *PassTruck   SlowsForConstruction   WaitForSchoolBus   (FollowBus=NavigateCircle)   BusTurnsOff* *ArriveAtWork LeaveHome JoinCarpool* }

This set of orders illustrates how even after all the temporal relationships known from the set of event-relation combinations have been accounted for and temporal semantic constraints have been applied, there is still variation between orders, for example, some of the orders begin with a different event. This indicates that there remains an opportunity for further review of orders in O_1 with respect to plausibility.

In addition to the step of refining orders based on temporal relation semantics, therefore, additional methods of refinement may also be applied. Up to this point, only temporal relationships between events have been considered for ordering and order refinement. In some cases, however, there will be spatial information available about the occurrence of events. Spatial distributions or patterns of events, for example, may hold and these can be exploited to increase further the plausibility of orders returned for consideration by a user. The following section discusses how an approach for automatically generating linear orders of events is extended to include spatial aspects relating to events.

3   Exploiting Spatial Information About Events for Ordering

The set of event-relation combinations may include information about the spatial location of events. For example, "LeaveHome [Unity] *before* JoinCarPool [Albion]." In order to incorporate event locations in the process of generating linear orders of events, all locations in a set of event-relation combinations must first be translated to the same level of detail or *granularity*. In this work, we assume that event locations are abstracted to one of three possible categories: street address, landmark, or city (including towns). We define a street address as consisting of a building number, street or road information, and city. This represents the finest granularity out of the three categories in our model, city being the coarsest abstraction. A landmark refers to a prominent or key feature that is used for orienting or navigating through space. For example, a park, a city's downtown area, or a geographic feature such as a mountain are commonly treated as landmarks in space. In this work, landmarks are modeled at a granularity that is coarser than a street address, but not as coarse as a city (Figure 1), and are defined as consisting of a landmark feature and a city. For spatial modeling tasks, locations that correspond to any of these three granularity classes are commonly modeled as point data.

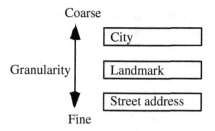

Fig. 1. Hierarchy of event location granularities

In a set of event-relation combinations, all event location granularities are coarsened to the *most common location granularity* in the set of event-relation combinations, that is, the *modal* granularity. The mode of a set of data refers to the most frequently occurring value within the set. Every event location in a set falls into one of the three potential categories with respect to granularity: street address, landmark, and city. Each of these granularities is assigned a variable: *street_address*, *landmark*, and *city*, respectively. The value of each of these variables is the number of occurrences of that granularity within the set. To determine the corresponding values for each of the n events in the set of event-relation combinations, the event location granularity is determined and the number of the appropriate granularity level variable is incremented.

The modal granularity is the level whose variable has the highest value at the end of this process. In cases where there is more than one modal granularity (i.e., multiple granularities are equally represented), event locations are translated to the coarsest granularity represented in the scenario. When the modal granularity has been determined, all event locations at a finer granularity than the mode are coarsened to match the modal granularity. Location granularities coarser than the modal granularity are not refined to the modal granularity, as the information necessary for this operation is unavailable. For example, a location given at the city level cannot be refined to the street address level, because the more detailed street address is not known. Consider the example scenario with a location associated with each event:

LeaveHome (116 Quaker Hill Rd, Unity) *before* JoinCarpool (Albion)
PassTruck (Bangor) *overlaps* SlowsForConstruction (Willow Beach, China)
WaitForSchoolBus (South China) *during* SlowsForConstruction (Willow Beach, China)
WaitForSchoolBus (South China) *meets* FollowBus (Threemile Pond, South China)
FollowBus (Threemile Pond, South China) *equals* NavigateCircle (Augusta)
BusTurnsOff (Augusta) *ends* FollowBus (Threemile Pond, South China)
NavigateCircle (Augusta) *before* ArriveAtWork (16 State House Station, Augusta)

In this example, there are $n= 9$ events and since every event has an associated location, there are also nine event locations in the set. Two of these locations – 116 Quaker Hill Rd and 16 State House Station – are at the **street address** granularity level. Two event locations, Willow Beach, China and Threemile Pond, South China, are at the **landmark** level. The remaining five event locations –

Albion, Bangor, South China, Augusta, and Augusta – are at the `city` granularity level. Thus, the final variable values are: *street_address=2*, *landmark=2* and *city=5*. The modal granularity in this set is `city`. The event locations at the `street address` and `landmark` granularities are subsequently coarsened to the `city` level. The result is a revised set of event-relation combinations:

LeaveHome (Unity) *before* JoinCarpool (Albion)
PassTruck (Bangor) *overlaps* SlowsForConstruction (China)
WaitForSchoolBus (South China) *during* SlowsForConstruction (China)
WaitForSchoolBus (South China) *meets* FollowBus (South China)
FollowBus (South China) *equals* NavigateCircle (Augusta)
BusTurnsOff (Augusta) *ends* FollowBus (South China)
NavigateCircle (Augusta) *before* ArriveAtWork (Augusta)

This coarsening of event location to the modal granularity can result in a loss of detail; however, analysis based on event locations requires that locations be at the same granularity level. In this work, linear regression and hierarchical clustering are performed on event locations after the location granularities have been coarsened to match the modal granularity. Event locations at a coarser granularity than the mode are treated as though they do not have an associated location. These events, as well as any events whose locations are unknown, do not contribute to the subsequent location-based filtering steps, i.e., there is no pruning of these events based on location information. These events, however, are still useful for generating linear orders based on the temporal relations that hold between them and other events in the event-relation combinations, and are included in the linear orders.

4 Determining the Spatial Relevance of Events Using Hierarchical Clustering

Given a set of event-relation combinations that include locations, we first determine whether any event locations can be considered spatial outliers (i.e., locations outside the geographic area represented by the majority of event locations in a set). Information about spatial outliers is used to determine whether any events in the set of event-relation combinations are superfluous such that these events can be pruned from the orders. By superfluous, we mean that an event is presumed to be irrelevant for an order if it exceeds a certain threshold distance from other events. This provides a mechanism for reducing the size of the orders with respect to the number of events they contain. In this work, a hierarchical clustering algorithm is applied to identify possible outlying event locations.

Hierarchical clustering is an agglomerative clustering technique. That is, each element is initially assigned its own cluster and sets of the two most similar clusters are systematically merged until a threshold value of clusters is reached. In single linkage clustering, the similarity between elements (event locations), is based on Euclidean distance between event locations [21-22]. For each pair of event locations $el1$ and $el2$, the Euclidean distance is referred to as $d[el1,el2]$. Two clusters a and b are considered most similar if the distance between them is the minimum distance between any two clusters, that is, $\forall el1,el2 \mid el1 \neq el2: d[(a),(b)] = \min d[el1,el2]$.

Distances between clusters are stored in an (n+1)×(n+1) proximity matrix P, which shows the Euclidean distance between each pair of event locations. Since the matrix P is symmetrical, only the top diagonal half of the matrix is populated. For the example introduced in the previous section, the matrix P is shown in Figure 2, with distances between locations in miles.

	Unity	Albion	Bangor	China	South China	South China	Augusta	Augusta	Augusta
Unity		8	34	13	20	20	40	40	40
Albion			42	6	13	13	25	25	25
Bangor				48	55	55	77	77	77
China					7	7	25	25	25
South China						0	13	13	13
South China							13	13	13
Augusta								0	0
Augusta									0
Augusta									

Fig. 2. The (n+1) × (n+1) proximity matrix P, showing the Euclidean distance between each pair of event locations

To begin the clustering algorithm, each of the n event locations is initially assigned its own cluster. Thus, every row and column in P represents a cluster. A search is conducted in the matrix P for the two clusters with the minimum distance between them, and these two closest clusters are merged. P is then updated to show this new cluster configuration, by merging the rows and columns of the two clusters that have been fused. The distance $d[old, new]$ between a previous cluster *old* and a newly formed cluster *new*, where *new* comprises the previous clusters *a* and *b*, is the smaller of the distances $d[(a)(old)]$ and $d[(b)(old)]$. The matrix P is then evaluated again for the shortest distance, and the two closest clusters are merged and P updated. The process is repeated until a *stopping rule* is reached.

A stopping rule is used to determine when the clustering process should be terminated. In this work, we use the stopping rule defined by Calinski and Harabasz [23-24] referred to as the *CH index*. The CH index is the ratio of the distance between clusters and the distance between events within clusters, accounting for the total number of events and the number of events within each cluster. The stopping rule indicates that a clustering process should be terminated when the CH index is at its maximum value. The equation for determining this ratio is based on D, the average distance between clusters; W, the average distance between events within clusters; n, the number of events; and k, the number of clusters (Equation 1).

$$CH\ index = D(n-k)/W(k-1) \tag{1}$$

At the end of the clustering process, the distribution of clusters is analyzed. If there is one single cluster, then all events are considered spatially relevant. If there is more than one cluster, however, the events in the largest cluster (the cluster containing the largest number of events) are considered spatially relevant, but events in other clusters are considered outlying events. Spatially relevant events are retained in the set O_l of linear orders, but outlying events are eliminated from the orders in O_l. Removing these events has the effect of shortening the orders.

When hierarchical clustering is performed on the current example, the result is that eight of the nine events are part of one cluster. Eight of the event locations fall in one cluster, where some events occur at the same location, and otherwise the average distance d between event locations is 19.8 miles. In contrast to this, the average distance d between the location of event PassTruck (Bangor) and the other event locations is 51.8 miles. The hierarchical clustering algorithm places the Bangor event location in its own cluster, and the remaining eight in another cluster.

With this result, the event that takes place in Bangor, PassTruck, is eliminated from the orders in O_l. The removal of this event results in several of the orders becoming identical; two pairs of orders become identical to each other, and three other orders are also now identical to each other. Filtering duplicate orders and retaining only one representative order for each case removes the redundancy resulting from identical orders. The set O_l is therefore reduced to only six orders:

$O_l =$
{*1 LeaveHome   JoinCarpool   SlowsForConstruction   WaitForSchoolBus   (FollowBus=NavigateCircle) BusTurnsOff ArriveAtWork*
2 LeaveHome   SlowsForConstruction   WaitForSchoolBus   (FollowBus=NavigateCircle)   BusTurnsOff JoinCarpool ArriveAtWork
3 LeaveHome   SlowsForConstruction   WaitForSchoolBus   (FollowBus=NavigateCircle)   BusTurnsOff ArriveAtWork JoinCarpool
4 SlowsForConstruction   WaitForSchoolBus   (FollowBus=NavigateCircle)   BusTurnsOff   LeaveHome JoinCarpool ArriveAtWork
5 SlowsForConstruction   WaitForSchoolBus   (FollowBus=NavigateCircle)   BusTurnsOff   LeaveHome ArriveAtWork JoinCarpool
6 SlowsForConstruction   WaitForSchoolBus   (FollowBus=NavigateCircle)   BusTurnsOff   ArriveAtWork LeaveHome JoinCarpool }

There may be cases where, instead of a single largest cluster, there are two or more clusters containing an equal number of event locations. In these cases, unless additional information is available, the clusters are assumed to be of equal importance, and all events contained in these clusters are included in the orders. Note also that although any outlying events are eliminated from the orders, the events are not eliminated from the set of event-relation combinations, because though spatial information may make an event less relevant to the orders, the temporal relations shared by that event with others are no less valid. For this reason, the event PassTruck can still be relevant to the ordering process due to the explicit temporal relation it shares with, for example, SlowsForConstruction and by extension, the relations it has with the other events in the set of event-relation combinations.

5   Filtering Orders Based on Event Locations

Spatial details of events are used as a foundation for eliminating outlying events from one or more orders in O_1, as well as filtering entire orders from the set. Given a set of event-relation combinations in which the locations of events are known, these locations will fall into one of two cases. The events occur in locations such that no pattern or order is apparent (Figure 3a), or the locations may be such that they exhibit a linear trend (Figure 3b), showing a *spatial order of locations*. In addition to linear trends, other patterns are also possible. These patterns include, for example, events occurring along a nonlinear roadway, such as a loop road, or events occurring in a circular pattern. We focus here on linear patterns of event locations and analysis of other possible patterns is considered as a topic for future work.

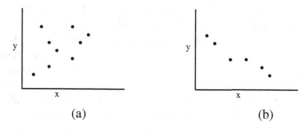

Fig. 3. Point locations in Euclidean space that show (a) no apparent spatial order and (b) a linear pattern of event locations

5.1   Detecting a Linear Pattern in Event Location Data

Detecting a linear pattern of event locations may be useful for the analysis and automatic derivation of plausible linear orders. Travel along a road, for example, may occur in a linear fashion. Weather events are another type of event that often occurs in a linear track through space. When the pattern of event locations is known to be linear, this information can be combined with the information about the orders of events generated from temporal information, such that orders that coincide temporally *and* spatially are retained as plausible orders. In this paper we examine the case where event locations are tested for linear trends, resulting in linear spatial orders of events. The example of the morning commute is treated once again; this time an analysis is performed to check whether the commute occurs along a linear track.

The first step in filtering orders of events from the set of plausible orders O_1 based on the spatial pattern of event locations is to determine the degree to which the event locations exhibit a linear trend. Linear regression is used to evaluate the degree of linearity shown by the event locations. Linear regression determines the equation of a line of best fit, based on the x and y values of a given set of points, as well as a correlation coefficient r, indicating the quality of the linear relationship between the x and y values. We apply least-squares regression, which minimizes the sum of the squares of the distance from each data point to the line in order to determine the equation of the line of best fit. The equation, in the form $y=mx+b$, gives the slope of the line m, and the y-intercept b, where x and y are in this case, the latitude and longitude of the locations.

The absolute value of r is always between 0 and 1, with values close to 1 indicating a very high linearity, and values close to 0 indicating that the linear relationship between the x and y values is very weak. In this work, linear regression is applied to each set of event locations, and if the absolute value of the correlation coefficient is greater than 0.7, then the linear trend exhibited by the locations is considered significant [25]. In cases where the correlation coefficient is less than 0.7, the linear trend is not considered significant, and therefore no subsequent analysis based on a linear trend of locations is performed.

5.2 Evaluating Orders for Correspondence with a Linear Trend

If linear regression has been applied to a set of event locations, and the value of r is greater than 0.7, then we conclude that the set of event locations shows a significant linear trend. The next step is to evaluate whether the linear trend apparent in the event locations corresponds to any of the temporal orders in the set O_1. Correspondence between the linear trend and any temporal orders in O_1 means that these orders in O_1 can be considered as *spatio-temporal orders*, supported by both spatial and temporal relations between events. Two lists, l_p and l_n, are formed by listing the events corresponding with the locations in positive and negative directions along the line of best fit. These lists show the two possibilities of spatial ordering exhibited by the event locations. As with events that are temporally equal, events with identical locations are listed together in l_p and l_n, and enclosed in parentheses. It seldom happens, however, that all events fall exactly along a line of best fit (that is, that $r=1$). In cases where $r \neq 1$, a method is needed to assign each location a position on the line before l_p and l_n can be generated. A perpendicular line is drawn between each event location and the line of best fit. The place at which a location's perpendicular line meets the line of best fit is that location's assigned position on the line (Figure 4).

The lists l_p and l_n are then compared with orders in the set O_1. If orders in O_1 match either l_p or l_n, then these orders are assumed to be more plausible than the other orders (i.e., orders which are supported only by temporal information). The set O_1 is then revised to include only those orders that correspond to either l_p or l_n. If linearity is confirmed but none of the orders in O_1 correspond with the linear pattern of event locations, then the spatial linearity of the events is assumed to be unrelated to the events' temporal relations to each other, and no updates are made to the set O_1. For example, in the case of crime events, several convenience stores may be located along a linear stretch of road, but if the stores are not robbed in order of location, the temporal order of events will not coincide with the linear order of locations.

In addition to the possibility of no matches between temporal and spatial orders of events, it is also possible that more than one temporal order corresponds to the spatial order of event locations. This can occur in cases where some events' locations are unknown, or when multiple events share the same location.

In the case where no event locations are known, reasoning about patterns of event locations is not possible and all orders are retained in O_1. In cases where location information exists for some events but not all, the available event locations are used in testing for linearity. Events without locations are automatically included in the set of linear orders, as discussed in Section 3. This means that in the case of a significant linear trend in event locations, the lists l_p or l_n will not contain all the events present in the orders within O_1. Instead of checking for orders that exactly match l_p or l_n, then, a search is performed for orders containing l_p or l_n as a subset. Searches for subsets of orders are performed using *longest common subsequence* methodologies, such as those used in genome-matching algorithms [26].

For the example of the morning commute, linear regression is applied to the locations of the n events in the set of event-location combinations, and the coefficient is computed as $r=0.9277$. This is well above the 0.7 significance threshold (Figure 5) and supports a linear trend for this set of events. The lists l_p and l_n, are:

$l_p=$ (NavigateCircle,   BusTurnsOff,   ArriveAtWork)   (WaitForSchoolBus, FollowBus) SlowsForConstruction JoinCarpool LeaveHome

$l_n=$ LeaveHome   JoinCarpool   SlowsForConstruction   (WaitForSchoolBus, FollowBus) (NavigateCircle, BusTurnsOff, ArriveAtWork).

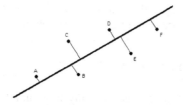

Fig. 4. Perpendicular lines from event locations to a line of best fit, showing each event's assigned position along the line. The lists formed from these locations are l_p = A, B, C, D, E, F and l_n = F, E, D, C, B, A.

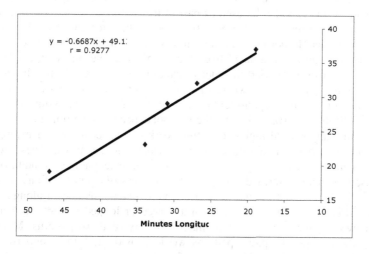

Fig. 5. Plot showing event locations and line of best fit

When the orders in O_1 are compared with l_p or l_n, one order, *LeaveHome JoinCarpool SlowsForConstruction WaitForSchoolBus (FollowBus= NavigateCircle) BusTurnsOff ArriveAtWork*, is found to correspond with l_n. As a result of this match, this order is retained in the set O_1, and the other orders are eliminated. The set O_1 now is reduced to only one order:

$O_1=\{LeaveHome$ *JoinCarpool*   *SlowsForConstruction*   *WaitForSchoolBus* *(FollowBus=NavigateCircle) BusTurnsOff ArriveAtWork* $\}$

This order is plausible based on the temporal knowledge available about the scenario, and the spatial order of the event locations corresponds with the route from Susan's home to her workplace.

In this case, only one order remains in O_1. Of course, for many other dynamic scenarios, the results are likely to be more varied and multiple orders are possible. This work shows how analysis of event locations can be applied usefully to augment the information derived by temporal relations that hold between events. Spatial information about events is used for both filtering events from orders (e.g., removing events that correspond to spatial outliers) as well as confirming the plausibility of orders of events that have occurred over a linear track in space.

6   Conclusions and Future Work

Information systems need to manage not only large numbers of events but also the relations that hold between events. An important aspect of managing events relates to the abstraction of events, where partial orders of events are simplified or summarized as linear orders. It is possible to automatically generate linear orders of events from partial orders based on temporal semantics that stem from the event interval relations that hold between events. In this work, we show how spatial aspects in the form of locations of events can also be used as part of the methodology to generate plausible linear orders. With respect to these spatial aspects, hierarchical clustering is used to identify events that do not meet a criterion of spatial relevance. For these cases, such outlying events can be eliminated from the set of plausible linear orders O_1. Linear regression can also be applied in order to capture linear trends in the distribution of event locations, and in cases where one or more orders in O_1 correspond with the spatial order of events, a spatio-temporal ordering of events is assumed and orders that do not display this spatio-temporal correspondence can be eliminated.

There are many related topics for future work. Event locations, for example, may correspond to other granularities, such as states or countries. These levels of abstraction may be modeled as area data and require techniques in addition to the point-based methods described above. Spatial patterns of event locations are also not always in the form of a straight line. For example, events may occur along a network such as a street or electrical network, in a curved or looped pattern, back and forth between two locations, or in a combination of any of these patterns. Methods for detecting these kinds of spatial patterns would provide additional opportunities for understanding spatio-temporal sequences of events.

Acknowledgments

This research has been supported by the National Geospatial-Intelligence Agency under grant number, NMA201-00-1-2009.

References

1. Alfonseca, E., Manandhar, S. A Framework for Constructing Temporal Models from Texts. Proceedings of the Fifteenth Florida Artificial Intelligence Conference, FLAIRS 2002, Pensacola Beach, FL, AAAI Press, (2002) 456-460.
2. Hornsby, K., Hall, S. Generating Linear Orders of Text-Based Events. Proceedings of the Workshop on Computational Lexical Semantics, Human Language Technology Conference 2004 (HLT/NAACL'04). Boston, MA, Association for Computational Linguistics, (2004) 92-99.
3. Allen, J., Ferguson, G. Actions and Events in Interval Temporal Logic. Journal of Logic and Computation 4 (1994) 531-579.
4. Raubel, M., Kuhn, W. Ontology-Based Task Simulation. Spatial Cognition and Computation 4 (2004) 15-37.
5. Yuan, M. Representing Complex Geographic Phenomena with Both Object- and Field-Like Properties. Cartography and Geographic Information Science 28 (2001) 83-96.
6. Galton, A. Desiderata for a Spatio-Temporal Geo-Ontology. In: Kuhn, W., Worboys, M. F., Timpf, S. (eds.) Spatial Information Theory: Foundations of Geographic Information Science, Proceedings of International Conference COSIT 2003, Kartause Ittingen, Switzerland, Springer, Lecture Notes in Computer Science, (2003) 1-12.
7. Galton, A. Fields and Objects in Space, Time and Space-Time. Spatial Cognition and Computation 4 (2004) 39-68.
8. Grenon, P., Smith, B. SNAP and SPAN: Towards Dynamic Spatial Ontology. Spatial Cognition and Computation 4 (2004) 69-103.
9. Worboys, M., Hornsby, K. From Objects to Events: GEM, the Geospatial Event Model. In: Egenhofer, M., Freksa, C., Miller, H. (eds.) Proceedings of GIScience 2004, Lecture Notes in Computer Science, 3234, Springer, Berlin (2004) 327-343.
10. Worboys, M. Event-Oriented Approaches to Geographic Phenomena. International Journal of Geographical Information and Analysis 19 (2005) 1-28.
11. Zacks, J., Tversky, B. Event Structure in Perception and Conception. Psychological Bulletin 127 (2001) 3-21.
12. Tversky, B., Zacks, J., Lee, P. Events by Hands and Feet. Spatial Cognition and Computation 4 (2004) 5-14.
13. Zacks, J. Using Movement and Intentions to Understand Simple Events. Cognitive Science 28 (2004) 979-1008.
14. Galton, A. Qualitative Spatial Change, New York, NY: Oxford University Press (2000).
15. Allen, J. Maintaining Knowledge About Temporal Intervals. Communications of the ACM 26 (1983) 832-843.
16. Knauff, M. The cognitive adequacy of Allen's interval calculus for qualitative spatial representation and reasoning. Spatial Cognition and Computation 1, (1999) 261-290.
17. Frank, A. Different Types of "Times" in GIS. In: Egenhofer, M. J., Golledge, R. G. (eds.) Spatial and Temporal Reasoning in Geographic Information Systems. Oxford, Oxford University Press (1998) 40-62.

18. Roddick, J., Mooney, C. Linear Temporal Sequences and Their Interpretation Using Midpoint Relationships. IEEE Transactions on Knowledge and Data Engineering 17 (2005) 133-135.
19. Freksa, C. Temporal reasoning based on semi-intervals. Artificial Intelligence, 54 (1992) 199–227.
20. Skiena, S. The Algorithm Design Manual. Santa Clara, CA, TELOS - the Electronic Library of Science (1998).
21. Johnson, S. Hierarchical Clustering Schemes. Psychometrika 32 (1967) 241-254.
22. Bailey, T., Gatrell, A. Interactive Spatial Data Analysis. Essex, Longman Group Limited (1995).
23. Milligan, G., Cooper, M. An Examination of Procedures for Determining the Number of Clusters in a Dataset. Psychometrika 50 (1985) 159-179.
24. Calinski, R. B., Harabasz, J. A dendrite method for cluster analysis. Communications in Statistics 3 (1974) 1-27.
25. Kiemele, M., Schmidt, S., Berdine, R. Basic Statistics. Colorado Springs, Air Academy Press (1997).
26. Smith, T., Waterman, M. Identification of Common Molecular Subsequences. Journal of Molecular Biology 147 (1981) 195-197.

Structural Salience of Landmarks for Route Directions

Alexander Klippel and Stephan Winter

Cooperative Research Centre for Spatial Information,
Department of Geomatics, The University of Melbourne, Australia
{aklippel, winter}@unimelb.edu.au

Abstract. This paper complements landmark research with an approach to formalize the structural salience of objects along routes. The aim is to automatically integrate salient objects—landmarks—into route directions. To this end, two directions of research are combined: the formalization of salience of objects and the conceptualization of wayfinding actions. We approach structural salience with some taxonomic considerations of point-like objects with respect to their positions along a route and detail the effects of different positions on the conceptualization process. The results are used to extend a formal language of route knowledge, the *wayfinding choreme theory*. This research contributes to a cognitive foundation for next generation navigation support and to the aim of formalizing geosemantics.

1   Introduction

This paper investigates the structural salience of objects along routes. The structural qualities considered are induced by embedding the route into a street network. Objects are called *structurally* salient if their location is cognitively or linguistically easy to conceptualize in route directions.

A generic and formal method of assessing the structural salience of objects with the goal of finding a landmark selection process for route directions is proposed. Such a measure of structural salience complements a formal model of salience recently developed (see, e.g., [1,2]), building on an earlier characterization of the nature of landmarks [3]. This model assumes visual, semantic and structural qualities of objects that contribute to their salience. Measures for structural qualities have been left out of this model so far, a gap which will be filled by this paper.

The proposed method in this paper utilizes a conceptual approach to spatial information exemplified by route direction elements (e.g., [4,5]), and extends the wayfinding choreme approach, i.e., a formal language for the specification of conceptual route knowledge, for the inclusion of salient features. This paper classifies structural aspects of landmarks, especially their location with respect to the re-orientation actions performed at the nodes of the underlying street network, which are frequently called decision points.

The linguistic complexity of characterizing the relative location of a landmark is discussed in relation to the conceptual complexity of realizing its wayfinding affordance. The hypothesis in this paper is that the conceptual complexity influences the selection of a landmark by direction givers, and hence, the salience of this landmark.

The described approach of the paper brings together two lines of research that have been unrelated so far: the formalization of salience and the conceptualization of

A.G. Cohn and D.M. Mark (Eds.): COSIT 2005, LNCS 3693, pp. 347–362, 2005.
© Springer-Verlag Berlin Heidelberg 2005

wayfinding actions. The paper is organized as follows: We start by reviewing work on landmarks and conceptualization processes relevant for interactions with spatial environments. For the conceptualization of landmarks as route elements we start by instantiating a taxonomy that classifies the location of landmarks with respect to routes. Building on this general classification scheme, we detail the different possibilities of locating landmarks at decision points. We investigate the conceptual complexity of using different types of landmarks, and derive a measure of structural salience from their ease of use. This measure can be combined with the existing measures for visual and semantic salience [1]. We briefly discuss the extension of the wayfinding choreme theory and the formalization of conceptual route knowledge by relying on the developed taxonomy, and conclude with a discussion.

2   Landmark Research and Conceptual Approaches to Spatial Information

Since the work by Lynch [6] on elements that structure our urban environmental knowledge, the concept of landmarks has inspired multiple research papers. There are some simple and straightforward facts that can be manifested from this research:

- Everything that stands out from a scene can be a landmark [7].
- In certain contexts, for example route following, even road intersections can be landmarks.
- Landmarks are pertinent for finding one's way.
- Landmarks are remembered/learned early on (i.e., landmark knowledge) [8].
- Landmarks structure environmental knowledge, for example, as anchor points [9].
- Landmarks are used to communicate route knowledge verbally and graphically [10].
- Landmarks are integrated in route directions to varying degrees, with greater quantities at origins, destinations and distinguished decision points [11,12].
- Landmarks at street intersections (decision points) are more pertinent when a change in direction is required [13].
- Landmarks generally work better than street signs in wayfinding (e.g., [14]).

While we understand how and why, *people* use landmarks in communication, and hence in memory and mental processes, the technical basis to *automatically construct* wayfinding directions with landmarks is still limited. Suitable formalisms for characterizing conceptual route knowledge that can be flexibly adapted to canonical and personal preferences are still missing.

Part of the problem is formalizing the concept of a landmark, such that a service can identify objects of some *landmarkness* or salience, i.e., objects that differ from their background [7]. According to a recent proposal [1,2,15] the salience of objects is determined by visual, semantic and structural qualities. These qualities can be characterized to provide an overall measure of salience. The approach proposes measures for visual qualities (such as the size, form, or texture of an object), and semantic qualities (such as their prominent or labelled use); measures for structural qualities have not been included so far. Objects that show large visual and semantic salience show good compliance with cognitively salient objects, i.e., landmarks chosen by people for their communication of routes.

The measure of salience can be adapted to context [15], and can be weighted by advance visibility along a route [16], which makes it route specific. The approach has also been adopted for data mining in topographic data sets [17,18].

An approach orthogonal to the work on the salience of objects is taken by wayfinding choreme theory [19,4,20]. This approach, however, can provide the missing aspect of structural salience. It is based on wayfinding events such as re-orientations and turns at street intersections as primitives; conceptual primitives of turns are derived from this.

The conceptualization of actions and events[1] and their formal specification is a recently well discussed area of research [21,22]. Complementary to other computational formalisms [23] the wayfinding choreme theory stresses the cognitive aspects of route knowledge by making (cognitive) conceptual primitives the basis of the formalism. Yet, the focus on conceptualization and the development of a formal language [4] offers many ties to recent discussions on the formalization of conceptual spaces [24], e.g., for landmarks [25], and the general approach of integrating cognitive semantics (sometimes referred to as conceptual semantics) into information systems [26]. The wayfinding choreme approach seeks conceptual primitives as a foundation for a formal language of space in which the number of basic elements are restricted and the combinatorial possibilities are constraint by the represented knowledge, in our case, the linear character of a route. Additionally, the focus on conceptual aspects of information creates a basis for multimodality in that the externalization of conceptualizations can be specified in various output formats (e.g., [12,27]): verbal, graphical or gesture.

The conceptualization of an action at a decision point, however, is not only dependent on the angle of the turn, i.e., the geometric representation of the trajectory of the movement, but also on the street structure in which the action is embedded [28]. Additionally, the possibility of relating the turning action to supplementary information—for example a landmark—has an influence on the conceptualization.

A formal specification of conceptual primitives of landmark locations should therefore allow the characterization of different layers of interaction with the environment and grasp the resulting conceptual primitive adequately. In this paper, we will further elaborate the conceptual approach and focus on landmarks, specifically their structural salience induced by the conceptualization of a wayfinding action. Based on this characterization we will extend the rules specified for the higher order route direction elements (HORDE) [29,30] to allow for different levels of granularity in route directions.

3   Conceptualizing Landmarks as Route Elements

Some of the observed characteristics of landmarks discussed in Section 2 concern their structural qualities with respect to a given route: the route structure (co-)determines which salient objects are selected to give route directions. Hence, this paper develops a classification schema for point-like landmarks depending on their location relative to a route and the route's structural embedding in the street network. For some ideas on other

---

[1] In this article we do not distinguish between actions and events.

types of landmarks and their relation to route directions see [20]. We further show that some locations are conceptually easier and conceptually less ambiguous than others, especially with regard to building complex route elements, i.e., combining conceptual primitives into HORDE.

3.1 Landmark Locations

The following taxonomy of landmarks induced by a route embedded into a street network is illustrated in Figure 1. Landmarks can occur (i) distant from the route (distant landmarks), (ii) somewhere along route segments (segment landmarks), or (iii) at specific route nodes (node landmarks). Segment landmarks and node landmarks can be grouped as either close to or on-route. Route nodes are also called *decision points*.

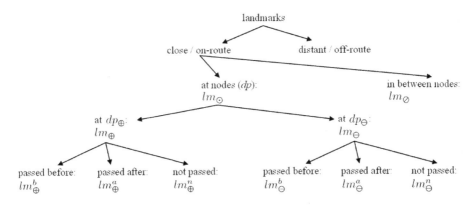

Fig. 1. A structural taxonomy of landmarks; the abbreviations are detailed in the text

Distant and On-Route Landmarks. With respect to a route, these categories of landmarks have different degrees of freedom regarding their location. A distant landmark is not determined in its exact location in two dimensions. The conceptualization details an area, a region in which the landmark is placed, or a line of sight. A segment landmark, on the other hand, has its location determined only within the one-dimensional interval between two route nodes; their exact location along the segment—the linear reference from start or end node—remains under-specified. In contrast, node landmarks are constrained in their location by a node of the street network or decision point. With respect to their function in route directions and the conceptualization of the action that has to be performed, both their location and their remaining degrees of freedom in location have to be reflected, for example, in the type of verbal reference.

Distant landmarks fulfill a variety of functions, for example, global orientation, reassurance and confirmation [31]. Their actual location (or distance) is irrelevant as long as the direction or visibility can be taken as reference. Distance generally is not a criterion for exclusion from route directions. Consider for example the route direction:

Follow the street until you see the castle distantly on your right.   (1)

Yet, the effect of distant landmarks on the conceptualization of route parts, for example, spatial chunking [32], is rather complex. Many parameters that influence the conceptualization of these distant landmarks are not primarily spatial (at least not planar spatial). They are therefore not the focus of this paper.

Within the category of landmarks that are close to or on-route (cf. [33]) segment landmarks [34] are used for reassurance and confirmation. The influence of segment landmarks on chunking is discussed in Section 3.3. In contrast, node landmarks may be used as an anchor for action (re-orientation and turning), and hence, their location with respect to the route is relevant. But node landmarks can also occur at decision points where no re-orientation is necessary. Within our taxonomy we write segment landmarks as $lm_{\oslash}$, and node landmarks as $lm_{\odot}$.

Route directions cannot neglect any necessary re-orientation, but they can neglect confirmations that may occur either between decision points or at decision points where no re-orientation is required. That means, node landmarks at decision points with re-orientation are essential, and other landmarks are less important or optional. This characterization establishes a first indication of structural salience.

Node Landmarks. A further common distinction is made between decision points with direction change, $dp_{\oplus}$, and decision points without direction change, $dp_{\ominus}$. This distinction has to be accounted for in the taxonomy of node landmarks. It has been shown that landmarks at $dp_{\oplus}$ are more pertinent to route knowledge [11,13]. Within our taxonomy we coin them $lm_{\oplus}$, and landmarks at $dp_{\ominus}$ are coined $lm_{\ominus}$, such that $lm_{\odot} = \{lm_{\oplus}, lm_{\ominus}\}$.

At a more detailed level of spatial granularity, it is of interest *where* with relation to any decision point, $dp_{\oplus}$ or $dp_{\ominus}$, a landmark is placed. Not every node landmark is equally suited to aid wayfinding and to be integrated into route directions and the conceptualization of route parts, respectively.

Note that this characterization is based on locational properties, i.e., the location of a landmark with respect to the physical layout of an intersection. It is not a characterization based on the visual or semantic salience of a landmark. This characterization is also a specification of the locational properties from the perspective of mental conceptualization processes, i.e., the conceptualization of an action performed in a spatial environment. In experimental settings (e.g., in [32]) it was made sure that at $dp_{\oplus}$ primarily those landmarks that can be integrated into a route direction conceptually easily are used. This integration is afforded by the landmarks' location with respect to the action at the decision point. More specifically, landmarks at $dp_{\oplus}$ were chosen that are passed immediately before a turning decision. These node landmarks may be located on the left or on the right side of the route, independent from the direction of the turn. A natural language example would be:

$$\text{Turn right after the post office.} \tag{2}$$

Based on the idea that the action performed (or imagined) at a decision point is the pertinent factor for the conceptualization process, we introduce further sub-concepts for landmarks, namely landmarks passed before the action is performed, lm^b, landmarks not directly passed, lm^n, and landmarks passed after decision, lm^a (see Figure 1, and

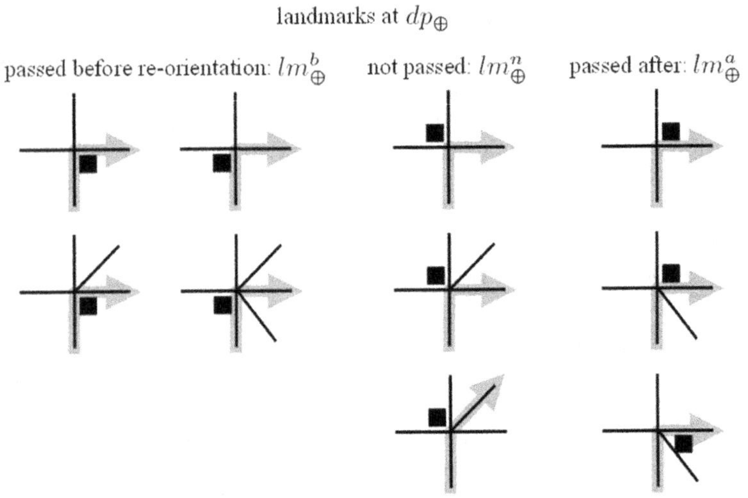

landmarks at $dp_\oplus$

passed before re-orientation: $lm_\oplus^b$ not passed: $lm_\oplus^n$ passed after: $lm_\oplus^a$

Fig. 2. Possible locations of node landmarks with re-orientation. The different locations result in different conceptualizations, and not every location of a landmark functions equally well as an identifier for the required decision.

also Figure 2 for more details). These concepts can be specified for decision points with a direction change, but also for decision points without a direction change.

At $dp_\oplus$, landmarks passed before decision, $lm_\oplus^b$, work equally well for all turning concepts. That is, they are straightforward to conceptualize as the turning occurs immediately after them and does not conflict with the overall branching structure of the decision point. Compare the use of a $lm_\oplus^b$:

$$\text{Make a sharp right turn after the post office.} \qquad (3)$$

with a $lm_\oplus^n$ or $lm_\oplus^a$:

$$\text{Make a sharp right turn at the intersection where the post office is.} \qquad (4)$$

which represents here a more precise, but also more complicated direction '*make a sharp right turn at the intersection where the post office is at the opposite corner*'. Especially at more complex intersections, where it is difficult to conceptualize the location of a landmark, a $lm_\oplus^b$ is the only unambiguously identifiable one.

3.2 A Route Direction Grammar with Node Landmarks

Having a taxonomy for the location of landmarks with respect to a route, we can integrate them into the wayfinding choreme route grammar [4]. Generally, two turning concepts have been differentiated, standard turning concepts ⟨STC⟩ and modified turning concepts ⟨MTC⟩ [4]. Both can be extended to incorporate the different landmark locations. To this end, the node landmark and its location is added to the wayfinding choreme grammar as an annotation. We exemplarily detail the notation for the wayfinding choreme of a right turn, wc_r, added with a landmark passed before the decision, $lm_\oplus^b$:

$$\langle wc_r{}^{lm_\oplus^b} \rangle \tag{5}$$

Likewise for other turning concepts.

3.3 Spatial Chunking with Node Landmarks

This section exemplifies the influence of structural aspects of landmark positions within yet another aspect of route directions, the change in granularity by applying higher order route direction elements (HORDE). We discuss here the possibilities of chunking with node landmarks at decision points without a direction change, $lm_\ominus$ preceding a $dp_\oplus$. $lm_\ominus$ have two functions: First, they are used to identify a decision point resulting in a verbalization such as *'go straight at the intersection where the McDonald's is'*. Second, they are used in a way analogous to a segment landmark $lm_\oslash$, such as *'pass the McDonald's and turn right at the Shell gas station'*. Here it is not specified whether the $lm_\ominus$ is placed at a decision point or between two decision points. The second case might be an example of spatial chunking.

Two distinctions are pertinent for $lm_\ominus$ that determine their function in spatial chunking. First, whether the landmark is passed before ($lm_\ominus^b$) or after ($lm_\ominus^a$) the action of straight crossing the intersection. Second, whether a landmark is present at the chunk terminating $dp_\oplus$ ahead. We observe that $lm_\ominus$ only appear within small chunks (similar to segment landmarks $lm_\oslash$) if at least one and possibly both of the following conditions are met: (i) the $lm_\ominus$ is of the type $lm_\ominus^a$; (ii) there is an easy to conceptualize node landmark at the chunk terminating $dp_\oplus$. Figure 3 illustrates these assumptions. The following cases demonstrate some rules that can be distinguished and integrated into the wayfinding choreme route grammar:

– Consider the first example in Figure 3. $lm_\ominus$ is passed before orientation: $lm_\ominus^b$, and at the chunk terminating $dp_\oplus$ there is a landmark. The resulting concept is: PASS 'FIRST LANDMARK' AND TURN RIGHT AT 'SECOND LANDMARK'. This concept is over-specified when only two decision points are present; the first landmark should be left out, even if it is the somewhat more salient one.
– Consider the second example in Figure 3. $lm_\ominus$ is passed before re-orientation: $lm_\ominus^b$, but no landmark is present at the chunk terminating $dp_\oplus$. This situation has to be put in a less specific form: AFTER THE INTERSECTION WHERE THE 'LANDMARK' IS TURN RIGHT AT THE NEXT INTERSECTION, or alternatively, in a more complex concept: AFTER THE INTERSECTION WHERE THE 'LANDMARK' IS AT THE RIGHT CORNER TURN RIGHT AT THE NEXT INTERSECTION.
– Now turn to the third example in Figure 3. $lm_\ominus$ is passed after re-orientation: $lm_\ominus^a$, and a landmark is present at the corresponding $dp_\oplus$, for example, a $lm_\oplus^b$. The resulting concept is similar to the first case: PASS 'FIRST LANDMARK' AND TURN RIGHT AT 'SECOND LANDMARK'. When only two decision points are involved, the first landmark is left out.
– Finally, consider the fourth example in Figure 3. $lm_\ominus$ is passed after re-orientation: $lm_\ominus^a$, but no landmark is present at the corresponding $dp_\oplus$. Here, $lm_\ominus$ is used similarly to a $lm_\oslash$: TAKE A RIGHT TURN AFTER 'LANDMARK'.

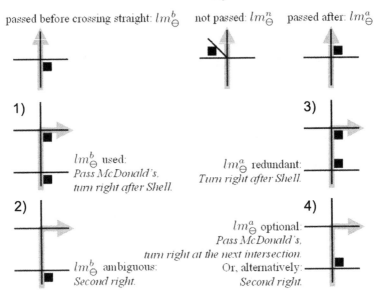

Fig. 3. Landmarks at a decision point without a direction change, and their influence on spatial chunking in examples 1-4

So far, distance has not been considered. There are, however, similarities to another approach to integrate distance in term rewriting as a method to extract conceptually connected primitives [35]. A more detailed characterization of spatial situations to differentiate, for example, the following two concepts is ongoing research: (a) MAKE A RIGHT AND THEN ANOTHER QUICK RIGHT, versus (b) MAKE A U-TURN.

4   Structural Salience of Landmarks

The previous sections discussed the conceptual approach to characterize the positions of landmarks with respect to their relevance for route directions. This section extends the salience model for landmarks to include structural aspects derived from the findings above.

4.1   The Salience Model

The salience model [1,2] provides a measure of salience for all identified objects within a street network. These measures enable choosing the most salient objects along a specific route, for example, at decision points with direction change, to enrich route directions. For static objects the measures can be calculated once, and stored as parameters of the objects.

The original model of salience was based on three qualities: visual, semantic and structural [3]. Each quality can be characterized by a normed measure of salience (with values $0 \ldots 1$), resulting in visual salience s_v, semantic salience s_s, and structural salience s_u. These measures can be combined to a weighted average of joint salience, s_o:

$$s_o = w_v s_v + w_s s_s + w_u s_u \qquad \text{with } w_v + w_s + w_u = 1 \qquad (6)$$

So far, structural salience has been considered in the model, but it was not developed. Saliences s_v and s_s are determined by comparing visual and semantic properties of objects with the properties of other objects in their neighborhood. The more distinct a property of an object is, in its neighborhood, the higher the object's salience measure is. By this way, the figure-ground relation mentioned in Section 2 is quantified. This means:

– Visual and semantic salience is dependent on the properties of objects nearby; it is a relative property of an object, not an absolute one. For instance, a red facade in an area where all facades are red will not stand out. But the same facade in a grey neighborhood stands out.
– Joint salience is quantitative, i.e., it can be represented by a real number between 0 and 1. It is not qualitative (e.g., 'landmark', 'no landmark'), as supposed so far in the previous sections.

It was shown further that weighting individual visual and semantic criteria differently allows for adapting the salience measure to different wayfinding contexts [15].

This original model of salience was investigated for one class of objects: facades of buildings. Consider for example Figure 4, which shows eight street facades of buildings at a street intersection, and their visual and semantic salience represented by grey intensity. According to the original salience model we would choose the object of highest salience (in Figure 4, one of the two facades of building c) as a landmark for route directions.

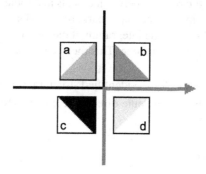

Fig. 4. A street intersection with eight street facades; each individual facade has a salience indicated by grey intensity

4.2 Advance Visibility

Visual and semantic salience characterize properties of objects that are independent from routes and the street network. Route dependent properties of objects can be considered additionally. Each object can be related to an infinite number of routes, but locally the number of combinations, i.e., the number of approaching directions, is small.

This means local route-dependent properties can be represented by a small number of fixed parameters. This number is rarely larger than four; extreme cases, such as the Arc d'Triomphe in Paris, extend already the concept of an intersection to a large circle which can be considered as comprising of several intersections.

The salience model can be extended by *advance visibility* [16]. This measure characterizes the visibility of the object from an approaching direction, and hence, is different for each approaching direction. The rationale for this additional component is that the most salient object at a decision point can form a poor reference in a route direction if it is not visible in advance but an alternative salient object is. Thus advance visibility s_a has to be balanced with joint salience to characterize total salience s_t, e.g., by multiplying the two measures:

$$s_t = s_o * s_a \tag{7}$$

Note that s_a is also a normed value between 0 and 1. Multiplication favors objects that are at the same time at least *to some extent* jointly salient and *to some extent* visible in advance, compared to salient but not visible, or visible but not salient objects. This behavior seems to be reasonable.

Consider for example Figure 4 again. The four facades facing towards the street the wayfinder is approaching are all visible to the wayfinder, but from the four facades facing the cross-road two are only partially visible (4(a) and (b)), and two are not at all visible (4(c) and (d)). Hence, the total salience s_t is largest for the facade of the building c that faces towards the street the wayfinder approaches. This facade should be used for route directions if we consider visual and semantic salience and advance visibility.

At this stage, we have a model that ranks objects by salience and advance visibility, but remains indifferent to the structural characteristics of the relation between objects and street network, or to the relation between objects and routes. However, in Section 3 we saw that the relationships between landmark, route and street network influence the selection process of landmarks. The integration of the structural properties of objects in the salience model still needs to be done. This means we have to develop

- normed salience measures for the identified structural properties of objects (s_u);
- an adaptation of higher order route direction elements (HORDE) for quantitative measures of landmarkness.

4.3 Structural Salience

The discussion of structural properties of landmarks in Section 3 showed that

1. structural properties of objects co-determine their suitability as a landmark;
2. structural properties are, if not quantitative, at least ordered, such that a specific weight of at least an ordered scale can be attached to each situation;
3. structural properties are determined by the structure of the underlying street network, and locally route dependent, which means they are countable and constant.

The set of weights should reflect the hierarchy of Figure 1, and the distinctions of Figures 2 and 3. The order reflected in these figures is motivated by the previous discussions, and partially validated by cognitive, behavioral or linguistic experiments. Presently, we convert the ordered scale measures of structural salience into ratio scale by matching an equally partitioned interval from 0 to 1. However, determining more realistic ratio scale weights needs careful human subject testing and is beyond the scope of this paper.

The third aspect—countable and constant measures—means that the measures can be stored as properties with each object. They are route- and street network dependent, and hence their combinatorial complexity is higher than, for example, for advance visibility. For node landmarks, for example, the complexity is $n(n-1)$, with n being the node degree of the street intersection, because the structure requires consideration of not only the incoming direction, but also the one going off. Note that this includes the distinction between node landmarks with re-orientation, $lm_{\oplus}$, and landmarks without, $lm_{\ominus}$. This means with street intersection degrees of rarely larger than four the complexity is rarely larger than twelve.

The measure of structural salience can be integrated into the original model of salience (Eq. 6). It is still one of three components that add up, i.e., an object is salient if it is visually, semantically *or* structurally distinct.

If we survey people for measures of structural salience with the one-dimensional configurational relationships, the results might be mixed up with expectations of advance visibility. However, in the motivation for structural distinctions we only argued for cognitive and linguistic simplicity. Hence, advance visibility is different and remains a component of total salience (Eq. 7).

For an illustration, consider the situation in Figure 5. The situation shows a route at a decision point with direction change, and some facades with their measures of salience. According to the discussion in Section 3, landmarks at decision points are more pertinent than those along route segments, and at decision points distinctions can again be made in relation to the action (here: turning right). The structural salience measures for the given facades reflect this hierarchy. Note that the structural salience measures recur for buildings (as point-like landmarks), not for facades individually. Advance visibility is assumed to be equal for all facades facing the street the wayfinder approaches to the decision point, less for the facade on the cross-road facing the advancing wayfinder, and zero (not visible in advance) for the facade on the cross-road facing away from the advancing wayfinder.

With Equation 7 we derive the (route- and network dependent) measures of total salience for the considered facades shown in Figure 6. In the given spatial configuration, and for the given visual and semantic salience, the facade with $s_t = 0.72$ is the most salient one. This is particularly interesting as it is not the most visually or semantically salient one. Hence taking into consideration the route- and street network dependent aspects can change the priorities significantly, a behavior that was sought for.

4.4   Selection Process in HORDE

The original salience model did provide a comparison between objects, but did not look into selection. It still assumed a superordinate selection process that exploits salience

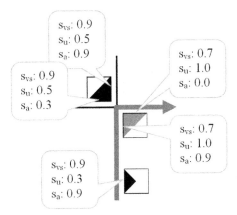

Fig. 5. A route at a decision point with direction change, and measures of salience for some facades (s_{vs}: visual and semantic salience, s_u: structural salience, s_a: advance visibility)

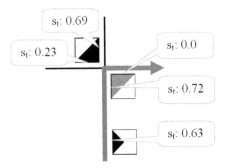

Fig. 6. The total salience s_t for the facades

measures to select references to salient objects where needed. In contrast, structural salience prioritizes visually and semantically salient objects at specific locations along a route. It establishes a selection process by weighting objects between decision points against objects at decision points and so on.

Compared to the discussion in Section 3.3 the situation at this stage has changed. Objects along the route now have more or less salience, and are no longer categorically considered as 'landmarks'. The measures of salience along a route form a distribution, which can further support selection. Let us study the distribution of values with two examples:

– Imagine a route through a green suburb of one-family houses. Salience measures of the objects (facades) along the route differ slightly, but no object stands significantly out. The distribution of salience measures has a small variance and no outliers. Selecting the most salient object along a route segment is possible, but not really helpful.

– Imagine a route along Vienna's *Ringstrasse*. There are frequent salient objects (the parliament, the *Burgtheater*, the city hall, the university, the stock exchange, and so on), and the measures of salience vary largely. The wayfinder is attracted, and if the route description only indicates to *'walk straight to the Danube'* and does not mention the attractions, she might feel uncomfortable and wonder if she is on the right track[2].

In other words, in an environment with one or a few outstanding object(s) these objects can be used as 'landmarks' in the categorical sense of Section 3.3. In an environment with no outstanding objects it is better to refer to other (structural) properties, such as the number of intersections. The appropriate method to distinguish these two cases is outlier detection, i.e., basing the decision on the standard deviation of salience in that environment, not on an absolute threshold value.

An object with a large salience has probably, but not necessarily, structural salience as well. This means that objects with large salience measures have a high probability of being at decision points $dp_\oplus$, or in another salient structural relationships to the route.

5   Conclusions and Outlook

In this paper we have combined two approaches on formalizing route knowledge relevant to the selection of landmarks and for integrating them into route directions: on the one hand the salience of landmarks as dominant objects in route knowledge and route directions, and on the other hand the conceptualization of wayfinding actions in relation to landmarks, i.e., the integration of landmarks in the formal specification of a conceptual route language, the wayfinding choreme theory.

Both approaches on their own are well established and the combination of them results in efficient formalisms addressing several unsolved research questions. Combined, they allow for the specification of structural salience and will complement the basis for an automatic, cognitive adequate generation of route directions in wayfinding assistance systems.

The approach of defining the structural salience of landmarks through the application of a conceptual approach also offers answers to research questions in the area of geosemantics; especially, their formalization, standardization and automatization, for example, for mobile navigation systems. The application of conceptual (cognitive) semantics for geographic information science has recently gained attention through research on ontologies [26].

Other approaches that aim to formally characterize spatial structures have to be considered in greater detail. Especially the approach of space syntax provides several concepts that relate to the topics discussed in this paper [36,37].

With the precisiation of location of landmarks at intersection the next step in the formalization of conceptual knowledge, especially with respect to different modi of externalization, has been achieved. The dual approach of a generic concept that in general specifies the presence of a landmark and the possibility of a more detailed analysis of-

---

[2] An option currently investigated relies on recursion to higher levels of abstraction, such as *'walk straight along the attractive Ringstrasse to the Danube'*.

fers a means to model different levels of granularity in route directions. It also offers a means to contribute the structural salience to models finding salient features by data mining in text documents [38,39].

Acknowledgements

This work has been supported by the Cooperative Research Centre for Spatial Information, whose activities are funded by the Australian Commonwealth's Cooperative Research Centres Programme. We would like to thank Kai-Florian Richter, Bremen, who saw an earlier version, and the anonymous reviewers for invaluable comments.

References

1. Raubal, M., Winter, S.: Enriching wayfinding instructions with local landmarks. In Egenhofer, M.J., Mark, D.M., eds.: Geographic Information Science. Volume 2478 of Lecture Notes in Computer Science. Springer, Berlin (2002) 243–259
2. Nothegger, C., Winter, S., Raubal, M.: Selection of salient features for route directions. Spatial Cognition and Computation **4** (2004)
3. Sorrows, M.E., Hirtle, S.C.: The nature of landmarks for real and electronic spaces. In Freksa, C., Mark, D.M., eds.: Spatial Information Theory. Volume 1661 of Lecture Notes in Computer Science. Springer, Berlin (1999) 37–50
4. Klippel, A., Tappe, H., Kulik, L., Lee, P.U.: Wayfinding choremes – A language for modeling conceptual route knowledge. Journal of Visual Languages and Computing (in press)
5. Klippel, A., Dewey, C., Knauff, M., Richter, K.F., Montello, D.R., Freksa, C., Loeliger, E.A.: Direction concepts in wayfinding assistance systems. In Baus, J., Kray, C., Porzel, R., eds.: Workshop on Artificial Intelligence in Mobile Systems 2004 (AIMS 2004). Volume Memo Nr. 84., Nottingham, UK, Sonderforschungsbereich 378 (2004) 1–8
6. Lynch, K.: The Image of the City. MIT Press, Cambridge (1960)
7. Presson, C.C., Montello, D.R.: Points of reference in spatial cognition: Stalking the elusive landmark. British Journal of Developmental Psychology **6** (1988) 378–381
8. Siegel, A.W., White, S.H.: The development of spatial representations of large-scale environments. In Reese, H.W., ed.: Advances in child development and behavior. Volume 10. Academic Press, New York (1975) 9–55
9. Couclelis, H., Golledge, R.G., Tobler, W.: Exploring the anchorpoint hypothesis of spatial cognition. Journal of Environmental Psychology **7** (1987) 99–122
10. Denis, M.: The description of routes: A cognitive approach to the production of spatial discourse. Current Psychology of Cognition **16** (1997) 409–458
11. Michon, P.E., Denis, M.: When and why are visual landmarks used in giving directions? In Montello, D.R., ed.: Spatial Information Theory. Volume 2205 of Lecture Notes in Computer Science. Springer, Berlin (2001) 292–305
12. Tversky, B., Lee, P.: Pictorial and verbal tools for conveying routes. In Freksa, C., Mark, D., eds.: Spatial Information Theory. Volume 1661 of Lecture Notes in Computer Science. Springer-Verlag, Berlin (1999) 51–64
13. Lee, P., Klippel, A., Tappe, T.: The effect of motion in graphical user interfaces. In Butz, A., Krüger, A., Olivier, P., eds.: Smart Graphics. Springer, Berlin (2003) 12–21
14. Tom, A., Denis, M.: Referring to landmark or street information in route directions: What difference does it make? In Kuhn, W., Worboys, M., Timpf, S., eds.: Spatial information theory. Volume 2825 of Lecture Notes in Computer Science. Springer, Berlin (2003) 384–397

15. Winter, S., Raubal, M., Nothegger, C.: Focalizing measures of salience for wayfinding. In Meng, L., Zipf, A., Reichenbacher, T., eds.: Map-based Mobile Services – Theories, Methods and Implementations. Springer Geosciences, Berlin (2005) 127–142

16. Winter, S.: Route adaptive selection of salient features. In Kuhn, W., Worboys, M.F., Timpf, S., eds.: Spatial Information Theory. Volume 2825 of Lecture Notes in Computer Science. Springer, Berlin (2003) 320–334

17. Elias, B.: Extracting landmarks with data mining methods. In Kuhn, W., Worboys, M.F., Timpf, S., eds.: Spatial Information Theory. Volume 2825 of Lecture Notes in Computer Science. Springer, Berlin (2003) 398–412

18. Elias, B., Brenner, C.: Automatic generation and application of landmarks in navigation data sets. In Fisher, P., ed.: Developments in Spatial Data Handling. Springer, Berlin (2004) 469–480

19. Klippel, A., Lee, P.U., Fabrikant, S.I., Montello, D.R., Bateman, J.: The cognitive conceptual approach as a leitmotif for map design. In Barkowsky, T., Freksa, C., Hegarty, M., Lowe, R., eds.: Reasoning with Mental and External Diagrams: Computational Modeling and Spatial Assistance. Volume TR SS-05-06., Stanford, CA, AAAI Press (2005) 90–94

20. Richter, K.F., Klippel, A.: A model for context-specific route directions. In Freksa, C., Nebel, B., Knauff, M., Krieg-Brückner, B., eds.: Spatial Cognition IV. Volume 3343 of Lecture Notes in Artificial Intelligence. Springer, Berlin (2005) 58–78

21. Casati, R., Varzi, A., eds.: Events. Dartmouth, Aldershot (1996)

22. Zacks, J., Tversky, B.: Event structure in perception and conception. Psychological Bulletin 127 (2001) 3–21

23. Worboys, M.: Event-oriented approaches to geographic information. International Journal of Geographical Information Science 19 (2005) 1–28

24. Gärdenfors, P.: Conceptual Spaces. The MIT Press, Cambridge, MA (2000)

25. Raubal, M.: Formalizing conceptual spaces. In Varzi, A.C., Vieu, L., eds.: Formal Ontology in Information Systems. Volume 114 of Frontiers in Artificial Intelligence and Applications., Amsterdam, NL, IOS Press (2004) 153–164

26. Kuhn, W.: Why information science needs cognitive semantics. In: Paper presented at Workshop on the Potential of Cognitive Semantics for Ontologies (FOIS 2004), Torino, Italy, http://musil.uni-muenster.de/documents/WhyCogLingv1.pdf (2004)

27. Sowa, J.F.: Knowledge Representation: Logical, Philosophical, and Computational Foundations. Brooks Cole, Pacific Grove (2000)

28. Klippel, A., Montello, D.R.: On the robustness of mental conceptualizations or the scrutiny of direction concepts. In Egenhofer, M., Miller, H.J., Freksa, C., eds.: GIScience 2004, University of Maryland, MD, Regents of the University of California (2004) 139–141

29. Klippel, A.: Wayfinding choremes. In Kuhn, W., Worboys, M.F., Timpf, S., eds.: Spatial Information Theory. Volume 2825 of Lecture Notes in Computer Science. Springer, Berlin (2003) 320–334

30. Dale, R., Geldof, S., Prost, J.P.: Using natural language generation in automatic route description. Journal of Research and Practice in Information Technology 37 (2005) 89–105

31. Couclelis, H.: Verbal directions for way-finding: Space, cognition, and language. In Portugali, J., ed.: The Construction of Cognitive Maps. Kluwer, Dordrecht (1996) 133–153

32. Klippel, A., Tappe, H., Habel, C.: Pictorial representations of routes: Chunking route segments during comprehension. In Freksa, C., Brauer, W., Habel, C., Wender, K.F., eds.: Spatial Cognition III. Volume 2685. Springer, Berlin (2003) 11–33

33. Lovelace, K.L., Hegarty, M., Montello, D.R.: Elements of good route directions in familiar and unfamiliar environments. In Freksa, C., Mark, D.M., eds.: Spatial Information Theory. Volume 1661 of Lecture Notes in Computer Science. Springer, Berlin (1999) 65–82

34. Herrmann, T., Schweizer, K., Janzen, G., Katz, S.: Routen- und Überblickswissen – konzeptuelle Überlegungen. Kognitionswissenschaft 7 (1998) 145–159

35. Kulik, L., Egenhofer, M.: Linearized terrain: Languages for silhouette representation. In Kuhn, W., Worboys, M.F., Timpf, S., eds.: Spatial Information Theory. Volume 2825 of Lecture Notes in Computer Science. Springer, Berlin (2003) 128–145

36. Davis, L.S., Benedikt, M.L.: Computational models of space: Isovists and isovist fields. Computer Graphics and Image Processing 11 (1979) 49–72

37. Batty, M.: Exploring isovist fields: space and shape in architectural and urban morphology. Environment and Planning B 28 (2001) 123–150

38. Tezuka, T., Yokota, Y., Iwaihara, M., Tanaka, K.: Extraction of cognitively-significant place names and regions from web-based physical proximity co-occurrences. In Zhou, X., Su, S., Papazoglou, M.P., Orlowska, M.E., Jeffrey, K.G., eds.: Web Information Systems (WISE 2004). Volume 3306 of Lecture Notes in Computer Science. Springer, Berlin (2004) 113–124

39. Winter, S., Tomko, M.: Translating the web semantics of georeferences. In Taniar, D., Rahayu, W., eds.: Web Semantics and Ontology. Idea Publishing, Hershey, PA (accepted for publication)

Expert and Non-expert Knowledge of Loosely Structured Environments*

Sylvie Fontaine[1,2], Geoffrey Edwards[1,2], Barbara Tversky[2,3], and Michel Denis[2,4]

[1] Centre de Recherche en Géomatique, Laval University, Quebec City, Canada
[2] The GEOIDE Network
[3] Department of Psychology, Stanford University, USA
[4] Groupe Cognition Humaine, LIMSI-CNRS, Orsay, France

Abstract. Three experiments investigated expert and non-expert knowledge of a familiar but loosely structured spatial environment as revealed through the production of sketch maps. In the first experiment, experts and non-experts in geomatics sketched maps of a well-known park. The analysis of the maps revealed that experts and non-experts used different drawing strategies that reflected different mental representations. In the second experiment, new participants identified good and poor examples from the previous maps. Expert and non-expert evaluators agreed, indicating that experts and non-experts alike agree on what constitutes a "good map". In the third experiment, people familiar and unfamiliar with the park were asked to remove non-essential features from a consolidated map that incorporated all the features drawn by the participants of the first experiment. Those familiar and unfamiliar with the environment retained the same features, notably, the paths in the park. Together, the research shows that experts produce superior maps to non-experts, but that people, irrespective of expertise and familiarity, concur on the features that make a map effective. Even for relatively unstructured environments like a large park, people seek structure in the configuration of paths. These findings have implications for the design of maps.

Keywords: Spatial cognition, maps, navigation, metacognitive knowledge, expertise, design, parks.

1 Introduction

To communicate environments, people commonly rely on descriptions or depictions, language or graphics. These two modes of externalization of spatial knowledge have been analyzed to reveal the content and structure of the mental representations of space. Studies have emphasized both the specificities of depictive and descriptive modes of representation, and also their intimate connections (e.g., Przytula-Machrouh, Ligozat, & Denis, 2004; Rinck & Denis, 2004; Taylor & Tversky, 1992a, 1992b). Tversky and Lee (1998, 1999) went as far as suggesting a common conceptual structure underlying depiction and description of familiar routes. They showed that people's spontaneous sketch maps and verbal directions were described

---

* The work reported in this paper was funded through two projects within the purview of the GEOIDE Network of Centers of Excellence, the DEC 30 project and the DEC/JON project. The authors are grateful to Ariane Tom for her help in the preparation of the manuscript.

A.G. Cohn and D.M. Mark (Eds.): COSIT 2005, LNCS 3693, pp. 363–378, 2005.
© Springer-Verlag Berlin Heidelberg 2005

by the conceptual structure, a structure Denis (1997) derived from a large corpus of spontaneous route directions. This suggests that both sketch maps and verbal directions are different externalizations of the same underlying mental representation. The core of that structure is a network of paths and nodes.

Corpora of spontaneous route directions have provided a rich source of information about effective directions (e.g., Allen, 2000; Denis, 1997; Denis, Pazzaglia, Cornoldi, & Bertolo, 1999; Golding, Graesser, & Hauselt, 1996; Michon & Denis, 2001; Schneider & Taylor, 1999). From these corpora, skeletal directions can be abstracted. To derive skeletal directions, first, all elements from all participants' directions are combined. Then, a group of judges selects those elements that are essential for navigation. Interestingly, judges familiar and unfamiliar with the environment tend to pick the same elements (Denis et al., 1999). The agreement of judges who do and do not know the environment suggests that selecting the crucial pieces of information in route directions is based on metacognitive knowledge that is to some extent independent of a specific environment. Similarly, participants familiar and non-familiar agreed on ratings of the communicative value of the original directions. The skeletal directions and the rated spontaneous directions were validated in studies using directions of varying judged goodness as well as the skeletal directions as navigation aids (Daniel, Tom, Manghi, & Denis, 2003; Denis et al., 1999). These studies confirmed that descriptions are variants of a core structure, a combination of links and nodes reflected in the skeletal directions (see also Fontaine & Denis, 1999; Michon & Denis, 2001). As noted, the core structure is expressed in sketch maps of routes as well as verbal directions (Lynch, 1960; Tversky & Lee, 1998, 1999). It has been applied to the design of computer algorithms that generate effective and popular route maps (Agrawala & Stolte, 2001).

Is this link/node core reflected in survey maps as well as route ones? Will it hold for environments that are not as highly structured as urban environments, environments that are used for recreation and wandering rather than for getting from place to place? Do maps produced by experts in map use and design differ from those produced by non-experts? And, finally, do people familiar and unfamiliar with an environment agree on the features that make for an effective map? In other words, do people have metacognitive knowledge of what is important and what is secondary in maps? We posed these questions in three studies. In the first, experts and non-experts in map production and use were asked to produce maps of a large park well-known to all of them. In the second study, those maps were evaluated by other participants, familiar or unfamiliar with the park. In the third experiment, new participants familiar or unfamiliar with the park selected the information they deemed important from an amalgamation of the information included in the original maps.

This procedure accomplishes two objectives simultaneously: it both reveals the mental representations people have of environments and establishes principles for designing effective maps to communicate those representations, thus creating a context for the development of new representational tools. Because the principles turn out to be the same for familiar and unfamiliar users, they can be broadly applied.

2 Experiment 1: Sketching Maps

The use of sketch maps as indices of spatial knowledge is not free of difficulties. These maps are generally incomplete and distorted, and they tend to mix metrics.

However, the distortions and omissions in sketch maps reflect people's underlying mental representations of environments by numerous other methods (e.g., Tversky, 1981, 1993). They are schematic and incomplete, often including blank spaces and unconnected networks. As a result, scoring for the purpose of assessment is a challenge. However, sketch maps have been shown to be reliable and preserve consistent information over time (e.g., Blades, 1990; Tversky, 1981). Moreover, they closely correspond to other indices of mental representations, such as descriptions, recall, and response times to answer questions about proximity and direction (e.g., Taylor & Tversky, 1992b; Tversky & Lee, 1998). As suggested by Davies and Pederson (2001), analyzing sketch maps can be challenging if the focus is on accuracy, but this does not preclude the value of sketch maps if the focus of the study is to exploring the knowledge elicited and the strategy followed by the people engaged in map drawing.

The construction of sketch maps is related to the organization of information in the mental representation of the described environment. Taylor and Tversky (1992a) analyzed the order in which elements of an environment were included in a map. Drawing order varied, and depended on cognitive features of the environments, over and above any constraints that might be imposed by the task of drawing. Taylor and Tversky found that the order of drawing reflected hierarchical organization of the environments, and that the hierarchy depended on both spatial and functional aspects of the environments. Subgroups were based on spatial proximity, spatial scale, and functional features. Walsh, Krauss, and Regnier (1981) used sketch maps to discover the structures people rely on to describe their neighborhoods. Most participants began their maps with some sort of street grid, and then filled in the pattern with landmarks and a few more streets.

Following these endeavors, the maps collected in the present experiment were first analyzed for their content and structure. We focused on the quality and quantity of information included, in particular landmarks and roads. Errors of location were also examined. As in the previous investigations, we recorded the order in which the different parts of the map were drawn, expecting to find evidence for a hierarchical organization of the maps. Spatial proximity and functional aspects were thought to be potential sources of influence on the structure of the map. Classic research on expertise generally attributes the memory superiority of experts to better organization of information in their knowledge base (e.g., de Groot, 1966). Therefore, the structuring of information in maps of experts should differ from that of non-experts.

2.1 Method

Environment. The environment selected for the study was the major park of Quebec City, the Plains of Abraham. It lies over an extended space, covering about one hundred hectares, rather longer than wide. The park is delimited on the north side by the city and on the south by a steep hill overlooking St. Laurent River. The park presents a wide variety of relief. There are only a few roads in the park. Compared to a city or a campus, this environment is only loosely structured.

Participants. Two groups of people participated in the experiment. The first group was composed of 9 graduate students in geomatics at Laval University (8 men, 1 woman). They were considered as experts in the domain of map processing. The second group was composed of 27 graduate students in other disciplines (13 men, 14 women). They were considered as non-experts as regards map processing. The criterion for including the participants in the study was their knowledge of the park of which they would draw the map. Participants of both groups had been living in Quebec City for more than 15 years and reported to experience the park frequently, at least once a month on the average, both during winter and summer. In this and subsequent studies, the effect of gender was examined; there were no reliable effects, so these analyses are not included.

Materials. White sheets of paper, legal size, were made available to participants to draw the maps.

Procedure. Participants were asked to draw a map of the Plains of Abraham. The map was intended to provide information necessary to navigating the park and finding the major points of interest to those unfamiliar with the park. Sessions were video recorded. At the end of the experiment, participants filled in a questionnaire on how they perceived the task just completed.

2.2 Results

Map Content. For each map, the number of landmarks, road segments, and road intersections were tallied; these appear in Table 1 for expert and non-expert participants. An analysis of variance (ANOVA) was conducted on each group of items. Experts reported more landmarks, $F_{(1, 34)} = 5.70$, $p < 0.05$, road segments, $F_{(1, 34)} = 17.12$, $p < 0.001$, and intersections, $F_{(1, 34)} = 21.32$, $p < 0.001$, than non-experts. Overall, experts reported an average of 52.0 items, while non-experts reported an average of 25.4 items, $F_{(1, 34)} = 15.64$, $p < 0.001$.

Table 1. Average number of items reported (standard deviations are in parentheses)

	Experts	Non-experts
Landmarks	20.4 (9.8)	13.2 (7.2)
Road segments	17.7 (8.8)	7.4 (5.5)
Intersections	13.9 (7.4)	4.8 (4.2)

Errors were categorized as "global" or "local". To this effect, the area of the park was divided into six sub-areas. For a given sketch map, we considered as a global error every occurrence of an object (a landmark, for instance) which was drawn in a wrong sub-area, and as a local error every occurrence of an object wrongly positioned in its correct sub-area. The average number of errors is shown in Table 2. There were overall very few global errors, but non-experts made more such errors than experts, $F_{(1, 34)} = 4.55$, $p < 0.05$. There was no difference between experts and non-experts in local errors.

Debriefing revealed that all experts but one reported having seen a map of the park, but only half the non-experts had (13 had and 14 had not seen a map). Those who had seen a map produced more landmarks, 16.0 (sd = 7.9), than those who had not, 10.5 (sd = 5.5), $F_{(1, 23)} = 4.74$, $p < 0.05$.

Table 2. Average number of errors (standard deviations are in parentheses)

	Experts	Non-experts
Global errors	0.1 (0.3)	0.8 (1.0)
Local errors	2.1 (1.4)	2.0 (1.4)

Questionnaire. In the post-experimental questionnaire, participants rated several aspects of the task on a 1-5 rating scale: confidence in the information contained in the map, confidence in the location of items on the map, ease of map drawing, self-rated knowledge of the park, and self-rated sense of direction. Only the first measure differed between the groups, with experts expressing more confidence in the information they included in their maps than non-experts, 4.1 (sd = 1.0) and 3.5 (sd = 0.9), respectively, $F (1, 34) = 3.85$, $p < 0.05$.

Orientation of Maps. As revealed in Table 3, experts tended to orient their maps north-up, but non-experts did not, $Chi2 (1) = 14.48$, $p < 0.001$. Non-experts preferred to orient maps with the park entrance at the bottom, as though one could walk into the map, a strategy observed in previous work (e.g., Taylor & Tversky, 1992; Tversky, 1981).

Table 3. Frequency of placement of north at the top or bottom of the sheet by experts and non-experts

	Experts	Non-experts
North at the top	8	5
North at the bottom	1	22

Order of Drawing Roads and Landmarks. We selected the first 20 items (roads and landmarks) drawn by each participant and, among these, those produced by at least half the participants. A value was given to each item, corresponding to the rank order of drawing of this item. The median rank was then calculated for each item. These computations revealed differences between the two groups. Experts drew the structure of the roads earlier than non-experts. Significantly, the first item drawn by experts, but not non-experts, was the Grande Allée, the street which runs along the park and marks the border between the city and the park. This street orients the park in the surrounding environment. Both experts and non-experts drew roads prior to landmarks; roads ranked 6.5 and landmarks 11.5. Thus, maps are structured first by roads or links, and these are used for locating landmarks.

Order of Drawing Landmarks. We selected the 10 major landmarks drawn by all participants in order to determine whether these were hierarchically organized. Following Taylor and Tversky (1992), we conducted cluster analyses on these landmarks. In their research, the clusters were an excellent index of hierarchical organization. Recall order has been used as an index of hierarchical organization at least since Tulving (1962). For each map, we calculated the recall interval for every pairwise combination of landmarks, that is, the number of other landmarks recalled between the two items of the pair. The median recall interval for each pair of landmarks was calculated and represented in a half matrix. We used this matrix to compute the cluster analysis for both groups of participants (ADDTREE; Sattath & Tversky, 1977).

Figure 1 shows the clustering of landmarks for experts. Two groups of items emerged. The first one included the Jogging Loop, the Grey Terrace, the Garden, and the Museum. The second one included the Bandstand, the Loews Hotel, the Martello Tower, and the Citadel. Landmarks from the first group were mostly in the west part of the park and those from the second group were mostly in the east part. The further two landmarks (the Kiosk and the Promenade) were at the eastern limit of the park. This structure thus confirmed the progression from west to east in map drawing and showed that the construction of the experts' maps was mainly based on the principle of spatial proximity.

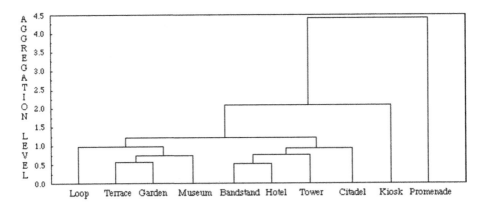

Fig. 1. Cluster diagram for landmarks identified by experts. The ordinal variable is Aggregation Level.

Figure 2 shows the clustering of landmarks for non-experts. The clustering is quite different than for the experts. Two groups of items emerged. The first included the Jogging Loop, the Loews Hotel, the Grey Terrace, and the Citadel. The Jogging Loop is at the western end of the park; the Loews Hotel is on a border of the park, equidistant from the western and eastern extremities; the Grey Terrace is in the west part of the park, south of the Jogging Loop; and the Citadel is at the eastern extremity. These items are all located on the borders of the park and their positions provide a rectangle-like frame. Once these items were drawn, the resulting virtual rectangle was filled in with the items located inside the park. Thus, the elaboration of the maps by the non-experts followed a strategy consisting in drawing items on the borders first, then filling in the structure. Spatial proximity was not used as a governing rule in the construction of the maps.

To summarize, while experts seemed to rely on spatial proximity to draw the landmarks, non-experts seemed to rely primarily on the functional properties of the landmarks. Because landmarks were located on the borders, they became functionally significant to enclose the space of the park.

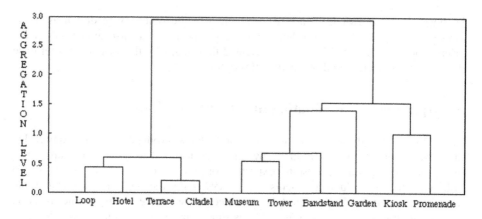

Fig. 2. Cluster diagram for landmarks identified by non-experts. The ordinal variable is Aggregation Level.

2.3 Discussion

Experts' maps of a familiar, loosely structured environment differ from those of non-experts. Experts included more information than non-experts, an effect not due to different exposure as the groups reported equal knowledge and frequency of visiting. More likely, the superior performance of experts is connected to their greater acquired capacity to manipulate spatial information, read and use cartographic materials, which helps them to better organize spatial information. Internal organization of information thus facilitates the retrieval of items to be included in the map. The marked reference to road information is another indication that experts' knowledge is more strongly structured than that of non-experts.

The analysis of errors revealed an interesting finding. Even if we condition recall of location of landmarks on overall recall of landmarks, experts were locating landmarks better. This suggests that for experts, memory for landmark and memory for location were tightly linked, but for non-experts, they were more independent. When non-experts remembered the location, they were as accurate as experts (the number of local errors was the same).

The maps of both experts and non-experts were hierarchically structured, but differently. Experts' maps were primarily structured by roads. The roads constitute a framework with respect to which landmarks are located. Non-experts relied less on roads. They constructed their maps from the borders inside. In addition, the representations of non-experts were less structured than those of the experts.

Expertise had also an effect on the orientation of the maps. The experts followed cartographic convention by placing north at the top of the map. They also demonstrated greater ease in adopting a survey perspective to externalize their spatial knowledge. By contrast, the orientation of the maps suggested that non-experts did not adopt a consistent survey perspective, but rather mixed survey and route perspectives. Taylor and Tversky (1996) reported that people often mix perspectives when they have to produce descriptions of environments. A similar process may be at work in the construction of maps. Inspection of non-expert maps revealed that some landmarks were drawn from a bird's eye view, while others were drawn as if the drawer took a frontal view on them. A route perspective was also evidenced by the orientation of the maps. Non-experts oriented their maps by the way they experience the park when entering and proceeding through it.

3   Experiment 2: Evaluating the Quality of Maps

The maps produced by experts are superior to those produced by non-experts. Do their evaluations of maps produced by others correspond to their own maps, or is there general agreement despite expertise on the qualities of a "good map"? Following the procedures of Denis et al. (1999) for the analysis of verbal route directions, experts and non-experts were asked to assess the quality of the maps on several rating scales. Because this task was time-consuming, we randomly selected a subset of 25 maps from the 36 collected in Experiment 1. Cartographers use explicit criteria for the generation of maps and if these criteria are applied, the quality of the resulting map is assured. The question here was whether non-experts would adopt the same or different criteria.

Based on the literature in graphic semiology and cartography (e.g., Bertin, 1967), we selected two classes of criteria that seemed to be important to experts: those related to the physical qualities of the maps, and those related to their functional qualities. For the physical qualities, three aspects pertain to structures (i.e., roads and landmarks): identifying the structures, preserving their proportions, and preserving their relative positions. Another three aspects pertain to the map itself: amount of information included, homogeneity of scale, and aesthetic qualities of the map. For functional qualities, three aspects pertain to the processing of the map: ease of reading, ease of locating structures, and ease of recognizing structures. Another three aspects are related to using a map: ease of locating oneself, ease of selecting a goal, and ease of constructing a route.

If people have metacognitive knowledge of what constitutes a good map, judgments of experts and non-experts, those familiar with the environment and those not, should be similar. If, on the other hand, such shared knowledge does not exist, we would expect experts, who rely on a set of cartographic rules, to give more importance to these criteria than non-experts. Additionally, experts might be harsher in their evaluations. Moreover, not knowing the described environment could make the judges more demanding, so that they might give lower evaluations than judges familiar with the park. On the other hand, those unfamiliar with the environment might be more forgiving of the inclusion of landmarks and of the accuracy of their locations simply because their knowledge is incomplete.

3.1   Method

Participants. Twelve people participated in this experiment. Four of them were experts according to the criterion used in Experiment 1, and eight were non-experts. In each group, half were familiar with the park (visiting it at least once a week), and the other half had never visited it or had done so just once. Within these categories, there was an equal number of men and women.

Materials. A subset of 25 of the maps collected in Experiment 1 were used, 9 from experts and 16 from non-experts, presented on separate sheets of paper.

Procedure. Participants evaluated the overall quality of the maps and then used 7-point scales to judge them on 12 criteria.

3.2 Results

Overall Scores. An ANOVA did not reveal any significant differences between judgments of experts and non-experts, nor between participants who were familiar or unfamiliar with the environment. Furthermore, the correlation matrix among the scores given by the 12 judges revealed that all 66 correlation values were positive, with 55 significant at a probability level of 0.05 or less. Intra-class coefficients amounted to 51.3% for the whole set of judges; 52.6% and 48.9% for experts and non-experts, respectively; and 45.2% and 53.7% for familiar and unfamiliar judges, respectively. These data suggest a common conception of what is a good map, and of implicit criteria shared by the experts and the non-experts.

Scores on Individual Criteria. ANOVAs were conducted on scores given to the maps for each of the 12 criteria considered in turn. Expertise and familiarity did not affect the scores on any of these criteria. We also wanted to estimate the relative weight of the criteria in the global evaluation expressed by the overall score. This was done by using an analysis of stepwise regression on the overall score. The analysis proposed a model with 8 of the 12 criteria, with $R^2 = 0.8455$. The results showed that 81% of the variance of the overall scores was explained by three criteria (in decreasing order): ease of locating oneself; amount of information included; and ease of recognizing structures. These three criteria were also found in the models calculated for experts and non-experts separately, and for familiar and unfamiliar participants, separately. The model obtained for the experts also included the aesthetic qualities of the map.

"Good" Versus "Poor" Maps. Three maps received average overall scores of 5.00 or more; two of these were produced by experts, and one by a non-expert. The three maps had similar profiles over the 12 individual criteria. The three maps rated poorest (below 2.00) were drawn by non-experts. When examining their scores across the 12 criteria, there was in fact less homogeneity in their profiles than for the best maps.

Drawing Expertise. The maps produced by experts received higher overall scores than those produced by non-experts, 4.0 and 3.2, respectively, $F (1, 284) = 19.01$, $p < 0.001$. Experts' maps were rated higher on many of the criteria for a good map: preserving proportions among structures; preserving relative positions of structures; amount of information included; homogeneity of scale; ease of locating structures; ease of locating oneself; and ease of constructing a route (in all cases, $p < 0.001$). The criteria receiving the highest scores in experts' maps were related to the spatial properties of the maps. Thus, what differentiates expert from non-expert maps is spatial adequacy and veracity. These, of course, are the first requisites of a map, and point to the difficulties encountered by non-experts in accurately representing spatial relations among structures.

3.3 Discussion

This study, in which experts and non-experts rated maps produced by experts and non-experts, provides clear evidence for shared conceptions of what constitutes a good map. The ratings of map quality were strongly correlated across participants irrespective of expertise and familiarity, echoing previous work on route directions (Denis et al., 1999). Shared knowledge and criteria create a context conducive to easier communication, whether that communication is by maps or language.

Three criteria for a good map were especially strong in the regression analysis. A good map must, first of all, help users position themselves in an environment; next, it must contain an adequate amount of information; and finally, the structures drawn on the map should be recognizable.

4 Experiment 3: Constructing a Skeletal Map

The aim of Experiment 3 was to construct a "skeletal map" of the environment considered, by following a procedure paralleling a similar procedure used in building "skeletal directions" (Denis et al., 1999; Fontaine, 2000). As a first step, we built a "mega-map" containing all information provided by all the participants in Experiment 1. Participants in the present experiment selected the items that they thought should be present in a map intended to provide necessary and sufficient information to users. As before, both people familiar and people unfamiliar with the environment participated, allowing assessment of effects of familiarity. By comparing the responses from people familiar or unfamiliar with the described environment, we expected to uncover whether common implicit knowledge is available for people, independent of their knowledge of the environment. If the responses of familiar and unfamiliar participants are similar, then it is likely that this is because they share knowledge of the criteria of good maps.

4.1 Method

Participants. Thirty-two participants were recruited, half of them familiar and the other half unfamiliar with the park, according to the criteria used for the previous two experiments. In both groups, there was an equal number of men and women.

Materials. A mega-map of the environment was generated on a computer from a geo-referenced database. A total of 114 informational items, drawn from the responses of participants of Experiment 1, were positioned on the mega-map at their exact locations. For the roads and the major landmarks, existing locational data were used, but for many other landmarks, we had to measure their exact spatial coordinates with a GPS receiver. The map was then constructed using MapInfo™ software (see Figure 3).

Procedure. Participants were tested in groups. The experiment took place in a classroom. Participants faced two screens. On one screen, the mega-map was shown for the whole duration of the experiment. On the second screen, four successive enlargements of the mega-map were projected, each enlargement representing an area of the park. On each enlargement, information items were shown, then suppressed, then shown again. Instead of selecting or rejecting each item by all-or-none choice, the participants were invited to use a 5-point rating scale to estimate the extent to which they thought this item should be kept in the skeletal map. The map was said to allow a person who does not know the park to move efficiently without getting lost and to find every element that he or she could be interested in. With this purpose in mind, the participants were invited to give the value 1 to information items that should definitely be eliminated, 2 to items that should probably be eliminated, 3 to items that could be kept or discarded indifferently, 4 to items that should probably be kept, and 5 to items that should definitely be kept. This was done for all 114 information items in turn.

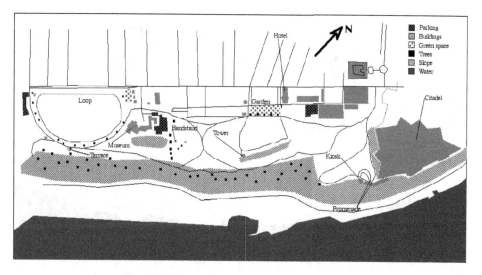

Fig. 3. Mega-map for the Plains of Abraham Park

4.2 Results

We classified the 114 information items of the mega-map into ten classes, which are listed below (with the number of items included):

- Roads within the park (13)
- Roads at the outside border of the park (28)
- Buildings within the park (large surface objects) (30)
- Buildings at the outside border of the park (10)
- Objects and monuments within the park (small surface objects) (15)
- Objects and monuments at the outside border of the park (3)
- Properties of the terrain (9)
- Specific indications (restrooms, points of view, services) (4)
- Indication of north (1)
- St. Laurent River (1)

For each information item, we computed the average rated value. Those items receiving a value equal to or above 4.0 were considered to be kept as items of the skeletal map (a total of 55 items were in such a situation). Not surprisingly, the single items of the last two classes were selected as skeletal items, namely, the reference to north, and the reference to St. Laurent River. Although the river was not part of the park itself, this remote landmark had a special status as a reference in the description of the park (see Figure 4).

The first four classes listed above contained items that were selected to be included in the skeletal map, but none of the items in the next four classes (objects and monuments of secondary importance, properties of the terrain, and specific indications) were rated to be included in the skeletal map.

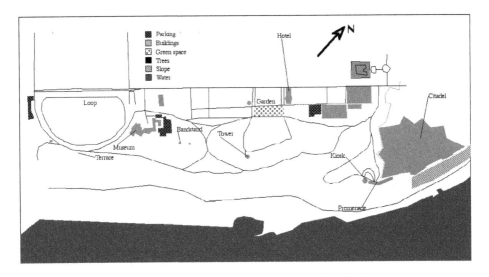

Fig. 4. Skeletal map for the Plains of Abraham Park

Table 4 shows the number of items of the first four classes kept in the skeletal map by the two groups of participants. Not surprisingly, more items within the park were maintained in the skeletal map than outside items, and roads were preserved more than buildings. The most interesting feature here was that the familiarity of the participants with the environment did not affect their perception of the importance of items. In other words, those items of primary importance for a guidance or navigation purpose were perceived as such even by those participants who had no knowledge of the environment. Based on the number of items kept by familiar and unfamiliar participants, the Chi2 value was not significant.

Following the procedure developed with route directions (Denis, 1997), we computed a measure of richness for the maps, that is, the proportion of skeletal items present in individual maps collected in Experiment 1. Here, we focused on the best three and the poorest three maps, according to the participants of Experiment 2. The first three maps had an average richness index of 69.1%, whereas the last three had an index of 16.4%. Thus, the richer a map is in items belonging to the skeletal map, the better it is judged in terms of quality.

Table 4. Number of items in the mega-map and in the skeletal map for participants familiar and unfamiliar with the environment

	Roads within the park	Roads at the outside border of the park	Buildings within the park	Buildings at the outside border of the park
Mega-map	13	28	30	10
Skeletal map (Familiar part.)	12	14	17	5
Skeletal map (Unfamiliar part.)	12	17	16	6

4.3 Discussion

The analyses reported above did not show any effect of familiarity on the judgment of the necessity of including items in the skeletal map. This lack of difference is highly compatible with the hypothesis of a common knowledge base. Being familiar or not with an environment does not appear to be crucial for determining the necessity of information on a map. Selecting essential elements in a map is based on knowledge that is independent of the specific environment.

The information that is preserved on the skeletal map essentially consists of roads and landmarks. The selected landmarks only consist of large-size buildings. This confirms visual saliency as a primary criterion of landmark selection (cf. Nothegger, Winter, & Raubal, 2004; Tom & Denis, 2003, 2004).

5 Conclusions

The three experiments reported here were conducted to investigate the mental representations of loosely structured environments by experts and non-experts, by those familiar and those unfamiliar with the environment. Implicit in this interest is the hope that mental representations of such environments will provide clues to the design of effective visualizations of environments. This double enterprise extends the efforts of Denis and his collaborators (Denis, 1997; Denis et al., 1999) and Tversky and her collaborators (Tversky, Agrawala, Heiser, Lee, Hanrahan, Stolte, & Daniel, in press; Tversky & Lee, 1998, 1999) from route directions and route maps of structured environments to area maps of loosely structured environments, in particular, a large urban park. This endeavor raises several questions. Is there a core structure underlying mental representations and visualizations of environments? The previous findings, discussed in depth in the introduction, indicate that there is core knowledge for route maps; here we have provided such evidence for the case of survey maps. Is there any metacognitive knowledge of what is important in a map and of what may be considered to be a good map?

To summarize, our results showed that experts' maps are different and better than those of non-experts. Experts begin by orienting the environment in the larger surroundings, continue to the basic framework of the environment, the structure of the roads, and then attach the landmarks to the framework. This structure and the order of drawing contradict some old notions of spatial cognition that claim that people construct mental representations of space first from landmarks and then paths, followed by survey representations (e.g., Siegel & White, 1975).

People who are not expert and not familiar with the environments prefer the maps that experts construct, a recurrent finding (see Tversky et al., in press) and the reflection of a lag between comprehension and production. We can appreciate and evaluate movies and books and meals that we cannot create. This is encouraging for design, as it says that design principles can be extracted from expert productions that will be successful for experts and non-experts alike. The techniques developed by Denis (1997; Denis et al., 1999) of extracting collective knowledge (mega-descriptions and skeletal descriptions) and judgments thereon are useful for finding design principles. The present research provides guidelines for constructing survey maps that are analogous to the guidelines for route directions produced by Denis (1997) and confirmed by Tversky and Lee (1998, 1999), namely, they provide the structure of the links and locate the landmarks with respect to these.

Design principles for constructing effective route maps growing out of the research of Denis (1997) and Tversky and Lee (1998, 1999) were implemented in an algorithm that generates thousands of route maps a day on demand (http://www.mappoint.com; cf. Agrawala & Stolte, 2001). These maps have been enthusiastically received by users (cf. Tversky et al., in press). The design principles for route maps include depicting the paths and turning points (links and nodes) clearly; exact distance and direction as well as links not on the path can be ignored. The present research suggests that these principles can be extended to designing survey maps. In the case of survey maps, the link and node structure will place additional constraints on distance and direction, increasing their accuracy.

The experiments reported here allowed us to situate the knowledge of experts with respect to the knowledge of non-expert map users, and hence to advance understanding of how spatial information is organized and presented as a function of expertise. There has been a longstanding interest in whether efforts should be made to structure map representations more "naively", closer to the way that non-expert users experience the environments. Our research suggests that experts' maps serve the needs sought by experts and non-experts alike, and hence justify the role that experts play in the process.

Furthermore, we have gained insight into how spatial information in loosely structured environments is organized and represented. By focusing on map knowledge of the space, our experiments confirmed what appears to be a shared knowledge core about the organization of spatial information for different tasks, different levels of expertise, and different levels of familiarity. It may be the case, likewise, that loosely structured environments which favor less goal-oriented navigation are more readily represented using survey knowledge, although our experiments did not lead to unequivocal results. It would be useful to test this further in other experiments.

The role of roads as organizing elements, even when these are not regularly structured, is an important result for representing loosely structured environments. One may speculate that hiking trails as well as roads are useful reference structures in large wilderness parks and that efforts should be made to include these in map representations. Topography was not extensively used in the representations of the Plains of Abraham Park. In larger unstructured environments, it may play a more important role, but representing topography in ways understandable to non-expert map users is still a challenge.

Overall, the experimental program shows that basic and applied research can be done at the same time, especially using generated external representations. The map sketches, when carefully analyzed, reveal the mental representations of their producers and, when evaluated by others for goodness and essential information, provide principles for designing effective visualizations for all.

References

Agrawala, M., & Stolte, C. (2001). Rendering effective route maps: Improving usability through generalization. *Proceedings of SIGGRAPH '01*, pp. 241-250.

Allen, G. L. (2000). Principles and practices for communicating route knowledge. *Applied Cognitive Psychology, 14*, 333-359.

Bertin, J. (1967). *Sémiologie graphique*. Paris: Gauthier-Villars.

Blades, M. (1990). The reliability of data collected from sketch maps. *Journal of Environmental Psychology, 10*, 327-339.

Daniel, M.-P., Tom, A., Manghi, E., & Denis, M. (2003). Testing the value of route directions through navigational performance. *Spatial Cognition and Computation, 3*, 269-289.

Davies, C., & Pederson, E. (2001). Grid patterns and cultural expectations in urban wayfinding. In D. R. Montello (Ed.), *Spatial information theory: Foundations of geographic information science* (pp. 400-414). Berlin: Springer.

de Groot, A. D. (1966). Perception and memory versus thought: Some old ideas and recent findings. In B. Kleinmuntz (Ed.), *Problem solving* (pp. 19-50). New York: Wiley.

Denis, M. (1997). The description of routes: A cognitive approach to the production of spatial discourse. *Current Psychology of Cognition, 16*, 409-458.

Denis, M., Pazzaglia, F., Cornoldi, C., & Bertolo, L. (1999). Spatial discourse and navigation: An analysis of route directions in the city of Venice. *Applied Cognitive Psychology, 13*, 145-174.

Fontaine, S. (2000). La cognition spatiale dans des environnements souterrains et urbains: Aides verbales et graphiques à la navigation. Unpublished doctoral dissertation, Université René-Descartes, Boulogne-Billancourt, France.

Fontaine, S., & Denis, M. (1999). The production of route instructions in underground and urban environments. In C. Freksa & D. M. Mark (Eds.), *Spatial information theory: Cognitive and computational foundations of geographic information science* (pp. 83-94). Berlin: Springer.

Golding, M. J., Graesser, A. C., & Hauselt, J. (1996). The process of answering direction-giving questions when someone is lost on an university campus : The role of pragmatics. *Applied Cognitive Psychology, 10*, 23-39.

Lynch, K. (1960). *The image of the city*. Cambridge, MA: The MIT Press.

Michon, P.-E., & Denis, M. (2001). When and why are visual landmarks used in giving directions? In D. R. Montello (Ed.), *Spatial information theory: Foundations of geographic information science* (pp. 292-305). Berlin: Springer.

Nothegger, C., Winter, S., & Raubal, M. (2004). Computation of the salience of features. *Spatial Cognition and Computation, 4*, 113-136.

Przytula-Machrouh, E., Ligozat, G., & Denis, M. (2004). Vers des ontologies transmodales pour la description d'itinéraires: Le concept de "scène élémentaire". *Revue Internationale de Géomatique, 14*, 285-302.

Rinck, M., & Denis, M. (2004). The metrics of spatial distance traversed during mental imagery. *Journal of Experimental Psychology: Learning, Memory, and Cognition, 30*, 1211-1218.

Sattath, S., & Tversky, A. (1977). Additive similarity trees. *Psychometrika, 42*, 319-345.

Schneider, L. F., & Taylor, H. A. (1999). How do you get there from here? Mental representations of route descriptions. *Applied Cognitive Psychology, 13*, 415-441.

Siegel, A. W., & White, S. H. (1975). The development of spatial representations of large-scale environments. In H. W. Reese (Ed.), *Advances in child development and behavior* (Vol. 10, pp. 9-55). New York: Academic Press.

Taylor, H. A., & Tversky, B. (1992a). Descriptions and depictions of environments. *Memory and Cognition, 20*, 483-496.

Taylor, H. A., & Tversky, B. (1992b). Spatial mental models derived from survey and route descriptions. *Journal of Memory and Language, 31*, 261-282.

Taylor, H. A., & Tversky, B. (1996). Perspective in spatial descriptions. *Journal of Memory and Language, 35*, 371-391.

Tom, A., & Denis, M. (2003). Referring to landmark or street information in route directions: What difference does it make? In W. Kuhn, M. F. Worboys, & S. Timpf (Eds.), *Spatial information theory: Foundations of geographic information science* (pp. 384-397). Berlin: Springer.

Tom, A., & Denis, M. (2004). Language and spatial cognition: Comparing the roles of landmarks and street names in route instructions. *Applied Cognitive Psychology, 18*, 1213-1230.

Tulving, E. (1962) Subjective organization in free recall of "unrelated" words. *Psychological Review, 69*, 344-354.

Tversky, B. (1981). Distortions in memory for maps. *Cognitive Psychology, 13*, 407-433.

Tversky, B. (1993). Cognitive maps, cognitive collages, and spatial mental models. In A. U. Frank & I. Campari (Eds.), *Spatial information theory: A theoretical basis for GIS* (pp. 14-24). Berlin: Springer.

Tversky, B., Agrawala, M., Heiser, J., Lee, P. U., Hanrahan, P., Stolte, C., & Daniel, M.-P. (in press). Cognitive design principles for generating visualizations. In G. L. Allen (Ed.), *Applied spatial cognition: From research to cognitive technology*. Mahwah, NJ: Erlbaum.

Tversky, B., & Lee, P. U. (1998). How space structures language. In C. Freksa, C. Habel, & K. F. Wender (Eds.), *Spatial cognition: An interdisciplinary approach to representation and processing of spatial knowledge* (pp. 157-175). Berlin: Springer.

Tversky, B., & Lee, P. U. (1999). Pictorial and verbal tools for conveying routes. In C. Freksa & D. M. Mark (Eds.), *Spatial information theory: Cognitive and computational foundations of geographic information science* (pp. 51-64). Berlin: Springer.

Walsh, D. A., Krauss, I. K., & Regnier, V. A. (1981). Spatial ability, environmental knowledge, and environmental use: The elderly. In L. S. Liben, A. H. Patterson, & N. Newcombe (Eds.), *Spatial representation and behavior across the life span: Theory and application* (pp. 321-357). New York: Academic Press.

Landmark Extraction: A Web Mining Approach

Taro Tezuka and Katsumi Tanaka

Graduate School of Informatics,
Kyoto University
{tezuka, tanaka}@dl.kuis.kyoto-u.ac.jp

Abstract. Landmarks play crucial roles in human geographic knowledge. There has been much work focusing on the extraction of landmarks from geographic information systems (GIS) or 3D city models. The extraction of landmarks from digital documents, however, has not been fully explored. The World Wide Web provides a rich source of region related information based on our understanding of geographic space. Web mining enables a new mean of extracting landmarks, differently from conventional vision oriented methods. Our approach is based on how geographic objects are *expressed* by humans, instead of how they are *observed*. We extend existing methods of text mining so that spatial context is considered. The results of the experiments showed that adopting spatial context into text mining improves the precision of extracting landmarks from web documents.

1 Introduction

Cognitive geography has interested many researchers from various fields, including civil engineering, geography, cognitive science, sociology, and marketing [7,12,35]. Researchers are interested in this subject because human spatial behavior is often based on a cognitive image of space, rather than on the actual physical structure. People act according to how they understand their environment. A pioneering work in this field is that of Lynch, a civil engineer, who uncovered basic elements of a city image from questionnaires collected from local residents [15].

From a theoretical viewpoint, Egenhofer and Mark described the characteristics of *naive geography*, a system greatly different from physical geography [8]. Mark and Frank discussed cognitive geographic space based on the recent achievements of cognitive linguistics [16].

Cognitive geography is increasingly important in the new applications of geographic information systems (GIS). Until recently, GIS has been a specialized tool for trained users such as scientists and city planners. Now more and more people use GIS for daily activities, including car navigation, pedestrian navigation, and a map service over the Internet. For these new applications, cognitive information plays an important role in making the map easier to understand for untrained users.

Conventional work on uncovering of cognitive geography, however, was mainly based on questionnaires. Such an approach is not directly applicable for practical purposes in landmark extraction, because collection and analysis of questionnaires are often cumbersome, labor-intensive tasks. In this paper, we determine the capability of extracting such information from digital documents collected from the World Wide

A.G. Cohn and D.M. Mark (Eds.): COSIT 2005, LNCS 3693, pp. 379–396, 2005.
© Springer-Verlag Berlin Heidelberg 2005

Web. The Web today contains a tremendous amount of region related document, and it is continuously expanding.

Although cognitive geography consists of a wide variety of elements, we limited our target to landmarks. The Oxford English Dictionary defines a landmark as follows [24]:

Landmark: An object in the landscape, which, by its conspicuousness serves as a guide in the direction of one's course (orig. and esp. as a guide to sailors in navigation); hence, any conspicuous object which characterizes a neighborhood or district.

Some of the important characteristics of landmarks are as listed below.

- Cognitively significant.
- Visually salient.
- Used in navigation tasks.
- Used for determining the direction.
- Has a specific location and is often abstracted as a point.

In this paper, we consider a landmark to be a cognitively significant geographic object that is geometrically categorized as a point. This is an abstraction. Some landmarks may have relatively large spatial extensions, for example Champs Elysee in Paris or the River Thames in London. However, we consider them as a point too. In a large scale Champs Elysee or the River Thames must be considered as regions, yet in a smaller scale, they can be considered as points.

The importance of landmarks in geographic cognition has been discussed in many literatures. Tom and Denis compared street and landmark information in giving directions, and concluded that landmark oriented directions are more effective in many cases [34]. Michon and Denis discussed in what situation landmarks become effective means of giving directions [19]. However, the importance of landmarks is not limited to the way findings.

Neisser pointed out that cognitive maps are useful tools for *memorizing* geographic knowledge [22]. Indeed, much of human geographic knowledge is said to be stored with respect to landmarks and other cognitively significant geographic objects, rather than by coordinates [15]. This is quite different from conventional GIS data structures. Figure 1 shows two models for storing geographic data. The one on the left is a coordinates-oriented model, on which most conventional GISs are based on. The one on the right is a landmark-oriented model, which we assume corresponds to most of human geographic knowledge. In our model, landmarks are linked to each other by *spatial relationships*. These relationships include topological ones such as *inside of*, geometrical ones such as *close to*, and directional ones such as *to the north of*. The location of each landmark is thus determined in relation to other landmarks. Landmarks have *neighborhoods*, which are areas considered to be close enough from the landmark. The criteria for the closeness vary among observers, yet the distance in physical space is one common factor. Locations of many insignificant geographic objects are memorized using the neighborhoods of the landmarks. Such hierarchical structure in cognitive geography has been discussed for example in anchor-point theory by Counclelis et al [5].

We propose an automated landmark extraction method based on the usage of landmarks in digital documents. We collected documents from the World Wide Web and evaluated different measurements that could be used for the landmark extraction.

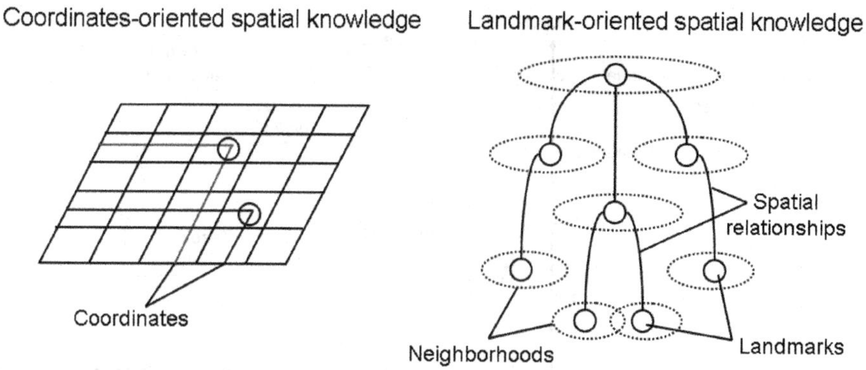

Coordinates-oriented spatial knowledge   Landmark-oriented spatial knowledge

Fig. 1. Coordinates- and landmark-oriented models for geographic space

2   Related Work

The extraction of landmarks from spatial data has interested many researchers (Figure 2). Burnett and May asked subjects to write descriptions of paths to (1) their familiar locations and (2) an unfamiliar location along a path shown by a video. From this collection of descriptions, they manually extracted landmarks and also asked the subjects why they chose these objects as landmarks [1]. Raubal and Winter developed a combined method employing a 2-D GIS, photographs of road intersections, and a prominent architecture database. Indicators such as the size of a building's facade, colors, architectural importance, and shape deviation from a rectangle were used to determine the overall significance of landmarks. Statistical tests were applied to judge whether the difference from the environment was large enough [25]. Brenner and Elias used cadastral maps and airborne laser scanning data to obtain layouts and height information for various buildings. They then applied data mining techniques, such as ID3 and clustering, and obtained visually significant objects and the sizes of the areas in which these landmark can be seen [3,9]. Koiso et al. extracted landmarks according to occupancy of the visual field and categorical differences from the environment [13]. This approach is based on a hypothesis that an object that is visually significant and that belongs to a different category from the surrounding environment is more likely to be a landmark. Moon et al., in dealing with robot navigation, pointed out that the vertical lines of objects can be used as a good indicator of landmarks, even though they are much smaller in scale than the typical geographic scale. These landmarks, incidentally, are used by a robot to navigate their way through a workspace [21]. Finally, Sorrows and Hirtle provided a good survey on what is necessary for a geographic object to become a landmark [30]. Their list of landmark characteristics included *singularity*, *prominence*, *accessibility*, *content*, and *prototypicality*.

There has been extensive research on the extraction of region related information from the Web. Most of the research, however, focused on providing users with a set of web pages related to certain area or theme [2,28,17,14]. For example, Georeferenced Information Processing SYstem (GIPSY) [37] is a system similar to ours in that it parses through documents and retrieves place names and their characteristics. MetaCarta is a

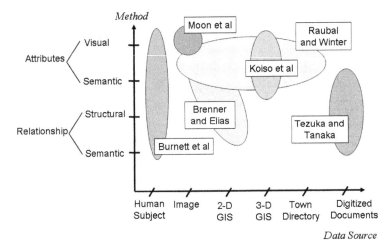

Fig. 2. Methods for landmark extraction

commercial geographic information retrieval system where documents from the Web and other sources can be searched based on the place names contain [18]. However, the aim of these systems was to assign geographical coordinates to documents. Our focus is on extracting information and obtaining new knowledge on cognitive geographic space.

Text mining research has dealt with extracting significant *terms* from documents. The text mining methods extract terms that are significant in the general sense [29,31]. In this paper, we discuss modifying existing text mining methods to include spatial context, in order to obtain spatially significant geographic objects from a very large corpus such as the Web.

3   Characteristics of Our Approach

Conventional methods for landmark extraction have focused mainly on modeling how landmarks are perceived by people. The basic idea was to model human perception and to implement a system that imitates the process of landmark cognition. Figure 3 shows the schema for this approach. The problem with this approach is that it is considering only a partial structure of landmark-human interaction. Much research has pointed out that the visual significance is not the only factor that determines landmarks. For example, despite their visual significance, skyscrapers do not always become landmarks. Another problem is that the process of modeling always encompasses selecting a limited number of attributes and ignoring the others.

In this paper, we focus on the usages of landmarks. We propose a model that landmarks are objects that are not only visually significant but also those that are frequently *used* by people. Landmarks have a variety of uses. First, they are used in organizing geographic knowledge, as described in the previous section. Second, they are used for finding one's way. Third, they are used for communication. People discuss certain locations by referring to nearby landmarks. Expressions such as *near A* are commonly used

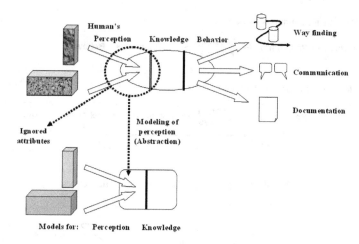

Fig. 3. Perception-based extraction of landmarks

Table 1. Uses of landmarks

Organizing concept	Organization of spatial knowledge
Way finding	Navigation to a destination
Communication	Description of regional knowledge to others
Symbol	Symbol of a region, city, or district

if *A* is a significant landmark. Lastly, landmarks are used as a symbol of either a city or a small district within a city. Table 1 lists some of the prominent uses of landmarks.

The emergence of a landmark is rather a circulatory process, in which perception is influenced by one's actions, as indicated in Figure 4. Evidence from cognitive science suggests that people are more likely to recognize objects that they expect [27]. Applying this to geographic level, objects are more likely to be recognized if the observer already knows them. Thus the objective properties such as visual significance are not the only factors that affect the emergence of landmarks. Familiarity with the object, the behaviors involved, and communication with other people play important roles. This is a *perceptual cycle*, as described by Neisser, where the significance of geographic objects increases as they are repeatedly used [22].

In extracting landmarks, not only their visual significance should be considered, but also their interaction with humans. Figure 5 illustrates the characteristics of our approach.

Because it is still difficult to trace all of human actions related to landmarks (barring drastic advancement in measurement technologies), we focus only the documentation activities of the landmarks.

Today, the World Wide Web provides a rich source of region related document. Our approach uses the Web for extracting significant landmarks to overcome the limit of perception-oriented landmark extraction methods. While most existing methods in landmark extraction are aimed at estimating how humans *observe* each geographic object, our method focuses on how people *express* landmarks.

Existing Methods for Landmark Extraction

Environment → Perception	Thinking Process → Judgment → Behavior

Proposed Method for Landmark Extraction

Action

| Environment → Perception → | Thinking Process → Judgment → Behavior | Documentation |

Communication

Fig. 4. Perception cycle in landmark emergence

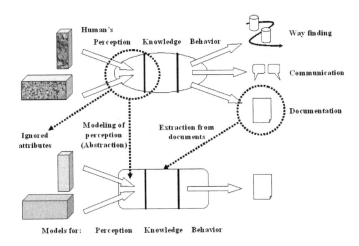

Fig. 5. Document-enhanced extraction of landmarks

In this paper, we employ text mining methods to extract significant landmarks from web documents. We have extended conventional measures in text mining so that the **spatial context** is considered.

Figure 6 shows a model that relates web, cognitive, and physical space. It asserts that the web content does not directly match the physical world. Rather, the web content is based on geographic knowledge owned by the creators of the contents.

4   Measures for the Cognitive Significance of Geographic Objects

In this section we describe the measures that we employ in extracting landmarks from web documents. First we discuss some of the general aspects of the measures, and then we describe each measure in more detail.

4.1   Landmark as a Relative Concept

Being a landmark is not a definite attribute. Whether a geographic object is observed as a landmark depends largely on the knowledge of the observer, his/her purpose, the

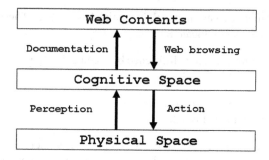

Fig. 6. Three-layer model: web content, cognitive space, and physical space

means of transportation that he/she employs, or even the time of the day. More importantly, it differs by the scale of the area being considered.

When discussing a small area, even a traffic sign or a grocery store sometimes acts as landmarks. On the other hand, there are highly significant landmarks that symbolize a city or even a state. In general, landmarks are objects that are significant *in respect to* the environment.

Thus, instead of obtaining static and absolute sets of landmarks, it is preferable to assign each geographic object a value that indicates its level of cognitive significance. In this way, whenever a range query is given, the system can return the objects that are most significant in the range query.

Before going into the discussion on the measures, however, we first compare two types of cognitive significance, general and spatial.

4.2 General and Spatial Significance

In text mining, the document frequency (DF) has been widely used as a measure for a word's commonness. The DF represents the number of documents in which a term appears. When applied to proper nouns, the popularity of an object can be estimated using the DF. However, this measure is insufficient in terms of measuring an object's significance in spatial context, for several reasons, since 1) well-known geographic objects are not necessary significant under the direct observation in the physical space, 2) the names of geographic objects often have ambiguities, where two different objects have a common name, 3) a single geographic objects may have more than one name, such as the official name and popular alternative names. In this paper, we introduce **spatial significance** in addition to **general significance** to cope with such problems. The difference between the two is as follows.

General significance: There is a class of geographic objects that are well-known in general sense but not as much in spatial sense. For example, although branches of enterprises and universities have specific locations in space, they are well-known not for its spatial properties but rather for other properties.

Spatial significance: Landmarks, nodes of traffic, significant paths, and characteristic regions are all significant in spatial sense. People know their locations, and they play significant roles in people's spatial knowledge. These are objects that local residents can easily locate on a map.

The examples below indicate a difference between general and spatial use of a term "McDonald's".

- Our store is next to a McDonald's.
- McDonald's says its beef is safe.

The former sentence indicates that the McDonald's is sometimes used as a spatially significant object in the area, while the latter sentence doesn't have such indication.

The aim of our research was to extract geographic objects that have high spatial significance by introducing spatial context into web mining techniques. We created four measures that consider spatial context, and compared them with a measure that does not consider spatial context.

From this point on, we use the term *place name* to indicate the names of geographic objects as they are expressed in web documents. In our method, the extraction of place names itself from sentences is not involved. Instead, our goal is to measure the level of significance for each given place name, when a set of place names is available from GIS.

Text mining methods can roughly be classified into two groups, **statistical** and **linguistic** [20]. The former takes a document as a set of terms, while the latter also uses the structures of the sentences. In our measures, the first three are statistical, and the latter two are linguistic. The characteristics of the five measures are listed in Table 2.

Table 2. Characteristics of the measures

Measure	Spatial context	Category
Document frequency (DF)	No	Statistical
Regional co-occurrence summation (RS)	Yes	Statistical
Regional co-occurrence variation (RV)	Yes	Statistical
Spatial sentence frequency (SF)	Yes	Linguistic
Case frequency (CF)	Yes	Linguistic

4.3 Definitions of Measures

In this subsection, we describe five measures that are expected to reflect significance of place names as landmarks. For each measure, we give an underlying hypothesis that advocates that the measure is suitable for obtaining landmark significance.

1) Document Frequency (DF)

Underlying Hypothesis: Landmarks are frequently mentioned in web documents.

The document frequency (DF), as described in the previous subsection, is defined for each term as the number of documents (web pages) in which the term appears. This measure is commonly used in text mining [29]. The formula for the DF is as follows:

$$d(p_i) = |\{d \in D | p_i \in s \land s \in d\}|$$

Here, p_i indicates the target place name (for which the DF is calculated), D is the document set, and s is a sentence.

2) Regional Co-occurrence Summation (RS)

Underlying Hypothesis: Landmarks are frequently mentioned in web documents together with place names in its surrounding.

A problem with the DF is that it does not examine whether a place name is used in a spatial context or not. Therefore, it is often the case that branches of enterprises, universities, or chain stores come highly ranked by the DF. In order to avoid such inappropriateness, we want to measure the frequency that a place name is actually used in spatial context. In calculating a regional co-occurrence summation (RS), we assume that when two neighboring place names appear in a same document, it is likely that these two place names were both used in spatial context. In terms of text mining, we consider a *co-occurrence* of two neighboring place names as an indicator of spatial context. Co-occurrence is a commonly used measure for term relationships in text mining [29,26].

The RS is defined as the total number of co-occurrences that the target place name has with its surrounding place names.

Before calculating the RS, we must first define the *surrounding place names* of the target place name. We call this set the **physical proximity** of the target place name. One way to define this is to use a threshold distance. The formula is as follows.

$$P'(p) = \{p_i | p_i \in P_{all} \land \delta(p, p_i) \leq R \land p_i \neq p\}$$

Here, p is the target place name, and P' is the threshold-based physical proximity. P_{all} is the original set of place names. The function δ gives the distance between place names, and R indicates the threshold distance.

This model becomes inappropriate if the target area contains both dense and sparse distributions of place names. In such case, there will be place names with a large number of neighboring place names, while some other place names have only few neighbors. As a result, the measure will have a low reliability.

Instead, we define the physical proximity as *the set of n-closest place names from the target place name*. Such a set can be obtained by sorting the place names according to their distance from the target place name. The formula for this definition is as follows.

$$P(p) = \{p_i | p_j \in P_{all} \land \delta(p, p_j) \leq \delta(p, p_{j+1}) \land 1 \leq i \leq n \land p_i \neq p\}$$

The formula for the RS, denoted by $r(p_i)$, is as follows.

$$r(p_i) = \sum_{p_j \in P(p_i)} \kappa(p_i, p_j)$$

Here, $\kappa(p_i, p_j)$ is the number of documents (web pages) containing both p_i and p_j. In other words, $\kappa(p_i, p_j)$ is the number of co-occurrences between p_i and p_j, in terms of documents.

The use of the RS reduces the effect of the ambiguities in place names. Suppose that a place name a indicates two different coordinates, x_a and $x_{a'}$, while place name b indicates coordinates x_b. Suppose also that the distances between the three coordinates follow the order $|x_a - x_b| < |x_{a'} - x_b|$. If a and b co-occur in document A, a in document A likely refers to coordinates x_a, rather than to $x_{a'}$. Because the DF does not account for such ambiguities, the RS is expected to perform better than the DF in extracting spatially significant objects.

Although various distances can be defined (i.e. network metric distance and time distance), we used Euclidean distance between the coordinates, since data necessary for calculating other distances are not as easily obtained for many target areas.

3) Regional Co-occurrence Variation (RV)

Underlying hypothesis: Landmarks are frequently mentioned in web documents together with a wide variety of place names in its surrounding.

The regional co-occurrence variation (RV) is another measure based on the co-occurrences between the target place name and its surrounding place names. Instead of using the total number of co-occurrences, the diversity in co-occurrences was used. The formula for the RV is as follows.

$$v(p_i) = |\{p_j \in P(p_i)|\kappa(p_i, p_j) \geq 1\}|$$

As with the RS, $P(p_i)$ is the physical proximity of the target place name p_i, and $\kappa(p_i, p_j)$ is the number of co-occurrences between place names p_i and p_j.

4) Spatial Sentence Frequency (SF)

Underlying Hypothesis: Landmarks are frequently mentioned in spatial sentences.

The spatial sentence frequency (SF) represents the frequency that the target place name is used in sentences that discuss spatial subjects. We estimate here that a sentence containing both a place name and also a *spatial trigger phrase* discusses spatial subject. We manually created a set of spatial trigger phrases. The formula for the SF is as follows.

$$s(p_i) = |\{d \in D|p_i \in s \wedge e \in s \wedge s \in d \wedge e \in E\}|$$

Here, D is the set of documents, d is a document, s is a sentence, p_i is the target place name, E is the set of spatial trigger phrases, and e is a spatial trigger phrase. Table 3 lists some of the spatial trigger phrases used in the extraction.

Table 3. Examples of spatial trigger phrases

Actions	walk, drive, turn at, go up, go down, arrive, stop at
Directions	right, left, front, back
Orientation	north, south, east, west
Spatial relationships	next to, at the corner of, behind, other side of
Spatial objects	intersection, road, street, railroad crossing
Means of transportation	car, bicycle, train

5) Case Frequency (CF)

Underlying Hypothesis: Landmarks are frequently used in spatial deep structure cases.

The case frequency (CF) focuses on the *case* that a place name accompanies. According to Fillmore's case grammar, each noun phrase in a sentence belongs a certain deep structure case, which means a specific role assigned to the noun phrase [11].

Some of the examples of deep structure cases are a *subject*, an *object*, a *location*, a *source*, a *method*, and a *goal*. In case grammar, the predicate is considered to be the central element in a sentence. The subject, the direct object, the indirect object, and prepositional phrases are all considered as noun phrases that modify the predicate. The deep structure cases are sometimes used for information retrieval [36].

Because a deep structure case indicates the phrase's role in a sentence, the frequency that the target place name is used in a spatial case is speculated to reflect the significance of the object in a spatial role.

Although the underlying deep structure of cases is common to all natural languages, the surface structure may vary. In isolated or inflective languages such as English and most other Indo-European languages, the case is expressed either by a preposition or word order. On the other hand, in agglutinative languages such as Japanese, Korean, Hungarian, and Finnish, the case is expressed by a suffix or a case particle.

One of the most common styles of spatial description in Japanese is as follows.

$$(w*) + pn + cp + (w*) + sp$$

Here, w is a term in general, pn is a place name, cp is a case particle, sp is a spatial predicate, and $*$ indicates an arbitrary number of repetition.

Because our target area is a city in Japan, we used case particles as the indicators of a deep structure case. We selected a set of Japanese case particles that often indicate spatial deep structure cases: *kara, made, yori, e, ni*, and *de*, which roughly correspond to the English prepositions *from, until, from, toward, to*, and *at*, respectively.

We define the case frequency (CF) as the frequency where the target place name is followed by the spatial case particle. The formula for the CF is as follows.

$$c(p_i) = |\{d \in D | p_i \in s \land c \in s \land s \in d \land c \in C_+ \land \alpha(p_i, c)\}|$$

Here, D is the set of documents, d is a document, s is a sentence, p_i is the target place name, C_+ is the set of case particles indicating a spatial deep structure case, c is a case particle, and $\alpha(p_i, c)$ indicates adjacency between p_i and c within a sentence (defined as true if p_i and c appear in this order).

5   Experiment

We performed a series of experiments to compare the validity of the proposed measures of landmark significance. In the experiment, we asked subjects to name a set of place names that they consider to be the landmarks of the target area. We compared the human judged sets of landmarks with the aforementioned five measures in terms of the recall and precision. This is a common evaluation method in information retrieval [29]. The recall and precision curves were graphed for the sets of landmarks extracted from the GIS data based on our five measures.

5.1   Data Set

The data set used in the experiment is as follows.

Subjects: 50 subjects consisted of 36 residents of the target area, Kyoto, and 14 people from outside the city. 40 were male and 10 were female.

Answer Set: Each subject was asked to name 20 of the most notable landmarks in Kyoto. A total of 1,000 entries consisted of 275 different place names. Table 4 lists the most frequently mentioned place names.

Table 4. Top 10 significant landmarks in Kyoto, collected from the subjects

Place name	# of answers
Kinkakuji (Golden Pavilion)	44
Ginkakuji (Silver Pavilion)	43
Kiyomizudera Temple	42
Kyoto Station	39
Kyoto Tower	34
Heian Shrine	32
Kyoto Imperial Palace	30
Kyoto University	29
Nijo Castle	29
Yasaka Shrine	25

GIS Data: The five measures for the cognitive significance were applied to the place names taken from a regular GIS, a digitized residential map provided by Zenrin, Ltd. [38]. This map data is divided into layers, including a "significant objects" layer that contains 7,109 place names. Although we can assume objects in this layer are mostly potential landmarks, their levels of significance vary. Famous temples and ordinary elementary schools alike are included in this layer. Thus, our goal in this experiment was to assign the level of significance to each of the place names included in the "significant objects" layer.

Web Documents: We collected 157,297 regional web pages for the web documents that were used to calculate our measures. Only the text part was used in the information extraction. The total file size was 2.45GB.

A focused web crawler was used for the collection of web pages. A focused web crawler is a special type of crawler that collects only the pages meeting a certain criterion [4,6]. The links are traced only when a page satisfies the criterion. In many cases, focused crawlers have greater efficiency in retrieving web pages under a certain topic, than regular web crawlers do. In this experiment, we used the place names taken from the GIS as the criterion of collection. Each page was guaranteed to contain at least one place name in the target area. The details of our implementation of a focused crawler are described in our previous paper [32].

Figure 7 is the architecture of the system that we implemented to evaluate our measures.

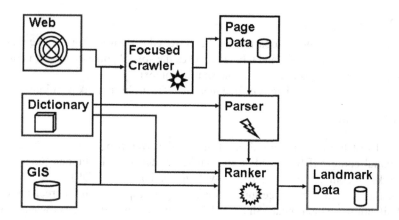

Fig. 7. System architecture for the measurement system

5.2   Evaluation Method

The definitions of the precision and recall are as follows.

$$\text{Precision} = \frac{\text{Retrieved Correct Objects}}{\text{Retrieved Objects}}$$

$$\text{Recall} = \frac{\text{Retrieved Correct Objects}}{\text{Correct Objects in the Population}}$$

When the rank position k is small, the set is likely to have high precision and low recall, while when k is large, the set tends to be the opposite. The precision and recall are functions of the rank position k. A precision-recall curve (P-R curve) is commonly used to visualize a series of the precision and recall pairs obtained by altering k [26].

The place names from the GIS in our evaluation were set in a decreasing order of the values calculated by each measure (DF, RS, RV, SF, and CF). The top k place names were selected, and the precision and recall were calculated with respect to a set of human-judged landmarks.

We consider only the points where the value of the recall increases to make the P-R curve smooth. By re-numbering the extracted pairs, we obtained the series of P-R pairs as a function of a new parameter j.

Then, we averaged the P-R pairs collected from different subjects for each j and obtained the averaged P-R curve, which is a function of j. This is called averaging by micro-evaluation [26]. In our case, the k ranged from 1 to 7,109 (= the number of the potential landmarks in GIS), j ranged from 1 to 20 (= the number of the "correct land-marks" given by each subject), and the number of the P-R pair series (to be averaged) was 50 (= the number of the subjects).

The evaluator consisted of a PostgreSQL database and approximately 1,100 lines of Perl scripts, including the part where each measure is calculated.

5.3   Results

Figures 8-11 indicate the comparison of the averaged P-R curves for the five measures. In these figures, RS, RV, SF, and CF, which uses the spatial context, are each compared with the DF, which does not employ the spatial context.

Table 5 compares precisions of five measures for different rank positions.

5.4   Discussion

The results of the experiments showed that in the overall performance the measures with spatial context (RS, RV, SF, CF) matched better with the human-judged sets of landmarks, in comparison to a measure without spatial context (DF).

The regional co-occurrence summation (RS) gave especially high precision for low recall situations, which means that the RS is the best measure to use when only the

Table 5. Precision of each measure

P. at rank	DF	RS	RV	SF	CF
5	0.016	0.368	0.132	0.272	0.152
10	0.140	0.212	0.186	0.276	0.130
15	0.096	0.259	0.245	0.241	0.147
20	0.106	0.239	0.192	0.251	0.152

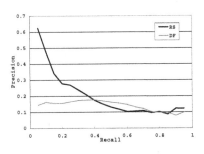

Fig. 8. RS and DF

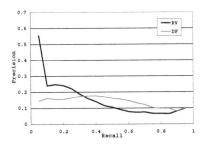

Fig. 9. RV and DF

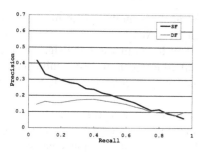

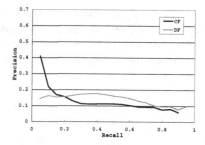

Fig. 10. SF and DF   **Fig. 11.** CF and DF

most significant landmarks are required. These are the cases showing only the typical landmarks of the region, such as in case of roughly abstracted route maps.

On the other hand, the spatial sentence frequency (SF) had relatively high precision for high recall situations. If a large set of landmarks is required, the SF is the preferred measure. For sightseeing maps where users want to see a large number of landmarks, the SF can be used.

In calculating the RS, all neighboring place names were treated equally. However, we could have used heterogeneous data sources, for example in different scales, and calculated the RS in a multi-layered manner. This approach may help avoid ambiguities in place names and improve the result.

For low recall situations, the linguistic approaches (SF and CF) did not perform as well as the statistical approaches (RS and RV). This is probably because of the relative scarcity of the obtained samples in comparison with the statistical approach.

Since linguistic approaches aggregates longer patterns of terms than the statistical approaches, with smaller size of documents, the signal-to-noise ratio increases. An increase in the number of collected web pages may change the situation better for the linguistic approaches.

Although using a specific measure proposed in this paper is the simplest solution for presenting landmarks, a combined measure could be used also.

6 Conclusion

In this paper, we discussed extracting cognitively significant geographic objects using web documents. Our main achievement is that we generated new geographic knowledge that is not present in conventional GIS, by aggregating distributed region related document on the Web.

Our approach has also introduced spatial context into text mining. The results of experiments showed that measures adopting spatial context match with human judged landmarks better when compared with the document frequency (DF).

In this paper, we did not consider how the effect of spatial tasks held by different users to the significances of landmarks. Such personalization is a difficult task, yet we are considering the extraction of such information by focusing on the grammatical aspects of the spatial sentences. For example, the types of the cases may provide clues. In calculating the CF, we only used the total frequency of spatial cases. However, their types are closely related to the spatial tasks. An analysis on the spectrum of the cases may unravel the preponderant spatial task for each landmark.

One advantage of our proposed method is that it can measure the significance of landmarks *quantitatively*. Although human map editors can choose significant landmarks when creating maps, giving each object its *level of significance* is often a difficult question.

The proposed method can be used in applications such as a progressively zoomable map interface, since place names shown on the map must be altered as the scale changes. The system can show the user the most significant and important place names on the map interface, without cramming too many characters on the screen.

Unlike questionnaire-based methods that are too expensive for collecting answers and analyzing results, our method is *scalable*, i.e. it can be extended simply by collecting more web pages. Because the size of the Web is growing continuously, the precision of our method is speculated to rise.

References

1. G. E. Burnett, D. Smith, and A. J. May, Supporting the navigation task: characteristics of 'good' landmarks, *Proceedings of the Annual Conference of the Ergonomics Society*, Taylor & Francis, 2001
2. O. Buyukokkten, J. Cho, H. Garcia-Molina, L. Gravano, N. Shivakumar, Exploiting geographical location information of web pages, *Proceedings of Workshop on Web Databases (WebDB99) held in conjunction with ACM SIGMOD99*, pp. 91-96, 1999
3. C. Brenner and B. Elias, Extracting landmarks for car navigation systems using existing GIS databases and laser scanning, *Proceedings of the ISPRS Workshop 'Photogrammetric Image Analysis,'* Munchen, Germany, 2003
4. S. Chakrabarti, M. van den Berg, B. Domc, Focused crawling: a new approach to topic-specific web resource discovery, *Proceedings of the 8th International World Wide Web Conference (WWW8)*, Toronto, Canada, 1999
5. H. Couclelis, R. Golledge, N. Gale and W. Tobler, Exploring the anchor-point hypothesis of spatial cognition, *Journal of Environmental Psychology*, Vol. 7, No. 2, pp. 99-122, 1987
6. M. Diligenti, F. M. Coetzee, S. Lawrence, C. L. Giles and M. Gori, Focused crawling using context graphs, *Proceedings of the 26th International Conference on Very Large Databases (VLDB 2000)*, Cairo, Egypt, 2000
7. R. M. Downs and D. Stea (Eds.), *Image and Environment : Cognitive Mapping and Spatial Behavior*, Aldine Publishing Co., Chicago, Illinois, 1973
8. M. J. Egenhofer and D. M. Mark, Naive geography, A. U. Frank and W. Kuhn (Eds.), *Spatial Information Theory: A Theoretical Basis for GIS*, Lecture Notes in Computer Science 988, pp. 1-15, Springer-Verlag, Berlin, 1995
9. B. Elias, Extracting landmarks with data mining methods, W. Kuhn, M. Worboys, and S. Timpf (Eds.), *Spatial Information Theory: Foundations of Geographic Information Science*, Lecture Notes in Computer Science 2825, Springer-Verlag, pp. 375-389, 2003
10. B. Elias and C. Brenner, Automatic Generation and Application of Landmarks in Navigation Data Sets, P. Fisher, (Eds.), *Developments in Spatial Data Handling*, pp. 469-480, Springer-Verlag, Berlin, 2004
11. C. J. Fillmore, The case for case, E. Bach and R. T. Harms (Eds.), *Universals in Linguistic Theory*, Holt, Rinehart and Winston, Inc., pp. 1-88, 1968
12. P. Gould and R. White, *Mental Maps*, Pelican Books, 1974

13. K. Koiso, T. Mori, H. Kawagishi, K. Tanaka, and T. Matsumoto, InfoLOD and LandMark: spatial presentation of attribute information and computing representative objects for spatial data, *International Journal of Cooperative Information Systems*, Vol. 9, No. 1-2, pp. 53-76, 2000

14. F. Lee, S. Bressan, and B. Ooi, Hybrid transformation for indexing and searching Web documents in the cartographic paradigm, *Information Systems*, Vol. 26, No. 2, pp. 75-92, 2001

15. K. Lynch, *The Image of the City*, MIT Press, Cambridge, Massachusetts, 1960

16. D. M. Mark and A. U. Frank, Experiential and formal models of geographic space, *Environment and Planning*, Vol. 23, No. 1, pp. 3-24, 1996

17. K. S. McCurley, Geospatial mapping and navigation of the web, *Proceedings of the Tenth International World Wide Web Conference (WWW10)*, pp. 221-229, Hong Kong, 2001

18. Metacarta : geographic text search engine, http://www.metacarta.com/technology/index.html#geoparse

19. P. E. Michon and M. Denis, When and why are visual landmarks used in giving directions?, D. R. Montello (Ed.), *Spatial Information Theory: Foundations of Geographic Information Science*, Lecture Notes in Computer Science 2205, Springer-Verlag, pp. 292-305, 2001

20. G. A. Mitra, C. Buckley, A. Singhal and C. Cardie, An analysis of statistical and syntactic phrases, *Proceedings of 5th International Conference on Computer-Assisted Information Searching on Internet (RIAO'97)*, pp. 200-214, Montreal, Canada, 1997

21. I. Moon, J. Miura and Y. Shirai, Automatic extraction of visual landmarks for a mobile robot under uncertainty of vision and motion, *IEEE International Conference on Robotics and Automation*, pp. 1188-1193, 2001

22. U. Neisser, *Cognition and reality: principles and implications of cognitive psychology*, W. H. Freeman and Company, San Francisco, 1976

23. C. Nothegger, S. Winter and M. Raubal, Selection of Salient Features for Route Directions, *Spatial Cognition and Computation*, Vol. 4, No. 2, pp. 113-136, 2004

24. *The Oxford English Dictionary*, Second Edition, Oxford University Press, 1989

25. M. Raubal and S. Winter, Enriching wayfinding instructions with local landmarks, M. Egenhofer and D. Mark (Eds.), *Geographic Information Science*, Lecture Notes in Computer Science 2478, Springer-Verlag, pp. 243-259, 2003

26. C. J. van Rijsbergen, *Information Retrieval - Second Edition*, Butterworth & Co. Publishers Ltd, 1979

27. D. E. Rumelhart, *Introduction to Human Information Processing*, John Wiley & Sons, Inc., 1977

28. T. Sagara, M. Arikawa, and M. Sakauchi, Spatial Information Extraction System Using Geo-Reference Information, *Information Processing Society of Japan Journal:Database, Vol. 41, No. SIG6 (TOD7)*, pp. 69-80, 2000

29. G. Salton, *Automatic Information Organization and Retrieval*, McGraw-Hill Inc., 1968

30. M. Sorrows and S. Hirtle, The nature of landmarks for real and electronic spaces, C. Freska and D. Mark (Eds.), *Spatial Information Theory: Cognitive and Computational Foundations of Geographic Information Science*, Lecture Notes in Computer Science 1661, Springer-Verlag, pp. 37-55, 1999

31. T. Strzalkowski, Natural language information retrieval, *Information Processing and Management*, Vol. 31, No. 3, pp. 397-417, 1995

32. T. Tezuka, R. Lee, H. Takakura, and Y. Kambayashi, Integrated model for a region-specific search system and its implementation, *Proceedings of the International Conference on Internet Information Retrieval (ICIIR 2003)*, pp. 243-248, Koyang, Korea, 2003

33. T. Tezuka, Y. Yokota, M. Iwaihara, and K. Tanaka, Extraction of cognitively-significant place names and regions from web-based physical proximity co-occurrences, X. Zhou, S. Su, M. P. Papazoglou, M. E. Orlowska, and K. G. Jeffery (Eds.), *Web Information Systems - WISE 2004*, Lecture Notes in Computer Science 3306, pp. 113-124, Springer-Verlag, 2004

34. A. Tom and M. Denis, Referring to landmark or street information in route directions: what difference does it make?, W. Kuhn, M. Worboys, and S. Timpf (Eds.), *Spatial Information Theory: Foundations of Geographic Information Science*, Lecture Notes in Computer Science 2825, Springer-Verlag, pp. 375-389, 2003

35. C. S. Yadav (Eds.), *Perceptual and Cognitive Image of the City*, Concept Publishing Company, New Delhi, India, 1987

36. E. B. Wendlandt and J. R. Driscoll, Incorporating a semantic analysis into a document retrieval strategy, *Proceedings of the Annual International ACM SIGIR Conference on Research and Development in Information Retrieval*, pp. 270-279, Chicago, 1991

37. A. G. Woodruff and C. Plaunt, GIPSY: automated geographic indexing of text documents, *Journal of the American Society for Information Science*, Vol. 45, No. 9, pp. 645-655, 1994

38. Zenrin Co.,Ltd, http://www.zenrin.co.jp/

Satellite Images – A Source for Social Scientists? On Handling Multiple Conceptualisations of Space in Geographical Information Systems

Anders Wästfelt

Department of Human Geography,
Stockholm University, 106 91 Stockholm, Sweden
anders.wastfelt@humangeo.su.se

Abstract. Current conceptualisations of space in the social sciences are detached from an absolute representation of space. Instead, space is seen as a relationship between subjects or actors. At the same time, the development of Geographical Information Systems (GIS) targeted the development of three-four-dimensional models in a realistic, absolute and relative mode of representation in which one single space is handled. Satellite images, on the other hand, are open to different interpretations because they can be perceived as pictures as well as spatial data in a GIS, and what can be read from these images is an open question. This article reviews current discussion on space and especially the way in which space is considered in GIS. It is proposed that, instead of perceiving one single space at a time, the social scientists' view of dealing with a social-relational space consisting of multiple actors or spaces can improve the usefulness of satellite images in the social sciences. A result of this study is that there is not necessarily a conflict between the physical and the social sciences in the use of satellite images; it is merely a question of epistemology. The proposed shift in the epistemological viewpoint means that satellite images used as primary sources in the social sciences make it possible to handle all kinds of questions that have a direct or indirect relationship to the environment and society.

1 Introduction

The conceptualisation of space in the social sciences and especially in human geography has become detached from physical space and is seen as relative and relational abstractions of space rather than as a realistic representation of it. Developments in these areas have been discussed by Perkins under the title "The Great Divide" [18]. Satellite images have at the same time become a common source in geographical and environmental studies, but there are very few studies that use remotely-sensed images in the social sciences. The use of satellite images as a source with a modern social-science perspective is limited by problems concerning different conceptualisations of space which have their roots at an epistemological level.

The purpose of this paper is to review different conceptualisations of space and particularly the way space has been handled in Geographical Information Systems (GIS). On the basis of criticisms of GIS and from general discussions about maps and images, a new approach is proposed for conceptualising space and handling

A.G. Cohn and D.M. Mark (Eds.): COSIT 2005, LNCS 3693, pp. 397–408, 2005.
© Springer-Verlag Berlin Heidelberg 2005

remotely-sensed image data in GIS. This approach harmonises more closely with the way social scientists today conceptualise space in a relational way. It is built on the idea that space must be seen as multiple, continuous and finite.

When a single connection between society and a map is in focus, as in locating positions or finding the way in unknown terrain, the map is a relatively unproblematic source. In other cases, when more abstract understandings of society are in focus, for example when dealing with environmental values in a village, or individual farmers perspectives and intentions regarding land use, the study immediately becomes very complex. This is because intentions and values are socially constructed: farms or villages constitute a single absolute spatial entity but are built on relations between entities such as inhabitants, farms, arable fields, woodland and the like using physical landscape in a specific way. In these cases, it is not possible to directly connect the image or map to the dimension under discussion in absolute space [2].

Satellite imageries are a source of information that can be used in Geographical Information Systems (GIS). They may also be seen as pictures of land and landscape. In contrast to a map, a picture is more open to different interpretations, but what may be interpreted from a satellite image is not presupposed [11]. This opens up opportunities to overcome some constraints on different conceptualisations of space, which is what I intend to discuss.

The bulk of research in GIS and remote sensing can still be seen as built on "non-critical" assumptions of representation and an absolute or relative conceptualisation of space. The focus has been on developing GIS within the framework of the idea of a physical representation of space and space-time. Recent developments have resulted in a more dynamic handling of geographical information in four dimensions (three dimension + time) and in studying relative aspects [13] [20]. However, it is problematical to base geographical modelling only on the conceptualisation of physical space. Not even nature behaves in a way that can be treated in a absolute four-dimensional model [13]. Handling satellite images in GIS means using the underlying absolute and relative conceptualisation of space and space-time as the basis for the implicit framework of analysis. This framework is in principle different from the conceptualisation of space in current human geography and other social-science disciplines. On the other hand, the creation of meaning and understanding of humans and society is based on a relational conceptualisation of space that combines a materialistic and an idealistic perception of space. In my opinion, if improved technologies such as remote sensing and GIS are to become more useful to society, they need new strategies for interpreting their source material that will bridge the wide epistemological gap between physical and social-scientific understandings of space.

Bridging this gap is not so much a question of the technology itself; it is more about the epistemological perspective. With an epistemological shift [8] it becomes possible to use GIS-rs technology to analyse and communicate directly and indirectly with the societal world. The argument I will focus on in this paper is that a new kind of geographical understanding and research will be possible if an anthroprocentric, geographical perception of multiple spaces is combined with remotely sensed image data seen as pictures that can be read as social signs representing multiple spaces.

2 Space in GIS Versus Human Geography

In human geography society is seen as consisting of both a social and a material part, the social part being based on relations between subjects or socially-constructed objects. This starting point can be useful for the understanding of values and qualitative aspects and for an understanding of cultural aspects and society. In many respects, this kind of knowledge cannot be reached in any other way.

The critics argue that this type of social-constructive worldview does not give any guidance for dealing with the concrete world. For example, Jonathan Raper states that, with an idealistic view as a starting point it follows "that 'representation' is merely an ideology and not a correspondence between a concept and a part of the external world". This is a perspective in which knowledge is produced rather than discovered [20]. Raper argues that GIS science needs to work with the social sciences and not dismiss them, but he does not present any examples of this himself.

Developments in Geographical Information "Science" have reached new levels by producing spatial representation in four-dimensional space-time models [20]. There is also new research on dealing with relative dimensions such as scale. The focus is at an epistemological level on improving methods and technology for dealing with all the complex aspects that exist in a satellite image or an existing dataset and which affect a "correct" corresponding representation of a landscape.

To develop this technology it has been necessary to deal with complexity in image data and uncertainty in concepts. Fuzzy logics and rough sets have been used fruitfully to deal with uncertainty in geometry and concepts, and the development of multiscale handling of image data makes it possible to manage complexity in data. This indicates that the discrete-object model is dissolved [20][5][1][7]. All these contributions to modelling important steps towards bridging the gap between different spatial worlds but they are still based on a realistic model of earth, but as mentioned above the conceptualisation of society and space in the social sciences is different.

I will argue that, regardless of how many new aspects we include to make a correct corresponding representation of the earth´s surface, the debate will sooner or later come back to the essential question of how this representation is related to the social and cultural world. This can be compared with Nadine Shuurman, who argues that, at some point, "a broader discussion of the sort initiated by critics from human geography will be necessary to position GIS in relation to the social"[24]. I mean that it is not enough merely to position GIS and remotely-sensed images in relation to society; it is also necessary to understand how to make use of these technologies in the social sciences.

Owing to the spherical shape of earth, spatial conceptualisations have from a "human" perspective of necessity always been built on abstractions. These abstractions have taken different forms in different historical periods. By using geometry and coordinate systems it is possible to produce images/maps of the surface of earth in three dimensions in GIS. Nevertheless, our personal experience is limited to what we can perceive from one single position, and the impressions we have at one single spot in time-space are much wider than can be represented on a map or in a GIS. Such a representation always implies a reduction and simplification that is based on our view of the world.

Accordingly, there has never been one single conceptualisation of space or one single consistent epistemology of the world, and none of the known representations can capture and deal with all aspects and conceptualisations. The world is too complex and the human world is too intricate. This does not mean that there is no point in dealing with a representation of earth, but the problems are much more complicated than we first imagine. For example, the Greek geographer Ptolemy worked with two different kinds of representations: firstly, the positioning of different places in a coordinate grid, which is thought of as the basis of geography, and secondly, chorography, which was some kind of qualitative description of places [14]. As a parallel to this reasoning, Openshaw has concluded that GIS deals with space but not place, which is a good illustration of the limitations of GIS [17].

The development of geographical information technology is primarily based on the concept that space is absolute and relative. This concept relates historically to Newton's work, ín which space is seen as external to humans and only one single space-time world exists in which empty space is allowed. Andrew Frank points out that this four-dimensional representation implies that there is one single time-space and that this excludes "parallel universes" [5].

There have been many other conceptualisations of space. For example, Newton's contemporaries, Descartes and Leibniz, suggested space in different ways. Descartes defined space as the extent of matter, while Leibniz argued that all possible relations between entities formed space [20]. Harvey explains Leibniz´s relational space "as being contained in objects in the sense that an object can be said to exist only insofar as it contains and represents within itself relationsship to other objects" [9]. A current interpretation of Leibniz´s relational space that I propose can include in principle all kinds of objects based on their relation to other objects. In both Descartes´s and Leibniz´s concepts of space, empty space cannot exist, which is an important aspect of social space and environmental issues in general.

GIS – RS community	Human Geography
<ul><li>One single space-time</li><li>Empty space exists</li><li>Realistic absolute and relative space</li><li>External reality</li><li>Generalisation is possible</li><li>Quantitative statistical analysis can easily display spatial variation</li></ul>	<ul><li>Multiple spaces</li><li>No empty space</li><li>Relations between subjects generate space</li><li>Context is socially as well as spatiall dependent</li><li>Concepts and values are socially constructed</li></ul>

Just as there have always been different conceptualisations of space in historical times, space is still conceptualised today in different ways, as has already been pointed out. The differences in the conceptualisations of space between the social and the natural sciences need to be understood if there is to be any use of satellite images in the social sciences.

In summary, the division between the GIS-rs community and the social-science perspective can be described in a diagram.

The development of new technologies like GIS resulted in criticism of GIS in the 1990s. Before discussing signs of changes and the possibility of dealing with other conceptualisations of space in GIS, the criticism of GIS in the 90s will be discussed in more detail.

3 Criticism and Limitations of GIS

In the 1990s there was criticism of GIS. [24][19]. The criticism was that GIS restricted the worldview and transformed the way in which geographers look at the world; it was seen as a biased perspective from a social-science point of view. In this criticism GIS was seen as implying a return to a deterministic and positivistic research method. The criticism was based on the recognition that either the perspective the GIS community proposed or the social contexts that these kinds of studies were inscribed inside were problematised [24].

The conclusion Nadine Shuurman draws is that the critics of GIS in the 1990s have not had any marked influence on the discipline; one example of this is the fact that increasing abstraction is still considered a problem that needs to be minimized [24].

One useful example in the search for a productive understanding of GIS limitations is Shuurman´s discussion of the concept of ontology. Shuurman argues that this concept is used in many different ways in the discussion of GIS. First, she claims that there is a deep gap between what philosophers mean by ontology, which is normally defined as the "essence of being", and what GIS researchers mean: that it refers to the conceptualisation and formalisation of the world. Shuurman argues that, in the conceptual stage, ontology refers to a stable external reality, and when the conceptual model is formalised, ontology refers to the formal system. She also points out that all entity relations in a GIS must always be formally inscribed [24].

What Shuurman mean is that epistemology in GIS always involves a break between conceptualisation and formalisation; in the formalisation process, all entities which are represented in a GIS become objects with an absolute extension in space.

"Formalisation thus involves an epistemological break between conceptualisation and formalisation, in the shift from realism to positivism" [24].

Shuurman believes that the formalisation process then guarantees that data will be handled in a positivistic way From a theoretical point of view this conclusion is correct, but Shuurman does not seem to offer any solution or way beyond these theoretical problems of principle.

In contrast, I mean that the primary problem does not concern formalisation and the epistemological question; instead, it is how we consider representation and the connection between social concepts and the empirical worldview [3]. An initial question is, What happens if we try to take control of the inscribed entity relations via the social concepts analysed within GIS? If we use primarily socially-constructed and essentially spatial concepts in a spatialised mode (for example, a village, which is both social and physical – a village can never be seen in a single pixel) that has indirect connections with the external world, is it possible to use GIS in a constructive way? This means that, instead of tackling the problem, we go behind it by applying a constructivistic approach to both the conceptualisation and the analytical step in the analysis.

In other words, we accept the formalisation and positivistic processing, but not the result, as a final truth about the external world. This presupposes that we see the final map as interpretable and as a subjective part of a constructional process [18][12].

There are signs in the debate about GIS that a change is about to take place. Starting from the idea that space is continuous, we always impose limits (borders) on this continuous space to make it possible to understand what we see. But the delineation of objects is seldom discrete.

These problems have occupied many researchers over the past decade. As a result, uncertainty in concepts and spatial representation can be dealt with through new forms of logic, for example, fuzzy logic and rough sets [1][5]. The interesting aspect for this paper regarding these types of logic is described by Openshaw: "In these types of logic the objectivity is represented by fuzzy sets and is simultaneously arbitrary, value-system dependent and multivalued" [17]. From this viewpoint Openshaw draws the conclusion that, in a fuzzy world, multiple truths are acceptable.

At the same time Doreen Massey has argued from a human geographical viewpoint concerning the relationship between physical and human geography that, "a more adequate understanding of spatiality for our times would entail the representation that there is more then one story going on in the world and that these stories have, at least, a relative autonomy" [13]. What does this mean in terms of dealing with data in GIS?

I mean that, on a conceptual level, it can be seen as the starting point for dealing with multiple trajectories and local spaces. This presupposes that the relations inside and between local spaces can be analysed. Taking this as a starting point, it will be possible to search for, and display, how different stories are developing in a region and in the world, which, of course, is part of the same world.

I will then argue that it is possible to investigate society through a representation in which socially-constructed concepts and situated knowledge can be analysed spatially. However, this presupposes a conceptualisation in which space is seen as multiple, continuous and socially finite, and where the interpreter of the space plays an active role. Before finally discussing different conceptualisations of space and the proposed approach, it will be necessary to take a close look at the differences between maps and images.

4 Maps

Geographical information systems are rooted in the construction of a system for handling maps in a digital environment. Much energy and large resources have been put into methods for converting maps into databases. Similarly, large resources have been put into converting information from satellite images into objects stored in GIS, which is the normal way of using remotely- sensed data.

I mean that these processes have restricted the use of GIS and especially the use of remotely-sensed data within GIS. Even if the process today in modelling GIS data is much more dynamic than previously, the primary idea is still to handle geographical information as "maps" in GIS. But maps themselves are no longer seen as unproblematic; today they are discussed as representations and the agenda has changed. The discussion of representation has implications for how we can look at satellite images.

During the past decade maps have been questioned as a neutral source. That a map describes summary knowledge of a territory is an old principle in the ontological assumption of cartography [23]. Maps have been criticised as representations of land that are very static and charged with power. Brian Harley has gone so far as to claim that "... the map is never the reality, in such ways it helps to create a different reality. But in either case, the map is never neutral" [8]. Harley argues that maps need to be contextualised if it is to be possible to understand the meaning of the sources behind the classical communication paradigm. He claims that external power links maps to the centre of political power, and the internal power creates power by creating a spatial panopticon [8]. A map can then be seen as an interface in which power relations are inscribed between different structural levels in society. Harley´s conclusion is that it is necessary to have a paradigm shift whereby maps are seen as social images [8]. This perspective has not yet been used in any studies but is close to my argument about how satellite images could be useful in the social sciences.

Mapping has traditionally been considered as driven by a communication paradigm. During the past decade some researchers have shown that this is a limitation, both on the interpretation of maps in a historical context and on the use of maps in current use, especially when maps have become a part of GIS [12]. The criticism of maps can also be illustrated by a quotation from David Harvey: "Any system of representation in fact is a spatialisation of sorts which automatically freezes the flow of experiences and in so doing distorts what it strives to represent [13].

In his search for a modern understanding of how maps work, Alan MacEachren has come to the conclusion that maps need to be seen from a representational perspective.

A representational perspective, in contrast, begins with an assumption that the process of representation results in knowledge that did not exist prior to that representation; thus mapping and map use are a process of knowledge construction rather then transfer [12].

This means that there is not merely a problem concerning different perspectives on power related to maps; what comes out of looking at a map with a representational perspective is also arbitrary. In contrast to Harley, MacEachren does not just analyse maps in a critical way, he elaborates maps in a constructive way, arguing that maps can work [22], which relates to the fact that MacEachren elaborates maps with a representational perspective.

MacEachren states that "semiotics considers the relationship between an "expression" and the "concept" to which that expression refers" [12]. The consequence is that "we arrive at the conclusion that maps do not refer to the real world, but to a concept about the world. We can then only know the world in the context of constructed categories that we apply to organise that world.

In conclusion, the critical discourse about maps has opened up a new window for the use of maps in the social sciences. However, in many ways this discourse is the same as the criticism of GIS. It can be associated with the fact that GIS, in the first instance, is a digital representation of maps. As I will be arguing, what is more interesting is that the full potential of GIS and remotely-sensed images will not be reached as long as the realistic view is prioritised in GIS analysis.

I will argue that it is possible to deal with socially-constructed concepts in GIS as representations, but it is necessary to establish a relationship both with the implemented "concepts" and with the interpreter's knowledge and impressions of these concepts.

Satellite imageries can, at first glance, be seen as maps. On the other hand, satellite images are not necessarily maps. However, they have important advantages as they are built on pixel values that can be characterised as representing a spatial continuum carrying social signs. There are not necessarily any predefined interpretations of a satellite image. Seen in this way, images allow more flexible interpretations and different interpretations.

5 Horizontal Versus Bird's Eye View

Horizontal images of the earth's surface are much more familiar to humans than a bird's-eye view. Traditional photographs have been in use for more than a century, and they have also been used in scientific work, but they were rarely criticised in the 70s.

During the last decade renewed interest in visual aspects has grown up in geography. However, visual analysis was considered to carry power relations inscribed in the image analysis [21]; the visual was seen as culturally constructed. Gillian Rose calls the framing of the seen a scopic regime. This scopic regime distinguishes true knowledge from perceived knowledge [21]. Perceiving images this way can be seen as a parallel to the criticism of maps discussed earlier.

Judith Oakley argues that criticism of the visual was a reaction to the insight that images are always charged with a defining perspective and that the research community should therefore abandon both visual and geographical representation. But Oakley believes that the visual does not need to be reduced to a controlling eye. She claims that "There may be a tension between the eye of the outsider and that of the inhabitants, but there is also possible interplay" [21]. She also points out that "to look at a landscape can be to receive it, to be 'disponible' (Breton 1937) to what it offers, rather than to spy upon it" [21]. In this statement Oakley point to a way forward, a sort of interaction whereby we do not just receive information.

It is easy to understand the kinds of internal and external power relations that can be associated with satellite images, since they are acquired from a bird's-eye view. However, it is more complicated to see how an analytical approach can be developed in order to acquire an emancipated approach to the "social" landscape through satellite images – instead of the concept of controlling land from space. I argue that this is possible by using the images in new ways. However, this presupposes that the images are perceived as a interpretable representation, in other words as a picture.

These images do not carry any meaning before interpretation; they consist only of non-sorted social signs. Rather, meaning is created in the relationship between different implemented concepts and the perceptions and knowledge of what can be seen.

The depicted landscape is perceived as many different social spaces, which can be interpreted in relation to each other. Thus, local configurations or contexts are important features in reading the image.

With the proposed perspective, there is no point in giving advice on the image itself. Its meaning exists in the possible interactions and interpretation of the image, which in turn are dependent on the interpreter's knowledge and the concepts used. Hence, satellite images can be used to deal with quality dimensions in relation to landscape.

6 Handling Multiple Spaces in GIS – A Solution for the Use of Satellite Images in the Social Sciences

Up till now an absolute and relative conceptualisation of one single space-time has been common in dealing with remotely-sensed images in GIS. The information is normally transformed into objects in an absolute or relative mode.

In cases where remotely sensed pictures are processed in GIS, both restrictions and possibilities are inherent in the images. The remotely-sensed pictures presuppose a raster/grid model for analysis. This implies that it is a field model and not an object model. The field model is considered to be continuous, albeit discrete and no empty space exists inside the field studied.

In principle, space has inside GIS has been conceptualised as endless and discrete (empty space exists).

To make it possible to deal with multiple, anthroprocentric perspectives and socially-constructed concepts, space must be seen as multiple, continuous and finite (no empty space exists).

The field model itself makes it possible to deal with quality dimensions [6]. Thus, it is possible to manage qualitative differences between different spaces which belong to the same social concept at different places, and not just to separate them according to their relative position in space, which is one criterion for a relational space [20].

The proposed handling of space in this paper is close to how Andrew Frank argues for a five-tier ontology in GIS databases. Frank claims that it would be possible to deal with the principal problems in GIS: the integration of vector and raster data, the handling of continuous space and the integration of social reality. He says "An ontology constructed from tiers can integrate different ontological approches in a unified system" [5]. He proposes that a five-tier ontology should be implemented in GIS databases in order to achieve (1) a human independent reality, (2) observation of the physical world, (3) objects with properties, (4) social reality and (5) subjective knowledge. Frank also points out that consistent constraints may make it possible to link the socially-constructed terms to the physical object, but this link is only valid in a specific context. Therefore the database must model context if it is to be possible to make the semantic shift from the physical world to social reality [5].

I mean that Frank's argument can be implemented in satellite-image analysis using multiple spaces and the conceptualisation of social space as continuous and finite. Combined with an interpretation of images in a finite and continuous mode, as proposed, and including the idea that there is no social meaning or knowledge in the image itself, this means that image analysis must include:

- Applying knowledge and local concepts in image processing;
- Statistically assessing the visualised spatial variety of socially-defined concepts in a specific spatial context;
- Interpretating the results in a similar way to image reading, which means that knowledge is constructed;
- Still handling and processing data in a realist mode in absolute space in GIS;

7   Conclusion

The proposed conceptualisation of space seen as consisting of multiple, continiuous and finite spaces means a paradigm shift in favour of a social- science perspective and an anthroprocentric point of view. Thus, space is seen in principle as continiuous, non-empty and finite. In contrast to the conceptualisation of physical space and space in GIS, there cannot be any empty space in the discussion of social space. There must be at least two individuals, otherwise there cannot be any relationship. The essence of being inside a social space produces a relationship and a space. I mean that this is one of the crucial aspects of using remotely sensed images in the social sciences; it must deal with multiple spaces, because individuals and societies are created of multiple local spaces.

Exploring the relations and differences between local spaces related to each other, and the variations between them, means that an intrepretation of a relational space can be drawn.

Because of the flexible nature of a picture, the use of remotely–sensed pictures could open up a way for bridging the wide gap between different epistemological perspectives that exists in geography today. This, however, depends on the way the remotely sensed images are perceived. I mean that a satellite image can be seen as social signs as well as physical signs before interpretation. Taking this as the starting point, it is possible to interact with the image in new ways to produce meaningful information that has an indirect connection with the socially-conceptualised world and multiple spaces. In consequence, this means that the results of such a process cannot be considered without a connection with the social concepts that were initially used.

Thus, the meaning is constructed in the interpreter´s mind, and at the same time as the interpreter reads the image, it "works" [12]. This means that the broad complexity of the image and society is accepted, and that the image is dealt with as a representation which is understood in a social context.

This statement has a parallel in the creation and perception of art; artists "paint not the thing, but the effect it produces" [16]. In the case of satellite images, it is the opposite. We can see the social effects of land use in the image but initially have no idea what it means. Using a socially-constructed perspective and interacting with knowledge and local perspective will make it possible to extract spatial variations from the matrix of coordinates and the image, which in turn can be interpreted in a social context.

One result of this study is that there is not necessarily a constraint between the physical and the social sciences in the use of satellite images; it is only a question of epistemology. The proposed shift in the epistemological viewpoint means that satellite images used as primary sources, seen as a representation in the social sciences, make it possible to deal with all kinds of questions which have a direct or indirect relationship to the environment and society.

This means that the whole discussion about embedded realist/positivist framing in GIS must be seen as a concept and not necessarily as a restriction with a socially-constructed perspective. It all depends how we imagine the result; as true knowledge of the external world, or as a picture that carries information about many different social spaces that build up a social-relational space. In conclusion, there is not one single perspective or one story, there are many different stories. It is in the variations between these stories that the interpretation of society rests.

References

1. Ahlkvist, O., J Keukelaar, K. Oukbir.: Rough and Fussy Geographical Data Integration, in International Journal of Geographical Information Science, vol. 17, no. 3, (2003) 223–234.
2. Bibby, P. & J. Shepard.:GIS, Land Use, and Representation, in Environment and Planning B: Planning and Design, vol. 27, (2000) 583–598.
3. Couclelis, H.: Spatial Information Technologies and Societal Problems in M. Craglia & H. Onsrud (eds.): Geographic Information Research: Trans-Atlantic Perspectives. Taylor & Francis: London (1999)
4. Flowerdew, R.: Reacting to Ground Truth, in Environment and Planning A, vol. 30, (1998) 289–301.
5. Frank, A. U.: Tiers of Ontology and Consistency Constraints in Geographical Information Systems, in International Journal of Geographical Information scince, vol. 15, no. 7, (2001) 667–678
6. Gärdenfors, P.: Conceptual Spaces: The geometry of thought. A Bradford Book. MIT Press, Cambridge, Massachusetts (2000)

7. Hall, O.: Landscape from Space. Geographical Aspects on Scale, Regionalization and Change Detection. Meddelande nr. 116, Department of Human Geography, Stockholm University (2002)

8. Harley, J. B.: Deconstructing the Map in Cartographica, vol. 26, (1989) 1–20

9. Harvey, D.: Social Justice and the City. In The Spaces of Postmodernity, in M. J. Dear & S. Flusty (eds.): Readings in Human Geography. Blackwell. Oxford (2002)

10. Ingold, T. The Perception of the Environment. Essays in Livelihood, Dwelling and Skill. Routledge: London & New York (2000)

11. Liverman, D., et. al.: People and Pixels. Linking Remote Sensing and Social Science. National Research Council: Washington, D.C (1998)

12. Mac Eachren, A.: How Maps Work: Representation, Visualisation, and Design. Guilford Press. New York, London (1995)

13. Massey, D.: Space-time, ´Science´ and the Relationship Between Physical and Human Geography in Transactions of the Institute of British Geographers. Vol. 24:3, (1999) 261–276.

14. Nuti, L.: Mapping Places: Chorography and vision in the Renaissance, in D. Cosgrove (ed.): Mappings. Rektion. London (1999)

15. Oakley, J.: Visualism and Landscape: Looking and Seeing in Normandy, in Etnos, vol. 66:1, (2001) 99–120

16. Olsson, G.: Constellations. Sistema Terra, in Remote Sensing and the Earth, vol. VIII, no. 1–3. (1999)

17. Openshaw, S.: Towards a More Computationally Minded Scientific Human Geography, in Environment and Planning A, vol. 30, (1998) 317–332

18. Perkins, C.: Cartography: Mapping Theory. Progress Report in Progress in Human Goegraphy, vol. 27, no. 3, (2003) 341–351.

19. Pickles, J.: Ground Truth. The Social Implications of Geographical Information Systems. The Guilford press: New York & London (1995)

20. Raper, J.: Multidimensional Geographic Information Science. Taylor & Francis: London & New York (2000)

21. Rose, G.: Visual Methodologies: an Introduction to the Interpretation of Visual Materials. Sage: London (2001)

22. Röd, J. K.:Geographical Information Processing. Towards Transparent Statistical Mapping. Dr. Polit. Thesis. Department of Geography, Norwegian University of science and Technology (2002)

23. Smith, R. G.: Baudrillard's Nonrepresentational Theory: Burn the Signs and Journey without Maps, in Environment and Planning D: Society and Space, vol. 21, (2003) 67–84

24. Shuurman, N.: Critical GIS: Theorizing an Emerging Science, in Cartographica, vol. 36, no. 4. (1999)

3D Topographic Data Modelling: Why Rigidity Is Preferable to Pragmatism

Friso Penninga

Delft University of Technology, OTB, section GIS Technology,
Jaffalaan 9, 2628 BX Delft, The Netherlands
F.Penninga@otb.tudelft.nl

Abstract. In this paper two concepts for modelling 3D topography are introduced. The first concept is a very pragmatic approach of 3D modelling, trying to model as much as possible in (less complicated) 2.5D and use 3D modelling only in exceptional cases. The idea is to use a constrained TIN in 2.5D and place 3D TENs on top or below this surface. As both data structures use the same simplexes (nodes, edges, triangles) this integration should be very well possible. At a conceptual level this approach seems suitable, but at design level serious problems occur. To overcome these a rigid approach is developed, modelling all features in a 3D TEN, including the air above and earth beneath these topographic features. This model is stored and maintained within a spatial database. Despite its more advanced concept, it is shown that this approach offers huge advantages compared to the initial pragmatic approach.

1   Introduction

This paper describes two modelling concepts developed within the research on 3D Topography as carried out within the GIS Technology group at Delft University of Technology. Initially the aim was a pragmatic approach, in which applicability was one of the keywords. However, during the research we came to the conclusion that our pragmatic approach seemed suitable at conceptual level, but that it causes some serious modelling problems at design level. Based on the identified strengths of this initial model a rigid approach is developed, which turned out to actually simplify most of our problems at design level, although it is more advanced at conceptual level.

This paper starts with an introduction on the backgrounds of the 3D Topography research in Paragraph 2, including a short overview of relevant available data sets. Next the concepts of the initial pragmatic modelling approach are introduced in Paragraph 3. Its drawbacks are addressed and some preliminary conclusions on the initial modelling approach described. Based on these conclusions the concepts of the new rigid approach are presented in Paragraph 4, followed by some remarks on the implementation of this approach. The paper ends with final conclusions and future research in Paragraph 5.

A.G. Cohn and D.M. Mark (Eds.): COSIT 2005, LNCS 3693, pp. 409–425, 2005.
© Springer-Verlag Berlin Heidelberg 2005

2 Backgrounds of the 3D Topography Research

2.1 The Need for the Third Dimension

Most current topographic products are limited to representing the real world in only two dimensions. As the real world exists of three dimensional objects, which are becoming more and more complex due to increasing multiple land use, accurate topographic models have to cope with the third dimension. The overall goal of this research is to extend current topographic modelling into the third dimension. Applications of 3D modelling are not limited to the terrain surface and objects built directly on top or beneath it, as geological features and air traffic or telecommunication corridors can be modelled too.

Most initiatives on developing 3D GIS focus on supporting visualisation, often in Virtual Reality-like environments. One of the objectives of this 3D modelling research is to enable 3D analysis as well, as this traditional GIS-strength lacks until now in most 3D GIS approaches. Another important assumption within this research follows from the required wide variety of applications of topographic data. As topography is ranked high in the spatial data infrastructure hierarchy, one cannot optimize the data model for one specific purpose. One has to be able to serve the complete range of user applications, regardless whether these applications require for instance optimal visualisation capabilities or optimal analytical capabilities.

In 3D modelling one needs a 3D primitive (a volume) beside points, lines and faces to represent 3D objects accurately. Earlier research proposed amongst others using simplexes (point, line, triangle, tetrahedron) (Carlson 1987), points, lines, surfaces and bodies (3D Formal Data Structure (FDS)) (Molenaar 1990a, Molenaar 1990b, Molenaar 1992), combining Constructive Solid Geometry (CSG) and a B-rep. (de Cambray 1993) and integrating a 2.5D Triangulated Irregular Network (TIN) with 3D FDS (Pilouk 1996). In applications polyhedrons are often used as 3D primitive (Zlatanova 2000, Stoter 2004). These publications on 3D modelling concepts are often limited to a conceptual description of the use of a 3D primitive, without addressing any of the actual problems concerning the use of these 3D primitives as in analysis. As a result true implementations (besides some very small experiments) are rare, thus not proving actual usefulness of the concepts.

2.2 Current Available Data Sets

The initial approach to extend topographic modelling into 3D is both supply and demand driven. The required data sets -both topography and height data-become available in a growing number of countries. 2D Topographic data sets are available and are being converted into object-oriented models, as this offers huge advantages in digital processes such as GIS analysis. Due to the growing popularity of airborne laser scanning high resolution height data becomes available. Combining both types of data can lead to full 3D data. Within the presented research Dutch data sets will be used in implementation tests. The height information is available in a Digital Elevation Model (DEM) called the AHN (Actual Height model the Netherlands). As large parts of the Netherlands are situated below sea level accurate large scale height data is of great importance, thus leading to the development of the AHN, a height model with one height point for on average every 16m2. The most relevant topographic data set (1:10.000) is currently being converted by the Dutch Topographic Survey into an object-oriented structure called TOP10NL.

At the demand side developments as increasing multiple land use and rising awareness of the importance of sustainable urban development (both caused by urban space scarcity), increase the need for real 3D topographic data sets. At dataset level the major shortcomings of the current 2D products lie in the absence of height information for buildings and other constructions, which is essential for for instance noise and odour modelling and in problems at viaducts with crossing (rail)roads on different levels.

3   Initial Modelling Concept

Based on the shortcomings at product level of the traditional topographic map the initial idea was that only in some specific cases true 3D modelling would be necessary, whereas in the majority of cases modelling in 2.5D would be sufficient. I define 2.5D modelling as using a single height value at every x,y-coordinate. Sometimes this is referred to as 'strict 2.5D modelling', as some people define 2.5 modelling as using 2D-simplexes (faces) in 3D space (Pilouk 1996), thus enabling several height values on one x,y-coordinate.

As 2.5D modelling is far less complex than 3D modelling, this has lead to the concept of combined 2.5D/3D modelling. The basic assumption is that the earth's surface can be modelled in 2.5D and that some more complex situations like buildings, viaducts or tunnels can be placed on top or below this surface. Apart from the intention to extend topographic models from 2D into 3D another important characteristic of the new modelling approach is to introduce the use of a foundation data structure. Within a data structure redundant data storage (geometry) can be avoided and the relationships between objects enable validation. In 2D one might require for instance that all objects form a planar partition, thus banning empty spaces between objects. The availability of topological relationships can also improve query performance during analyses.

3.1   Concepts of the Integrated TIN/TEN Approach

This leads to the concept of a topographic terrain representation in an integrated TIN/TEN model (TIN: Triangulated Irregular Network / TEN: Tetrahedronized Irregular Network). Four types of topographic features can be determined: 0D (point features), 1D (line features), 2D (area features) and 3D (volume features). For each type of feature simplexes of corresponding dimension are available to represent the features with, i.e. nodes, edges, triangles and tetrahedrons. A great advantage of using these simplexes is the well-defined character of the mutual relationships: a kD simplex is bounded by k+1 (k-1)D-simplexes (Pilouk 1996). This means that for instance a 2D simplex (a triangle) is bounded by three 1D simplexes (edges) and a 3D simplex (tetrahedron) is bounded by four 2D simplexes (triangles). The second important advantage of simplexes is the flatness of the faces, which enables one to describe a face using only three points. The third advantage is that every simplex, regardless its dimension, is convex, thus making convexity testing unnecessary. This quality simplifies point-in-polygon test significantly. The price for this comes with increased modelling complexity. Compared to for instance using polyhedrons as 3D primitive it will be clear that there exists a 1:1 relationship between a 3D feature (for instance a building) and its representation (the polyhedron), but that there will be a 1:n relationship between this 3D feature and its tetrahedrons. However, as long as one is able to hide this complexity from the average user, the advantages will overcome this drawback. To further illustrate the strength of using well-defined primitives, consider a real estate tax application that determines the tax assessment based on the volume of the building. In order to automate this process, a formula for determining volumes is required. Designing a formula capable of determining a polyhedron's volume is more complex due to the unlimited variation in shape. Contrarily, implementing a formula for the volume of a tetrahedron is straightforward, it only has to be applied several times as a building will be represented as a set of tetrahedrons. This repetition is however exactly what computers are good for.

The concept of the integrated TIN/TEN model is to represent 0D-2D objects in a TIN and 3D objects as separate TENs, that will be placed on top or below the TIN. This principle is illustrated in Figure 1. Note that the TIN for simplicity reasons is shown as a 2D TIN but that the model uses a 2.5D TIN. As both TINs and TENs are using triangles they can be 'put together' by making sure that they both contain the corresponding triangles.

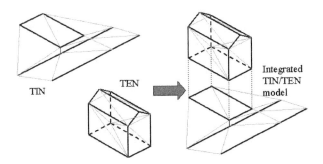

Fig. 1. Principle of modelling in an integrated TIN/TEN model

The underlying data structure can be designed quite straightforward, as the model only consists of nodes, edges, triangles and tetrahedrons. In Figure 2 an initial UML diagram is given of the structure. It consists of the four primitives, linked as described by the definition of Pilouk: a kD primitive is bounded by k+1 (k-1)D primitives. Flags (booleans) indicate whether a node/edge/triangle is part of the TIN, TEN or both. However, it shows that in this simple structure some problems occur as some attributes apply only on TINs or on TENs. Furthermore these attributes are in fact associations and have to be modelled accordingly. For example: within a TIN an edge has two neighbouring faces (left/right), within a TEN the number of associated faces in unbounded. An issue that will not be addressed in this section is whether to store only the node geometry and compute all other geometries based on the these nodes or to store the geometry for every simplex, as it is currently modelled in the class diagram.

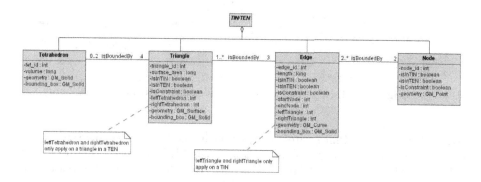

Fig. 2. Initial UML class diagram of the data structure. Note that some UML class diagram modelling rules are ignored in this diagram.

The attempt of the first conceptual model to store and integrate TIN and TENs at the most fundamental level was futile. Conceptually a TIN edge is not the same as a TEN edge and the same holds for TIN triangles and TEN triangles. Its geometries might be identical, but TIN and TEN edges and triangles have different mutual relationships. Therefore it is necessary to model the TIN and TEN separately (and where appropriate link or even merge its components). However a close relationship exists between for instance a TIN edge and a TEN edge, as they both are 1-simplexes. The UML class diagram in Figure 3 illustrates this relationship explicitly. An optional 'isEquivalentTo' relationship is defined between TIN and TEN nodes, edges and triangles. Note that this relation could indicate redundancy in the model, depending the actual implementation at storage level. This relationship is also visible in Figure 1, where the floor of the building is modelled in the TEN with nodes, edges and faces, equivalent to the nodes, edges and faces in the footprint in the TIN. Parallel to the Formal Data Structure (Molenaar 1990a) the initial modelling approach is feature-oriented. The model is able to represent point, line, area and volume features. In order to integrate TIN and TEN (and thus the 2.5D world with the 3D world) also semantically, the footprint of a volume feature is also integrated in the TIN. As a result the TIN can be considered to form a representation at

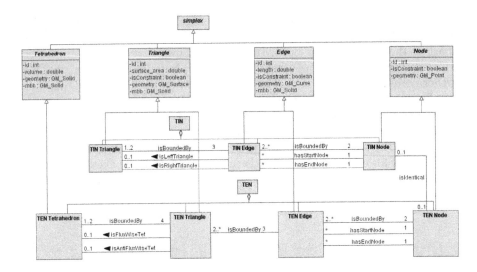

Fig. 3. UML class diagram of the integrated TIN/TEN model

terrain level. Figure 4 integrates the features in the UML class diagram. In this figure the connection between the TIN and the TEN at data structure level is only available at node level (and not at edge or triangle level); the reasons for this will be clarified in the next section. Three different types of linking can be distinguished:

- link at triangle level (thus 'glueing' the TENs on top of the TIN)
- link at edge level (thus 'stitching' the TENs on top of the TIN)
- link at node level (thus 'nailing' the TENs on top of the TIN)

Linking at the node level is the most fundamental one of this three, as edges and (indirect) triangles are defined by their nodes. However in order to optimize analytical capabilities one would prefer to have the link available at triangle level (which implies also the relationship on edge and node level).

3.2   Drawbacks Initial Modelling Concept

Although our initial modelling approach seems to make sense from a practical topographic point in a UML class diagram, it has some serious hidden problems. The first problem lies within the integration of the TIN with the TENs. At a conceptual level the 'isEquivalentTo' relationship is an appropriate way of linking both models. However, in order to optimize analytical capabilities one would prefer this link to exist on node as well as on edge and triangle level. In order to do so, one needs an implementation that takes care of ensuring the 1:1 relationship between TIN surface and TEN bottom. If one considers the example of a building placed on top of the terrain (as illustrated in Figure 1), not only the footprint of the building should match in TIN and TEN, but also the internal

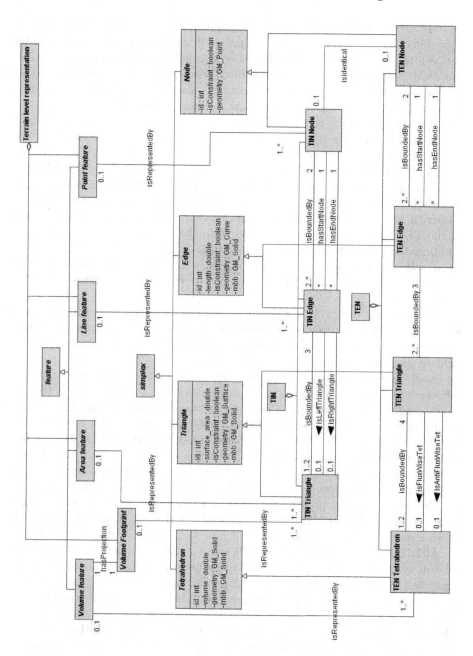

Fig. 4. UML class diagram of the feature-based integrated TIN/TEN model

edges in the shared face. The building will be represented as a rectangle in the TIN and this rectangle is identical to the floor boundaries in the TEN. However, in order to ensure the 1:1 relationship also on edge and face level, one needs the guarantee that the internal edge that triangulates this rectangle is the same diagonal in both data structures. Unfortunately it is not possible to ensure a match between the TIN and TEN triangles, as constrained triangulations and tetrahedronizations are only capable of handling constrained edges (Shewchuk 2004). This implies that not only the outer boundaries should be handled as constraints, but also this internal edge. As a result the outcome of the TIN triangulation can be transferred as input into the TEN triangulation or vice versa. This dependency wouldn't be a problem if one could be sure that these constraints will be triangulated without problems. However additional Steiner points are often required in order to enable tetrahedronization (as Schönhardts polyhedron (Schönhardt 1928) can't be tetrahedronized without additional Steiner points) or to improve the quality of a constrained Delaunay triangulation (and thus the numerical stability (Shewchuk 1997)). As Steiner points are inserted for instance in the TIN, they need to be transferred into the TEN, accompanied with the additional edges created by the Steiner points. At the same time the TEN algorithm might insert additional Steiner points, which should be transferred back to the TIN, thus resulting in the threat of a never-ending exchange of Steiner points.

This problem can probably be solved (partially) by handling Steiner points in a different way. Within algorithms used in GIS Steiner points are almost always used to split long edges into smaller ones. In the more general research field of meshing, where amongst others triangulation and tetrahedronization are used in order to simplify complex objects to enable appliance of partial differential equations, Steiner points are also added in the interior of a triangle or tetrahedron (Shewchuk 1997). In particular most refinement algorithms select skinny triangles and add the centrepoint of the circumcircle as a Steiner point. In TENs the centrepoint of the circumsphere can be used. Within GIS adding internal nodes is rather unusual, probably due to the fact that this data is collected by surveying techniques as GPS or photogrammetry, which are point measurements. As a result a node usually represents a measurement, which is not the case for Steiner points in the interior of an object. However, the question whether every polyhedron can be triangulated or tetrahedronized without adding Steiner points at the boundary of the object is theoretically not answered yet.

Although the modelling approach is quite straightforward ("model in 2.5D, only in exceptional cases model in 3D"), the questions how to link both models and when to switch between the two representations are not easy to answer (and thus to implement in practice). An important design question is whether both TIN and TEN should exist in case of more complex situations or that only the TEN should be available in these situations (implying a 'hole' in the TIN), thus creating 2.5D representation that is not a surface partition. Both approaches (TIN+TEN and TIN-TEN) are suitable in certain situations. One can imagine a situation in which a single building is represented in a TEN. At this location it will be quite easy to include the footprint in the TIN, thus using a TIN+TEN approach. However, if one considers the complex situation in Figure 5, in which a highway and railroad tracks are planned in tunnels, with a station and offices

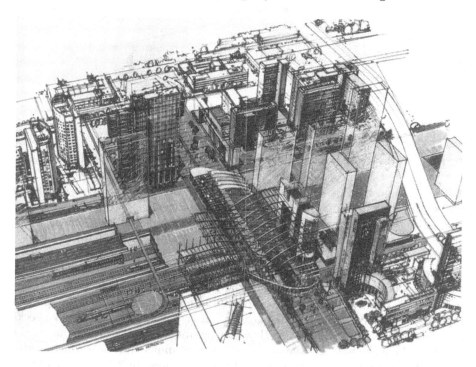

Fig. 5. Impression of the plans for the renewal of Amsterdam WTC Station, with several tunnels, offices and a station build on top of each other

build on top, the question is how to give a meaningfull definition of the 'terrain level' and thus what has to be included in the 2.5D representation. In such a situation it would make more sense to model the entire complex situation in one TEN, which shares its borders with the surrounding TIN, thus using a TIN-TEN approach. Virtual closure surfaces (Kolbe, Gröger & Plümer 2005) can be added to obtain a closed surface, although these surfaces have no relationships with actual features.

As a result from both examples one cannot select either the TIN+TEN or the TIN-TEN approach alone. It makes sense to use both approaches, but how to satisfactory define a general rule when to apply TIN+TEN and when TIN-TEN? This criterion adds more complexity to the initial simple modelling concept. In order to further contribute to the confusion, let's concentrate at the question which feature types are modelled in the TIN and which in a TEN. If one considers the simplified viaduct in Figure 6, applying the initial modelling approach would imply that only the viaduct itself will be modelled in 3D and the ascent and descent in 2.5D. Suppose that both on and under the viaduct a highway exists. In this case the bottom highway is represented in a TIN and the upper highway in a TEN. As a result the upper highway will have a thickness, while the bottom highway has not. One can also decide to only label the top triangles of the TEN as highway, but this will result in a meaningless volume, acting as a 'carrier object'

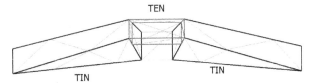

Fig. 6. Simplified viaduct, partially modelled as TEN

for the highway object. Regardless the chosen approach, it will still be an issue how to define the thickness of the TEN, as this data will not be available. One could use some kind of standard thickness, based on prior knowledge of some civil engineering rules of thumb. Still, why should one include this thickness or a 'carrier object' for highways on a viaduct and not include the standard thickness of the foundation of the bottom highway? This inconsistency is hard to accept.

The earlier observation that it will be difficult to define a 2.5D terrain surface everywhere implies that it is apparently not possible to extend all characteristics of a 2D representation into 2.5D, especially when this 2.5D representation has to fit with 3D TENs. In 2D a topographic representation can be considered as a topologically closed surface. This plays an important role in consistency checks of the data. However in 2.5D this rule no longer applies, for instance if one thinks of a tunnel entrance. In 2D the road stops at the tunnel entrance, which is also the border of the terrain feature lying above the tunnel. In 2.5D there will be a vertical gap between these two features, resulting in a non-watertight surface.

So far it is clear that - after solving a number of practical problems as described above - it will be possible to create a model in which TIN and TENs are combined, thus enabling a representation in which both surface and volume objects are present. Now the question arises whether this representation also offers the required functionality. For instance, Figure 6 illustrated that it will be possible to model a viaduct as a combination of a TIN and a TEN, but does this model enable the user to analyse the clearance under the viaduct? In a geometrical sense a relationship can be discovered between the TIN triangle of the bottom highway and the TEN lying above, but only after some serious calculations. Wouldn't it be better if the air between the viaduct and the underlying terrain was also modelled, thus enabling the user to calculate the height of these 'air' tetrahedrons in order to solve his query? The same holds for a tunnel. In the TIN+TEN case the tunnel TEN is only attached to the TIN at both tunnel entrances, but would this enable a quick analysis of which buildings are located on top of the tunnel? If the earth between the tunnel and the buildings is also modelled, this would simplify the query. Again this raises the question whether it is possible to satisfactory define a general rule which 'air' or 'earth' tetrahedrons should be included and which not.

3.3   Conclusions on the Initial Modelling Approach

Our initial modelling approach has some strengths and weaknesses. At a conceptual level the TIN and TEN have very close relationships and can be combined in an integrated data structure. The combination of TIN and TEN fits with the important observation that large parts of a country can be considered to be 2.5D and therefore there is no need for more complicated 3D modelling in these areas. However at the lower level problems occur regarding the actual integration and connection of the TIN and TEN models. Maintaining a 1:1 relationship between a TIN triangle and a TEN triangle is difficult due to the addition of Steiner points. Another problem in linking the TIN and TEN lies within the limited analytical capabilities of the combined model, as 'empty' space between the TIN and TEN is not present in the model. Therefore the link is only clearly present at the shared boundaries of TIN and TEN. Another problem is the non-existence of a topologically closed surface in 2.5D in the combined 2.5D/3D model.

4   Proposed New Modelling Approach

As the initial modelling approach had several drawbacks, a better approach was looked for. However, it was intended to retain the strengths of the initial modelling approach. These identified strengths were the point of departure for further developments:

- triangulation in 2D/2.5D/3D is a powerful data structure and offers computational advantages
- triangles correspond to well defined (flat) faces
- triangulations are suitable for storing objects in several dimensions
- large parts of a country can be represented in 2.5D due to the absence of true 3D shapes

4.1   Fundamental Concepts

Besides the strengths from the initial modelling approach the new approach is based on two fundamental observations:

- The ISO 19101 Geographic information - Reference model defines a feature as an 'abstraction of real world phenomena'. These real world phenomena have by definition a volumetric shape. In modelling often a less-dimensional representation is used in order to simplify the real world. Fundamentally there are no such things as point, line or area features; there are only features with a point, line or area representation (at a certain level of abstraction/generalization).
- The real world can be considered to be a volume partition. A volume partition can be defined (analogously to a planar partition) as a set of non-overlapping volumes that form a closed modelled space. As a consequence objects like 'air' or 'earth' are explicitly part of the real world and thus have to be modelled.

These two observations contain no shocking new insights; the first principle even goes back as long as to ancient cartography. However, combining both observations leads to an important method of treating less dimensional representations.

Based on the first observation one might wonder whether less-dimensional representations are even allowed in the new modelling approach, for instance using a face instead of a volume. The answer is positive, but only in special cases. Looking at the real world one can see that the features that are represented by faces are actually marking a border between two volume objects. For instance an area labelled as 'forest' might still be represented as a face, as it represents not only 'forest' but also implicitly the earth's surface, thus marking the transition between 'air' above and 'earth' beneath the surface. A lot of common modelling approaches are in its ways to define objects actual some kind of boundary representations. As in 2D a building is often represented by its walls; these walls actually mark the transition from the building's interior and the outside world. In the new modelling approach these volumes play a central role. The faces marking the borders between volumes might still be labelled, for instance as 'wall' or 'roof', but semantically they do not bound the building anymore, as the building in itself is represented by a volume, with neighbouring volumes that represent air, earth or perhaps another adjacent building.

At this point it might seem that also modelling 'air' and 'earth' in addition to all common topographic features is a very rigid approach of modelling, more serving the abstract goal of 'clean' modelling than an actual useful goal. This is however not the case. These air and earth objects do not just fill up the space between features of the other types, but are often also subject of analyses, such as noise and odour modelling. Another great advantage is the flexibility introduced by these features, as they enable future extensions of the data model. For instance, if one wants to model air traffic corridors, they can simply be inserted in the model. The space that is now simply labelled as 'air' can be subdivided in much more types, such as air traffic or telecommunication corridors. The same holds for the 'earth' tetrahedrons. In a later stage this general classification can be replaced by a more accurate one, for instance based on geotechnical and geological layers or polluted regions. Another advantage of modelling these 'empty' spaces is that it enables very pragmatic solutions for short term problems. If one thinks again of the viaduct in Figure 6, the problem was that feature instances of the same type were sometimes represented in 2.5D and sometimes in 3D, even though the thickness was not known. As long as no real data is available at the viaduct, one might model both crossing roads as faces. The upper highway will be represented as a set of flat faces on the border of two volumes, this time both 'air' volumes. This might not be a desirable definite solution, but until the availability of real 3D data of viaducts this is a pragmatic solution. This 3D data will become available in the future as terrestrial laser scanning will become more and more common practice (see Figure 7 for a terrestrial laserscan of a bridge). As a last advantage the simplicity of the concept in itself can be mentioned: modelling 'everything' in a TEN is easier than modelling sometimes in a TIN, sometimes in a TEN. Immediately it should be mentioned that modelling in a TEN is more complex than modelling in a TIN, thus questioning whether modelling everything in a TEN will be also easier at implementation level.

Fig. 7. Terrestrial laserscan of a bridge gives 3D information

Another interesting concept that is loosely related to the new modelling approach is an idea that still enables a user to work with a 2.5D TIN, even though the topographic model is stored as a full TEN. Up till now no distinction has been made between the way in which features are presented to the user and the way these representations are stored. One of the goals of the 3D Topography research is to implement the model in a spatial database, as this will improve manageability of the data. Databases have a rather nice feature and that is that one can work with views instead of tables. Whereas a table is physically stored in the database, a view can be seen as some kind of a virtual table. The user has all functionality as if he is working with a table, but the view is only a certain filter on top of one or more tables. Within our topography model one can think of node, edge, triangle and tetrahedron tables as basic storage structure of the 3D TEN. As stated at the beginning of this paragraph, the observation is that large parts of a country can be considered to be 2.5D, thus it still would simplify some applications if a TIN surface would be available. The idea is now that it should be possible to define a view on top of the triangle table of the TEN in such a way that this view consist of TIN triangles. As seen in the previous paragraph linking the TIN with the TENs was very difficult due to several problems, amongst others with Steiner points. The concept of using a view on a TEN actually does not try to solve the integration problem of the TIN and TEN, but it actually avoids the problem completely. Using a view in queries and visualisation makes the user believe that there is a TIN, while it's actually just a subset of the TEN triangles. The idea is quite simple, but with all earlier integration problems in mind it is an ingenious solution. It is almost surrealistic, as on user level the TIN is available for all kind of applications, including navigating to neighbours via edges. It resembles (see Figure 8) a work of Rene Magritte, La Trahison des Images (the Betrayal of Images), with the caption 'Ceci n'est pas un pipe' (This is not a pipe). This is exactly what defining a view does with the average user: he is convinced that he is working with a TIN (in an analysis or visualisation), but actually there is no (physically stored) TIN, as he is presented with a subset of TEN triangles.

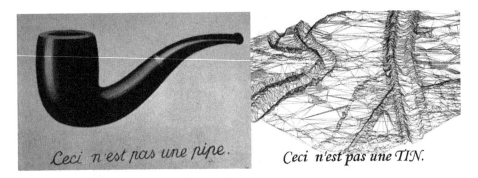

Fig. 8. La Trahison des Images, Rene Magritte, 1929 (left) and the TIN variant (right)

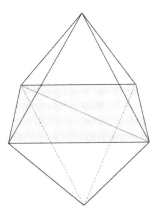

Fig. 9. Starting point for incremental insertion of topographic features: two 'air' tetra-hedrons and two 'earth' tetrahedrons. Note that these initial tetrahedrons will contain the entire topographic model and thus are extremely large.

4.2   Proposed Implementation

The new modelling approach is still triangulation-based, as one still wants to benefit from the well-defined character and its strong computational and ana-lytical capabilities. As a result the whole topographic model will be stored as one large TEN. The triangulation and tetrahedronization of newly inserted objects can be performed separately in order to improve performance; the resulting edges can be inserted into the TEN after the initial triangulation/tetrahedronization. In order to maintain a volume partition one does not start with an empty model, but with four initial tetrahedrons, see Figure 9. These four tetrahedrons (two 'air' tetrahedrons and two 'earth' tetrahedrons) will ensure that all space between features is modelled to without the need for explicitly surveying and maintaining 'air' or 'earth' objects. Note that in order to enclose the complete model these initial tetrahedrons are very large, as in theory a complete country or more can be modelled. The process of modelling topographic features consist of four discernable steps.

1. Start with four initial tetrahedrons, two 'air' and two 'earth' tetrahedrons;
2. Refine the earth's surface by inserting height information from a DEM;
3. Refine 'air' and 'earth' tetrahedrons in case of ill-shaped tetrahedrons by insertion of Steiner points;
4. Add real topographic features;

Within this last step both non-volumetric and volumetric features might be triangulated or tetrahedronized separately using a constrained triangulation/tetrahedronization algorithm. The outcome of these algorithms can be inserted incrementally in the full topographic model. As mentioned earlier there are no algorithms for the construction of constrained TENs capable of handling constrained faces. As a result one needs to 'translate' a constrained face into a set of constrained edges to preserve this constrained face in a tetrahedronization. This is done by first triangulating the boundary faces of an object and inserting all resulting edges as constrained edges into the tetrahedronization. As a post-processing step one needs to label the boundary faces explicitly as constrained faces.

Triangulating or tetrahedronizing the features one-by-one before insertion in the topographic model reduces computational complexity and thus saves computer time. The results need to be inserted into the full topographic model. This requires the use of an incremental algorithm to avoid recomputing the whole model. As the complete topographic model (the TEN) will be stored in a spatial database, it is necessary to implement the incremental algorithm within the database. As a result a full DBMS approach is required, instead of using the database just to store results of the computations.

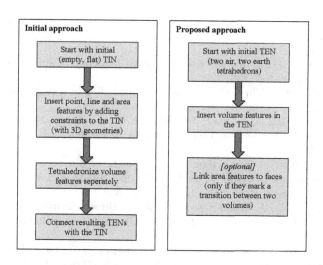

Fig. 10. Summary of the two proposed approaches

5   Conclusions and Future Research

This paper first introduced a very pragmatic approach of 3D modelling as it aims at modelling as much as possible in (less complicated) 2.5D, where full 3D modelling will be applied only in exceptional cases. Triangulations were selected as data structure due to their strong computational capabilities. The triangulation and tetrahedronizations can be integrated at a conceptual level, as both TIN and TEN use nodes, edges and faces. However the actual connection at design level appeared to be very difficult. As a result a more rigid approach was designed to solve the problems. Both approaches are summarized in 10. An important assumption is that the 3D model should form a volume partition and that preferably features are modelled in 3D. If one wants for some reason to use a less dimensional representation, this should only be done when these less dimensional features mark the transition between two 3D objects. For objects located on the earth's surface this is very clear: these objects might be modelled as faces instead of volumes, because one of the neighbouring volumes will represent 'air' and the other 'earth'.

As the rigid concept shows great promise, the following topics need further research:

1. Develop an UML class diagram of the rigid model
2. Implement required algorithms for constrained triangulation and tetrahedronization
3. Implement within the DBMS an incremental algorithm for tetrahedronization
4. Perform real tests with real world features
5. Define views within the TEN to offer TIN functionality on the earths surface

References

Carlson, E. (1987), Three-dimensional conceptual modeling of subsurface structures, *in* 'Auto-Carto 8', pp. 336–345.

de Cambray, B. (1993), Three-dimensional (3D) modelling in a geographical database, *in* 'Auto-Carto 11', pp. 338–347.

Kolbe, T., Gröger, G. & Plümer, L. (2005), Citygml: Interoperable access to 3d city models, *in* 'Geo-information for Disaster Management (GI4DM)', Springer, pp. 884–899.

Molenaar, M. (1990a), A formal data structure for three dimensional vector maps, *in* '4th International Symposium on Spatial Data Handling, Zürich', International Geographical Union IGU, Columbus, OH, pp. 830–843.

Molenaar, M. (1990b), A formal data structure for three dimensional vector maps, *in* 'Proceedings First European Conference on GIS (EGIS'90, Volume 2', Amsterdam, pp. 770–781.

Molenaar, M. (1992), 'A topology for 3D vector maps', *ITC Journal* **2**, 25–33.

Pilouk, M. (1996), Integrated Modelling for 3D GIS, PhD thesis, ITC Enschede, Netherlands.

Schönhardt (1928), 'Über die Zerlegung von Dreieckspolyedern in Tetraeder', *Mathematische Annalen* **98**, 309–312.

Shewchuk, J. R. (1997), Delaunay refinement mesh generation, PhD thesis, Carnegie Mellon University.

Shewchuk, J. R. (2004), 'General-Dimensional Constrained Delaunay and Constrained Regular Triangulations I: Combinatorial Properties', *To appear in: Discrete & Computational Geometry* . Available at http://www-2.cs.cmu.edu/ jrs.

Stoter, J. (2004), 3D Cadastre, PhD thesis, Delft University of Technology.

Zlatanova, S. (2000), 3D GIS for urban development, PhD thesis, Graz University of Technology.

Morse-Smale Decompositions for Modeling Terrain Knowledge

Lidija Čomić, Leila De Floriani, and Laura Papaleo

Department of Information and Computer Sciences (DISI),
University of Genova

Abstract. In this paper, we describe, analyze and compare techniques for extracting spatial knowledge from a terrain model. Specifically, we investigate techniques for extracting a morphological representation from a terrain model based on an approximation of a Morse-Smale complex. A Morse-Smale complex defines a decomposition of a topographic surface into regions with vertices at the critical points and bounded by integral lines which connect passes to pits and peaks. This provides a terrain representation which encompasses the knowledge on the salient characteristics of the terrain. We classify the various techniques for computing a Morse-Smale complexe based on the underlying terrain model, a Regular Square Grid (RSG) or a Triangulated Irregular Network (TIN), and based on the algorithmic approach they apply. Finally, we discuss hierarchical terrain representations based on a Morse-Smale decomposition.

Keywords: Terrain modeling, morphological representations, spatial knowledge modeling, critical net, Morse-Smale complexes.

1   Introduction

Terrain data consists of a finite set of points in a domain in the x-y plane at each of which an elevation value f is given. If the data points are regularly spaced in the domain, the terrain model is called a *Regular Square Grid* (RSG). Otherwise, the data points are connected to form a *triangle mesh* and a piecewise-linear interpolating function is defined on such mesh. The resulting model is called a *Triangular Irregular Network* (TIN) [7]. Regular grids can be encoded in very compact data structures, since only the elevation values need to be stored. TINs, on the other hand, better adapt to the shape of the terrain, since their vertices are irregularly and adaptively sampled.

A geometry-based description, such as RSG or TIN, provides an accurate representation of a terrain, but fails in capturing its morphological structure defined by critical points, like pits, peaks or passes, and separatrix lines, like ridges or valleys. Beside being compact, a morphological terrain description supports a knowledge-based approach for the analysis, visualization and understanding of a terrain dataset, as required for instance in visual data mining applications.

In the last decades, there has been a lot of research focused on extracting critical features (points, lines or regions) from images, or terrain data described

A.G. Cohn and D.M. Mark (Eds.): COSIT 2005, LNCS 3693, pp. 426–444, 2005.
© Springer-Verlag Berlin Heidelberg 2005

by regular grids, or triangulated surfaces. More recent work in computational geometry concentrates on representing the morphology of terrains through a decomposition of the terrain surface into regions bounded by critical points and integral lines, called a *Morse-Smale decomposition (or complex)* [10,25]. These techniques are rooted in Morse theory and try to simulate the Morse-Smale decomposition defined for C^2-differentiable functions in the discrete case [13,23]. Moreover, there has been quite a lot of research in image analysis on the watershed transform in the discrete case, which allows segmenting a gray-level image in regions, called *catchment basins*, bounded by lines, called *watershed lines*. Since terrain models can be interpreted as gray-level images, such techniques can be applied to morphological terrain representation.

The objective of this paper is to provide an overview and an analysis of existing techniques for extracting knowledge from a terrain, by decomposing the terrain surface into critical points, separatrix lines and regions defined by the Morse-Smale complex. We classify methods for morphological analysis of terrain data into methods based on an RSG, and methods based on a TIN. Then, we further classify such techniques for extracting a Morse-Smale decomposition into *boundary-based* and *region-based* depending on whether they compute the critical points and the separatrix lines, or they directly extract the regions forming the decomposition.

The remainder of this paper is organized as follows. In Section 2, we introduce some background notions on cell complexes, on Morse-Smale complexes and on the watershed transform in the continuum. In Section 2.2, we classify methods for extracting critical points and a Morse-Smale decomposition from RSG and triangle mesh, respectively, pointing out differences between boundary- and region-based methods. in Section 4 we review methods for extracting critical points starting from an RSG or a TIN. In Section 5, we present, analyze and compare methods for extracting a Morse-Smale complex from an RSG, while in Section 6 we consider methods which work on a TIN. Section 7 discusses hierarchical morphological representations of scalar fields. Finally Section 8 draws some concluding remarks.

2   Background Notions

In this Section, we present some background notions on cell complexes, Morse theory, and on the watershed transform in the continuum, that we will use in the rest of the paper.

2.1   Cell Complexes

A *k-dimensional cell* (*k*-cell) in E^d, $0 \le k \le d$, is a subset of E^d homeomorphic to a closed *k*-dimensional ball $B^k = \{x \in E^d : ||x|| \le 1\}$. A cell complex is a finite set of cells Γ in E^d, such that the interiors of the cells are disjoint, and the boundary of each *k*-dimensional cell is made of cells of Γ of dimension less than *k*. A cell γ' on the boundary of a cell γ is called a *face* of γ. The maximum of dimensions of cells γ over all cells of a complex is called the *dimension*, or the *order*, of the complex.

A subset Λ of Γ which has the structure of a cell complex is called a *subcomplex* of Γ. The *k-skeleton* of a cell complex Γ is the subcomplex of Γ which consists of all the cells of Γ having dimension less than or equal to k.

2.2   Morse-Smale Complexes

We review here the basic notions of Morse theory in the case of 2-manifolds. A 2-manifold is a topological space in which each point p has a neighborhood which is homeomorphic to the space R^2, or to the halfspace $R_+^2 = \{(x, y) \in R^2 : y \geq 0\}$ (see [13,23] for more details on Morse theory).

We consider a C^2-differentiable real-valued function f defined over a domain $D \subseteq R^2$. We denote by S the surface in R^3 which is the graph of function f. A point $p \in R^2$ is a *critical point* of f if and only if the gradient ∇f of f vanishes on p, i.e., if and only if $\nabla f(p) = 0$. Function f is said to be a *Morse function* when all its critical points are non-degenerate. i.e., when the Hessian matrix $Hess_p f$ of the second order derivatives of f at p is non-singular (its determinant is $\neq 0$). This implies that the critical points of f are isolated.

The number of negative eigenvalues of $Hess_p f$ is called the *index* of a critical point p. In 2D, there are three types of non-degenerate critical points. A critical point p is a *minimum (pit)*, a *saddle (pass)*, or a *maximum (peak)* when p has index 0, 1 or 2, respectively. An *integral line* of a function f is a maximal path which is everywhere tangent to the gradient vector field ∇f of f. The classical Taylor formula shows that integral lines follow the gradient directions in which the function has the maximum increasing growth. Integral lines cannot be closed, nor infinite and they cover S. An integral line is emanating from a critical point, or from the boundary of S, and it reaches another critical point, or the boundary of S.

An integral line which connects a minimum to a saddle, or a saddle to a maximum is called a *separatrix*. From each pass p there are two separatrices (also called *ridge lines*), which connect p to peaks, and two separatrices (also called *valley lines*) which connect p to pits. Integral lines that converge to (originate from) a critical point p of index i form an *i-cell* ((2-i)-cell) form a *stable (unstable) manifold* of p. The stable (unstable) manifolds are pairwise disjoint and decompose surface S into open cells which form a complex, since the boundary of every cell is the union of lower-dimensional cells. Such complexes are called *stable* and *unstable Morse complexes*, respectively.

A Morse function f is called a *Morse-Smale function* if and only if the bounaries of the stable and unstable complexes intersect only transversally. This means that such boundaries cross when they intersect, and that the crossing point is a saddle. Cells that are obtained as the intersection of the stable and unstable complexes of a Morse-Smale function f decompose surface S into a *Morse-Smale complex*. Cells of dimension 0, 1 and 2 in this complex are called *vertices*, *edges* and *regions*, respectively. The 1-skeleton of this complex consists of the critical points and the separatrix lines connecting them, and it is called the *critical net*.

Figure 1 (a) and Figure 1 (b) show an example of an unstable and of a stable Morse complex, respectively. Figure 1 (c) shows the corresponding Morse-Smale decomposition.

The combinatorial structure of the critical net is called a *surface network* [18,26]. It is a graph in which the nodes correspond to the critical points in the critical net and the arcs to the separatrices lines. A generalization of the surface network is provided by the *Critical Point Configuration Graph (CPCG)*[14], which has been defined for a Morse function, which does not necessarily satisfy the Morse-Smale condition.

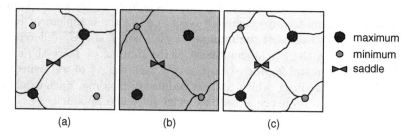

Fig. 1. (a) An unstable Morse complex, (b) a stable Morse complex, (c) the critical net

2.3   Watershed Transform in the Continuum

The watershed transform has been first introduced for gray-scale images. Several definitions exist in the discrete case [12,28]. The watershed transform has also been defined for C^2-differentiable functions over a connected domain D for which the critical points are isolated, and, thus, for Morse functions.

Catchment basins and watershed lines are basic notions in the watershed transform. They can both be defined in terms of topographic distance [12,20]. If f is a function which has a gradient ∇f everywhere except possibly at some isolated points, then the *topographic distance* $T_D(p,q)$ between two points p, q belonging to the domain D of f is defined as

$$T_D(p,q) = \inf_P \int_P |\nabla f(P(s))| ds$$

where P is a path (smooth curve) inside D, such that $P(0) = p$, $P(1) = q$, and $|\cdot|$ denotes the magnitude (norm) of a vector.

The topographic distance is defined in this way in order to ensure that the path which minimizes the topographic distance between two points p and q in D is the path of steepest slope, if it exists. In other words, if p and q are two points in D and if there exists an integral line which reaches both p and q, then the topographic distance between these two points is equal to the difference in elevation between them ($T_D(p,q) = |f(p) - f(q)|$). Otherwise, if such an integral line does not exist, the topographic distance between p and q is strictly greater than the difference in elevation, between p and q ($T_D(p, q) > |f(p) - f(q)|$). If p and q are points which belong to the same flat surface (plateau) then the topographic distance between them is zero. This means that the topographic distance is unable to distinguish among points belonging to the same plateau, and it is not a distance function, since the separation condition is not satisfied.

Let m_i be one of the minima of the function f, the *catchment basin* $CB(m_i)$ is defined as the set of points which are closer (in the sense of topographic distance) to m_i than to any other minimum. Thus, $CB(m_i) = \{f(p) \in S : f(m_i) + T_D(p, m_i) < f(m_j) + T_D(p, m_j), j \in I - \{i\}\}$ (where I is some index set).

Watershed (or *watershed lines*) $WS(f)$ of f is defined as the set of points in S which do not belong to any catchment basin, i.e., as the complement in S of the set of catchment basins of the minima of f. When f is a C^2-differentiable Morse function, then the catchment basins of the minima of f are the closure of the 2-cells in the unstable Morse complex of f, and the set of watershed lines forms a subset of ridge lines, which connect saddles to maxima. Each catchment basin is bounded by a sequence of saddles, ridge lines and maxima.

3   Issues in Approximating a Morse-Smale Complex

In this section, we characterize and analyze the methods proposed in the literature for decomposing the graph of an elevation function f into an approximation of a Morse-Smale complex and related issues. Such an approximation is obtained either fitting a C^1- or C^2-differentiable surface on a discrete terrain dataset, or by simulating a Morse-Smale complex on a piecewise-linear interpolation of the terrain. A terrain dataset consists of a finite set of elevation data given at a set of points not necessarily sampled from a Morse function, thus, not necessarily with isolated critical points. Note that information on the gradient and higher order derivatives are not generally available.

An approximation of the Morse-Smale complex in the discrete case can also be computed by applying the discrete watershed transform. The watershed transform in the C^2-differentiable case provides a decomposition of the graph of the elevation function f into the unstable manifolds of the minima, and thus an unstable Morse complex. Through a change in the sign of the elevation function, the stable manifolds of the maxima, and thus we can obtain the stable Morse complex for the original function. The overlay of the two complexes generates the Morse-Smale complex.

Note that, if f is a C^2-differentiable function, then the integral lines of f do not intersect, and, for every point p in the graph of f, there is only one integral line that passes through p. This is not true for piecewise-linear functions. Let us consider a piecewise-linear function f defined over a triangulated domain, i.e., a TIN, and two points p and q on the graph of f. Then, if the line of steepest descent from p contains q, then the line of steepest ascent from q does not necessarily contain p. This means that the lines of steepest ascent and descent may intersect, as noted also in [22]. One possible intersection is illustrated in

Figure 2, where the line of steepest descent (from point q at elevation 23 to point p at elevation 10) intersects the line of steepest ascent (from point s at elevation 17 to point r at elevation 30) at point t of elevation 20.

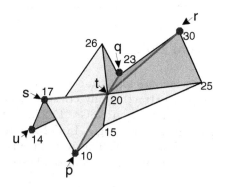

Fig. 2. Lines of steepest ascent connects the points s, t, r and descent connects the points q, t, p. Numbers indicate elevation values.

The stable (unstable) manifold of a point p in a Morse complex is defined as the set of points q on the graph S of f, such that the ascending (descending) integral lines from q reach p. As a consequence of this definition, the boundaries of stable (unstable) 2-cells are integral lines (or separatrix lines). These latter two conditions cannot be both guaranteed in the piecewise-linear case. If the definition of the stable (unstable) manifold is *region-based*, i.e., as the set of points q such that lines of steepest ascent (descent) reach the same critical point, then it cannot be guaranteed that the boundaries of stable (unstable) 2-cells are lines of steepest descent (ascent) that emanate from saddle points. If, on the other hand, the definition is *boundary-based*, i.e., as the set of points in regions bounded by descending (ascending) separatrix lines that emanate from saddles, then it cannot be guaranteed that a stable (unstable) 2-cell of a critical point p contains all points, and only those points, for which the line of steepest ascent (descent) ends in p. For example, in Figure 2, a point s at elevation 17 is a saddle point. One of the lines of steepest ascent starts from s goes through a regular point t at elevation 20, and ends in a maximum r at elevation 30. This line separates the domain into unstable components of the two minima u and p at elevation 14 and 10, respectively if the boundary-based definition is adopted. Let us consider now a point q at elevation 23. It belongs to the unstable component of u, but the line of steepest descent from q reaches the minimum p, and not u.

Most of the algorithms proposed in the literature use a boundary-based definition, that is, they extract and classify critical points and then trace the integral lines starting from saddle points, until a maximum or a minimum is reached. Some of these algorithms are based on RSGs, such as [1,21,22], others are based on TINs, such as [26,10,2,5,16]. The region-based approaches in [8], and watershed algorithms based on simulated immersion [28], topographic distance [12] and on rain-falling simulation [11] are the exceptions. Watershed algorithms by topographic distance and by rain-falling simulation adopt the definition of a catchment basin of a minimum, which is analogous to the definition of an unstable component of a Morse complex, i.e., as a set of points for which the path (integral line) of steepest descent ends in the same minimum.

4   Extracting Critical Points

In this section, we review techniques for extracting only critical points, or labeling terrain data sets (without extracting any critical lines). Several techniques focus on extracting critical points from RSGs [17,27,29], most of which have been developed initially for gray-scale images. In [3,15], a technique is proposed for extracting critical points from a TIN. This technique is used by most of the algorithms which compute the Morse-Smale complex based on a TIN (described in Section 6).

In both cases (RSG or TIN), a point p is classified as a critical point (maximum, minimum, saddle), or as a point which belongs to a critical line, based on the values of the elevation function f at p and at the points in some neighborhood of p. This means that the characterization and classification of characteristic points is done with a *local* criterion and, thus for instance, can be performed in parallel. This is similar to the C^2-differentiable case, where a critical point p is characterized and classified based on the partial derivatives of f at p, which are also local information.

On the other hand, it is not possible to characterize the points which belong to separatrix lines based only on local considerations. These methods, which are based on an RSG, can only extract points which belong to crest and course lines, since these features can be characterized locally (for example by considering principal normal curvature k_1 and k_2 of f at p). A line c is a crest (course) line if each point p on c is a local maximum (minimum) for the restriction of f to the line obtained as the intersection of the surface S associated with f and the vertical plane at p normal to the projection of the tangent vector of c at p to the xy plane. The set of ridge and valley lines is the subset of the set of crest and course lines, but the reverse is not true.

The method in [17] uses the relative elevation values at points in a 4- or 8-connected neighborhood [24] of a point p, and considers the number of the sign changes of the (signed) difference between the elevation at p and elevation at neighboring points. The neighborhood of p is scanned in a clockwise, or counter-clockwise, order and the number of sign changes from the points at lower elevation to the points at higher elevation and viceversa is counted, as well as the number of points in each contiguous sequence of points at lower (higher) elevation.

The method in [27] uses more elaborate concepts for classifying a point p in an RSG, thus producing a set of uniformly labeled regions. It introduces two quantities, namely the *connectivity number* CN and the *coefficient of curvature* CC, which are local in nature, since they are computed from elevation values at p

and at the points in some neighborhood of p. Intuitively, the connectivity number CN at a point p measures how many connected components, with an elevation greater than the elevation at p, exist in the neighborhood of p. The coefficient of curvature CC measures the sum of angle changes, in the neighborhood of p, of the contour line with elevation $f(p)$, measured counterclockwise. Characteristic regions are extracted as the connected components of points belonging to the same class.

The method in [29], uses approximations of the initial data points through either a generalized B-spline or a discrete cosine transformation. These approximations, which are C^0-differentiable inside the neighborhood of p and which are C^2-differentiable at p, are built based on the elevation values at p and at its neighboring points.

Since all the three methods are heuristic in nature, the quality of the output they produce (especially when they classify a point as belonging to a critical line) depends on the input data set. The first two methods [17,27] do not assume any approximation function, but also the method in [29] uses a globally discontinuous approximation.

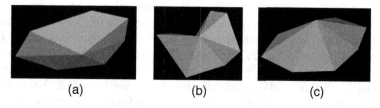

(a)   (b)   (c)

Fig. 3. (a) a minimum. (b) a saddle point. (c) a maximum.

The technique described in [3,15] work on a TIN by considering the link of each vertex p of the TIN. The *link* of a given vertex p in a TIN, denoted $Lk(p)$, is the set of edges in the TIN which are on the boundaries of the triangles incident in p and are not incident in p. $Lk(p)$ can be decomposed into the union of three different subsets of edges, namely $Lk^+(p)$, $Lk^-(p)$ and $Lk^\pm(p)$. $Lk^+(p)$, called the *upper link*, consists of the set of edges edges $[p_i, p_j] \in Lk(p)$ such that $f(p_i) > f(p), f(p_j) > f(p)$. $Lk^-(p)$, called the *lower link*, consists of the edges $[p_i, p_j] \in Lk(p)$ such that $f(p_i) < f(p), f(p_j) < f(p)$, while $Lk^\pm(p) = [p^+, p^-] : f(p^+) > f(p) > f(p^-)$ is the set of mixed edges. In this way, if $Lk^-(p) = 0$, a point p is a minimum, if $Lk^+(p) = 0$, p is a maximum, if $Lk^\pm(p) = 2$, p is a regular point, while if $Lk^\pm(p) = 2 + 2n$, p is a saddle with multiplicity n (n-fold saddle). Figure 3 shows examples of a minimum (a), a saddle (b) and a maximum (c), respectively.

5   Extracting a Morse-Smale Complex from an RSG

In this Section, we describe methods that, starting from an RSG, extract an approximation of the Morse-Smale complex. These methods can be subdivided into two main categories:

- methods which compute the critical net on an approximating surface defined on the grid (*boundary-based methods*)
- methods for computing the discrete watershed transform, which are applied to compute the stable and unstable Morse complexes (*region-based methods*).

5.1   Boundary-Based Methods for Computing the Critical Net

In this subsection, we present three boundary-based methods [1,21,22]. They first extract critical points, and then compute the separatrix lines as lines of steepest ascent (descent) starting from the saddle points. All three methods try to fit a surface of a certain degree of continuity to the input data.

The interpolating surface used in [1] is a globally C^1-differentiable Bernstein-Bèzier bi-cubic function. This interpolating function does not remove any critical point of the initial input data. In order to ensure that the number of newly introduced critical points is small, a "damped" central differencing scheme for computing the coefficients of the interpolating function is devised. The basic idea of "damping" is to keep the interpolant monotone inside the grid cells whenever possible. Integral lines are computed through a Runge-Kutta technique [19]. Four separatrix lines are traced, from each saddle point, in the direction of the appropriate eigenvectors. Computation of a separatrix line ends when the line reaches a neighborhood of another critical point, or the boundary of the RSG.

The method proposed in [21,22] uses a bilinear C^0-differentiable interpolating function $f(x,y) = axy + bx + cy + d$. Coefficients a, b, c, d are obtained unambiguously for each 2-cell from the elevation values of the vertices of the 2-cell. This interpolating function cannot introduce additional minima or maxima (for $a \neq 0$), so minima and maxima can only occur at the vertices of the RSG, but it may introduce additional saddles inside cells at a point with coordinates $(-\frac{c}{a}, -\frac{b}{a})$. A grid point p is classified by considering only the elevation of its 4-adjacent neighbors, while a 2-cell, which contains a pass, can be detected by considering the elevation of its four vertices. Separatrix lines are traced and constructed point by point and they can follow grid edges, or go through 2-cells. When a separatrix line crosses a grid cell, it can be approximated with small (linear) steps, or computed exactly, by solving a linear system of differential equations. The exact solution (integral line) is a hyperbolic function inside a grid cell.

The second method described in [21] uses a bi-quadratic approximation $f(x,y) = ax^2 + by^2 + cxy + dx + ey + f$ to the input data. This approximating function is constructed by fitting a bi-quadratic polynomial to the 8-coneected neighbors to each point of the RSG. The method produces a globally discontinuous approximation, formed by local surface patches. In this approach, all critical points are constrained to lie on the grid vertices. On the basis of the coefficients of the bi-quadratic polynomial, each surface patch is classified as *elliptic, parabolic* or *hyperbolic*, if $4ab - c^2$ is > 0, $= 0$, or < 0, respectively. The center and semi-axes of each conic section (patch) are determined, and the point p is classified, based on the type of the corresponding conic section and the number of intersections of the semi-axes with a circle around p with user defined radius r (see [21] for details). This classification follows the approach of [17].

Unlike the method in [1], the methods in [21,22] compute the first and second derivatives analytically. The second method proposed in [21] uses this information to trace the separatrix lines, while the first method in [21] proceeds in a step-by-step numerical manner.

5.2   Region-Based Methods

Watershed methods are region-based techniques. All watershed-based approaches start by first extracting the minima and then assigning points of the terrain to catchment basins related to the minima. Points that are not assigned to any catchment basin belong to watershed lines. A similar procedure can be applied to the same RSG and the elevation function $-f$, starting from the maxima. By computing the overlay of the two segmentations we obtain an approximation of the Morse-Smale complex. Figure 4 (left) shows the catchment basins of two minima and the related watershed lines.

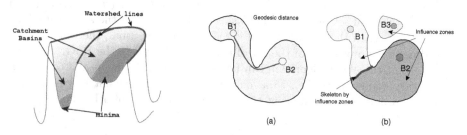

Fig. 4. The catchment basins (green) of two minima and the related watershed lines (red) are depicted on the left. (right) (a) The geodesic distance between two points in A. (right) (b) The influence zones and the skeleton by influence zones of B with respect to A.

Basically two techniques have been developed for computing the watershed transform in the discrete case starting from an RSG, namely watershed methods based on *simulated immersion* [28], and on *discrete topographic distance* [12]. The method in [12] extends the idea of topographic distance from the continuous to the discrete case, while the concept of simulated immersion is defined only in the discrete case [28].

The idea of simulated immersion can be described in an intuitive way. Let us consider a terrain and assume to drill holes in place of local minima. We assume to insert this surface in a pool of water, building dams to prevent water coming from different minima to merge. Then, the watershed of the terrain is described by these dams, and the catchment basins of minima are delineated by the dams. The method, that belongs to this class, uses the concept of skeleton by influence zones [24] in order to define catchment basins and watershed lines.

To understand the concept of *skeleton by influence zones*, we can imagine a set $A \subseteq R^n$ (or $A \subseteq Z^n$) and a set $B \subseteq A$ composed of n connected components $B_1, ..., B_n$. The skeleton by influence zones is the set C of points in A which are equally close (in the sense of geodesic distance) to at least two connected components of B. We recall that the geodesic distance between two points p and q in A is the length of a minimal path which connects p to q and stays within A. The *influence zone* of a component $B_i \in B$ is the set of points in A which are closer to B_i than to any other connected component B_j of B. Note that the skeleton by influence zones C of B within A is the complement of the union of influence zones of B_i within A (see Figure 4 on the right (a) and (b)).

The method in [28] recursively extracts catchment basins and watershed lines, starting from the minimal value of the elevation function f and going up. At each level of recursion, new minima can be found, or already created catchment basins are be expanded. The expansion process continues until, at a given level h, a potential catchment basin CB_h (related to level h) contains at least two catchment basins (for example CB_{ih-1}, CB_{jh-1}) already is present at level $h-1$.

This is the case in which the definition of skeleton by influence zones comes up: CB_h is partitioned into three elements, the two influence zones of CB_{ih-1} and CB_{jh-1} and the set of points in CB_h equally distant from CB_{ih-1} and CB_{jh-1} (skeleton by influence zones). The influence zones of CB_{ih-1} and CB_{jh-1} will be part of the final set of catchment basins in the output of the algorithm. The process stops when the maximal level is reached and, as usual, the watershed is defined as the complement of the set of catchment basins in the RSG.

Note that the above algorithm can be extended to 3D images and to a mesh of arbitrary connectivity, i.e., to a mesh in which each element is a node of a graph and a node can be adjacent to an arbitrary number of other nodes.

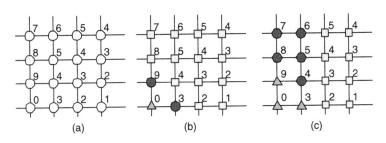

Fig. 5. (a) The initial RSG, (b) The result obtained by applying simulated immersion, (c) The result obtained using a watershed technique by topographic distance. Both use a 4-connected neighborhood [20]. Circles belong to the watershed, while squares and triangles identify catchment basins of the minima at elevation 1 and 0, respectively.

The topographic distance between two points p and q in the discrete case is defined as the minimum sum of cost among all the possible path from p to q in the RSG [12]. A cost is associated with each edge in the RSG, computed on the basis of the *lower slope* $LS(p)$, defined as the maximal slope linking p to any of its neighbors of lower elevation. As in the continuous case, a path P from p to q is a path of steepest descent when the topographic distance between p and q is equal to the difference in elevation between p and q. Otherwise, the topographic distance between p and q is greater than the difference in elevation between p and

q. The definition of a catchment basin in an RSG is similar to the definition in the differentiable case. In the discrete case, the continuous topographic distance is replaced with the (cost-based) discrete topographic distance. The problem with this approach is that watersheds can be thick. This is due to the fact that this topographic distance between two points that belong to the same plateau is zero. Figure 6 illustrates the computation of the topographic distance along a path $P = (a, b, c, d, e, f, g, h)$. In each of the regions on the surface (graph of the elevation function), the intensity of the gradient is assumed to have a constant value. The topographic distance between points a and b is computed as $T_f(a, b) = \alpha d_P(a', b')$, where α is the intensity of the gradient in the region which contains the subpath (a, b), and $d_P(a', b')$ is the geodesic distance between the projections a' and b' of a and b (the length of the path (a', b')).

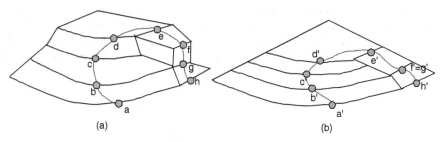

(a)   (b)

Fig. 6. Computation of the topographic distance between points a and h along a path $P = (a, b, c, d, e, f, g, h)$. (b) is the projection on the xy plane of (a).

In [20] it is shown that the two approaches based on topographic distance or on simulated immersion produce different results on the same RSG. Figure 5 (a) shows the initial RSG, while Figure 5(b) reports the result obtained by applying simulated immersion approach and Figure 5 (c) the result obtained by using a watershed technique by topographic distance. In both examples, we use a 4-connected neighborhood, for simplicty.

6   Methods for extracting a Morse-Smale complex from a TIN

As for RSGs, we can classify the methods for extracting an approximate Morse-Smale complex from an TIN as boundary-based and region-based methods.

6.1   Boundary-Based Methods

All boundary-based methods start by extracting the critical points. The methods in [26,10,5,16] use the technique described in Section 4, while the method in [2] uses the normals of the triangles incident at point p in order to classify p. Then, all the methods extract separatrix lines. Starting from the saddle points, two lines of steepest descent and two lines of steepest ascent are traced. The methods in

[26,10,2] extract the separatrix lines along the edges of the TIN. The methods in [5,16] compute the gradient along edges and triangles, and create the lines that may cross the triangles of the TIN.

The methods in [26,10], after extracting the critical points, unfold the saddles in such a way that a k-fold saddle p is replaced with k simple saddles $p_1, \ldots, p_k$. The procedures in [10,26] differ in the way they connect the neighboring points of a multiple k-fold saddle p to k simple saddles into which p is decomposed.

The difference between the path tracing procedure of [26] and [10] is that, in [26] the neighboring point of any given point with highest elevation is chosen, while in [10] the steepest edge is chosen at every point. Note that the two approaches produce two different critical nets, as illustrated in Figure 7. According to the algorithm in [10], the point at elevation 2 is connected to the point at elevation 20, while in the algorithm in [26], such point is connected to the point at elevation 21 by assuming that the distance between points at elevation 2 and 20 is (12) smaller than the distance between points at elevation 2 and 21 (20), so that the corresponding slope is greater. In [10], the resulting Morse-Smale complex is further modified by using a sequence of local transformations, called *handle slides*.

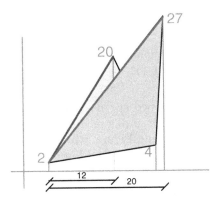

Fig. 7. The point at elevation 2 is connected to a points at elevation 20 in the method of [10] (red line), or to the point at elevation 21 in the method of [26] (blue line).

To overcome the limitation of implementing handle slides, the algorithm described in [5,16] extracts an approximation of a critical net from a TIN by crossing the triangles so that separatrix lines of the Morse-Smale complex are not constrained to be edges of the TIN, but are computed along the actual paths of steepest ascent (descent).

Finally, the method in [2] uses a similar procedure to [10,26] for tracing separatrix lines, but it adopts a different technique to classify the critical points. A point p in the TIN is classified based on the normal vectors to the triangles incident at p. The gradient at p, which is undefined in the piecewise-linear case, is assumed to take a range of values based on the normal vectors of adjacent triangles: p is considered a critical point when this range of values includes a vector $(0, 0, 1)$. A critical point p is classified as minimum, maximum, or saddle point, depending on whether the gradient flow is away from, towards, or towards-and-away from p, respectively.

6.2   Region-Based Approaches

In this subsection, we present two region-based methods [8,11] for extracting an approximation of a Morse-Smale complex starting from a TIN. The method in [11] uses a watershed approach, which is based on a rain-falling simulation technique. The method in [8] applies a region-growing technique on the faces of the TIN based on the gradient.

In [8], a constructive algorithm is described to compute an approximation of a Morse-Smale complex, obtained by extracting the stable and unstable Morse complexes separately and then overlaying them. It extracts the minima for the unstable complex and the maxima for the stable complex.

First, all triangles incident at a minimum (maximum) p are assigned to the unstable (stable) component of p. This component is extended considering the triangles adjacent to its boundary and their gradients. Let t be a triangle in the component C of p and let e be the edge of t, on the boundary of C. Let t' be the triangle which shares e with t: if the gradient of t' has the same orientation as the gradient of t, t' is added to the unstable (stable) component of p (see Figure 8). The gradient field inside a triangle can have one of five possible configurations: if it is incident to a minimum (see Figure 8 (a)) or a maximum (see Figure 8 (b)), then a bundle of integral lines is emanating from the minimum or it is converging to the maximum. The gradient can be parallel to one edge of the triangle (see Figure 8 (c)) or enter the triangle along one (two) edge and leave it along two (one) edges (see Figure 8 (d)-(e)). Flat triangles, having all three vertices at the same elevation, are marked and not considered by the algorithm.

In [11], a discrete approach extending the morphological watershed algorithm by rain-falling simulation is applied. This algorithm does not assume any interpolant to be defined on the TIN. In the rain-falling approach, watersheds are seen as the divide lines of the domains of attraction of rain falling over the region. Thus, a point p is assigned to a catchment basin of the minimum m_i if there is a

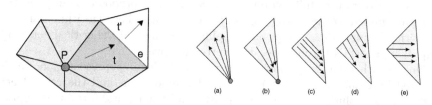

Fig. 8. (left) t is a triangle in the component of p, e is the edge of t shared with the triangle t'. The gradient of t' is concordant to the gradient of t, thus t' is added to the unstable component of p (it will become green). The gradient is depicted with arrows. (right) The five configurations of bundles of integral lines traversing a triangle. Green point is a minimum, blue point is a maximum.

path of steepest descent from p to m_i. The algorithm extracts all local minima as the points which have the lowest elevation compared with all their neighboring points, and assigns a unique label to each minimum. Then, portions of surface at the same elevation (plateaus) are identified by considering connected components of the vertices of the TIN having the same elevation. Beginning at the lowest vertex adjacent to a plateau, a descending path P is computed on the TIN until a labeled vertex q is encountered. The path is constructed by considering, at each of its vertices p, the vertex adjacent to p with minimum elevation. The vertices of P and the vertices of the portions of surface at the same elevation are labeled with the same label as q. After the plateaus have been processed, descending paths are constructed in a similar manner, starting from each unlabeled vertex. The set of the vertices of the TIN is segmented into catchment basins which correspond to minima. Catchment basins are formed by considering maximal connected components of the vertices of the TIN with the same label. Finally, a region-merging process is applied, which is based on the depth of the basins, that is the difference in the elevation between the highest and the lowest vertex in each basin.

The method in [28] which has been originally designed for RSG can be extended to a mesh of arbitrary connectivity, i.e., to a mesh in which each element is a node of a graph and a node can be adjacent to an arbitrary number of other nodes. Thus, this watershed algorithm can be applied also to a TIN.

The method described in [8] extracts maxima and minima and computes the saddle points via intersection of the stable and unstable Morse complexes. The method in [11], if applied to both the function f and the function $-f$, extracts also the minima and maxima but it does not identify saddles. Moreover, the method in [11] is a discrete approach that uses a discontinuous surface, while the method in [8] uses an approximation which is C^0 continuous.

7   Hierarchical Representation of Terrain Morphology

A hierarchical representation of the terrain morphology is critical for interactive analysis and exploration of a terrain in order to maintain and analyze characteristic features at different levels of resolutions. Current multi-resolution terrain models are just based on a progressive simplification process applied to a TIN describing a terrain at full resolution. In [8], we have proposed a simplification algorithm guided by the morphology of the critical net. The algorithm maintains the critical points at each step, but simplifies the edges of the critical net, while simplifying the TIN through iterative vertex removal.

On the other hand, a concise and powerful terrain representation can be obtained by combining the Morse-Smale decomposition (and the critical net) with a multi-resolution approach, thus providing a structural terrain description as a hierarchy of complexes.

Generalization of a critical net is the process of transforming one critical net into another, with fewer number of vertices [5,10,25,30]. This is accomplished by applying an elementary generalization operation, usually called *cancellation of critical points*. This generalization operation consists of removing a pair of adjacent critical points and in reconnecting the remaining points of the critical net. The pair of removed critical points consists of a saddle and a maximum, or a saddle and a minimum. The two critical points are removed from the critical net together with all the lines incident in them. The pair of critical points to be removed must be selected in such a way that the resulting decomposition is still a Morse-Smale complex. When a saddle s together with an adjacent minimum (maximum) p is removed, the remaining saddles s_i can be connected to the other minimum (maximum) q adjacent to s, if the elevation of s_i is higher (lower) than the elevation of q. Figure 9 shows an example of generalization operation that removes the join (s, p), where s is a saddle and p is a minimum.

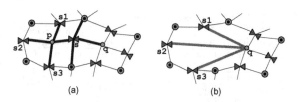

(a)   (b)

Fig. 9. (a) A surface network $S = (C, A)$. The arcs to be removed are in bold. (b) The surface network S_0 obtained from S by removing the join (s, p), where s is a saddle and p is a minimum. The new arcs inserted are in bold.

In [30], a minimum (maximum) p is chosen for cancellation together with its lowest (highest) adjacent saddle s, the order in which minima and maxima are chosen is not specified. In [25], a pair of critical points (p, s) is chosen such that the difference in elevation between p and s is minimal among all (unsigned) differences in elevation between a saddle and an adjacent minimum, or a saddle and an adjacent maximum. In the approach described in [10], and later in [5] (where geometric considerations are taken into account), a saddle s is chosen together with its adjacent maximum at lower elevation, or its adjacent minimum at higher elevation. The order in which the pairs of points are canceled is determined based on the notion of persistence (see [10] for details). Finally, in [22] a hierarchical critical net is proposed based on an entirely different approach, which identifies hierarchies of ridges and valleys inside the critical net at maximum resolution.

8   Concluding Remarks

In this paper, we have described, analyzed and compared existing methods for extracting an approximation of a Morse-Smale complex from a terrain model. We have classified the algorithms based on the terrain model used namely a Regular Square Grid or a Triangulated Irregular Network and based on the algorithmic approach used for computing the Morse-Smale decomposition. Tables 1 and 2 summarize the analysis and comparisons performed in the paper of the different methods.

Table 1. Classification of the analyzed methods for extracting critical points

Methods for Extracting Critical Points				
Method (ref)	Input (RSG/TIN)	Continuity	Output	Points on vertices?
[17]	RSG	local	labeling RSG	Points ON
[29]	RSG	local	labeling RSG	Points ON
[27]	RSG	patches	labeling RSG	Points ON
[3]	TIN	C^0	max, min, saddles	Points ON
[15]	TIN	C^0	max, min, saddles	Points ON

Table 2. Classification of the examined methods for extracting a Morse-Smale complex

Methods for computing a Morse-Smale Complex						
Method (ref)	RSG/ TIN	Cont.	Output	Points on vertices? Lines crossing?	B/R	Notes
[1]	RSG	C^1	critical points, lines	Points ON, lines CROSS	B	Bezier function
[22]	RSG	C^0	critical points, lines	Points ON, lines CROSS	B	bilinear
[21]	RSG	patches	critical points, lines	Points ON, lines CROSS	B	bicubic
[26]	RSG/TIN	C^0	critical points, lines	Points ON, lines ON	B	
[10]	TIN	C^0	critical points, lines	Points ON, lines ON		
[5]	TIN	C0	critical points, lines	Points ON, lines CROSS	B	
[2]	TIN	C^0	critical points, lines	Points ON, lines ON	B	
[6]	TIN	C^0	critical points, lines, regions	Points ON, lines ON	R	
[8]	TIN	C^0	critical points, lines, regions	Points ON, lines ON	R	up tp nD
[12]	RSG	local	watershed	Points ON, lines ON	R	
[28]	RSG/TIN	local	watershed	Points ON, lines ON	R	
[11]	TIN	C^0	watershed	Points ON, lines ON	R	

Challenging research issues are developing techniques for computing the Morse-Smale decomposition for $3D$ scalar fields. The algorithm proposed in [9] is theoretical correct and its implementation [4] is efficient only for small size datasets. Region-based techniques seem to be more promising for extensions to higher dimensions.

Acknowledgments

This work has been partially supported by the project founded by the Italian Ministry of Education, University and Research (MIUR) on Representation and Management of spatial data in the web ($SPADA@WEB$) and by the European Network of Excellence $AIM@SHAPE$ under contract number 506766. We also want to thank Emanuele Danovaro and Maria Vitali for all the helpful discussions.

References

1. C. L. Bajaj, V. Pascucci, and D. R. Shikore. Visualization of scalar topology for structural enhancement. In *Proceedings IEEE Visualization'98*, pages 51–58. IEEE Computer Society, 1998.
2. C. L. Bajaj and D. R. Shikore. Topology preserving data simplification with error bounds. *Computers and Graphics*, 22(1):3–12, 1998.

3. T. Banchoff. Critical points and curvature for embedded polyhedral surfaces. *American Mathematical Monthly*, 77(5):475–485, 1970.

4. P. Bremer, V. Pascucci, and B. Hamann. Maximizing adaptivity in hierarchical topological models. In *Proceedings of International Conference on Shape Modeling and Applications*, 2005.

5. P.-T. Bremer, H. Edelsbrunner, B. Hamann, and V. Pascucci. A multi-resolution data structure for two-dimensional Morse functions. In G. Turk, J. van Wijk, and R. Moorhead, editors, *Proceedings IEEE Visualization 2003*, pages 139–146. IEEE Computer Society, October 2003.

6. E. Danovaro, L. De Floriani, and M. M. Mesmoudi. Topological analysis and characterization of discrete scalar fields. In T.Asano, R.Klette, and C.Ronse, editors, *Theoretical Foundations of Computer Vision, Geometry, Morphology, and Computational Imaging*, volume LNCS 2616, pages 386–402. Springer Verlag, 2003.

7. L. De Floriani, P. Magillo, and E. Puppo. Data structures for simplicial multi-complexes. In R. H. Guting, D. Papadias, and F. Lochovsky, editors, *Advances in Spatial Databases*, volume 1651 of *Lecture Notes in Computer Science*, pages 33–51. 1999.

8. E.Danovaro, L. De Floriani, P.Magillo, M. M. Mesmoudi, and E. Puppo. Morphology-driven simplification and multiresolution modeling of terrains. In E.Hoel and P.Rigaux, editors, *Proceedings ACM GIS 2003 - The 11th International Symposium on Advances in Geographic Information Systems*, pages 63–70. ACM Press, 2003.

9. H. Edelsbrunner, J. Harer, V. Natarajan, and V. Pascucci. Morse-Smale complexes for piecewise linear 3-manifolds. In *Proceedings 19th ACM Symposium on Computational Geometry*, pages 361–370, 2003.

10. H. Edelsbrunner, J. Harer, and A. Zomorodian. Hierarchical Morse complexes for piecewise linear 2-manifolds. In *Proceedings 17th ACM Symposium on Computational Geometry*, pages 70–79. ACM Press, 2001.

11. A. Mangan and R. Whitaker. Partitioning 3D surface meshes using watershed segmentation. *IEEE Transaction on Visualization and Computer Graphics*, 5(4):308–321, 1999.

12. F. Meyer. Topographic distance and watershed lines. *Signal Processing*, 38:113–125, 1994.

13. J. Milnor. *Morse Theory*. Princeton University Press, 1963.

14. L. R. Nackman. Two-dimensional critical point configuration graph. *IEEE Transactions on Pattern Analysis and Machine Intelligence*, PAMI-6(4):442–450, 1984.

15. X. Ni, M. Garland, and J.C. Hart. Fair morse functions for extracting the topological structure of a surface mesh. *ACM Trans. Graph.*, 23(3):613–622, 2004.

16. V. Pascucci. Topology diagrams of scalar fields in scientific visualization. In S. Rana, editor, *Topological Data Structures for Surfaces*, pages 121–129. John Wiley and Sons Ltd, 2004.

17. T. K. Peucker and D. H. Douglas. Detection of Surface-Specific Points by Local Parallel Processing of Discrete Terrain Elevation Data. *Computer Graphics and Image Processing*, 4:375–387, 1975.

18. J. L. Pfaltz. Surface networks. *Geographical Analysis*, 8:77–93, 1976.

19. W. Press, S. Teukolsky, W. Vetterling, and B. Flannery. Numerical recipes in c - second edition. Cambridge University Press, 1992.

20. J. Roerdink and A. Meijster. The watershed transform: definitions, algorithms, and parallelization strategies. *Fundamenta Informaticae*, 41:187–228, 2000.

21. B. Schneider and J. Wood. Construction of metric surface networks from raster-based DEMs. In S. Rana, editor, *Topological Data Structures for Surfaces*, pages 53–70. John Wiley and Sons Ltd, 2004.

22. Bernhard Schneider. Extraction of hierarchical surface networks from bilinear surface patches. *Geographical Analysis*, 37:244–263, 2005.

23. S. Smale. Morse inequalities for a dynamical system. *Bulletin of American Mathematical Society*, 66:43–49, 1960.

24. P. Soille. *Morphological Image Analysis: Principles and Applications*. Springer-Verlag, Berlin and New York, 2004.

25. S. Takahashi. Algorithms for extracting surface topology from digital elevation models. In S. Rana, editor, *Topological Data Structures for Surfaces*, pages 31–51. John Wiley and Sons Ltd, 2004.

26. S. Takahashi, T. Ikeda, T. L. Kunii, and M. Ueda. Algorithms for extracting correct critical points and constructing topological graphs from discrete geographic elevation data. *Computer Graphics Forum*, 14(3):181–192, 1995.

27. J. Toriwaki and T. Fukumura. Extraction of structural information from gray pictures. *Computer Graphics and Image Processing*, 7:30–51, 1978.

28. L. Vincent and P. Soile. Watershed in digital spaces: an efficient algorithm based on immersion simulation. *IEEE Transactions on Pattern Analysis and Machine Intelligence*, 13(6):583–598, 1991.

29. L. T. Watson, T. J. Laffey, and R. M. Haralick. Topographic classification of digital image intensity surfaces using generalized splines and the discrete cosine transformation. *Computer Vision, Graphics, and Image Processing*, 29:143–167, 1985.

30. G. W. Wolf. Topographic surfaces and surface networks. In S. Rana, editor, *Topological Data Structures for Surfaces*, pages 15–29. John Wiley and Sons Ltd, 2004.

2D-3D MultiAgent GeoSimulation with Knowledge-Based Agents of Customers' Shopping Behavior in a Shopping Mall

Walid Ali *,** and Bernard Moulin *,**

* Computer Science Department, 3904 Pav. Pouliot, Laval University,
Ste Foy, Québec G1K 7P4, Canada
** Research Center on Geomatics, Laval University, Ste Foy, Quebec G1K 7P4, Canada
{walid.ali, bernard.moulin}@ift.ulaval.ca

Abstract. In this paper we present a simulation prototype of the customers' shopping behavior in a mall using a knowledge-based multiagent geosimulation approach. The shopping behavior in a shopping mall is performed in a geographic environment (a shopping mall) and is influenced by several shopper's characteristics (internal factors) and factors which are related to the shopping mall (external or situational factors). After identifying these *factors* from a large literature review we grouped them in what we called "*dimensions*". Then we used these dimensions to design the knowledge-based agents' models for the shopping behavior simulation. These models are created from empirical data and implemented in the MAGS geosimulation platform. The empirical data have been collected from questionnaires in the *Square One shopping mall* in Toronto (Canada). After presenting the main characteristics of our prototype, we discuss how mall's managers of the *Square One* can use the Mall_MAGS prototype to make decisions about the mall spatial configuration by comparing different simulation scenarios. The simulation results are presented to mall's managers through a user-friendly tool that we developped to carry out data analysis.

Keywords: Knowledge-based agent, MultiAgent System, GeoSimulation, Shopping mall, Shopping behavior, Spatial characteristics, Spatial behavior.

1 Introduction

Geosimulation (Benenson and al., 2004) and more specifically the simulation of human behavior in space is an extremely interesting and powerful research method to advance our understanding of human spatial cognition and the interaction of human beings with their spatial environment. MultiAgent systems provide a computing paradigm which has been recently used to create such simulations (Frank and al., 2001). Several researchers used this paradigm to develop simulation applications that simulate different behaviors in spatial environments. For example, (Raubal, 2001) (Frank and al., 2001) presented an application which simulates a wayfinding behavior in an airport. (Dijkstra et al., 2001) simulated, using cellular automata, pedestrian

A.G. Cohn and D.M. Mark (Eds.): COSIT 2005, LNCS 3693, pp. 445–458, 2005.
© Springer-Verlag Berlin Heidelberg 2005

movements in a shopping mall. (Koch, 2001) simulated people movements in a large scale environment representing a town. These applications successfully simulated certain kinds of behaviors, but they have some limitation related to the capabilities of the agents used in the simulation. For example, the agents of (Raubal, 2001) and (Frank and al., 2001) perceive and memorize their environment using the concept of information and affordance (Frank and al., 2001). (Dijkstra et al., 2001) and (Koch, 2001) use a message passing technique between the agent and the environment. These perception mechanisms do not allow agents to perceive the geographic characteristics of the environment which are important to take into account certain spatial behaviors. Furthermore, in these applications the agents do not have a memorization capability which can used to memorize the elements perceived in the environment. To develop more realistic applications that simulate human behavior in spatial environments, we need agents equipped with a more *"accurate perception"* and able to perceive all the components included in the environment and the changes that may occur to objects in the geographic space. Agents also need richer cognitive capabilities such as memorization.

In addition, the applications of (Raubal, 2001), (Frank and al., 2001), (Dijkstra et al., 2001) and (Koch, 2001) are only used to display on screens the behaviors to be simulated. To be more useful, simulation applications must be used beyond the mere visualization function: They should generate simulation output data which can be used by users in order to make decisions. In MAGS project (Moulin et al., 2003) we develop a method and a geosimulation platform in order to overcome some of the aformentionned limitations of exisiting geosimulations.

In this paper we present a multiagent geosimulation application which simulates customer's shopping behavior in a shopping mall in order to understand how shoppers interact with a mall and how they react to the changes of the mall's configuration and atmosphere. We use a *multiagent geosimulation* approach for the following reasons:

- The shopping behavior in a mall is an activity which is performed by a large number of shoppers (hundreds or thousands of shoppers). Hence, the use of a multi-agent approach and a realistic simulation should take into account individual behaviors (Frank et al., 2001).

- It is, essentially, a spatial behavior carried out in an environment (shopping mall) in which the geographic and spatial characteristics are very important.

- Furthermore, we are interested in the behavior of customers who perceive, memorize, decide and navigate in the shopping environment. Hence, our agents should be equipped with some of these cognitive capabilities (Frank et al., 2001).

Furthermore, in our prototype we feed the simulation with data using specific tools that we developed for our work. In addition, our proptotype generates output data using software agents called *Observers* and we analyse this data using non-spatial and spatial analysis techniques. This analysis is carried out through a data analysis tool that we developed in *Microsoft Visual Basic 6.0*. Using our application, mall' managers (i) can visualize and understand how shoppers interact with the mall's environment (spatial or atmosphere aspects of the mall), (ii) can change the configuration or the atmosphere of the mall and see how virtual shoppers react to these changes and (iii) can identify the parts of the mall on which they need to concentrate their efforts in order to make their mall more confortable for shoppers and to seduce them.

This paper is organized as follows. In Section 2 we discuss the main properties of the shopping behavior in a mall. In Section 3 we present the characteristics of some agents' models that we designed to simulate the customers' shopping behavior. In Section 4 we present our multiagent geosimulation prototype of the shopping behavior. In the same section we present how mall' managers can use the prototype in order to evaluate different configurations and atmospheres of the mall. Finally, in Section 5 we discuss some related works and in Section 6 we present future works and conclude the paper.

2 The Shopping Behavior in a Shopping Mall

The shopping behavior that we consider in this research consists of all the activities that shoppers can carry out in a mall. These activities depend on the mall's characteristics. Hence, it is relevant to first study these characteristics.

(Roberts et al., 2000) presented five stages of the evolution of shopping malls. Furthermore, they identified the four critical components of malls: retail, services, entertainment and social components. These elements are discussed in the following points.

- *Retail*: The retail component is the most dominant element of a shopping mall. It includes the stores and kiosks, the department stores, etc. Some shopping malls focus on a specific retail format to suit their particular area of specialisation. In this component we find the spatial configuration (layout and positions of the stores, kiosks, doors, etc.) and the atmosphere (music, lighting, odor, temperature, etc.) of the shopping mall.

- *Services*: The provision of services is a key element in shopping malls. Some malls provide outlets for the post office, laundromat, pharmacy, medical services, bank, wheelchair access, free child care services, community halls, a courtesy bus and parcel pickup, etc.

- *Entertainment*: Entertainment is a critical element in a mall and is perhaps the aspect that is remembered most by shoppers. Here are some examples of this component: fashion parades, live television broadcasts, cinema. Entertainment sometimes depends on the particular focus of the mall.

- *Social*: The social aspect of the mall relates to the idea of being a community meeting space. It allows people to participate in a community recreational activity, to seeing and being seen, to meet and passively enjoy the atmosphere. Ambient music, landscape and a wide variety of places to eat support this component. The social element is very important and a large number of people come to the mall not to purchase but to socialize.

Our literature review confirmed that the shopping behavior in a mall is affected by the four shopping mall's components discussed by (Roberts et al., 2000). These components represent, what is called in the literature, the *"external factors"* which come from the environment and affect the shopping behavior. Other studies present factors called *"internal factors"* (they are related to the shopper). Here are some examples of these *internal factors* : demographic variables (gender, age group, marital status, occupation, etc.), personality, values, attitudes, culture, social class, goals, preferences, habits, temporal factors, the financial status, the knowledge of the shopper, etc. (Duhaime et al., 1996).

After identifying these factors, we grouped them into several categories. Some of these categories belong to the shopper and the others belong to the environment (shopping mall). In this paper we refer to these categories by the term "*dimensions*".

- *The Shopper dimensions:* These dimensions represent a large number of factors belonging to the shopper. In our work we consider the following dimensions: The shopper Characteristics, the Shopper Knowledge, the Shopper Behavior dimensions.
- *The Shopping mall dimensions:* These dimensions represent several factors belonging to the environment (shopping mall) and affect the shopping behavior. In our work we consider the following dimensions: Space, Ambiance and the Information dimensions of the shopping mall.

All these dimensions are used to design the knowledge-based agents' models that are used to develop the geosimulation prototype of the shopping behavior in a mall. The detail of these dimensions and the agents' models are discussed in Section 3.

3   The Characteristics of the MultiAgent Geosimulation Models of the Shopping Behavior in a Shopping Mall

In our work we use knowledge-based agents to simulate the shopping behavior. This section aims to present the agent-based simulation models that we used to develop the geosimulation prototype. The first agent model simulates the shopper. This model contains the shopper's dimensions which affect the shopping behavior in a mall. Furthermore, since we use this model to simulate human behavior, we integrate spatial and non-spatial capabilities that simulate those possessed by people such as perception, memorization, decision making, navigation, etc. In the second agent model we integrate the dimensions that belong to the mall and which affect the shopping behavior.

3.1   The Characteristics of the Shopper Agent's Model

In the Shopper agent model the dimensions take into account several factors that can affect the shopping behavior in a mall and different processes that compose this behavior.

Characteristics Dimension: In this dimension we distinguish two kinds of characteristics: Non-spatial and spatial characteristics. The *non-spatial characteristics* represent the Shopper agent's attributes. They can be either static or dynamic. The static characteristics represent the attributes that can affect the shopping behavior and do not change during the simulation. Here are some examples: *gender, age group, marital status, sector of employment* of the Shopper, etc. The dynamic characteristics represent the attributes that affect the shopping behavior and can change during the simulation. Here are some examples: *level of hunger, level of thirst, level of fear or stress, need to go to the restroom, emotional states*, etc. The *spatial characteristics* represent the physical attributes of the Shopper agent. We have both 2D and 3D representations of the agents. In a 2D mode the Shopper agent can be represented by a point, a triangle, a square (Fig. 1.a presents these representations). The 2D representations (point, triangle or square) can be chosen by the user of the simulation in order to distinguish between the simulation agents. In 3D it is represented by slightly animated figures of a man or a woman that can be young or old (Fig. 1.b).

Fig. 1a. The 2D spatial characteristics of the Shopper agent

Fig. 1b. The 3D spatial characteristics of the Shopper agent

Knowledge Dimension: The Shopper agent's knowledge obviously influence its shopping behavior and decision making such as choosing the type of stores or kiosks to visit, the path or corridors to follow during the shopping trip, etc. In our model we distinguish non-spatial and spatial knowledge. *Non-spatial knowledge* represents what the Shopper agent knows about the non-spatial characteristics of the environment (the mall). Here are some examples: The name and speciality of stores or kiosks, the products found in stores or kiosks. *Spatial knowledge* represents what the Shopper agent knows about the spatial characteristics of the environment. Let us mention for example the locations of stores or kiosks, of doors, of restrooms, of corridors.

Behavior Dimension: Shopper agents are autonomous and can perform different spatial aor non-spatial behaviors in the mall.

Non-Spatial Behavior Contains the Following Processes:

-The Perception process: The Shopper agent can perceive the non-spatial information of the environment such as the types of stores or other messages that can be broadcasted in the mall (ads messages or others).

-The Memorization process: The Shopper agent can memorize the non-spatial information that comes from the environment such the type of stores or the kiosks.

-The Decision making process: The Shopper agent makes decision about its movements inside the mall. On the bases of its own characteristics, of what it perceives in the environment, of what it memorizes (its knowledge). For example, the Shopper agent can decide to go to a store, to go to a place where to eat, to go to the restroom, to play, to socialise with other people, to leave the mall.

Spatial Behavior Contains the Following Processes:

-The Perception process: The Shopper agent can perceive the spatial characteristics of the environment. The perceived elements (stores, obstacles, other agents, etc.) are used in the decision making process.

-The Memorization process: The Shopper agent uses this process in order to memorize the spatial elements perceived in the environment. In our simulation the Shopper agent can memorize the locations of stores, doors, corridors, and use them in the decision making process.

-The navigation process: The Shopper agent can move from one point to another in 2D and 3D modes (Fig. 2.a and 2.b).

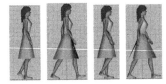

Fig. 2a. The 2D spatial behavior of the Shopper agent (navigation)

Fig. 2b. The 3D spatial behavior of the Shopper agent (navigation)

3.2   The Characteristics of the Shopping Mall Agents' Model

The model of the shopping mall contains several dimensions. Unlike the Shopper's model which is represented by an agent, the shopping mall model is represented by different software agents. Each important element of the shopping mall such as stores, kiosks, restrooms, doors, is represented by a software agent. These agents cannot move, but they have all the other characteristics of agents. These agents represent the limits of the simulation, in the sense that we do not simulate in details what happens inside the stores. The behaviors associated with these agents enable us to model, using probabilistic models, the outcomes of the shopper agent's visits in the stores. For example, a behavior of a store computes the duration and the results of the visit of each agent entering in the store based on the agent's charactristics and some probabilities. Another store behavior broadcasts to agents located in it information about the items and ads displayed in the store window. In the shopping mall's model we consider *Space*, *Ambiance* and *Information* dimensions.

Space Dimension: It represents the geographic and spatial characteristics of the mall, and is taken into account by the spatial processes of the Shoppers agents (perception , memorization, navigation and movement decisions, etc.). Fig. 3.a and Fig. 3.b

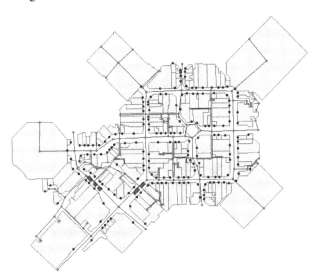

Fig. 3a. The 2D spatial structure of Square One shopping mall

Fig. 3b. The 3D spatial structure of Square One shopping mall

represent respectively 2D and 3D spatial representations of *Square One shopping mall* in which we carry out our simulation. To feed this dimension we use data manipulated with Intergraph's Geomedia Geographic Information System (http://www. intergraph.com).

***Ambiance* Dimension:** It contains the elements of the ambiance or the atmosphere of the shopping mall which can affect the shopping behavior. As example of elements we can cite the *temperature, colors, music, lighting,* etc.

***Information* Dimension:** It represents the Non-Spatial Information of the Mall.

4 The Geosimulation Implementation: The Case of the Square One Shopping Mall (Toronto)

4.1 The MAGS: The MultiAgent GeoSimulation Platform

The simulation models presented in the previous section are used to develop a multi-agent geosimulation prototype using a geosimulation platform called MAGS (Multi-Agent Geo-Simulation) (Moulin et al., 2003). It is a generic platform that can be used to simulate, in real-time, thousands of knowledge-based agents navigating in a 2D or 3D virtual environment. MAGS agents have several knowledge-based capabilities such as perception, navigation, memorization, communication and objective-based behavior which allow them to display an autonomous behavior within a 2D-3D geographic virtual environment. The agents in MAGS are able to perceive the elements contained in the environment, to navigate autonomously inside it and react to changes occurring in the environment. These agents have several knowledge-based capabilities.

- *The agent perception process*: In MAGS agents can perceive (1) terrain characteristics such as elevation and slopes; (2) the elements contained in the landscape surrounding the agent including buildings and static objects; (3) other mobile agents navigating in the agent's range of perception; (4) dynamic areas or volumes whose shape changes during the simulation (ex.: smoky areas or zones having pleasant odors); (5) spatial events such as explosions, etc. occurring in the agent's vicinity; (6) messages communicated by other agents (Moulin and al., 2003).

- *The agent navigation process*: In MAGS agents can have two navigation modes: *Following-a-path-mode* in which agents follow specific paths which are stored in a bitmap called ARIANE_MAP or *Obstacle-avoidance-mode* in which the agents move through open spaces avoiding obstacles. In MAGS the obstacles to be avoided are recoded in specific bitmap called OBSTACLE_MAP.

- *The memorization process*: In MAGS the agents have three kinds of memory: *Perception memory* in which the agents store what they perceive during the last few simulation steps; *Working memory* in which the agents memorize what they perceive in one simulation and *Long-term memory* in which the agents store what they perceived in several simulations (Perron et al., 2004).

- *The agent's characteristics*: In MAGS an agent is characterized by a number of variables whose values describe the agent's state at any given time. We distinguish *static states* and *dynamic states*. A static state does not change during the simulation and is represented by a variable and its current value (ex.: gender, age group, occupation, marital status). A dynamic state is a state which can possibly change during the simulation (ex.: hunger, tiredness, stress). A dynamic state is represented by a variable associated with a function which computes how this variable changes values during the simulation. The variable is characterized by an initial value, a maximum value, an increase rate, a decrease rate, an upper threshold and a lower threshold which are used by the function. Using these parameters, the system can simulate the evolution of the gents' dynamic states and trigger the relevant behaviors (Moulin et al., 2003).

- *The objective-based behavior*: In MAGS an agent is associated with a set of objectives that it tries to reach. The objectives are organized in hierarchies which are is composed of nodes that represent composite objectives and leaves that represent elementary objectives which are associated with actions that the agent can perform. Each agent owns a set of objectives corresponding to its needs. An objective is associated with rules containing constraints on the activation and the completion of the objective. Constraints are dependent on time, on the agent's states, and the environment's state. The selection of the current agent's behavior relies on the priority of its objectives. Each need is associated with a priority which varies according to the agent's profile. An objective's priority is primarily a function of the corresponding need's priority. It is also subject to modifications brought about by the opportunities that the agent perceives or by temporal constraints (Moulin and al., 2003).

- *The agent communication process*: In MAGS agents can communicate with other agents by exchanging messages using mailbox-based communication.

The spatial characteritics of the environment and static objects are generated from data stored in Geographic Information System and in related databases. The spatial characteristics of the environment are recorded in raster mode which enables agents to access the information contained in various bitmaps that encode different kinds of information about the virtual environment and the objects contained in it. The *AgentsMap* contains the information about the locations of agents and the static objects contained in the environment. The *ObstaclesMap* contains the locations of obstacles, the *ArianeMap* contains the paths that can be followed by mobile agents, the *HeightMap* represents the elevations of the environment, etc. The information contained in the different bitmaps influences the agent's perception and navigation. In MAGS the simulation environment is not static and can change during the simulation. For example, we can add new obstacles, or gaseous phenomena such as smoke, dense gases and odors which are represented using particle systems, etc. (Moulin and al., 2003).

4.2 The Mall-MAGS Prototype

Using the MAGS platform we developed a multiagent geosimulation prototype that simulates customers' shopping behavior in a mall. As a case of study we use the *Square One shopping mall* in Toronto (Canada). To feed the simulation models with data we carried out a survey in October 2003 and collected 390 questionnaires filled by real shoppers in the *Square One* shopping mall. This data belongs to two categories: Non-spatial data such as demographic information (gender, age group, marital status, occupation, preferences, habits, etc.) and spatial data such as preferred entrance and exit doors, habitual itineraries, well-known areas in the mall, etc.

In Fig. 4.a and Fig. 4.b we display 2D and 3D screenshots of a simulation that involved 390 software Shoppers agents navigating in the virtual shopping mall.

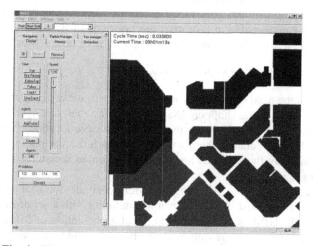

Fig. 4a. The 2D simulation in MAGS platform (Square One mall

In the simulation prototype the Shopper agent comes to the mall to visit a list of specific stores or kiosks that are chosen before the simulation on the basis of the agent's characteristics. It enters by a particular door and starts the shopping trip. Based on its position in the mall, its knowledge (memorization process), what it perceived in the mall (perception process), it makes decision about the next store or kiosk to visit (decision making process). When it chooses a store or kiosk, it moves in its direction (navigation process). Sometimes, when it is moving to the chosen store or kiosk, it perceives another store or kiosk (perception process) that is in its shopping list and that it did not know it before. In this case, the Shopper agent moves to this store or kiosk and memorizes it (memorization process) for its next shopping trips. The shopper agent accomplishes this behavior continually until it visits all the stores or kiosks or until it has not time left for the shopping trip. If the shopper agent has still time for shopping and some stores or kiosks of its list are in locations unknown by the agent, it starts to explore the shopping mall to search for stores or kiosks. When the shopper agent reaches the maximum time allowed to the shopping trip, it leaves the mall.

The Shopper agent can also come to the mall without a specific list of stores or kiosks to visit: It comes to the mall to explore it, to see people, or to make exercice, etc. In the exploration mode the Shopper agent takes its preferred paths in the shopping mall. In this mode the moving action of the Shopper agent to the stores, kiosks, music zones, odor zones, lighting zones, is directed by its habits and preferences. For example, if the Shopper agent likes *cars* and it passes in front of a car exhibition, it can move to this exhibition. To extend our simulation prototype we can simulate the shopper reactions to the mall's atmosphere. We can insert special agents that broadcast music, lighting or odor. If the shopper agent is in the exploration mode and likes the music or the lighting or the odor broadcasted by these special agents, the shopper agent can move toward them and possibly enter the store.

During its shopping trip the Shopper agent can feel the need to eat or to go to the restroom (simulated by a dynamic variable reaching a given threshold). Since these needs have a bigger priority than the need to shop or to play, the agent suspends temporarily its shopping trip and goes to the locations where it can eat something or to restrooms. In our geosimulation prototype the priorities of the activities of the shopping behavior are defined based on Maslow's hierarchy of needs (Maslow, 1970).

4.3   The Use of the Mall_MAGS Prototype

Mall_MAGS can be used by shopping mall managers to make decisions related to the spatial configuration of the shopping mall. A shopping mall manager can change the spatial configuration of the shopping mall (change a store location, close a door or a corridor, etc.). For each change the manager can launch the simulation and collect the results. By comparing these results he can make informed decisions about the impact of spatial changes in the mall.

To illustrate the use of the Shopping behavior geosimulation tool we used 2 simulation scenarios. In the first one we launch a simulation with a set of input data about the shopping mall (GIS) (see Fig. 5.a) and about a population of 390 shoppers. This first scenario generates for us output data about the itineraries that the Shoppers agents take in the shopping mall. In scenario 2 we exchange the location of a two department stores: *Wal-Mart* and *Zellers* (Fig. 5.b), we launch the simulation again and we generate the output data about the itineraries of the same population of Shoppers agents. By comparing the output data of the two scenarios we notice the difference of the paths that the Shopper agents followed to attend the department stores *Wal-Mart* and *Zellers* stores. The simulation output analysis shows us that corridor X is less frequented in scenario 2 than in scenario 1 (Fig. 6.a). However, corridor Y is more frequented in scenario 2 than in scenario 1 (Fig. 6.b). In these figures the flow of the agents Shoppers which pass through a corridor is represented by a line which is attached to this corridor. The width and the color o f this line are proportional to the flow of Shoppers agents that pass through the corridor. If this flow grows, the width of the line grows and its color becomes darker. By a data analysis on the characteristics' dimension of the Shopper agent we can see that in scenario 2, most of the Shoppers agents that go through corridor Y are female and they come to the mall to visit female cloth stores. If the mall manager chooses the mall configuration of scenario 2, he may think of renting the spaces along corridor Y to female cloth stores. It is important to note that:

- The simulation output data are generated using software agents called *Observers*. The mission of these agents is to gather data about the Shoppers agents which enter their perception area. This data is recorded in files and analysed after the simulation.

- The data analysis of the geosimulation output (non-spatial and spatial data) is implemented in an analysis tool that we developped using *Microsoft Visual basic 6.0.* This user-friendly tool uses the data generated by the *Observers* agents in order to make multidimensional non-spatial and spatial analysis using an OLAP (OnLine analytical Processing) approach (Bédard et al., 2001).

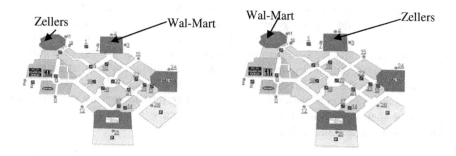

Fig. 5a. The simulation environment in Scenario 1   **Fig. 5b.** The simulation environment Scenario 2

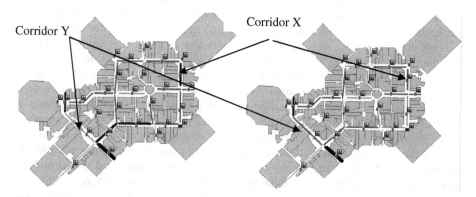

Fig. 6a. The spatial data analysis in Scenario 1   **Fig. 6b.** The spatial data analysis in Scenario 2

5 Related Works

(Bandini and al., 2002), (Ulicny and Thalmann, 2001), (Sung and al., 2004) and (Batty, 2003) developed some applications that simulate pedestrian behavior in geographic environments using the agent technology. These works do not focus on the individual features and behaviors of each agent but aim to study the emergent behavior of crowds in the simulation. They also focus on the animation and gestural movements of the agents rather than their internal behavior. The shopping behaviors are very much influenced by the agents' internal factors (Duhaime et al., 1996). In order to simulate such behaviors, we need to deal with the individual structure and behavior of the agents. So the structure of our agent must contain the majority of the factors that influence the shopping behavior. The agent's behavior must be enough developed to contain the processes that compose the shopping behavior. We also need to take into account the collective level of the simulation (the crowd). Furthermore, the applications of (Ulicny and Thalmann, 2001) and (Bandini and al., 2002) are used to merely visualize the simulation in 2D and 3D. A simulation application is often used to make decision about the system to be simulated. Hence, it must generate output data that can be manipulated for different purposes. These outputs should be easily used by users to make decisions. In our simulation we take into account this important aspect of a simulation and we generate, using what we call observer agents, output data about the simulation which is used thanks to a user friendly interface, by the user to make decision about the system to be simulated (the shopper) or about the configuration of the simulation environment (the mall). In addition, our simulation is based on real data about the shopper and the simulation environment (the mall), which improve the realism of the simulation. This is not taken into account by the simulation applications mentionned above because they use data generated randomly or automatically using algorithms such as genetic algorithms.

6 Conclusion and Future Works

In this paper we presented how we can simulate the spatial shopping behavior of customers in a mall using knowledge-based multiagent geosimulation. We presented the factors that influence this behavior. Based on these factors we proposed different relevant dimensions that belong to the shopper and to the spatial environment (the shopping mall). Based on these dimensions, we discussed the agents' models that we designed in order to simulate the shopping behavior in a shopping mall. Third, we presented the multiagent geosimulation prototype that we developped based on these agents' models. This prototype is called Mall_MAGS whitch refers to Mall (shopping mall) using the MAGS simulation platform. Finally, we presented how the shopping mall's managers can use our prototype in order to make decisions about the assessment of the shopping mall configuration in order to make it more confortable for the shoppers. This decision making process is based on a non-spatial and spatial multidimensional analysis of the geosimulation output data. It is important to note that the simulation output data is generated using agents called *Observers* which use their perception capabilities to observe the simulation and store data about it in file structures.

In few months we plan:

- To enhance our prototype to simulate customers' entertainment activities if a Shopper agent feels the need to play or to entertain, it can move toward an entertainment zone such as merry-go-round in order to satisfy this need.

- To simulate the social aspect of the shopping behavior. So, if the Shopper agent does not like crowdy environments it can avoid them. If it perceives a store to be visited and it sees a large number of shoppers agents in front of this store or kiosk, it can decide to abandon its goal to visit to this store. However, if the Shopper agent likes to be in a crowd it can look for crowded areas around it and move toward them.

- To extend the usage of the simulator in order to help mall managers to make decisions about marketing strategies related to the changes of music or odor in a corridor, change of temperature, or wall colours in certain areas, etc. For each change they would execute the simulation and collect results. By comparing these results they can make decisions about the optimal marketing strategy to adopt. How to propose a systematic way to carry out these comparisons is still an open research area.

- To validate our geosimulation models, document our prototype and deliver a final version of the Mall_MAGS prototype to the managers of the *Square One* shopping mall in Toronto.

References

(Anu, 1997) Anu Maria. Introduction to modeling and simulation. Proceedings of the 29th Conference on Winter Simulation, p.7-13. December 07-10, Atlanta, Georgia, 1997.

(Bédard and al., 2001) Bédard Y., Rivest S., Marchand P. Toward better support for spatial decision making: Defining the characteristics of Spatial On-Line Processing (SOLAP). Geomatica, the Journal of the Canadian Institute of Geomatics, Vol. 55, No. 4,2001, p. 539,555. 2001.

(Benenson and al., 2004) Benenson, I. & Torrens, P.M. (2004). Geosimulation: Automata-Based Modeling of Urban Phenomena. London: John Wiley & Sons. On sale in April 2004.

(Dijsktra ad al., 2001) Dijsktra J., Harry JP., Bauke U. Virtual reality-based simulation of user behavior within the build environment to support the early stages of building design. In M. Schreckenberg and S.D. Sharma (ed.): *Pedestrian and Evacuation Dynamics*. Springer-Verlag, Berlin. pp. 173-181. 2001.

(Duhaime et al., 1996) Duhaime, Kindra, Laroche, Muller. Le comportement du consommateur. 2ème édition. Gaétan Morin, éditeur. 1996.

(Frank and al., 2001) Frank A.U., Bittner S., Raubal M. (2001), Spatial and cognitive simulation with multi-agent systems, in D. Montello (edt.), *Spatial information Theory: Foundations for Geographic Information Science*, Springer Verlag, LNCS 2205,124-139.

(Koch, 2001) Koch A. Linking Multi-Agent Systems and GIS- Modeling and simulating spatial interactions-. Department of Geography RWTH Aachen. Angewandte Geographische Informationsverarbeitung XII, Beiträge zum AGIT-Symposium Salzburg 2000, Hrsg.: Strobl/Blaschke/Griesebner, Heidelberg, p. 252-262. 2001.

(Maslow, 1970) Abraham Maslow, *Motivation and Personality*, 2nd ed., Harper & Row, 1970.

(Moulin and al., 2003) Moulin B., Chaker W., Perron J., Pelletier P., Hogan J. MAGS Project: Multi-agent geosimulation and crowd simulation. In the proceedings of the COSIT'03 Conference, Ittingen (Switzerland), Kuhn, Worboys and Timpf (edts.), *Spatial Information Theory*, Springer Verlag LNCS 2825, p. 151-168. 2003.

(Perron and al., 2004) Perron J., Moulin B. Un modèle de mémoire dans un système multi-agent de géo-simulation. *Revue d'Intelligence Artificielle*, Hermes. 2004.

(Bandini and al., 2002) Bandinin S., Manzoni S. Simone C. Heterogeneous agents situated in heterogeneous spaces. Applied Artificial Intelligence. 16(9-10): 831-852. 2002.

(Ulicny and Thalmann, 2001) Ulicny B. & Thalmann D. Crowd Simulation for interactive virtual environments and VRtraining systems. Proceedings of Eurographics workshop on Animation and Simulation. p. 163-170, Springer-Verlag. 2001.

(Sung and al., 2004) Sung M., Gleicher M., Chenny S. Scalable behaviours for crowd simulation. Computer graphics Forum 23, 3. 2004.

(Batty, 2003) Batty M. Agent-Based Pedestrian Modeling. CASA UCL Working papers series 61. 2003.

(Raubal, 2001) Raubal M. Agent-Based Simulation of human wayfinding. A perceptual model for unfamiliar building. PhD thesis. Vienna University of Technology. Faculty of Sciences and Informatics. 2001.

(Roberts and al., 2000) Roberts J., Merrilees B. Shopping centre concept evolution: An historical analysis of the Australian experience. ANZMAC 2000 doctoral colloquium proceedings. November 2000.

(Ruiz and al., 2004). Ruiz J.P., Chebat J.C., Hansen P. Another trip to the mall: A segmentation study of customers based on their activities. Les cahiers du GERAD. January 2004.

(Wakelfield and al., 1998) Wakelfield Kirk L., Baker J. Excitement at the mall: Determinants and effects on shopping response. Journal of retailing, 74(4): 515-539. 1998.

Biography

Walid Ali obtained a Master degree in Computer Science from Laval University (Quebec). He is now a PhD Student (supervised by Bernard Moulin since 2001). He is mainly interested in using multi-agent systems to simulate behaviors and phenomena in virtual spatial environments. He is working on a project whose goal is to simulate shopping behavior in shopping malls using geosimulation and multi-agent systems.

Bernard Moulin obtained an Engineering degree from Ecole Centrale de Lyon (France), a degree in Economics (University Lyon2) and a PhD (applied mathematics-computer science, University Lyon1). He is now a full professor at Laval University in the Computer Science Department. He is also one of the main researchers of the Research Center on Geomatics at Laval University. He has been working on software engineering and artificial intelligence for more than 20 years. He is supervising research projects mainly subsidized by the Natural Sciences and Engineering Research Council of Canada, the FQRNT fund of Quebec, the Network of Centers of Excellence in Geomatics GEOIDE and the Defence Research and Development Center at Valcarier. His main interests are: multi-agent systems, agent-based geosimulation, applications of AI techniques to geomatics, modeling and simulation of software agents' conversations.

Memory for Spatial Location: Influences of Environmental Cues and Task Field Rotation

Sylvia Fitting, Douglas H. Wedell, and Gary L. Allen

Department of Psychology, University of South Carolina,
Columbia, SC 29208 USA
fitting@sc.edu

Abstract. We developed a formal theoretical extension of the category-adjustment model (Huttenlocher, Hedges, & Duncan, 1991) to incorporate the potential impact of external reference cues in spatial memory. This extension was tested in an experiment in which individuals remembered locations within a circular task field, with different numbers of peripheral cues available. Orientation of the task field was dynamic in the sense that it was rotated on the majority of trials. By modeling the angular bias shown in observers' estimates on unrotated trials, we sought to distinguish between cue-based categories and viewer-based geometric categories as the source of such bias. Results were consistent with our fuzzy boundary extension of the category-adjustment model in which observers generate prototypes based on available reference cues. Both memory accuracy and bias were affected by the number of cues in the task field as predicted.

Keywords: spatial memory; bias; category effect; spatial cues; mental rotation.

1 Memory for Spatial Location

Upon first consideration, memory for spatial location seems a simple matter. The ease with which a stationary observer can recall where a flash appeared in the night sky, where a "blip" disappeared from a radar screen, or where a familiar person stood in a crowd might suggest a straightforward, uncomplicated process. However, the literature on spatial cognition indicates that, theoretically speaking, spatial memory is a rather complicated phenomenon involving different types of memory, spatial frames of reference, and coding processes.

The purpose of this investigation was to examine how characteristics of a spatial memory task, specifically the availability of environmental cues surrounding the task field and potential changes in task field orientation, influence performance and thus shape our inferences about spatial memory processes. Central to our approach were three considerations. First, we selected a memory task that included a stationary observer remembering a location in a small circular task space. This task provided a straightforward set of requirements, thus providing a clear picture of the influences of environmental cues and task field rotation. Second, we developed a specific quantitative model of the memory coding processes

A.G. Cohn and D.M. Mark (Eds.): COSIT 2005, LNCS 3693, pp. 459–474, 2005.
© Springer-Verlag Berlin Heidelberg 2005

that would illustrate clearly the presence and magnitude of the effects of interest. Third, we focused on the memory-based performance of individuals rather
than of groups. This case study approach provided a rigorous test of the application of the model while avoiding potential problems arising from averaging
across subjects who may employ different strategies. Our aim then was to develop a generally applicable theory-grounded framework for examining spatial
processing of environmental cues within a dynamic task field.

1.1   Types of Spatial Memory and Frame of Reference

Distinctions between different kinds of spatial memory, frames of reference, and
types of memory coding central to each are ubiquitous in the literature (Morris &
Parslow, 2004; Nadel, 1990; Newcombe & Huttenlocher, 2000). How might these
common theoretical distinctions apply to remembering the location of a point
in a circular field? One possibility is that an observer could rely on a response-
learning mechanism (Nadel, 1990) grounded in an egocentric frame of reference
(Klatzky, 1998). Such a mechanism would be consistent with the workings of
a perception-action system as posited by Proffitt and his colleagues (Creem &
Proffitt, 1998). Spatial behavior based on this system would be consistent and
unbiased by contextual information, although it would be severely disrupted by
changes in task demands brought about by field rotation.

A second possibility is that an observer could respond on the basis of a cue-
learning mechanism (Nadel, 1990) involving an allocentric object-based frame of
reference (Newcombe & Huttenlocher, 2000). An object-based reference system
in this case brings with it some ambiguity. Certainly, a discrete landmark within
the circular field would provide the basis for cue-guided responding (Nadel,
1990). However, the circular field itself could possibly be interpreted as an object
with intrinsic axes. Past research has demonstrated that observers tend to impose
horizontal and vertical axes on a two-dimensional circular field in remembering
locations (Huttenlocher, Hedges, & Duncan, 1991). Memory would be expected
to show bias reflecting the implicit or explicit structure of task objects or the
task field (Newcombe & Huttenlocher, 2000). Furthermore, contextual information and task demands would significantly impact performance.

A third possibility is that an observer's memory for location in this task
could be the product of a place learning mechanism (O'Keefe & Nadel, 1978;
Tolman, 1948), incorporating an allocentric coordinate frame of reference (Morris & Parslow, 2004; Nadel, 1990). However, there are different ways in which
behavior could be based on a coordinate system. Huttenlocher et al. (1991)
suggested that stationary observers tend to impose an implicit polar coordinate system on a circular task field in the absence of visual cues. However, it
is also possible to use cue-based axes extending from visible peripheral cues to
organize the field, as rats evidently do in the Morris Water Maze task (Morris,
Garrud, Rawlins, & O'Keefe, 1982). In theory, place memory is consistent and
unbiased (O'Keefe & Nadel, 1978), although its reliability is strictly dependent
upon stable spatial relations among environmental objects. In fact, there has

been relatively little empirical work examining the effects of task demands and contextual information on place memory per se.

Closely related to the concept of different types of spatial memory is the theoretical distinction between different spatial coding processes (Allen & Haun, 2004; Kosslyn et al., 1989). A well specified and empirically verified distinction of this type involves categorical and fine-grain coding (Huttenlocher et al., 1991). Categorical coding is posited as the robust product of a relatively rapid process in which location is remembered in terms of its being within a particular categorical region of the response space. A central or salient location within this category acts as a prototype, biasing remembered locations toward it. In the specific task of a stationary observer remembering a location with a circular field, observers appear to impose implicit horizontal and vertical axes on the circle, thus creating quadrants that function as categories (Huttenlocher et al., 1991), with centrally located points within these quadrants serving as prototypes. Remembered locations are "pulled" toward these prototypes.

Fine-grain coding is posited as the product of the somewhat more time-consuming process of remembering a location in terms of a geometric coordinate system imposed upon the response field. Fine-grain coding yields metric accuracy and less susceptibility to bias. Previous research has shown that stationary observers remembering a location within a circular field encode fine-grain information with respect to a polar coordinate system imposed on the response field (Huttenlocher et al., 1991). Fine-grain memory may also be more fragile and hence decay more rapidly than categorical information, resulting in a greater reliance on categorical coding with increased memory demands (Haun, Allen, & Wedell, 2005).

1.2 Spatial Memory as a Multifaceted Phenomenon

The preceding considerations suggest that a stationary observer remembering a location in a circular field could rely on an egocentric, an allocentric-intrinsic, or an allocentric-coordinate frame of reference and presumably could switch from one to the other in the face of task demands. Memory in this task could involve categorical coding, fine-grain coding, or both. Accordingly, it seems advisable to approach spatial memory as a multifaceted phenomenon involving an array of cognitive tools that can be flexibly enacted in response to task situations.

Based on this view, we sought to establish how well-documented findings across different tasks could be integrated into the current task. With respect to cue availability, it is well established that observers tend to impose implicit quadrants on a circular task space to code spatial location categorically (Huttenlocher et al., 1991), and it is well established that animals use cues peripheral to a circular response area to organize their memory of that space (Morris et al., 1982). However, it is not clear how observers in our memory task would use peripheral cues outside the circular response field. They could ignore them, relying instead on the implicit category structure imposed on the circular field, or they could use them as spatial prototypes for categorical coding.

With respect to field rotation, it is well established that observers can track spatial location when response fields are rotated relative to the observers' initial view. Thus, it was assumed that they could accommodate the rotation of the circular field used in this work, given an adequate cue for orientation. However, it is not known how the possibility of task field rotation per se affects such coding (as when memory trials requiring rotation are intermixed with trials requiring no rotation).

We addressed these issues by developing an extension of Huttenlocher et al.'s (1991) category-adjustment model, which features a combination of fine-grain and categorical coding. First, we formally examined how categories reflecting viewer-based horizontal and vertical axes would bias memory and then contrasted that prospective outcome with how categories reflecting peripheral cues in the task field would bias memory. Subsequently, we conducted an experiment designed to show which of the two versions of categorical organization was a better predictor of actual memory bias. In the next section, we present our modeling framework in some detail.

2 Fine-Grain and Categorical Coding in Place Memory

2.1 The Category-Adjustment Model

Our model builds on the category-adjustment model of Huttenlocher et al. (1991), according to which spatial location is coded at two levels, fine-grain and categorical. Each of these sources involves unbiased error. Accordingly, fine-grain information is centered on the actual location, whereas categorical information is centered on the category prototype. Bias in memory arises from combining information from these two sources to produce an estimate. Although this combination produces a predictable pattern of bias, it also results in an overall reduction in error (Huttenlocher et al., 1991).

Because bias in estimation is central to our investigation, we present a series of formal models that will be used to generate testable predictions. According to the category-adjustment model, the expected value of the response in an estimation task, $E[R]$, can be characterized as a weighted average of fine-grain and categorical information described by the following equation:

$$E[R] = \lambda\mu + (1 - \lambda)p, \tag{1}$$

where μ is the mean of the distribution of fine-grain memory values for the object, assumed to be unbiased and hence equated with the true location of the object. Similarly, p is the mean of the distribution of prototype locations for the relevant category. The parameter λ, which varies from 0 to 1, represents the relative weight of the fine-grain information. Bias is determined by subtracting the actual value from the response, and thus the expected bias is characterized by the following equation:

$$E[Bias] = E[R] - \mu = \lambda\mu + (1 - \lambda)p - \mu. \tag{2}$$

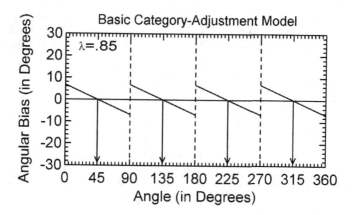

Fig. 1. Category-adjustment model of Equation (2) applied with boundaries fixed along the axes (i.e., at 0°, 90°, 180°, and 270°) and prototypes at the midpoint of each quadrant (i.e., at 45°, 135°, 225°, and 315°).

Within Huttenlocher's framework the weighting of fine-grain memory depends on the relative uncertainty of the fine-grain memory value and the prototype memory value. In general, the greater the uncertainty concerning fine-grain information, the less it is weighed (i.e., λ decreases). In Equations (1) and (2), λ is a constant, reflecting an assumption that uncertainty does not vary across locations within a category.

Figure 1 illustrates the model's prediction for remembered locations within a circular field. Following Huttenlocher et al. (1991), we impose a polar coordinate system that locates the dot in terms of its angle and radial distance from the center. In our examples, we focus on the angular component of the prediction, although similar predictions can be made concerning radial distance. As with Huttenlocher et al. (1991), we assume an implicit categorical structure that divides the circle into four quadrants along vertical and horizontal axes. As shown in Figure 1, estimates are unbiased when stimuli are located at the category protypes. Conversely, bias is maximized near the boundaries of each quadrant, where the deviation from the prototype is maximal.

One shortcoming with the model of Equation (2) is that it does not account for nonlinear bias that can appear in the data. For example, Haun et al. (2005) found that bias decreased near extreme angles of 0° and 90° that were the category boundaries in their azimuth and incline estimation tasks. They argued that this debiasing near the borders could be explained in terms of greater fine-grain memory discrimination, consistent with the bowing effect observed in perceptual judgments (Luce, Nosofsky, Green, & Smith, 1982). Increased discrimination for fine-grain information near the border would be reflected in corresponding higher weighting of λ. They used a simple quadratic function to capture this relationship as follows:

$$\lambda = a + b(\mu - p)^2, \tag{3}$$

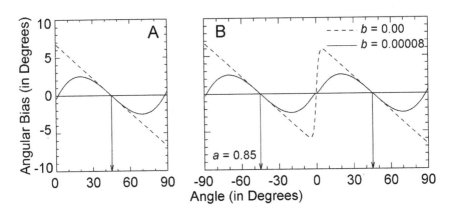

Fig. 2. Model prediction of bias under conditions of constant weighting (dashed line) of fine-grain information ($b = 0$) or a curvilinear relationship (solid line) between weight and angle ($b > 0$), with a representing baseline fine- grain weighting. Panel A: Application of Equations (2) and (3) to estimation within a single quadrant assuming a prototype value of 45°. Panel B: Application of Equations (2) through (4) to estimation across adjacent quadrants assuming prototypes at -45° and 45°.

where a represents the baseline weighting of fine-grain memory and b indexes the increment to this value as the stimulus deviates from the prototype and thus appoaches a boundary or border. In the Haun et al. model (2005), the expression for λ from Equation (3) was substituted into Equation (2) to model the increase in weighting of fine-grain memory values near the border. The pattern of bias produced from this model is presented in Panel A of Figure 2.

In Haun et al.'s (2005) Experiment 2, motor estimates of azimuth across a range of angles that spanned two quadrants were not well fit by the combination of Equations (2) and (3), primarily because estimates near the border appeared to be influenced by both prototypes. To accommodate this possibility, Haun et al. (2005) proposed a fuzzy boundary version of the category-adjustment model in which stimuli near the border beween two categories were sometimes misclassified and hence recruited the wrong category prototype. For the two category case involving a single boundary, the probability of retrieving a given prototype can be described as a logistic function of the difference between the actual angle and the category boundary location as follows:

$$Pr(p_1|\mu) = \frac{1}{1 + exp(-c(\mu - t))}, \tag{4}$$

where p_1 represents one of two prototypes, c is a sensitivity or scaling parameter, and t is the threshold or boundary dividing the two categories. Within this framework, the probability of retrieving the other prototype is simply the complementary probability. The pattern of bias produced from a model that combines Equations (2), (3) and (4) is presented in Panel B of Figure 2.

2.2   Further Developments

Potentially, the category-adjustment model is widely applicable. Thus far, it has been applied in its basic form to memory for locations in two-dimensional fields (Huttenlocher et al., 1991; Newcombe & Huttenlocher, 2000) and in modified form to memory for incline and azimuth (Haun et al., 2005). To apply the fuzzy boundary model of Haun et al. to the location memory task, however, requires some modification. One problem is that the task implies the use of four categories rather than two so that the logistic probability function cannot be used. To meet this problem, we propose a probability recruitment function that is based on the relative similarity of the stimulus angle to the midpoints of all the possible categories. A frequently applied similarity function is the exponential decay function (Nosofsky, 1986; Shepard, 1987), in which similarity falls off very rapidly with increased distance. We first apply this similarity function to the fixed four quadrant case. Accordingly, the probability of prototype recruitment can be described as follows:

$$Pr(p_j|\mu) = \frac{exp(-c|\mu - .5(t_{MIN,j} + t_{MAX,j})|)}{\sum exp(-c|\mu - .5(t_{MIN,k} + t_{MAX,k})|)}, \tag{5}$$

where similarity is calculated relative to the midpoint of the category, using the average of the lower boundary, t_{MIN}, and the upper boundary, t_{MAX}. By using the midpoints of boundaries, Equation (5) represents a fixed threshold (fuzzy boundary) model. Note that because of the polar coordinate system, angles that are in the first and fourth quadrants are incorrectly seen as distant from each other (i.e., 1° and 359° are seen as 358° apart rather than 2° apart). Thus, to properly apply this model we include two additional midpoints corresponding to the lowest valued midpoint plus 360° (i.e. 45° + 360° = 405°) along with the highest valued midpoint minus 360° (i.e. 315° - 360° = -45°). These virtual midpoints are necessary to allow the model to recruit the quadrant 1 prototype for quandrant 4 angles, and vice versa. In short, recruitment of prototypes from either clockwise or counter-clockwise rotation requires the addition of these "virtual" category midpoints.

Figure 3 shows the predicted pattern of bias from a model that combines Equations (2) and (5) for the location memory task. Note that as with the model of Figure 1, fixed boundaries are assumed at vertical and horizontal axes. While bias at first increases as the angle moves away from the prototype location, it later decreases as the boundary is approached. This phenomenon occurs because near the boundary the angle is increasingly likely to recruit the prototype from the adjacent category, and hence bias is added in the opposite direction. Although the prototype locations are shown at the midpoints of the categories, the model combining Equations (2) and (5) requires only that the prototype lie somewhere within the category boundaries.

The focus of our study is on the potential influence of cues peripheral to the task field on estimation bias. The most straight-forward extension of the fuzzy boundary category adjustment model is to assume that the cues serve as

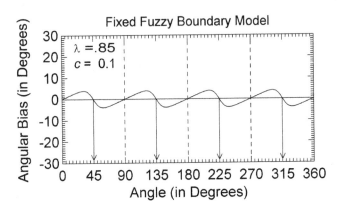

Fig. 3. Fuzzy boundary model of Equations (2) and (5) with fixed boundaries at $0°$, $90°$, $180°$, and $270°$ dividing the circle into four quadrants (prototypes located at $45°$, $135°$, $225°$, and $315°$). Bias is reduced near the boundaries due to recruitment of prototypes from adjacent categories. Note λ = weighting of fine-grain information and c = sensitivity parameter.

prototypes. Note that this assumption means that we can no longer use the fixed boundaries described by viewer-based vertical and horizontal axes. Instead, we infer boundaries by assuming they fall at equal distances from the category prototypes. Accordingly, we alter the prototype recruitment equation to be based on the similarity of the stimulus to the prototype rather than to the midpoint of the category as follows:

$$Pr(p_j|\mu) = \frac{exp(-c|\mu - p_j|)}{\sum exp(-c|\mu - p_k|)}. \tag{6}$$

Once again we include "virtual" prototypes for the lowest and highest categories so that recruitment may be conducted in a clockwise or counter-clockwise fashion. Note that even when there is only one prototype, there will be a "virtual" boundary created by the inclusion of these virtual prototypes. For example, consider the case in which the cue and, accordingly, the category prototype is located at $305°$. Virtual prototypes will be created at $665°$ and $-55°$. Given that similarity falls off exponentially, the virtual prototype at $665°$ will have no impact. However, the prototype at $-55°$ will have an impact. Indeed, angles in quadrant 1 ($0°$ to $90°$) will recruit this virtual prototype and hence exhibit a negative bias. The point halfway between the prototype at $305°$ and at $-55°$ (i.e., at $125°$) will then represent the virtual boundary in this single prototype case.

Panel A of Figure 4 describes the predictions of this cue-based fuzzy boundary version of the category-adjustment model for the one prototype case (with the prototype at $305°$). As shown, the bias function crosses $0°$ at two points, once at the prototype value of $305°$ and once at the virtual boundary value of $125°$. This striking pattern of bias is quite different from the usual pattern shown in Figures 1 and 3 and thus should be easily detected in observers' behavior.

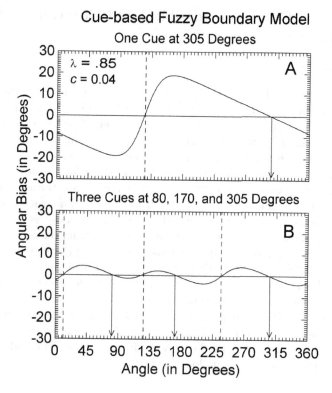

Cue-based Fuzzy Boundary Model

Fig. 4. Predictions of the fuzzy boundary model of Equations (1) and (6), with prototypes equated with cue location, probabilistic recruitment of prototypes across categories, and inferred boundaries located halfway between adjacent prototypes. Panel A: One cue case with the prototype at 305° and the boundary at 125°. Panel B: Three prototypes located at 80°, 170°, and 305°, with inferred boundaries at 237.5°, 12.5° and 125°. Note λ = weighting of fine-grain information and c = sensitivity parameter.

Panel B of Figure 4 describes the predictions of the cue-based fuzzy boundary model for the three-cue case with prototypes located at $p = 80°$, $p = 170°$, and $p = 305°$. Once again two virtual prototypes are added, one at $p = 440°$ and the other at $p = -55°$. These prototypes produce inferred boundaries at 237.5°, 12.5° and 125°. The bias pattern generated by three prototypes (Panel B) is thereby highly distinguishable from the pattern produced by one prototype (Panel A). It is also quite distinct from the quadrant-based four-category case shown in Figures 1 and 3.

In summary, the developments here show how the category-adjustment model can be modified to account for fuzzy boundaries and cue-based prototypes in a task requiring memory for location in a circular field. The patterns of bias have been shown to be quite different, depending on the number and location of cues. We may therefore pose the question of whether actual observers show bias in memory reflecting cue-based categories.

3   Effects of Cues and Possible Field Rotation on Spatial Coding

3.1   Rationale

In a previous study, Fitting (2005) tested different versions of the category-adjustment model in a memory for location task similar to the one conducted by Huttenlocher et al. (1991). In this study, number of peripheral cues was varied, and the task field was in a fixed orientation. Results from all conditions were consistent with the modified category-adjustment model shown in Figure 3. Four prototypes located near the middle in each of four quadrants were implicitly imposed by observers, with quadrant boundaries determined by vertical and horizontal axes. The lack of any significant interactions with cue condition provided evidence for a cue-independent spatial coding process within a static task field. Observers simply ignored the available external reference cues.

The lack of cue effects in this previous experiment prompted us to consider the possible influence of a dynamic context generated by adding a preponderance of trials in which the task field was rotated. Thus, we designed a study in which the number of potential reference cues was varied and there was the possibility of task field rotation. In this paper, we consider only the unrotated trials because they afford the use of either coding strategy, that is, categories based on the vertical and horizontal axes of the viewer (Fitting, 2005; Huttenlocher et al., 1991) or categories based on cue locations.

3.2   Method

In our task, participants attempted to reproduce the location of 32 dots in a circular field presented on a two-dimensional computer display, with 16 dots located at a short radius of 92 pixels and 16 dots located at a long radius of 168 pixels. In each of these two sets, four different angles ($3°$, $25°$, $43°$, and $75°$) were presented in each of the four quadrants (based on cardinal directions). In order to create a dynamic task field setting, four equally occurring rotations were included in this task (rotations of $0°$, $30°$, $90°$, or $160°$). Participants were tested under two conditions of cue availability. In the one-cue condition, an external cue appeared at $305°$ just outside the circular field. In the three-cue condition, unique external cues appeared at $80°$, $170°$, and $305°$ just outside the circular field. Thus, each cue set consisted of 128 trials (4 rotations of 32 dot locations). Note that rotation consisted of moving the locations of the external cues around the field by the prescribed angle.

All materials and instructions were presented on computers with 15-in. (38-cm) monitors at a resolution of 640×480 pixels. The circular region was identical to the white background and separated by a 20 pixel black circle (radius $= 212$ pixels). A red dot, 5 pixel in radius, was the target.

Participants were administered a series of training trials with and without rotation. These example trials were followed by the two actual test sets with 128 different dot locations. Each dot appeared on screen for 1 s, and then was covered by a dynamic checkerboard mask for 1.5 s followed by a blank circle. After the first set, there was a 3-min break. In the second set, the cue condition was changed. Subjects moved a cross-hairs cursor to the remembered location using a mouse and clicked a mouse button to record the response (measured in pixel units on the screen).

The dependent variable we focused on was 'angular bias', calculated by subtracting the angle of the actual location from the angle of the reproduced location. A negative value indicated a clockwise bias, whereas a positive value indicated a counter-clockwise angular bias.

3.3   Results

This paper presents results for angular bias scores of two individual participants whose pattern of responding was representative of the group data. Our case-study approach is powerful because it focuses on a within-subject design and shows how participants' estimates depend on available reference cues. Subject A experienced the one-cue condition first and the three-cue condition second; whereas Subject B experienced these conditions in the reverse order. One outlying data point from Subject A was deleted as it was displaced beyond 40°.

Figure 5 presents the data for Subjects A and B under the one-cue and three-cue conditions. A cursory glance at the pattern of data clearly supports two contentions. First, the pattern of data is inconsistent with the simple category-adjustment model that assumes fixed boundaries along the geometric axes of the circle. Thus, these results differ dramatically from those of Fitting (2005), in which there was no prospect of field rotation. Second, the patterns of bias for one and three cues strongly resemble the corresponding patterns shown in Figure 4, as predicted by the fuzzy boundary model with cue-determined prototypes.

To provide a more formal analysis, we fit different models to the data and used the change in proportion of variance explained (R^2) as a main criterion for assessing the model fit. The angular bias scores for the two subjects were fit separately to the original category-adjustment model by Huttenlocher et al. (1991) and the different versions of the fuzzy boundary model presented here.

The fit of the category-adjustment model of Equation (2) is presented in Table 1 for both subjects in the one-cue and three-cue condition, along with estimated parameter values. One feature of the Huttenlocher et al. model (1991) is the assumption of one prototype in each of the four quadrants. This assumption implies four downward sloping functions that cross the 0° bias line in the graphs at the locations of the prototype in each of the four quadrants. This clearly is not the case for the one-cue condition in which the bias functions appear to cross 0° bias line only twice. It also does not appear to be the case for the three-cue functions. Another feature of the original model is that it assumes a linear bias function without taking into account the possibility of fuzzy boundaries. Our data are not consistent with that outcome. Also, the original model predicts little or no bias, as fine-grain weighting values are close to 1.0 (as shown in Table 1). Inconsistent with this value of λ are the large observed deviations from the 0° bias line, especially for the one-cue condition. Finally, the inferred value

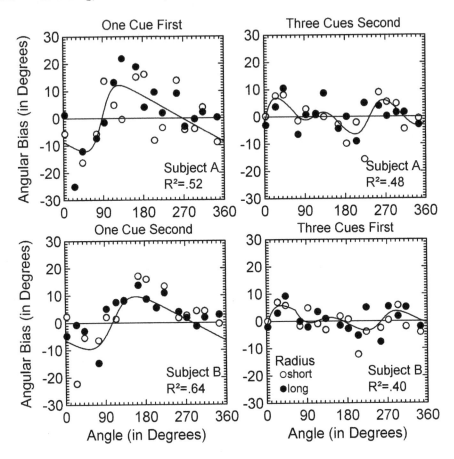

Fig. 5. Empirical angular bias scores plotted separately for one-cue and three-cue conditions along with model prediction functions. Panel A: Subject A. Panel B: Subject B.

of the λ parameter exceeds 1.0 for both subjects in the one-cue condition, a value that contradicts the spirit of the model. In conclusion, the category-adjustment model with geometrically determined boundaries provides a poor fit to these data.

The cue-based fuzzy boundary model of Equations (2) and (6) overcomes the shortcoming of the four-quadrant version of the model. Versions of the fuzzy boundary model were generated by successively freeing parameter values and using the iterative nonlinear regression procedure within SYSTAT (Wilkinson, 1989), with a least squared error criterion. Parameters were freed only if doing so led to a significant increment in R^2. The estimated parameter values of the best-fit versions of the model are presented in Table 2 and the predictions from these models are plotted in Figure 5. Even though the proportions of explained variance are not high in all four cases, there is clear evidence that the subjects

Table 1. Parameters for the Fit of the 5-Parameter Category-Adjustment Model of Equation (2)

Subject	Cue condition	Parameter					R^2
		p_1	p_2	p_3	p_4	λ	
A	One cue first	80°	123°	200°	340°	1.117	0.252
B	One cue second	80°	98°	183°	315°	1.114	0.315
A	Three cues second	78°	169°	183°	350°	0.991	0.053
B	Three cues first	83°	145°	183°	345°	0.950	0.269

Note: p = prototype value and λ = weight of fine-grain memory.

Table 2. Parameters for the Fit of Fuzzy Boundary Model of Equations (2) and (6)

Subject	Cue condition	Parameter					R^2
		p_1	p_2	p_3	λ	c	
A	One cue first	268°	-	-	0.906	0.043	0.523
B	One cue second	286°	-	-	0.911	0.027	0.635
A	Three cues second	66°	156°	310°	0.807	0.035	0.478
B	Three cues first	64°	171°	344°	0.913	0.035	0.399

Note: p = prototype value, λ = weight of fine-grain memory, c = sensitivity parameter

are influenced by the available reference cues. In the one-cue condition, both individuals are less biased near the location of the external cue, with the estimated prototype location being near the actual external cue location of 305°. Furthermore, freeing parameter values to include more prototypes did not significantly increase the fit of the model. In contrast to the one-cue condition, both observers evidently relied on three prototypes when three cues were available. Again the locations of the three prototypes estimated by the model are near the actual values of the external reference cues, 80°, 170°, and 305°. Here, the model fitting procedure clearly rejected models based on one or two prototypes, and the inclusion of a fourth prototype did not significantly increment the fit of the model.

In order to test for changes in accuracy across cue conditions, within-subject analyses of variance (ANOVA's) were conducted on the absolute values of angular bias scores, with cue condition as a repeated measure and using the three-way interaction term as an error term. Both participants were significantly less accurate when just one cue was available, $F(1, 8) = 26.17$, $p < .001$ and $F(1, 9) = 8.31$, $p = .018$, for Subjects A and B respectively. These results support the basic notion behind the category-adjustment model that including categorical information from multiple categories can reduce overall error.

4   Conclusions

Our results and those of related studies indicate that spatial memory, even in tasks involving nothing more than specifying a location seen moments earlier, is a multifaceted phenomenon that can involve a number of different processes in response to task demands and available information. Previous studies have established that memory for location involves coding two types of information, fine-grain and categorical (Allen & Haun, 2004; Huttenlocher et al., 1991). Fine-grain information is metrically veridical and based on a coordinate frame of reference (for example, polar coordinates). In contrast, categorical information is based on properties of the task field. Prototypes within categories draw remembered locations toward them, resulting in an unmistakable pattern of bias.

When observers remember locations within a circular field in the absence of peripheral cues, their memory is biased by implicit categories imposed on the circle itself (Huttenlocher et al., 1991). Specifically, observers tend to impose viewer-based horizontal and vertical axes on the circle, with the resulting quadrants acting as categories and an interior point within each quadrant serving as a category prototype. The presence of peripheral cues exerts no influence when field rotation is not a possibility (Fitting, 2005). In other words, observers tend to ignore the cues and impose viewer-based horizontal and vertical axes. The result is the same four-category structure that influences memory in the cue-free version of the task.

When the task includes both rotated and unrotated trials, however, the unrotated trials, which pose the same prima facie task demands as trials in the Fitting (2005) study with a static field, show a very different categorical structure. On these trials, the cues that are necessary for orientation during rotated trials are also used on unrotated trials. In short, the categorical structure imposed on the task field is cue-based, with the number of cues dictating the number of categories and corresponding prototypes.

In addressing the influence of environmental cues on categorical coding of location information, we developed a fuzzy boundary extension of Huttenlocher et al.'s (1991) category-adjustment model. Our aim was not only to develop a way of addressing this issue in its inherent complexity, but also to offer a more general theoretical tool for investigating the interplay between fine-grain and categorical coding of spatial relations in a variety of tasks. For example, recent research showed that this type of model provides a good account of memory for incline and azimuth, with categories based on implicit quadrants imposed on the response field (Haun et al., 2005). Our current findings suggest that this outcome may be limited to response fields that lack cues. Such cues in the visual environment may afford the viewer a different categorical structure.

Another potential application of our approach involves classic place learning tasks, such as the Morris Water Maze. In the current study, we employed a task that was comparable in some ways to the Morris Water Maze, but clearly there were differences. Navigation is involved in the Morris Water Maze, but not in our task. The Morris Water Maze typically involves a single place to be learned; our task involved hundreds of trials with different locations to be remembered. Thus, the results from our experiment are more suggestive than conclusive when addressing the question of whether categorical bias is present in animal and human behavior in this classic place learning paradigm. We hope to obtain more conclusive evidence with humans using two different approaches. One approach involves a place learning task performed by humans in an actual arena in which navigation is required. In a study using this approach, Fitting and Allen (2004) found some evidence that participants distorted remembered locations toward implicit axes extending from peripheral cues, a finding consistent with cue-based categorical coding. A different approach to the same issue involves a computerized version of the water maze task that requires navigation. We are designing studies using this approach in which participants guide a virtual rat in discovering and subsequently navigating to a safe platform in the maze. As with the location memory task we used in the current study, reliance on cue-based categorical coding should be revealed in memory bias.

Another question that needs further investigation is the applicability of the model in a large-scale space. How do the spatial memory processes proposed by the fuzzy boundary model come into play when location memory is tested in a large-scale space? This question raises interesting issues regarding the process by which observers parse the environment into categorical regions, the role of environmental geometry and peripheral cues in this parsing process, and the role of human information-processing limitations in cue selection and use.

Given our view of spatial memory as a multifaceted phenomenon involving an array of cognitive tools, we suspect that different contextual and task constraints may lead individuals to apply cue-based categories in large-scale tasks in ways similar to those found in the present small-scale study. From this perspective, a key issue for future research is the determination of conditions leading to the use of different spatial representations and cognitive tools across spatial environments.

References

Allen, G. L, & Haun, D. B. M. (2004). Proximity and precision in spatial memory. In G. Allen (Ed.), *Human Spatial Memory: Remembering where* (pp. 41-61). Mahwah, NJ: Lawrence Erlbaum Associates.

Creem, S. H., & Proffitt, D. R. (1998). Two memories for geographical slant: Separation and interdependence of action and awareness. *Psychonomic Bulletin and Review, 5*, 22-36.

Fitting, S. & Allen, G. L. (November, 2004). *Remembering places in spaces: Memory load, retention interval and environmental cues.* Presented at the 45th annual meeting of the Psychonomic Society. Minneapolis, MN.

Fitting, S. (2005) *Memory for spatial location: Cue effects as a function of field rotation.* Master's Thesis, Department of Psychology, University of South Carolina, Columbia, SC.

Haun, D. B. M., Allen, G. L., & Wedell, D. H. (2005). Bias in spatial memory: A categorical endorsement. *Acta Psychologica, 118*, 149-170.

Huttenlocher, J., Hedges, L. V., & Duncan, S. (1991). Categories and particulars: Prototype effects in estimating spatial location. *Psychological Review, 98*, 352-376.

Klatzky, R. (1998). Allocentric and Egocentric Spatial Representations: Denitions, Distinctions, and Interconnections. In C. Freksa, C. Habel, & K. Wender (Eds.). *Spatial cognition: An interdisciplinary approach to representation and processing of spatial knowledge* (pp. 1-17). Berlin: Springer-Verlag.

Kosslyn, S. M., Koenig, O., Barrett, A., Cave, C. B., Tang, J., & Gabrieli, J. D. (1989). Evidence for two types of spatial representations: Hemispheric specialization for categorical and coordinate relations. *Journal of Experimental Psychology: Human Perception and Performance, 15*, 723-735.

Luce, R. D., Nosofsky, R. M., Green, D. M., & Smith, A. F. (1982). The bow and sequential effects in absolute identification. *Perception and Psychophysics, 32*, 397-408.

Morris, R. G. M., Garrud, P., Rawlins, J. N. P., & O'Keefe, J. (1982). Place navigation impaired in rats with hippocampal lesions. *Nature, 297*, 681-683.

Morris, R. G., & Parslow, D. M. (2004). Neurocognitive components of spatial memory. In G. Allen (Ed.). *Human Spatial Memory: Remembering where* (pp. 217-247). Mahwah, NJ: Lawrence Erlbaum Associates.

Nadel, L. (1990). Varieties of spatial cognition: Psychobiological considerations. In A. Diamond (Ed.), *Annals of the New York Academy of* Sciences (Vol. 608; pp. 613-636). New York: New York Academy of Sciences.

Newcombe, N. S., & Huttenlocher, J. (2000). *Making space: The developmental of spatial representation and spatial reasoning.* Cambridge, MA: MIT Press.

Nosofsky, R. M. (1986). Attention, similarity, and identification-categorization relationship. *Journal of Experimental Psychology: General, 115*, 39-61.

O'Keefe, J., & Nadel, L. (1978). *The hippocampus as a cognitive map.* Oxford: Clarendon Press.

Shepard, R. N. (1987). Toward a universal law of generalization for psychological science. *Science, 237*, 1317-1323.

Tolman, E. C. (1948). Cognitive maps in rats and men. *Psychological Review, 55*, 189-208.

Wilkinson, L. (1989). *SYSTAT: The system for statistics.* Evanston, IL: SYSTAT, Inc.

Network and Psychological Effects in Urban Movement

Bill Hillier and Shinichi Iida

Bartlett School of Graduate Studies,
University College London,
1-19 Torrington Place, Gower Street, London WC1E 6BT, UK
{b.hillier, s.iida}@ucl.ac.uk
http://www.bartlett.ucl.ac.uk/

Abstract. Correlations are regularly found in space syntax studies between graph-based configurational measures of street networks, represented as lines, and observed movement patterns. This suggests that topological and geometric complexity are critically involved in how people navigate urban grids. This has caused difficulties with orthodox urban modelling, since it has always been assumed that insofar as spatial factors play a role in navigation, it will be on the basis of metric distance. In spite of much experimental evidence from cognitive science that geometric and topological factors are involved in navigation, and that metric distance is unlikely to be the best criterion for navigational choices, the matter has not been convincingly resolved since no method has existed for extracting cognitive information from aggregate flows. Within the space syntax literature it has also remained unclear how far the correlations that are found with syntactic variables at the level of aggregate flows are due to cognitive factors operating at the level of individual movers, or they are simply mathematically probable *network effects*, that is emergent statistical effects from the structure of line networks, independent of the psychology of navigational choices. Here we suggest how both problems can be resolved, by showing three things: first, how cognitive inferences can be made from aggregate urban flow data and distinguished from network effects; second by showing that urban movement, both vehicular and pedestrian, are shaped far more by the geometrical and topological properties of the grid than by its metric properties; and third by demonstrating that the influence of these factors on movement is a cognitive, not network, effect.

1 Introduction

A fundamental proposition in space syntax (Hillier and Hanson 1984[1]) is that, with the kinds of exceptions noted in Hillier et al. (1987)[2], Hillier et al. (1993) [3], Chang and Penn (1998)[4] and Penn et al. (1998)[5], the configuration of the urban street network is in itself a major determinant of movement flows: that is, the number of people observed moving along the street segments, without regard to the origins or destinations, or to the reasons for choosing to move

A.G. Cohn and D.M. Mark (Eds.): COSIT 2005, LNCS 3693, pp. 475–490, 2005.
© Springer-Verlag Berlin Heidelberg 2005

along that segment.[1] The research which supports this proposition is based on representing the street system as a network of the fewest lines that cover the system, translating the network into a graph in which lines are nodes and intersections are links (similar to what is known as 'line graph' as in Harary (1969)[8], reversing the more common 'primary' representation), and measuring configuration through topological distances in the graph, without metric weighting. The fact that strong correlations are commonly found between observed flow and such configurational measures (as in Hillier et al. 1993[3], Penn et al. 1998[5]) suggests that geometric (from the use of lines) and topological (from the use of metric free graph measures) factors are critically involved in how people navigate urban grids. However, because the reported results are about aggregate human behaviour, it has always been unclear how far they depended on individual spatial decisions, and how far they are simply mathematically probable *network effects*, that is emergent statistical effects from the structure of line networks, relatively independent of the psychology of navigational choices.

The apparent involvement of grid complexity in navigation has also brought space syntax into conflict with orthodox urban modelling, where it has always been assumed that insofar as spatial factors play a role in navigation, it will be on the basis of metric distance. However, in recent years, research results have accumulated in cognitive science which suggest that the metric distance assumption is unrealistic, not perhaps because we do not seek to minimise travel distance, but because our notions of distance are compromised by the visual, geometrical and topological properties of networks. For example, estimates of distance have been shown to be affected by the division of routes into discrete visual chunks (Golledge 1992[9]; Montello 1997[10]; Kim 2001[11]; Kim and Penn 2004[12]), by a tendency to correct bends to straight lines and turns to right angles (Allen 1981[13]),and even by the direction in which the estimate is made (Sadalla et al. 1980[14]; Montello 1992[15]; Golledge 1995[16]). As a consequence, much current cognitive work on spatial complexity explores how far route choices reflect the frequency (Duckham and Kulik 2003[17]) or degree (Conroy-Dalton 2001[18]; 2003[19]; Hochmair and Frank 2002[20]) of directional change, rather than metric distance. An obstacle to a more definitive resolution within the urban research community of how concepts of distance shape human movement—or even whether or not a general definition exists—is that no method exists to extract cognitive information from the aggregate flows in street networks, and distinguish this from emergent statistical effects of the network itself. In this paper, we seek to resolve these questions through a two stage argument. First we develop a theory of why network effects on movement flows

---

[1] For example, in the study of human cognition, a distinction is often made between *navigation* (a certain knowledge on the route to be followed is assumed) and *wayfinding* (no predetermined criteria is defined and therefore it involves search and exploration) (Golledge and Gärling 2002[6]), although two terms are also used interchangeably (Duckham et al. 2003[7]). Throughout this paper, the discussion of movement will only refer to the observed aggregate numbers.

are to be expected in spatial configurations in general, and why the syntactic measures of *closeness* and *betweenness* can be expected to capture them.

We then ask what remains for the psychology of navigation and suggest the answer can be found in the concepts of *distance* that must underlie the use of measures like closeness and betweenness. By using different concepts of distance in configurational analysis of urban networks, and correlating the results with real movement flows, we show how cognitive inferences can be made from aggregate movement data, and distinguished from network effects. By using this method in a study of movement in four areas of London, we also show that movement in cities reflects the geometrical and topological structure of the network configuration far more than metric distance.

2   Network Effects: Theoretical Motivation

Why and how, then, should we expect street networks to shape movement in cities? First, we must be clear about network effects—that is emergent statistical effects on aggregate movement from the structure of the network itself—and why they are to be expected in movement in urban systems.[2] First, we remind the reader of a simple yet motivating example (Hillier 1999[22]) that makes network effects intuitively obvious. Figure 1 shows a notional grid with a main street, a cross street, some side streets and a back street. Suppose all streets are equally

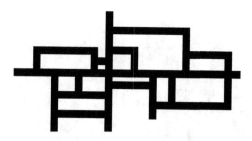

Fig. 1. Notional grid showing how the network will shape movement flows

loaded with dwellings, and that people move over time from all dwellings to all others using some notion of shortest (least distance) or simplest (fewest turns) routes. It is clear that more movement will pass through the horizontal main street than other streets, and that more will pass the central parts of the main

---

[2] The actual urban systems may contain many different levels of movements all of which may have a compound effect on route choice and navigation, as suggested, for example, by Timpf et al. (1992)[21]. In addition, each street segment may have a various non-physical constraints in access (for example, a restriction of entry by the direction). However, we will focus on the simplest level of movement (either pedestrian or vehicular), that is, the simple count of people or vehicles moving along a simple street system where all the streets are treated equally.

street than the more peripheral parts. This effect follows from the structure of the network, since no one needs a plan to pass through the spaces, and would hold under either assumption about distance. It is also clear that the main street considered as a whole is more *accessible* than other spaces, on either definition of distance. It will then be more advantageous to locate a shop on the main street, since it will be both easier to get to and also where people are likely to be when moving between locations. Although locating a shop is an individual decision, it is clear that the decision will be shaped first and foremost by the properties of the network. This simple example shows in a common-sense way why we should expect network effects on movement.

Now consider a more complex theoretical example. On the top left of Figure 2 is a notional arrangement of blocks with something like the degree of linear continuity between spaces that we expect in urban space. Visibility graph analysis (Turner et al. 2001[23]) shows an emergent warm colour pattern which looks a bit like a main street, with side streets and back streets, although of a rather a irregular kind. On the right we retain exactly the same blocks but move some marginally to 'just about' block lines of sight, so in effect, we change nothing but the linear relations between some of the spaces. The visibility graph analysis, shown on the same scale, shows not only a substantial loss of visual integration but also a totally transformed pattern, with little in the way of continuous structure. Spatial network seems to have lost structure as well as integration.

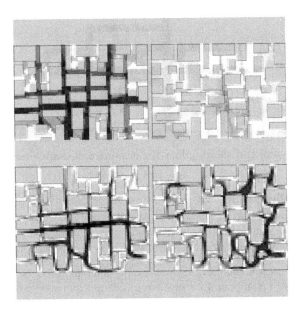

Fig. 2. Two layout showing how slight shifts in the positioning of blocks to allows or break lines of sight has major effect on degree and structure of visual integration (top row) and the traces of 10000 randomly moving sighted agents.

This is of course the well-known syntactic property of *intelligibility* (Hillier et al. 1993[3]). The r^2 between the *visual connectivity* and *visual integration* of point is 0.714 in the left case and 0.267 on the right. But this mathematical change in the network structure has consequences even for theoretical movement. In the bottom of Figure 2, we use the 'agent' facility in Turner's Depthmap software (Turner 2001[24]) to show the traces of 10000 sighted agents with 170 degrees of vision, who select a point within their field of view randomly, move towards it three pixels and repeat the process. Even with the sight and distance parameters set close to randomness (in that vision is diffuse rather than focused and distance between destination selections are short), the results are strikingly different. In the left case the highest density of traces follows the space structure to a remarkable degree, while in the right case, it spirals all over the system, reflecting the local scaling of spaces rather than overall configuration. Since the 'cognitive' structure and behaviour of agents is identical in the two cases, the differences between the trace patterns are clearly *network effects*.

3   Network Effects in Real Urban Systems

But what of real human subjects in real urban systems? Are there reasons for expecting network effects at this level? It is useful here again to distinguish between the *structure* of the graph, that is, the pattern of nodes and links, and how *distance* between nodes is to be calculated. There are (as noted in Hillier 2002[25]) in principle strong mathematical reason to expect network effects from the structure of any graph on movement, in that random one step movement in a non-bipartite graph will lead to the number of visits per node going to a limit of the degree of the node as the number of iterations goes to infinity (Norris 1997[26]; Batty and Tinkler 1979[27]). If movement is random and one step, then, the number of visits to each node will be wholly determined by the structure of the graph.

But of course human movement is neither random nor one step. For the most part it is both planned and n-step. Are there also mathematical reasons why flows arising from this kind of movement will be shaped by the network configuration? First let us consider the nature of human movement. It has two aspects: the selection of a destination from an origin; and the selection of the intervening spaces that must be passed through to go from one to the other. The former is about *to-movement*, the latter *though-movement*.

First, consider *to*-movement. From any origin, say, someone's house, we must expect that over time a range of trips will be made to various destinations, and these will be a matter of individual decision. But, over time, the choice of destinations from any origin would be expected to show some degree of statistical preference for closer rather than more remote destinations—say, by going more often to the local shop than to visit an aunt in Willesden. It does not have to be that way, but in most cases it probably will be. This is no more than an instance of what geographers have always called *distance decay*. But if there is any degree of distance decay in the choice of destinations, then it has the formal consequence that locations which are closer to *all* others in the network will be featured as destinations more often than those that are more remote—that is, more accessible locations will be theoretically more attractive as destinations than less accessible ones simply as a result of their configurational position in the complex as a whole. The bias towards more accessible locations for *to*-movement is then a network effect, due to the configurational structure of the network, even though it is driven by the accumulation of individual decisions.

This of course is to say no more than that central locations in a system are more accessible than others. But the argument becomes more interesting if we consider *variable radius* integration, that is the accessibility of nodes to its neighbouring network up to a certain graph or metric distance away. A node which is more integrated than others in its region at a given radius will also become more theoretically attractive as a destination to the degree that movement at that graph scale is preferred in the system. In other words, the network properties measured by variable radius integration can be *theoretically* expected to attract more movement to some destinations than others purely as an effect of the structure of the network, and this is of course what we find in urban reality.

But the effect of the network does not end there, since there will also be network effects on *through*-movement, more obvious than those for *to*-movement. The sequences of nodes that are available between origins and destinations will often vary with different definitions of distance (as we will see below) but whichever definition of distance we choose the available sequences are defined by the structure of the graph itself, so again we are dealing with network effects. However we define distance, in effect, the choice of routes is defined by the network.

So network effects must exist for both *to*- and *through*-movement. There will also be an interaction between them. First, every trip is made up of a pair of origin-destination, or *to*-movement nodes, and a variable number of *through*-movement nodes. But with increasing length of trip the *to*-movement pair will remain constant at two nodes while the *through*-movement node count will increase. It follows that the longer the trip, the higher will be the ratio of *through*-movement spaces to the origin-destination pairs, which of course always remain constant. We may expect then that the greater the graph length of the trip, the more it will reflect the choice, or betweenness, structure of the graph, rather than the integration, or closeness, structure.

Second, any *to*-movement bias towards integrated locations will also have an effect on *through*-movement, since routes leading to those locations will be more likely to be used than those leading to less integrated locations. It would be a simple matter to reflect this by weighting the choice measure for the integration value of destinations (see below). This combined measure should then theoretically measure both aspects of simplest path *n*-step movement in a system, with some adjustment for the mean length of trips. So in the 'state of nature' the graph should already have a tendency towards a certain pattern of movement reflecting the spatial configuration of the graph, and the configurational properties which produce this are exactly reflected in the syntactic measures of variable radius integration and choice, and by the relations between them.

This then is the theory of urban movement in a network considered as a graph. But how people actually move will clearly be affected by how distance is conceptualised. In what follows, we show how cognitive information on how distance concepts can be extracted from information on real flows in urban networks.

4 Varying Distance Concepts in Line Networks

The technique we propose to extract cognitive information from real flows is to take urban street networks and subject them to different mathematical interpretations according to how distance is defined, then to explore how well the different interpretations correlate with real movement patterns. The basis of the different interpretation is a disaggregated line-network model which is an extension of fewest line-network model ('axial map', originally developed by Hillier and Hanson (1984)[1]) for introducing fractional weights instead of a constant one (for example, see Turner (2001)[28], Dalton (2001[29], 2003[30]), Winter (2002)[31], Dalton et al. (2003)[32] and Asami et al. (2004)[33]). [3] We describe the construction of the model below.

We start from the existing fewest line map and represent the street network as its graph of line segments between intersections. Figure 3(a) is an example of the unweighted line network with three lines, and its graph representation is shown in Figure 3(b).

Figure 3(c) shows how the line network is disaggregated at intersections to form a segment network, and Figure 3(d) shows its graph representation. Each line segment is represented as a node in the graph and links between nodes are intersections. The distance cost between two line segments is measured by taking a 'shortest' path from one to the other, so the cost of travel between **S** and **a** can be given as $w(\pi - \theta) + w(\phi)$, while the cost between **S** and **b** can be $w(\theta) + w(\pi - \phi)$. Three different weight definitions are then used to represent different distance cost relations between adjacent segments:

Least length (metric) The distance cost of routes is measured as the sum of segment lengths, defining length as the metric distance along the lines between the mid-points of two adjacent segments. The distance of two adjacent line segments is thus calculated as half the sum of their lengths.

Fewest turns (topological) Distance cost is measured as the number of changes of direction that have to be taken on a route. In the example shown in Figure 3(c) and (d), $w(\theta) = w(\pi - \theta) = w(\phi) = w(\pi - \phi) = 1$ (however, $w(0) = 0$).

---

[3] The model also has a close relevance to the street-based network models such as Jiang and Claramunt (2002)[34], Thomson (2003)[35], Porta et al. (2004)[36] or Rosvall et al. (2005)[37].

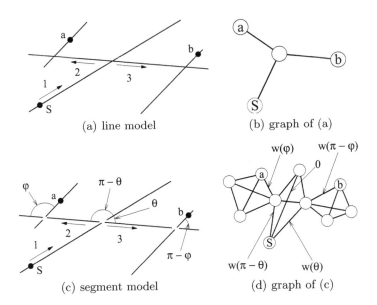

(a) line model

(b) graph of (a)

(c) segment model

(d) graph of (c)

Fig. 3. Line-network model and its disaggregated model with graph representation for each model

Least angle change (geometric) Distance cost is measured as the sum of angular changes that are made on a route, by assigning a weight to each intersection proportional to the angle of incidence of two line segments at the intersection. The weight is defined so that the distance gain will be 1 when the turn is a right angle. In other expression, $w(\theta) \propto \theta$ $(0 \leq \theta < \pi)$, $w(0) = 0$, $w(\pi/2) = 1$.

These definitions reflect different conjectures in the literature as to how distance is conceptualised in human navigation. Paths between all segments and all others can then be assessed in terms of least length, fewest turns, and least angle paths. Least length paths are the shortest metric distances, fewest turns paths the least number of direction changes, and least angle paths the smallest accumulated totals of angular change on paths, between all pairs of nodes. Note that the original lines of the fewest line map will emerge as sequences of zero-change weights for fewest turns and least angle change paths. In this sense the model also allows a test on how far the focus on linearity of syntactic axial mapping is justified.

We then apply the common centrality measures of 'closeness' and 'between-ness'. Closeness, as defined by Sabidussi (1966)[38], is:

$$C_C(p_i) = \left(\sum_k d_{ik} \right)^{-1}$$

where d_{ik} is the length of a geodesic (shortest path) between node p_i and p_k. Betweenness, as defined by Freeman (1977)[39], is:

$$C_B(p_i) = \sum_j \sum_k g_{jk}(p_i)/g_{jk} \quad (j < k)$$

where $g_{jk}(p_i)$ is the number of geodesics between node p_j and p_k which contain node p_i, and g_{jk} the number of all geodesics between p_j and p_k.

This gives six different mathematical interpretations of a street system: closeness and betweenness measures applied to least length, least angle change and fewest turns weightings of relations between adjacent segments in the system. In addition, closeness measures can be assigned for every radius, defining radius in terms of the number of segments distant from each segment treated as a root. This permits experimentation with the scale at which measures operate, from the most local, to the global level.

5   Empirical Studies

We then take four areas of London (Barnsbury, Clerkenwell, South Kensington and Knightsbridge)[4] on which earlier studies (Penn et al. 1998[5]) had established dense vehicular and pedestrian movement flows at the segment level throughout the working day for a total of 356 observation 'gates'.[5] The street network for each observation area was embedded in a contextual network of 3–3.5km radius, and analysed using the segment network representation. Closeness measures were calculated for every third radius, that is up to 3, 6, 9, ... segments distant from a root segment, up to the maximum radius of the system.

Translating the numerical results of the analysis into images of the network, with segments coloured in bands of value for each measure, from black for least distance through to pale gray for most, Figure 4 shows that the different interpretations give different pictures of the 'structure' of the network, some slight, others more substantial. The upper row of figures show closeness (a: geometric, b: topological, c: metric) and the figures at the bottom betweenness (d: geometric, e: topological, f: metric). In the closeness maps (a-c), line segments in dark tone are those with the least mean distance cost from that line segment to all others,

---

[4] Two of the areas (South Kensington and Knightsbridge) had originally been selected to pose problems for a purely configurational analysis, in that they had large movement attractors at their heart, one a complex of national museums adjacent to a tube station, and the other a leading department store. Both could be expected to distort correlations with purely network measures, but the original study by Penn et al. (1998)[5] found consistent agreement between vehicular flows and the local closeness measure, regardless of the existence of attractors.

[5] Observation points were set up on each street segment and the number of pedestrians and vehicles that pass each observation point was counted for ten minutes in each of five time peiriods during the working day, totalling fifty minutes of observation for each point. The hourly average volume of flows were then used to see the correlation with the measures.

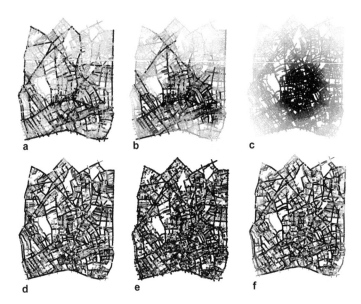

Fig. 4. Disaggregated line network models of Barnsbury area in London, each coloured up by a different measure

and shift to lighter tone as the degree of closeness declines. In the betweenness maps (d-f), line segments in dark tone are those which lie on most shortest paths between pairs of segments, declining towards lighter tone for those which lie on least. The network consists of 10897 line segments and the area covered by this model is roughly 3km in radius. The observation area was near the centre of the

Table 1. Adjusted R^2 values for correlations between vehicular flows and least sum of length, least sum of angle change and least number of turns analysis applied to closeness and betweenness measures

Area name	Gates	Measure	Least length	Least angle	Fewest turns
Barnsbury	116	closeness	0.131 (60)	0.678 (90)	0.698* (12)
		betweenness	0.579	0.720*	0.558
Clerkenwell	63	closeness	0.095 (93)	0.837* (90)	0.819 (69)
		betweenness	0.585	0.773*	0.695
S. Kensington	87	closeness	0.175 (93)	0.688 (24)	0.741* (27)
		betweenness	0.645	0.629	0.649*
Knightsbridge	90	closeness	0.084 (81)	0.692* (33)	0.642 (27)
		betweenness	0.475	0.651*	0.580

* Best correlation.
† Numbers in round brackets indicate best radius in segments for closeness measures.

Table 2. Adjusted R^2 values for correlations between pedestrian flows and least sum of length, least sum of angle change and least number of turns analysis applied to closeness and betweenness measures

Area name	Gates	Measure	Least length	Least angle	Fewest turns
Barnsbury	117	closeness	0.119 (57)	0.719* (18)	0.701 (12)
		betweenness	0.578	0.705*	0.566
Clerkenwell	63	closeness	0.061 (102)	0.637 (39)	0.624* (36)
		betweenness	0.430	0.544*	0.353
S. Kensington	87	closeness	0.152 (87)	0.523* (21)	0.502 (27)
		betweenness	0.314	0.457	0.526*
Knightsbridge	90	closeness	0.111 (81)	0.623* (63)	0.578 (63)
		betweenness	0.455	0.513	0.516*

* Best correlation.
† Numbers in round brackets indicate best radius in segments for closeness measures.

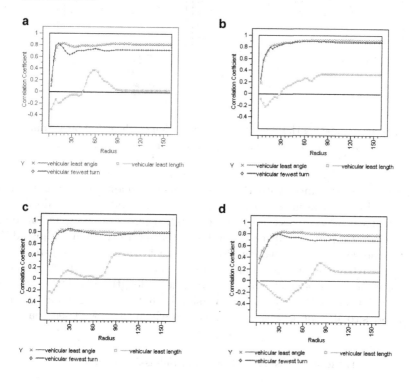

Fig. 5. Line charts showing the change in the correlation coefficient of vehicular movement with all three types of closeness measures with increasing radius

map (around 600m in radius), where vehicular and pedestrian flow data was collected. Of two measures, closeness shows a marked difference between different

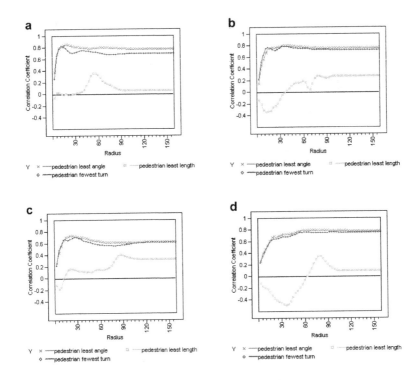

Fig. 6. Line charts showing the change in the correlation coefficient of pedestrian movement with all three types of closeness measures with increasing radius

interpretations of distance, with the metric version yielding only a concentric picture with highest closeness in the centre of the map and a smooth decline towards the edges. Betweenness shows more stability across the weights, although a more concentric picture can be observed in the metric version. Adjusted R^2 values for the correlations between closeness and betweenness measures range between 0.505–0.807 for the least angle interpretation, between 0.599–0.853 for fewest turns, and between -0.0067–0.0764 for least length. The reason for the lack of correlation for the least length version is that least length for closeness from all nodes to all others must give a result in which the closest segment to all

Table 3. Adjusted R^2 from axial radius-3 integration

Area name	Vehicular	pedestrian
Barnsbury	0.765**	0.706
Clerkenwell	0.627	0.570
South Kensington	0.819**	0.514
Knightsbridge	0.560	0.467

others is at the geometric centre, with a more or less smooth fall-off from centre to periphery, as can be seen in the concentric patterns shown in Figure 4c.

The different sets of values for segments were then correlated with observed vehicular and pedestrian flows averaged for the whole working day between 8 a.m. and 6 p.m. For vehicular movement correlations were confined to streets with unrestricted two-way flows, but for pedestrian movement the whole urban network was included (though observations made inside a college and a housing estate were not). The pattern of adjusted R^2 values for vehicular movement is shown in Table 1 and for pedestrian movement in Table 2 with a total of 48 correlations and therefore 16 possible best correlations. Correlations will be in general negative in closeness centrality (but see below) since movement should increase with less metric, angular or directional change.

The results give a consistent picture. In 11 out of 16 cases, (5 vehicular and 6 pedestrian) least angle correlations are best. In the remaining five cases fewest turns is best, but in each case only marginally better than least angle. In no case is a metrically based measure best, and in no case is a least angle measure worst. On average correlations based on least length measures are markedly lower than the other two. The pattern of correlation for the metric, least angle and fewest turns interpretations of the closeness measure with increasing radius are shown from left to right in Figure 5 and Figure 6. Each line chart corresponds to the vehicular (Figure 5) and pedestrian (Figure 6) movement data of Barnsbury (a), Clerkenwell (b), South Kensington (c) and Knightsbridge (d). Plots for least length (metric) are shown in gray, least angle change (geometric) in crossed dotted line, and fewest turns (topological) in dotted line. In all cases the correlation coefficient stabilises above a certain radius, but not necessarily at its optimal level, suggesting natural limits to the radius within which configurational measures operate. The plots show that the superiority of the least angle model is more marked than in the tabulated results in two of the four cases. We conjecture that the weakly negative correlations at low radii for metric (meaning that it is positively correlated with movement) are due to the fact that higher connectivity for a segment will both tend to produce more movement and have a higher sum of lengths, while higher sums of length at the larger scale will mean greater distance from the rest of the system and so less movement.

6   Discussion

How then are these results to be interpreted? From a cognitive point, it is clear that, unlike previous results from axial maps, the *differences* in movement correlations with the different definitions of distance cannot be network effects, since in each case the representation of the street network and its graph are identical, and all that differs is the mathematical interpretation by varying the concept of distance. The differences in correlation can then only be due to differences in the degree to which each mathematical interpretation coincides with the interpretations made by individuals moving in the system. It is then an unavoidable inference that people are reading the urban network in geometrical and topological rather then metric terms. Although it is perfectly plausible that people *try* to minimise distance, their concept of distance is, it seems, shaped more by the geometric and topological properties of the network more than by an ability to calculate metric distances. In general we might say that the structure of the graph governs network effects on movement and that how distance is defined in the graph governs cognitive choices.

These results have three implications. First, they show that it is the geometrical and topological architecture of the large scale urban grid that is, as space syntax has always argued, the most powerful shaper of urban movement patterns. These factors are not currently represented in most of the models currently in use to predict urban movement. The effect is that the design of movement systems does not take account of the primary factors which shape urban movement. Clearly, this situation cannot continue.

Second, the results show that axial graphs in their present form are in most circumstances a perfectly good approximation of the impact of spatial configuration on movement. In two of the eight cases reported, the correlation between movement and radius-3 integration is better than any of the segment analyses, and in general the spread of R^2 values for axial graph mirrors the pattern of correlation for the new, more disaggregated segment based measures.

Third, these results are the strongest demonstration to date that the architecture of the street network, in both geometrical and topological sense, can be expected, through its effect on movement flows, to influence the evolution of land use patterns and consequently the whole pattern of life in the city. This most powerful feature of the urban system can surely not continue to be sidelined in urban modelling, and the architectural effects of the large scale street network cannot be discounted.

References

1. Hillier, B., Hanson, J.: The Social Logic of Space. Cambridge University Press, Cambridge, UK (1984)
2. Hillier, B., Burdett, R., Peponis, J., Penn, A.: Creating life: or, does architecture determine anything? Architecture and Behaviour **3** (1987) 233–250
3. Hillier, B., Penn, A., Hanson, J., Grajewski, T., Xu, J.: Natural movement: or configuration and attraction in urban pedestrian movement. Environment and Planning B: Planning and Design **20** (1993) 29–66
4. Chang, D.K., Penn, A.: Integrated multilevel circulation in dense urban areas. Environment and Planning B: Planning and Design **25** (1998) 507–538
5. Penn, A., Hillier, B., Banister, D., Xu, J.: Configurational modelling of urban movement networks. Environment and Planning B: Planning and Design **25** (1998) 59–84
6. Golledge, R.G., Gärling, T.: Spatial behavior in transportation modeling and planning. In Goulias, K., ed.: Transportation Systems Planning: Methods and Applications. CRC Press (2002)
7. Duckham, M., Kulik, L., Worboys, M.F.: Imprecise navigation. Geoinformatica **7** (2003) 79–94
8. Harary, F.: Graph Theory. Addison-Wesley, Reading, Mass. (1969)

9. Golledge, R.G.: Place recognition and wayfinding: making sense of space. Geoforum **23** (1992) 199–214

10. Montello, D.R.: The perception and cognition of environmental distance. In Hirtle, S.C., Frank, A.U., eds.: Spatial Information Theory: Theoretical Basis for GIS. Number 1329 in Lecture Notes in Computer Science. Springer-Verlag, Berlin (1997) 297–311

11. Kim, Y.O.: The role of spatial configuration in spatial cognition. In: Proceeding of the Third International Space Syntax Symposium, Atlanta, GA, Georgia Institute of Technology (2001) 49.1–49.21

12. Kim, Y.O., Penn, A.: Linking the spatial syntax of cognitive maps to the spatial syntax of the environment. Environment and Behavior **36** (2004) 483–504

13. Allen, G.L.: A developmental perspective on the effects of "subdividing" macrospatial experience. Journal of Experimental Psychology: Human Learning and Memory **7** (1981) 120–132

14. Sadalla, E.K., Burroughs, W.J., Staplin, L.J.: Reference points in spatial cognition. Journal of Experimental Psychology: Human Learning and Memory **6** (1980) 516–528

15. Montello, D.R.: The geometry of environmental knowledge. In Frank, A.U., Campari, I., Formentini, U., eds.: Theories and Methods of Spatial Reasoning in Geographic Space. Number 639 in Lecture Notes in Computer Science. Springer-Verlag, Berlin (1992) 136–152

16. Golledge, R.G.: Path selection and route preference in human navigation: a progress report. In Frank, A.U., Kuhn, W., eds.: Spatial Information Theory: A Theoretical Basis for GIS. Number 988 in Lecture Notes in Computer Science. Springer-Verlag, Berlin (1995) 182–199

17. Duckham, M., Kulik, L.: 'simplest' paths: automated route selection for navigation. In Kuhn, W., Worboys, M.F., Timpf, S., eds.: Spatial Information Theory: Foundations of Geographic Information Science. Number 2825 in Lecture Notes in Computer Science. Springer-Verlag, Berlin (2003) 182–199

18. Conroy Dalton, R.: Spatial navigation in immersive virtual environments. PhD thesis, Bartlett School of Graduate Studies, University of London (2001)

19. Conroy Dalton, R.: The secret is to follow your nose: route path selection and angularity. Environment and Behavior **35** (2003) 107–131

20. Hochmair, H., Frank, A.U.: Influence of estimation errors on wayfinding-decisions in unknown street networks — analyzing the least-angle strategy. Spatial Cognition and Computation **2** (2002) 283–313

21. Timpf, S., Volta, G.S., Pollock, D.W., Frank, A.U., Egenhofer, M.J.: A conceptual model of wayfinding using multiple levels of abstraction. In Frank, A.U., Campari, I., Formentini, U., eds.: Theories and Methods of Spatial Reasoning in Geographic Space. Number 639 in Lecture Notes in Computer Science. Springer-Verlag, Berlin (1992) 348–367

22. Hillier, B.: The hidden geometry of deformed grids: or, why space syntax works, when it looks as though it shouldn't. Environment and Planning B: Planning and Design **26** (1999) 169–191

23. Turner, A., Doxa, M., O'Sullivan, D., Penn, A.: From isovists to visibility graphs: a methodology for the analysis of architectural space. Environment and Planning B: Planning and Design **28** (2001) 103–121

24. Turner, A.: Depthmap: a program to perform visibility graph analysis. In: Proceedings of the Third International Space Syntax Symposium, Atlanta, GA, Georgia Institute of Technology (2001) 31.1–31.9

25. Hillier, B.: A theory of the city as object. Urban Design International **7** (2002) 153–179

26. Norris, J.R.: Markov Chains. Cambridge University Press, Cambridge (1997)

27. Batty, M., Tinkler, K.J.: Symmetric structure in spatial and social processes. Environment and Planning B: Planning and Design **6** (1979) 3–28

28. Turner, A.: Angular analysis. In: Proceedings of the Third International Space Syntax Symposium, Atlanta, GA, Georgia Institute of Technology (2001) 30.1–30.11

29. Dalton, N.: Fractional configuration analysis and a solution to the Manhattan Problem. In: Proceedings of the Third International Space Syntax Symposium, Atlanta, GA, Georgia Institute of Technology (2001) 26.1–26.13

30. Dalton, N.: Storing directionality in axial lines using complex node depths. In: Proceedings of the Fourth International Space Syntax Symposium, London, University College London (2003) 63.1–63.10

31. Winter, S.: Modeling costs of turns in route planning. GeoInformatica **6** (2002) 345–360

32. Dalton, N., Peponis, J., Conroy Dalton, R.: To tame a TIGER one has to know its nature: Extending weighted angular integration analysis to the description of GIS road-centerline data for large scale urban analysis. In: Proceedings of the Fourth International Space Syntax Symposium, London, University College London (2003) 65.1–65.10

33. Asami, Y., Kubat, A.S., Kitagawa, K., Iida, S.: A three-dimensional analysis of the street network in Istanbul: an extension of space syntax using GIS. In Okabe, A., ed.: Islamic Area Studies with Geographical Information Systems. Routledge Curzon, London (2004)

34. Jiang, B., Claramunt, C.: Integration of space syntax into GIS: new perspectives for urban morphology. Transactions in GIS **6** (2002) 295–309

35. Thomson, R.C.: Bending the axial line: smoothly continuous road centre-line segments as a basis for road network analysis. In: Proceedings of the Fourth International SpaceSyntax Symposium, London, University College London (2003) 50.1–50.10

36. Porta, S., Crucitti, P., Latora, V.: The network analysis of urban streets: a dual approach. Available online at `http://arXiv.org/abs/cond-mat/0411241` (2004)

37. Rosvall, M., Trusina, A., Minnhagen, P., Sneppen, K.: Networks and cities: An information perspective. Physical Review Letters **94** (2005) 028701

38. Sabidussi, G.: The centrality index of a graph. Psychometrika **31** (1966) 581–603

39. Freeman, L.C.: A set of measures of centrality based on betweenness. Sociometry **40** (1977) 35–41

Probabilistic Techniques for Mobile Robot Navigation

Wolfram Burgard

Institut für Informatik,
Albert-Ludwigs-Universität Freiburg,
Georges-Koehler-Allee, Geb. 079
D-79110 Freiburg
Germany
burgard@informatik.uni-freiburg.de

Abstract. In recent years, probabilistic techniques have enabled novel and innovative solutions to some of the most important problems in mobile robotics. Major challenges in the context of probabilistic algorithms for mobile robot navigation lie in the questions of how to deal with highly complex state estimation problems and how to control the robot so that it efficiently carries out its task. In this talk I will discuss both aspects and present techniques currently being developed in my group regarding the problem of controlling a robot to efficiently learn a map of an unknown environment. I then will describe how a team of mobile robots can be coordinated to effectively explore unknown environments. Additionally, I will present probabilistic approaches to learn three-dimensional models from range data as well as techniques for classifying places based on range and vision data. For all algorithms I will present experimental results, which have been obtained with mobile robots in real-world environments as well as in simulation. I will conclude the presentation with a discussion of open issues and potential directions for future research.

A.G. Cohn and D.M. Mark (Eds.): COSIT 2005, LNCS 3693, p. 491, 2005.
© Springer-Verlag Berlin Heidelberg 2005

Spatial Language, Spatial Thought: Parallels in Path Structure

Barbara Landau

Department of Cognitive Science,
241 Krieger Hall,
Johns Hopkins University,
Baltimore, MD 21218 USA
landau@cogsci.jhu.edu

Abstract. A hallmark of human cognition is our capacity to talk about what we see– especially, the objects, motions and events around us. How is this accomplished? Given that language and spatial representations are likely to be quite different in kind, the challenge is to understand how such apparently different systems of knowledge map onto each other, and how these mappings are learned. One solution is to seek core communalities in the structure of linguistic and non-linguistic representations, which could afford learners and mature users a means for talking about their spatial experience. I will discuss this possibility with respect to the structure of paths within events. Over a broad range of event types, formal linguistic analyses, computational models, and experimental studies show that a fundamental property of path representation-an asymmetry between source and goal paths-is pervasive in our non-linguistic understanding of events as well as our encoding of these in language. This suggests that there are deep structural homomorphisms between spatial language and non-linguistic spatial representation. Such structural parallels may provide a partial solution to the problem of mapping dissimilar domains onto each other, and may even yield insight into how some aspects of language evolved.

A.G. Cohn and D.M. Mark (Eds.): COSIT 2005, LNCS 3693, p. 492, 2005.
© Springer-Verlag Berlin Heidelberg 2005

Author Index

Lecture Notes in Computer Science

For information about Vols. 1–3603

please contact your bookseller or Springer

Vol. 3658: V. Matoušek, P. Mautner, T. Pavelka (Eds.), Text, Speech and Dialogue. XV, 460 pages. 2005. (Subseries LNAI).

Vol. 3655: A. Aldini, R. Gorrieri, F. Martinelli (Eds.), Foundations of Security Analysis and Design III. VII, 273 pages. 2005.

Vol. 3654: S. Jajodia, D. Wijesekera (Eds.), Data and Applications Security XIX. X, 353 pages. 2005.

Vol. 3653: M. Abadi, L. de Alfaro (Eds.), CONCUR 2005 – Concurrency Theory. XIV, 578 pages. 2005.

Vol. 3652: A. Rauber, S. Christodoulakis, A M. Tjoa (Eds.), Research and Advanced Technology for Digital Libraries. XVIII, 545 pages. 2005.

Vol. 3649: W.M.P. van der Aalst, B. Benatallah, F. Casati, F. Curbera (Eds.), Business Process Management. XII, 472 pages. 2005.

Vol. 3648: J.C. Cunha, P.D. Medeiros (Eds.), Euro-Par 2005 Parallel Processing. XXXVI, 1299 pages. 2005.

Vol. 3646: A. F. Famili, J.N. Kok, J.M. Peña, A. Siebes, A. Feelders (Eds.), Advances in Intelligent Data Analysis VI. XIV, 522 pages. 2005.

Vol. 3645: D.-S. Huang, X.-P. Zhang, G.-B. Huang (Eds.), Advances in Intelligent Computing, Part II. XIII, 1010 pages. 2005.

Vol. 3644: D.-S. Huang, X.-P. Zhang, G.-B. Huang (Eds.), Advances in Intelligent Computing, Part I. XXVII, 1101 pages. 2005.

Vol. 3642: D. Ślezak, J. Yao, J.F. Peters, W. Ziarko, X. Hu (Eds.), Rough Sets, Fuzzy Sets, Data Mining, and Granular Computing, Part II. XXIII, 738 pages. 2005. (Subseries LNAI).

Vol. 3641: D. Ślezak, G. Wang, M. Szczuka, I. Düntsch, Y. Yao (Eds.), Rough Sets, Fuzzy Sets, Data Mining, and Granular Computing, Part I. XXIV, 742 pages. 2005. (Subseries LNAI).

Vol. 3639: P. Godefroid (Ed.), Model Checking Software. XI, 289 pages. 2005.

Vol. 3638: A. Butz, B. Fisher, A. Krüger, P. Olivier (Eds.), Smart Graphics. XI, 269 pages. 2005.

Vol. 3637: J. M. Moreno, J. Madrenas, J. Cosp (Eds.), Evolvable Systems: From Biology to Hardware. XI, 227 pages. 2005.

Vol. 3636: M.J. Blesa, C. Blum, A. Roli, M. Sampels (Eds.), Hybrid Metaheuristics. XII, 155 pages. 2005.

Vol. 3634: L. Ong (Ed.), Computer Science Logic. XI, 567 pages. 2005.

Vol. 3633: C. Bauzer Medeiros, M. Egenhofer, E. Bertino (Eds.), Advances in Spatial and Temporal Databases. XIII, 433 pages. 2005.

Vol. 3632: R. Nieuwenhuis (Ed.), Automated Deduction – CADE-20. XIII, 459 pages. 2005. (Subseries LNAI).

Vol. 3631: J. Eder, H.-M. Haav, A. Kalja, J. Penjam (Eds.), Advances in Databases and Information Systems. XIII, 393 pages. 2005.

Vol. 3630: M.S. Capcarrere, A.A. Freitas, P.J. Bentley, C.G. Johnson, J. Timmis (Eds.), Advances in Artificial Life. XIX, 949 pages. 2005. (Subseries LNAI).

Vol. 3629: J.L. Fiadeiro, N. Harman, M. Roggenbach, J. Rutten (Eds.), Algebra and Coalgebra in Computer Science. XI, 457 pages. 2005.

Vol. 3628: T. Gschwind, U. Aßmann, O. Nierstrasz (Eds.), Software Composition. X, 199 pages. 2005.

Vol. 3627: C. Jacob, M.L. Pilat, P.J. Bentley, J. Timmis (Eds.), Artificial Immune Systems. XII, 500 pages. 2005.

Vol. 3626: B. Ganter, G. Stumme, R. Wille (Eds.), Formal Concept Analysis. X, 349 pages. 2005. (Subseries LNAI).

Vol. 3625: S. Kramer, B. Pfahringer (Eds.), Inductive Logic Programming. XIII, 427 pages. 2005. (Subseries LNAI).

Vol. 3624: C. Chekuri, K. Jansen, J.D.P. Rolim, L. Trevisan (Eds.), Approximation, Randomization and Combinatorial Optimization. XI, 495 pages. 2005.

Vol. 3623: M. Liśkiewicz, R. Reischuk (Eds.), Fundamentals of Computation Theory. XV, 576 pages. 2005.

Vol. 3622: V. Vene, T. Uustalu (Eds.), Advanced Functional Programming. IX, 359 pages. 2005.

Vol. 3621: V. Shoup (Ed.), Advances in Cryptology – CRYPTO 2005. XI, 568 pages. 2005.

Vol. 3620: H. Muñoz-Avila, F. Ricci (Eds.), Case-Based Reasoning Research and Development. XV, 654 pages. 2005. (Subseries LNAI).

Vol. 3619: X. Lu, W. Zhao (Eds.), Networking and Mobile Computing. XXIV, 1299 pages. 2005.

Vol. 3618: J. Jedrzejowicz, A. Szepietowski (Eds.), Mathematical Foundations of Computer Science 2005. XVI, 814 pages. 2005.

Vol. 3617: F. Roli, S. Vitulano (Eds.), Image Analysis and Processing – ICIAP 2005. XXIV, 1219 pages. 2005.

Vol. 3615: B. Ludäscher, L. Raschid (Eds.), Data Integration in the Life Sciences. XII, 344 pages. 2005. (Subseries LNBI).

Vol. 3614: L. Wang, Y. Jin (Eds.), Fuzzy Systems and Knowledge Discovery, Part II. XLI, 1314 pages. 2005. (Subseries LNAI).

Vol. 3613: L. Wang, Y. Jin (Eds.), Fuzzy Systems and Knowledge Discovery, Part I. XLI, 1334 pages. 2005. (Subseries LNAI).

Vol. 3612: L. Wang, K. Chen, Y. S. Ong (Eds.), Advances in Natural Computation, Part III. LXI, 1326 pages. 2005.

Vol. 3611: L. Wang, K. Chen, Y. S. Ong (Eds.), Advances in Natural Computation, Part II. LXI, 1292 pages. 2005.

Vol. 3610: L. Wang, K. Chen, Y. S. Ong (Eds.), Advances in Natural Computation, Part I. LXI, 1302 pages. 2005.

Vol. 3608: F. Dehne, A. López-Ortiz, J.-R. Sack (Eds.), Algorithms and Data Structures. XIV, 446 pages. 2005.

Vol. 3607: J.-D. Zucker, L. Saitta (Eds.), Abstraction, Reformulation and Approximation. XII, 376 pages. 2005. (Subseries LNAI).

Vol. 3606: V. Malyshkin (Ed.), Parallel Computing Technologies. XII, 470 pages. 2005.

Vol. 3605: Z. Wu, M. Guo, C. Chen, J. Bu (Eds.), Embedded Software and Systems. XIX, 610 pages. 2005.

Vol. 3604: R. Martin, H. Bez, M. Sabin (Eds.), Mathematics of Surfaces XI. IX, 473 pages. 2005.